TRANSPORT PHENOMENA IN AQUEOUS SOLUTIONS

TRANSPORT PHENOMENA
IN
AQUEOUS SOLUTIONS

TIBOR ERDEY-GRÚZ

Member of the Hungarian Academy of Sciences
Professor of Physical Chemistry
Eötvös L. University, Budapest

A HALSTED PRESS BOOK

JOHN WILEY & SONS
New York Toronto

© Akadémiai Kiadó, Budapest 1974

Translated by

I. RUFF C.Sc.

Department of Inorganic and Analytical Chemistry
Eötvös L. University, Budapest

Translation revised by

I. EGYED Ph.D.

Research Institute for Organic Chemical Industry
Budapest

Published in the U.S.A. and Canada by
Halsted Press, a Division of
John Wiley & Sons, Inc., New York

Library of Congress Cataloging in Publication Data

Erdey-Grúz, Tibor.
  Transport phenomena in aqueous solutions.

  "A Halsted Press book."
  Includes bibliographies.
  1. Solution (Chemistry)   2. Transport theory.

I. Title.
QD543.E7513 1974   541′.34   74-13613
ISBN 0-470-24371-6

Published
as a co-edition of
AKADÉMIAI KIADÓ, BUDAPEST
and
ADAM HILGER LTD
Rank Precision Industries Ltd
29 King Street
London WC2E 8JH
Printed in Hungary at the Akadémiai Nyomda

CONTENTS

Preface 9

List of main symbols 11

1. Introduction 13

1.1 General 13
1.2 The structure of normal liquids 16
References 24
1.3 The structure of water 25
1.3.1 The structure of the water molecule 25
1.3.2 The structure of ice 27
1.3.3 The structure of liquid water 31
1.3.3.1 Special properties of liquid water 31
1.3.3.2 The quartz–tridymite theory 33
1.3.3.3 Theory of the bent bond 35
1.3.3.4 Theory of structural cavities 38
1.3.3.5 The imperfection theory 43
1.3.3.6 The 'water hydrate' theory 43
1.3.3.7 Association theories 45
1.3.3.8 The flickering cluster model 47
1.3.3.9 Theory of the unified structure 52
1.3.3.10 Further theories 53
References 56
1.4 Effect of solutes on the structure of water 59
1.4.1 Effect of solute molecules on the structure of water 59
1.4.2 Effect of dissolved ions on the structure of water 64
References 77

2. Viscosity 79

2.1 Basic empirical relationships of the viscosity of liquids 79
2.2 Fundamentals of the theories of viscosity 84
2.2.1 General 84
2.2.2 Holes and free volume in liquids 86
2.2.3 The Eyring theory of viscosity 90
2.2.4 Activation energy of viscous flow 95
2.2.5 The role of holes in the liquid with respect to viscosity 99
2.2.6 Activation entropy of viscous flow 103
2.2.7 Volume viscosity 104
2.3 Viscosity of liquid mixtures and non-electrolyte solutions 105
2.4 Viscosity of electrolyte solutions 111

2.4.1 Effect of the interaction of dissolved ions and solvent on the viscosity of solutions   113

2.4.2 Effect of the size and shape of the dissolved particles on viscosity (The Einstein effect)   122

2.4.3 Effect of electrostatic interaction of dissolved ions on viscosity   128

2.4.4 Viscosity of concentrated solutions   130

2.4.5 Effect of electrolytes on the viscosity of liquids at high pressures   134

2.4.6 Viscosity of ternary electrolyte solutions   135

References   137

3. Diffusion   141

3.1 Diffusion of non-electrolytes   141

3.1.1 Fundamental relationships of diffusion   141

3.1.2 Fundamentals of the theory of diffusion in liquids   149

3.1.2.1 Hydrodynamic theory of diffusion   149

3.1.2.2 Kinetic theories of diffusion   156

3.1.2.3 Thermodynamic theory of diffusion   170

3.1.2.4 Diffusion in concentrated solutions   175

3.1.2.5 Diffusion without activation energy   181

3.2 Diffusion of electrolytes   181

3.2.1 Diffusion of binary electrolytes in dilute solutions   184

3.2.2 Effects of association or incomplete dissociation on diffusion in dilute solutions   196

3.2.3 Diffusion of binary electrolytes in concentrated solutions   199

3.2.4 Effect of a magnetic field on the diffusion of electrolytes   202

3.3 Diffusion of mixtures of solutes   203

3.3.1 Fundamentals of the general theory   204

3.3.2 Diffusion of electrolyte mixtures   207

3.3.3 Diffusion of electrolytes in solutions of non-electrolytes   212

3.4 Self-diffusion and tracer diffusion   216

3.4.1 Fundamentals of the general theory of self-diffusion   217

3.4.2 Self-diffusion of water   219

3.4.3 Self-diffusion in mixtures   225

3.4.4 Self-diffusion of water in electrolyte solutions   231

3.4.5 Tracer diffusion of ions   235

3.5. Electron diffusion in solutions   245

References   247

4. Electrolytic conduction   253

4.1 Macroscopic relationships of electrolytic conduction   253

4.1.1 General   253

4.1.2 Electrolytic conductivity, transference number, and ionic mobility   265

4.2 Fundamentals of the theory of electrolytic conduction   266

4.2.1 Hydrodynamic theories of electrolytic conduction, I   266

4.2.2 Kinetic theories of electrolytic conduction, I   273

4.2.2.1 Absolute reaction rate theory of the transport of electricity   274

4.2.2.2 Theory of prototropic ('anomalous') conduction   278

4.2.3 Kinetic theories of electrolytic conduction, II. Theory of electrostatic interaction of ions   289

    4.2.3.1 Weak and strong electrolytes   290

    4.2.3.2 Fundamentals of the Debye–Hückel–Onsager theory of strong electrolytes; the variation of conductivity with concentration; the limiting law   294

    4.2.3.3 Deviations from the limiting law; conductivity of more concentrated solutions   302

    4.2.3.4 Dependence of conductivity on ionic association   307

    4.2.3.5 Variation of the transference number of strong electrolytes with concentration   314

    4.2.3.6 Variation of conductivity under high field strength (the Wien effect)   317

    4.2.3.7 Dispersion of conductivity (the Debye–Falkenhagen effect)   321

    4.2.3.8 Effect of magnetic field on conductivity   322

    4.2.3.9 Conductivity of concentrated electrolyte solutions   323

    4.2.3.10 Conductivity of solutions of electrolyte mixtures   325

References   326

4.3 Temperature dependence of electrolytic conductivity   331

References   338

4.4 Pressure dependence of electrolytic conductivity   339

4.5 Activation energy of ionic migration   341

References   346

4.6 Conductivity of weak electrolyte solutions   347

References   351

4.7 Effect of non-electrolytes on the conductivity of aqueous electrolyte solutions   351

    4.7.1 General   351

    4.7.2 Effect of changes in viscosity due to non-electrolytes on hydrodynamic migration   353

    4.7.3 Effect of non-electrolytes on prototropic conduction   372

    4.7.4 Changes in the activation energy of the transport of electricity due to the effect of non-electrolytes   388

4.8 Electronic conduction in aqueous solutions   390

References   391

5. Appendix   395

5.1 Interaction of dissolved ions (theory of strong electrolytes)   396

    5.1.1 Ratio of free ions in solution   397

    5.1.2 Distribution of ions in dilute solutions of strong electrolytes   399

    5.1.3 Chemical potentials and activity coefficients of dissolved electrolytes   407

        5.1.3.1 Calculation of the activity coefficient from the interaction between ions   409

        5.1.3.2 Variation of the activity coefficient due to the interaction between ions and solvent molecules   416

    5.1.4 Activity coefficients of more concentrated electrolyte solutions   420

    5.1.5 Association of ions   424

        5.1.5.1. Association and dissociation   424

5.1.5.2 Theory of ion pair formation 426

5.1.5.3 Association of three and four particles 434

5.1.5.4 Correlation between association and complex ion formation; localized hydrolysis 436

References 439

5.2 Interaction of dissolved ions with the solvent (The fundamentals of the theory of hydration) 441

5.2.1 Classification of hydration phenomena 441

5.2.2 Primary hydration 445

5.2.3 Secondary hydration 455

5.2.4 Experiments indicating hydration 458

5.2.4.1 Effect of hydration on the mobility and diffusion of ions 460

5.2.4.2 Correlation between transference number and hydration 468

5.2.4.3 Effect of hydration on the conductivities and activity coefficients of strong electrolytes, and on the solubilities of non-electrolytes 470

5.2.4.4 Effect of hydration on the compressibility and ultrasonic absorption of solutions 472

5.2.4.5 Optical consequences of ionic hydration 475

5.2.4.6 Thermodynamic aspects of hydration 479

References 489

Author index 493

Subject index 503

PREFACE

Transport processes occur in almost every chemical phenomenon and industrial process, and are significant factors in several cases. Thus, a physical–chemical review of their practical and theoretical relations is helpful to enable chemists and other experts dealing with chemical procedures, to gather information about the individual transport processes without detailed study.

This work covers the phenomena of viscous flow, diffusion, and electric conduction in water and aqueous solutions, since these are the most important transport processes in practice. Heat conduction is less significant in theoretical physical chemistry, and a systematic extension of the present treatment to non-aqueous solvents would considerably increase the size of the work. In some relations, however, other solvents and non-aqueous solutions will also be mentioned.

Within the topics given above, I could not aim for completeness in the scope of the phenomena covered or in the depth of the treatment, or in the list of the references given. The requirement of completeness would have unreasonably enlarged the field to be surveyed and, furthermore, I have limited the range of the topics involved in order that the reader may get a clearer view of the rather intricate and complicated relationships of the transport phenomena. In order to facilitate this, only the main aspects of the theories have been discussed. For those who take an interest in further details, I have referred to the literature cited. Experimental methods were not considered to be in the scope of this book.

I have tried to discuss the phenomena and problems with respect to their physical–chemical importance, giving a relatively detailed discussion of my research and that of my co-workers. The proportion of the various parts treated is, of course, a question of individual opinion, and many readers might well have proposed other proportions.

In solutions, the transport phenomena are closely related to the structure of liquids and to the changes in this structure caused by the solute. Hence, I have found it advisable to review first the main aspects of our knowledge on the structure of liquids — above all, of water. The complexity of the structure of liquids, the uncertainty of our knowledge of it, as well as the extensive polemics about it may serve as excuses for the exceptional length of the Introduction. The Appendix should facilitate the discussion of the topics in that it gives a review of the theory of the electrostatic interaction of ions in solution, and of the phenomena of hydration and its relationships which appreciably influence transport processes.

In spite of comprehensive and extensive studies on the structure of water and the transport phenomena proceeding in it, no uniform picture corresponding to reality—in all its details and consequences—has been achieved up to now. The directly observable facts are usually the result of a complexity of interactions in which the effects of the individual factors can hardly be separated unequivocally; a separate analysis of the particular interactions and their relationships, and, consequently, a reliable reconstruction of reality by a theoretical treatment are extremely difficult. This is accounted for mainly by the fact that different experiments concerning the same property can lead to different, and often contradictory conclusions, and sometimes even the same results can be explained in different ways. In other words, in independent experiments on the same property, appreciable effects can be produced by factors or interactions which escape notice or which are still unknown. In such cases the experiments may be related to properties other than those being studied.

The uncertainty in the interpretation of the results is also reflected in this work. It would not give a satisfactory view of the state of our knowledge if the treatment were limited to the theoretical conclusions which are accepted unequivocally—or believed to be such by me. The contradictory explanations also belong to our present state of knowledge about transport phenomena. Owing to the deficiencies in our knowledge, in several cases none of the explanations can be preferred, especially when a satisfactorily objective standpoint is sought.

In such a situation, the question may arise regarding the advisability of writing a work about transport processes in aqueous solutions. I think the answer can be positive. On one hand, several facts and correlations are quite reliably known from many points of view, and they are applicable in practice. On the other hand, by confronting the contradictory results and opinions and interpreting the experiments, the still unsolved problems will be brought to light, the complex nature of the phenomena emphasized, and the direction of further research pointed out. The uncertainties in our knowledge and the contradictions in the theoretical interpretation necessitate careful judgement of the various viewpoints. It helps one to avoid attributing too much importance to explanations of points of detail, and calls one's attention to the possibility of the existence of interactions not accounted for because of the extremely complicated relations in solutions.

As a consequence, it will not be surprising if the reader does not agree with some parts of my book. I am interested in having the opinions of my readers, and would very much appreciate having my attention drawn to any mistakes, misprints or other deficiencies.

I am very grateful to Prof. E. Berecz for his valuable advice during reading the manuscript and to Mrs. B. Janko for compiling the author and the subject index.

Tibor Erdey-Grúz

LIST OF MAIN SYMBOLS

a	mean 'ionic diameter' in the theory of strong electrolytes
a_i	activity of the ith component
a_{ca}	mean activity of electrolyte
c_i	concentration of the ith component, mole/unit volume
d	diameter of a molecule or ion
D	diffusion coefficient (defined in the appropriate parts of the text)
$D*$	self-diffusion coefficient
e	the elementary electric charge
E	energy per mole
f	force
F	distribution function in the ground state
$F\ddagger$	distribution function in the activated state
F	Faraday number, 96 494 coulomb/g equivalent
G	free enthalpy referred to one mole
$\Delta G\ddagger$	free enthalpy of activation
h	Planck's constant
h	hydration number (the number of water molecules bound to one ion)
H	enthalpy referred to one mole
$\Delta H\ddagger$	enthalpy of activation
I	ionic strength
J_i	flux of the ith component in g cm^{-2} sec^{-1} or mole cm^{-2} sec^{-1} units
k	rate constant (specific rate)
k	Boltzmann constant
K	equilibrium constant
L_i, L_{ii}, L_{ik}	conduction (phenomenological) coefficients
m	mass of one molecule
m_i	molality of the ith component
M	molecular weight
N	the Avogadro number
p	pressure
r	radius of a molecule or ion
R	universal gas constant
S	entropy referred to one mole
$\Delta S\ddagger$	entropy of activation
t_c, t_a	transference number of the cation and anion, respectively
T	absolute temperature on the Kelvin scale
u	molecular velocity
U	absolute mobility
U^0	limiting value of the absolute mobility in infinitely dilute solutions
v	free volume in liquids

V	molar volume
w_i	weight fraction of the ith component
x_i	mole fraction of the ith component
X_i	thermodynamic force acting on the ith component
z_i	valence or ionic charge of the ith component
γ_i	activity coefficient of the ith component
γ_{ca}	mean activity coefficient of the electrolyte
Δ^2	the Laplace operator $= \dfrac{\partial^2}{\partial x^2} + \dfrac{\partial^2}{\partial y^2} + \dfrac{\partial^2}{\partial z^2}$
ε	energy per molecule
ε	dielectric constant
η	viscosity coefficient
$\varkappa$	specific conductivity
$\varkappa$	reciprocal value of the 'thickness' of the ionic atmosphere in the theory of strong electrolytes
$\varkappa$	the ratio of the specific heat of gases measured at constant pressure and volume, respectively
λ	distance between two adjacent equilibrium positions of molecules in liquids
$\lambda_1, \lambda_2, \lambda_3$	distance between molecules in liquids in the three directions of space
λ_c, λ_a	relative mobility (equivalent conductivity) of the cation and anion, respectively
λ_c^0, λ_a^0	limiting value of the equivalent conductivity of the cation and anion, respectively, in infinitely dilute solutions
Λ	equivalent conductivity of an electrolyte
Λ^0	limiting value of the equivalent conductivity of an electrolyte in infinitely dilute solutions
μ_i	chemical potential of the ith component
$\tilde{\mu}_i$	electrochemical potential of the ith component
ν	number of moles of ions formed from 1 mole of salt ($\nu = \nu_c + \nu_a$)
φ	fluidity
Φ_i	volume fraction of the ith component
χ	packing factor
ψ	electrostatic potential

1. INTRODUCTION

1.1 GENERAL

Under the name of transport phenomena are summarized processes involving transfer of momentum, energy, or mass between different regions of an inhomogeneous material system (usually forming one phase, which is non-uniform in its properties). In *viscous flow* momentum transfer takes place between regions of a material system containing particles of different impulses; energy transfer occurs in *heat conduction* between places of different thermal energy; mass is transferred in *diffusion** between places with different chemical potentials (usually having different concentrations), and mass is also transferred in *electric conduction* between places of different electric potential. Transport phenomena occur in every state of matter, but their rate is especially high in fluid phases. In electrochemistry, transport processes are also essential, since mass and charge transfer accompany almost all electrochemical processes. Knowledge of the general relationships of transport phenomena is indispensable to their understanding.

In common liquids, close to the equilibrium state, the rate of the transport processes is proportional to gradients of impulse, temperature, concentration (more precisely: chemical potential), and electric potential. The corresponding proportionality factors are the *viscosity coefficient* (viscosity), the *heat transfer coefficient* (heat conductivity), the *diffusion coefficient* (diffusion constant), and the *electric conduction coefficient* (electric conductivity). If transport is caused by local differences in not only one, but two (or more) properties, the so-called cross-effects should also be taken into account (e.g., when local differences exist in both the concentration and temperature of a solution). In this work, the discussion is limited to viscous flow, diffusion, and electric conduction of aqueous solutions. Heat conduction, being less important in theoretical physical chemistry, is considered to be out of the scope of this discussion.[+]

Transport processes, in common with other phenomena, can be treated on the basis of a phenomenological, as well as a molecular theory. The phenomenological theory describes the observable phenomena by means of

---

* A special limiting case of diffusion is self-diffusion which is related to the displacement of molecules, atoms, or ions due to the thermal motion within a phase uniform in its properties. Under these conditions, the molecules move at random in a uniform phase.

[+] In connection with this, see, e.g. R. P. Tye, *Thermal Conductivity*, Vols 1 and 2 (London, 1969).

Literature on page 24

correlations between macroscopically measurable quantities. In this respect, these methods are somewhat similar to those of thermodynamics, and they are based on the thermodynamics of irreversible processes. Phenomenological theories usually meet the direct requirements for applications in engineering and other fields; however, they do not help much the detailed understanding of the mechanism of the processes. The molecular theories of transport phenomena, on the other hand, endeavour to discover just the mechanism of the processes. They try to deduce the laws of the transport processes from the properties of the molecular structure of the given medium, applying the kinetic-statistical theory of matter. In a certain sense, hydrodynamic theories exhibit some characteristics of both the phenomenological and molecular theories.

The phenomenological theories, which can already be considered as classical, lead to reliable relationships–within the experimental errors–for many purposes on the basis of experimental data. The rapid development of industrial and other practices, however, requires a more thorough knowledge of the details of the processes, i.e. their molecular mechanism is also required. Consequently, besides its great theoretical importance, the practical application of the molecular theory of transport phenomena is also increasing. In accordance with the fact that phenomenological and molecular theories are different approximations of the same objective reality, in the following chapters both will be dealt with to a certain degree, but with emphasis on the molecular theories; in this field there are several areas still to be investigated.

The main task of the molecular theory of transport processes is to interpret the transport coefficients defined in the phenomenological theory and determined experimentally, on the basis of molecular processes, and to calculate them from molecular parameters (e.g. from the potential energies between adjacent molecules).

Until the middle of the 1940's, the theory of transport phenomena was, on one hand, mainly a description of the consequences of the experimentally determined relations, and on the other hand, was based on a very simplified picture of the liquid state that contained a lot of arbitrary assumptions. However, it proved to be adequate for determining the fundamental relationships, in spite of its hypothetical nature. On the basis of non-equilibrium statistical thermodynamics, originally used in the investigations of Onsager, Born, and Green [1], as well as by Kirkwood and his co-workers [2], the development of a new period has started in the theory of transport phenomena. These theories deal with 'normal' liquids, and their structural model is based on rigid spherical molecules. In order to reduce the complexity of the mathematical treatment of the relationships to a suitable degree, several further simplifications have been introduced into the theories.

The newer theories explain several relationships in detail, but none of them can be regarded as satisfactory in all respects. Much research is still needed to get a detailed understanding of the transport phenomena in liquids. Essentially, the difficulties arise from the fact that the transport processes are related to dynamic and static properties of several interacting molecules. These properties, however, depend on the shape and nature of the molecules, and on the forces exerted by them on their environment; these forces are influenced by the nature of this environment, and are distance– and direction–dependent. A detailed molecular theory, resulting in satisfactory agreement with the experimental data, could be applied only to a special case where the potential corresponding to the force existing between two neighbouring molecules depends only on the distance between these molecules. This can be valid, however, only for monoatomic liquids. In the statistical-mechanical theory of monoatomic liquids, a further simplification is generally introduced by assuming that the force acting on the molecules is the resultant of the gradients of the intermolecular potential energy of molecule pairs formed by neighbouring molecules.

In the case of water (and other 'associating' solvents) the simplifying assumptions applied in the theory of monoatomic liquids are not valid even approximately. The non-spherical shape and electric dipole character of the water molecule, the various interactions with its environment (e.g. dipole–dipole interactions, hydrogen bridges), as well as the directional dependence of these effects, make the structure of water and the mechanism of the transport phenomena complicated.

The investigations on the structure of water are very extensive, but they still have not led to satisfactory results. Several models have been suggested for the structure of water which explain some properties more or less adequately; however, some discrepancies arise with respect to others. One of the greatest difficulties in this problem is that no experimental proof has been found which would unequivocally support the acceptance of one of the suggested models and reject all the others. Some models have been found that are in agreement with a group of experiments, but it cannot be proved that this is the only model suitable for the interpretation of this group of experimental data. Moreover, no suitable model to explain all of the properties of water and aqueous solutions has been developed up to now.

Although none of the theories succeed in the general interpretation of the structure of liquids and particularly that of water, the theories of the structure of liquids still provide some information on the mechanism of transport phenomena. Hence, it seems to be advisable to give a general review of the structure of liquids and water before discussing the transport processes, even if the current knowledge is not satisfactory to form a unified theory and is not free from (in some respects antagonistic) contradictions. This information is very useful, although some properties of electrolytic solutions (e.g. the activity coefficient of a dissolved electrolyte, its conductivity) can reliably be described within certain limits (e.g. by the Debye–Hückel theory), even when taking into account the mutual interaction of the solute particles only and regarding the solvent as a structureless continuum.

Literature on page 24.

1.2 THE STRUCTURE OF NORMAL LIQUIDS

The aim of the molecular theories of liquids, in common with those of gases and the solid state, is to describe the structure of a material system with satisfactory accuracy and to explain how it depends on the structure and other properties of molecules and atoms forming the system, as well as to establish a quantitative correlation between the macroscopical behaviour and the individual properties of molecules and atoms. This aim could satisfactorily be approached for gases, where molecules move at random and independently of one another for a rather long time, and also for crystalline solids where the particles are located in a rigorous order, with only few exceptions. In liquids, however, none of these conditions simplify the relations (for liquids there is no limiting state analogous to that of a perfect gas or a perfect crystal); thus the theories regarding liquids are much more complicated and their results are far less satisfactory than those of gases and crystals.

The theories of liquids [3] can be divided into two groups with respect to their main characteristics. Those in the first group are based on the description of the structure using some simplifying assumptions, and on the basis of experimental details, deal only subsidiarily with the question of how the structure is determined by the molecular properties. Since the assumed structure is similar, to a certain extent, to the regular lattice structure of crystals, these theories can be thought of as *lattice theories* of liquids [4]. The other group of theories of liquids includes the *distribution function theories* [5], concerning the forces arising from interactions between molecules, and the determination of how these forces influence the structure. In the description of a state by these theories, the probabilities of different molecular distributions are investigated, resulting in the possible actual configuration. The lattice theory is based on the successful theory of crystals, while the distribution function theory is based on the successful theory of gases.

Both groups of theories give rather important results, but neither group is satisfactory. The two types of theories approach the real state from different sides, so they supplement each other to a certain extent. From the physical-chemical point of view, the advantage of the lattice theories is that they provide an easier way to understand and explain the properties of solutions. For the time being, however, this usually means the explanation of experimental facts rather than the prediction of relationships.

The mutual position of molecules. Liquids with the most simple structure are those consisting of single atoms ('monoatomic molecules'), in this respect considered to be rigid spheres. Liquid argon is one such to a good approximation. However, these theories on the structure of the most simple, so-called *normal liquids* are not satisfactory either, since the conclusions deduced require empirical treatment to make them comparable with the observations [6]. The relationships for liquids consisting of polyatomic, nonspherical molecules are very complicated and make any conclusions about the structure merely qualitative or semi-quantitative in nature.

The most important experimental data on the structure of liquids can be obtained from X-ray and neutron diffraction investigations, measurements on thermodynamic equilibrium properties, (density, compressibility, thermal effects, vapour pressure) and from studies on non-equilibrium transport phenomena (viscosity, diffusion, electric conductivity). On the basis of diffraction experiments, the *radial distribution function* of molecules can be obtained from the first intensity maximum of the scattered radiation. This function gives the probability of the presence of another molecule as a function of the distance from a given molecule. On the basis of the theory of liquids consisting of structureless, spherically symmetrical molecules, the distribution function of pairs of molecules can be calculated and compared with the experimental results; it gives the probability of finding two molecules at a distance R, as a function of the mutual separation R. A knowledge of the distribution function is the minimum information needed to develop a picture of the structure of liquids.

For liquids with configurational energy obtainable by simple summation of the potential energies of molecular pairs, rather far-reaching conclusions can be drawn not only for transport phenomena, but also for thermodynamic properties which may be deduced from the distribution function or the pair-potential. In the case of liquids other than this simplest type, however, much less unequivocal conclusions can be drawn from the distribution function. Nevertheless, it is also still important in this case to determine the distribution function experimentally, because the necessary–but not sufficient–condition for the correctness of the theory on the structure of liquids is the agreement of the distribution function deduced from it with the one obtained experimentally.

The radial distribution function calculated from the first maximum of the X-ray or neutron diffraction functions–although, in general, not unequivocally–makes it possible to obtain the average distance of the adjacent molecules, i.e. the radius of the *coordination sphere* and the number of nearest neighbours, i.e. the *coordination number*. These two parameters regarding the positions of the molecules, however, are not enough to characterize the structure of the liquid. Structure means not only the relative position of the molecules, but also the forces due to their interaction, which should be taken into account by a third parameter at least. A more extensive analysis of the other maxima in the scattering function will surely lead to a more detailed understanding of the structure of liquids.

The properties of material systems in different states depend on both the *internal structure* of the atoms and molecules and on the structure of the *given state of matter as a unified system*. With respect to the structural state, molecules are relatively fixed structural units. According to experimental evidence, the internal structure of the molecules does not essentially change with changes in the state; in most cases only the bond angles are somewhat modified. There is no reason to suppose that in the liquid state the internal structure of the molecules is essentially altered by external conditions (unless the liquid is under very high pressure). Thus the depen-

Literature on page 24

2

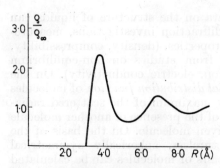

Fig. 1.1 Radial distribution function of liquid argon at 84 °K. Density p at a distance R from the given molecule is related to the average density pd

dence of the properties of liquids on temperature and pressure is mainly due to changes in the structure of the liquid. The liquid structure itself is determined, of course, by the internal structure of the molecules.

The analysis of the radial distribution functions calculated on the basis of X-ray and neutron diffraction measurements indicated that in liquids of simple structure, the primary result of an increase in temperature is not an increase in the distance between adjacent atoms, but a reduced average number of neighbours. (Namely, the first peak of the radial distribution curve is not altered much by the temperature, but the area under the peak—which is proportional to the number of immediate neighbours—is decreased.) The formation of vacancies of atomic size or an increase in the number of vacancies plays a significant role in increases of volume.

The experimental data obtained for liquids of the simplest structure already show that—at least far from the critical state—the disorder is not complete in liquids; the molecules are ordered to a certain extent within smaller or larger regions. For example, Henshaw's neutron diffraction experiments [7] have given evidence of a maximum in the radial distribution of the molecules in liquid argon (Fig. 1.1), which is proof of a certain order of the molecules. In liquids consisting of polyatomic molecules, the order is more extended, particularly if the molecules also have permanent dipole moments and intermolecular connections (mainly hydrogen bonds) are formed.

On the basis of our knowledge of liquids, it is not doubted that there is a certain ordered structure of the species in them. In order to describe their arrangement, a properly chosen crystal lattice should be considered first. In this sense the structure of liquids is somewhat similar to that of crystals, but only with respect to small regions and to temporary positions of the molecules. In the solid state, the lattice-like order is permanent and, in the ideal case, extends over the whole crystal. Of course, in real crystals, this order is disturbed in some places by different imperfections in the lattice, besides the thermal motion. In addition, the particles vibrating about their equilibrium positions sometimes jump to an adjacent equilibrium position. These disturbances *(lattice imperfections)*, however, are relatively rare, and, in general, they do not destroy the main features of the lattice structure extended over all the crystal, though they markedly influence some of its properties. In liquids, however, the more or less lattice-like order of the particles is extended only to *small regions* of molecular dimensions, and these regions are constantly changing in time.

The lattice theory of liquids assumes that the molecules (atoms or ions) vibrate about their equilibrium positions. The equilibrium positions, however, are not fixed, and it may often happen that an unoccupied place (vacancy) is formed due to fluctuations in the vicinity of a molecule, and this molecule jumps into the vacancy. This results in an altered equilibrium position of the molecular vibration. Since viscosity is much lower, and diffusion and ionic migration are much greater in liquids than in solid crystals, these jumps must be relatively frequent.

The lattice-like ordered regions rapidly break up because of the frequent displacement of the molecules, and others will develop elsewhere instead. Thus, on the statistical average a definite short-range order prevails in liquids, but the individual ordered regions are permanently changing. The equilibrium properties of liquids determined by time-averaged values are not influenced too much by the jump-like movement of the molecules, but they are very important with respect to transport phenomena.

The mechanical model of the structure of liquids. Bernal suggested a 'poly-hedral vacancy theory' [8a] to interpret the structure of normal liquids. This theory considers the liquid with respect to its structure, not as a crystal of destroyed lattice or a condensed gas, and investigates directly the arrange-ment of closely but irregularly packed spherical molecules in a continuous and homogeneous medium. Varying coordination is characteristic of the liquid state, i. e. the immediate neighbours are constantly changing and varying in number. Self-diffusion and fluidity are also consequences of this. Considering short time intervals (e. g. in the case of neutron or ultrasonic diffraction measurements) the structure of the liquid is similar to that of crystals. This similarity, however, is geometrical rather than physical. Liquid, in fact, differs from the ordered state of crystals simply by its disorder. In liquids, molecules pass through irregular series of disordered situations. There are many such energetically equivalent irregular complexes, and there are few regular ones, but they are in permanent exchange with each other. Fluidity is due to transitions between different regular and irregular arrangements caused by very small forces. The question whether there is any common structural feature in the irregular arrangements of the molecules has been studied by Bernal [8a] by means of model experiments.

In order to study the structure of liquids, Bernal put small steel spheres on to an irregular basis in complete disorder. This agglomeration of spheres can be considered in first approximation as an enlarged model of a liquid consisting of spherical molecules. By pouring dye on to this agglomeration and drying it, the number of the immediate neighbours of each sphere could be determined (the contact points of the spheres remained undyed). By a cumbersome study of the agglomeration of the spheres, the forms and sizes of the cavities between the spheres could also be determined, in addition to the number of the immediate neighbours (the coordination num-ber). According to this experiment, the number of the immediate neigh-bours of the spheres proved to be variable (5–10), and the frequency of spheres having given numbers of neighbours showed a definite statistical distribution. Bernal deduced from this that one of the characteristic proper-ties of liquid state is the varying coordination number.

Literature on page 25

2*

For a crystal-like regular lattice of spheres in contact and having
identical radii, only two types of polyhedral cavities could be found
(supposing that the spheres are in contact with as many neighbours as possible). These can have octahedral or tetrahedral shapes. In this model of
liquids, however, the regularity is not perfect, and the position of the
spheres continuing to build up the system is not defined accurately by their
neighbours. This model of liquids–according to its higher energy in comparison with crystals–is more voluminous and looser, having smaller cohesion
and larger volume. To develop this irregular arrangement characteristic
of liquids, the molecules need a volume about 15 per cent larger than
that required for regular close packing. It is an important feature of this
model of liquids that five types of polyhedral cavities are formed in which
there is no place for an additional sphere. These are tetrahedral, octahedral,
and three types of 'deltahedral' cavities. The latter ones are polyhedral
formations constructed from triangular planes which have no symmetry
suitable for a regular fitting. Besides these, there are a few 'incidental'
cavities in the model. The above five types of polyhedrons can form–with
small changes in their edge lengths–arrangements very different in volume;
the smallest volume of such a packing is, however, at least 15 per cent
larger than that of regular close packing.

The radial distribution curve calculated on the basis of Bernal's model
is in good agreement with that determined in liquid argon by means of
neutron diffraction. Thus, it can be assumed that this model reflects reality
fairly adequately, which suggests this will also be true for the further results.
It should be emphasized that the sphere model is not entirely irregular.
The predominating tetrahedral cavities occur in groups and in regular
arrangement. The tetrahedrons can fit to one another with a good utilization
of space, forming several types of structures, some of them are closed
(e.g. 20 tetrahedrons with three common planes each), or open (e.g. tetrahedrons with common planes in a triple helix). It should be pointed out
that in such local regions the fitting is of higher density than it could be in
the case of regular close packing. However, these ordered structures can
only occur locally, because they contain symmetry elements which prevent
their endless extension in the space. These structural regions have been
called *pseudonuclei* by Bernal, since–although they are somewhat similar
to crystal nuclei–they cannot be the starting points of a long-range arrangement. Locally, a structural order impossible in crystals because of the
symmetry properties of the ordering forces can also develop in liquids.
The structure of a liquid, in this way, cannot be deduced unequivocally
from that of the crystals producing it on melting. In liquids, owing to
fluctuations, regions of higher energy (i.e. of closer packing) and smaller
entropy (i.e. more regular) as compared with the average values can also
occur. From the assumption of pseudonuclei of higher density than the
average value, it also follows that there should be some local regions of
lower density. According to Bernal's model, the regions of higher density,
i.e. higher order, are surrounded by those of smaller density, i.e. higher
irregularity. It seems that in a mass in which ordered–disordered states
are in equilibrium, every locally ordered region is, in general, surrounded

by a more disordered zone non-conformed to its environment. It is possible that this circumstance also plays a role in the phenomena of solvation.

A complete description of the structure of liquids cannot be expected from the mechanical model of Bernal, particularly a treatment of the effect of temperature. Its general importance is the finding that there is not total disorder *even in the irregular agglomeration of spheres which are only in mechanical interaction with one another*. For a better understanding of the structure of liquids, it would be important to find some fundamental relationships for the 'statistical–geometrical' local arrangement. Nonspherical interactions between the molecules act as further ordering factors.

A novel model was elaborated by Pinsker [8b] for liquids consisting of identical spherical molecules, assuming that during the condensation of the vapour spatial regions are formed which are invariant with respect to a certain group of motions. According to this picture, the fluid is an indefinitely branching part of space filled up by a multitude of mirror image pairs of molecules.

Temperature dependence of structure of liquids. More detailed information on changes in the structure of normal liquids with temperature can be obtained from the Eyring theory [9], which takes into account both the crystal-like and gas-like properties of liquids. On one hand, this theory is in agreement with the fact, indicated by the scattering of X-rays on liquid argon, that near the melting point the coordination number is about 10–11, and with increasing temperature it continuously decreases to four, observed at 5 °C below the critical temperature. At the same time, the distance between the immediately adjacent molecules is almost unaltered in the whole temperature range. In the close vicinity of the critical temperature, the attracting force decreases, the distance between the immediate neighbours increases, and the coordination number becomes six. In the Eyring theory, the cavities are interpreted as 'fluidized vacancies', and it is assumed that these move as freely in the liquid as the molecules in a gas. This is in accordance with the observation that the average of the density of the liquid and the saturated vapour is almost independent of temperature and that it decreases slightly from the melting point up to the critical temperature according to a linear relationship (Cailletet–Mathias rule). The saturated vapour behaves in the opposite manner compared with the liquid, i.e. they 'reflect' each other. Of course, the movement of the vacancies is caused by the movement of molecules in the opposite direction. So the temporary lattice-like arrangement of the liquid molecules is not a consequence of some crystal-like structure of the liquid, but arises from the vibration of the molecules located in the 'potential well', produced by the neighbours, about their equilibrium positions.

Intermolecular forces and their potential energies. The structure of liquids is a result of the intermolecular forces. Although a detailed theory of intermolecular forces has not been successfully developed, several properties of these forces are known.

Literature on page 25

In liquids of neutral molecules (non-ions) the attractive forces keeping the molecules of the liquid together are of van der Waals character. Of these, the *dispersion forces* (the so-called London forces) which also act between neutral, non-polar molecules, are always efficient. Dispersion forces can be well characterized by the formation of a temporary dipole moment on fluctuation of the electron shell of the molecule; this polarizes the adjacent molecule, and attracts its induced dipole. Although the time average of the dipole moment produced by the fluctuation is zero (if the molecule has no permanent dipole moment), the average of the attracting force is still different from zero, because the force depends on the average of the squares of the dipole moments which is positive. The dispersion force decreases rapidly with distance and it becomes zero in a distance of a few Ångstrøms.

In the case of electrically asymmetrical molecules (possessing permanent electric dipole moments), the dispersion forces are accompanied by *dipole-dipole interactions* which cause a further attractive force, as well as orientating the molecules in a more or less ordered arrangement. The orientation effect of the permanent dipole moment is disturbed by the increase of thermal motion with temperature. In addition, the permanent dipole molecules polarize one another to a certain extent, i.e. they produce induced dipole moments. The energy of the interaction of these three types of effects between two molecules is:

$$E = -\frac{3}{4}\frac{\alpha^2 h \nu_0}{R^6} - \frac{2}{3kT}\frac{\mu^4}{R^6} - \frac{2\mu^2\alpha}{R^6},\tag{1.1}$$

where α is the average polarizability of the molecules, ν_0 is the zero-point frequency of the vibration related to the dispersion effect, R is the distance between the adjacent molecules, μ is the permanent dipole moment of the molecule, while the other letters have their usual meaning. In this expression, the first term represents the energy of the dispersion forces, the second one that of the dipole–dipole interaction, and the third one stands for the energy of the permanent dipole-induced dipole interaction. In general, the dispersion forces are predominant (except for liquids of very polar molecules), and their characteristic feature is that they are additive, i.e. the dispersion force arising from the interaction between several molecules can be calculated as the sum of the interactions existing in the individual pairs of molecules [10a].

If *hydrogen bonds* or any other intermolecular connections are formed between the molecules of the liquid, these also contribute to the attraction forces acting between the species [10b].

Of the attraction forces, only the dispersion forces are centre-symmetrical, i.e. they have no favoured direction, unlike the forces arising from permanent dipoles or hydrogen bonds which are directional. Consequently, when these effects are operating, the attraction force depends rigorously both on the separation of the centres of the molecules and on their mutual orientation.

Between molecules approaching each other to a short distance, repulsive forces arise because of the overlapping of the electron shells of the neighbouring molecules and the mutual repulsion of the atomic nuclei. The effective distance of the repulsive force is smaller than that of the van der Waals attraction forces, but it increases more rapidly with decrease in distance. This feature of the repulsive forces is the reason why the molecules can be regarded as rigid particles in first approximation.

In the most simple cases, (e.g. two helium atoms) the potential energy of the interaction between molecules can be calculated by means of quantum mechanics. If the molecules are very near to one another, the potential energy of their interaction is high and positive. However, with the increase of the distance between them, the potential energy rapidly falls (Fig. 1.2) to a negative value and after passing through a minimum it starts to increase again and will become zero at 'large' distances. According to approximate calculations, the potential function between polyatomic molecules exhibits a similar shape. If the mutual separation is not too small, the potential energy of the interaction is approximately inversely proportional to the sixth power of the distance. No detailed quantitative information on the interactions between molecules containing several electrons has yet been given by quantum mechanics.

As for the potential energy of the interaction between polyatomic molecules, one has to rely on empirical correlations. In these equations, the numerical values of the constants can be obtained only on the basis of empirically determined characteristics such as the second virial coefficient and the viscosity of gases, the density, cohesion energy and specific heat of solids.

The empirical formula given by Lennard-Jones [10a, 11] regarding the potential energy of interaction U between two molecules proved to be valid to a good approximation, and it can widely be applied in several cases:

$$U = 4\varepsilon \left\{ \left(\frac{\sigma}{R} \right)^{12} - \left(\frac{\sigma}{R} \right)^{6} \right\} \tag{1.2}$$

(Lennard-Jones 12–6 potential), where ε and σ are constants with dimensions in energy and length, respectively. By a proper choice of the latter quantities, the properties observed can be described in a rather good approximation for noble gas atoms and diatomic molecules having no dipole moment. However, the 12–6 potential is not in perfect agreement with the real conditions for argon atoms. According to Guggenheim and McGlashan [6], more than two constants are needed to describe accurately even this simple case.

The potential energy of interaction between more complicated but spherically symmetrical molecules can be represented by an equation similar to equation (1.2):

$$U = 4\varepsilon \left\{ \left(\frac{\sigma}{R} \right)^{n} - \left(\frac{\sigma}{R} \right)^{m} \right\} \tag{1.3}$$

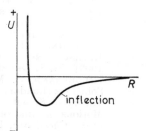

Fig. 1.2 Schematic representation of potential energy U of interaction of two molecules vs. their mutual distance R

Literature on page 25

(Lennard-Jones $n-m$ potential). For example, for the SF_6 molecule, $n = 28$, $m = 7$. The potential energy of interaction between non-spherical or polar molecules, however, cannot be obtained from an empirical expression involving only the distance between the centres of the molecules.

The mutual attraction of the molecules of the liquid results in a more or less loose connection between them which is often called as association. In this respect, however, association is not an unequivocal term, because several types of intermolecular connections can exist. Recently, it has been emphasized by Andersen [12] that there are two main types of molecular association in liquids; in some cases aggregates (complexes) consisting of several molecules are formed by the effect of intermolecular forces and they move as kinetic entities. In other cases, 'association' means only the existence of a correlation between the intermolecular forces and the position of the molecules, but it slightly affects the movement of the molecules. Andersen has supported this theory by NMR studies on chloroform.

A new method has been suggested by Lisnianskii [13] for the investigation of the association of liquid molecules by means of light absorption.

REFERENCES

to Sections 1.1 and 1.2

[1] L. ONSAGER, *Phys. Rev.*, 1931, **37**, 405; M. BORN and H. S. GREEN, *Proc. Roy Soc.*, 1946, **A.188**, 10; 1947, **A.190**, 455; H. S. GREEN, *Proc. Roy. Soc.*, 1947 **A.189**, 103.

[2] J. G. KIRKWOOD, *J. Chem. Phys.*, 1946, **14**, 180; J. G. KIRKWOOD, F. P. BUFF and M. S. GREEN, *J. Chem. Phys.*, 1949, **17**, 988; J. H. IRVING and J. G. KIRKWOOD, *J. Chem. Phys.*, 1950, **18**, 817; R. W. ZWANZIG, J. G. KIRKWOOD, J. OPPENHEIM and B. J. ALDER, *J. Chem. Phys.*, 1954, **22**, 783.

[3] As a review see, J. S. ROWLINSON, Liquids and Liquid Mixtures (London, 1969), K. SCHAEFER, Die Struktur der Flüssigkeiten, *Z. Elektrochem.*, 1948, **52**, 245; J. D. BERNAL, *Proc. Roy. Soc.*, 1964, **A.230**, 299; *Disc. Faraday Soc.*, 1967, **43**, 60; I. PRIGOGINE, The Molecular Theory of Solutions (Amsterdam, 1957); I. Z. FISHER, Statistic Theory of Liquids (Chicago, 1964); H. L. FRISCH and J. L. LEBOWITZ, The Equilibrium Theory of Classical Fluids (New York, 1964); S. A. RICE and P. GRAY, The Statistical Mechanics of Simple Liquids (New York, 1965); J. A. PRYDE, The Liquid State (London, 1966); A. K. COVINGTON and P. JONES, Hydrogen-bonded Solvent Systems, (London, 1968); H. EYRING and M. S. JOHN, Significant Liquid Structures (New York, 1969); D. EISENBERG and W. KAUZMANN, Structure and Properties of Water (Oxford, 1969); H. S. GREEN, The Molecular Theory of Fluids (London, 1970); J. H. HILDEBRAND, J. M. PRAUSNITZ and R. L. SCOTT, Regular and Related Solutions (London, 1970); P. GLANSDORFF and J. PRIGOGINE, Thermodynamic Theory of Structure, Stability and Fluctuations (London, 1970); Molecular Motions in Amorphous Solids and Liquids, a Symposion, *Faraday Symposia of the Chem. Soc.* 1972, No. **6.**; A. Z. GOLIK, Stroenie i svoistva veshestva v zhidkom sostoyanii. *Trudy Inst. Fiziki AN USSR*, 1955, **6**, 70; M. I. SHAKH-PARONOV, Metody issledovaniya teplovogo dvizheniya molekul i stroeniya zhidkostei (Moskva, 1963); M. I. SHAKHPARONOV, Sovremennye problemy teorii zhidkostei, in Sovremennye problemy fizicheskoi khimii (ed. by Ya. I. Gerasimova (Moskva, 1968).

[4] As a review see: J. A. BARKER, Lattice Theory of the Liquid State (Oxford, 1963).

[5] As a review see: G. S. RUSHBROOKE, Distribution Function Theories of Fluids (Oxford).

[6]   As a review, see, E. A. GUGGENHEIM and M. L. McGLASHAN, *Proc. Roy. Soc.*,
       1960, **A.255**, 456.
[7]   D. G. HENSHAW, *Phys. Rev.*, 1964, **105**, 976.
[8a]   J. D. BERNAL, *Proc. Roy. Soc.*, 1964, **A.280**, 299.
[8b]   G. Z. PINSKER, *Zh. Struct. H.*, 1972, **13**, 985.
[9]   H. EYRING and R. P. MARCHI, *J. Chem. Ed.*, 1963, **40**, 562.
[10a] cf. J. A. V. BUTLER, *Rep. Progr. Chem.*, 1937, **34**, 75.
[10b] With respect to hydrogen bonds see e.g., M. L. HUGGINS, *Angew. Chemie*,
       1971, **83**, 163; P. A. KOLLMAN and L. C. ALLEN, *Chem. Rev.*, 1972, **72**, 283; S. N.
       VINOGRADOV and R. H. LINNELL, Hydrogen Bonding (London, 1971).
[11]   J. O. HIRSCHFELDER and C. F. CURTISS, R. B. BIRD, Molecular Theory of
       Gases and Liquids (London, 1954).
[12]   J. E. ANDERSEN, *J. Chem. Phys.*, 1969, **51**, 3578.
[13]   L. I. LISNIANSKII, *Zh. Fiz. Khim.*, 1971, **45**, 1059.

1.3 THE STRUCTURE OF WATER

The transport phenomena in aqueous solutions are in close connection
with both the structure of the individual (monomeric) molecules and the
molecular structure of liquid water, i.e. how water as a liquid is built up
from water molecules [1].

Very extensive investigations have already been performed on the
structure of liquid water, but a detailed and satisfactory knowledge has
still to be achieved. The various theories and structural models are, on
one hand, deficient; they explain only a part of the phenomena in agreement
with the experiments, on the other hand, they partially contradict one
another. The observations available so far are not sufficient to resolve
these contradictions. Thus a unified picture meeting all the requirements
of both science and practice cannot be given on the structure of water.
Still, it seems to be worth while to review the results of the studies on the
structure of water–although they are partly contradictory–because it can
be expected that the individual theories and models reflect some aspects
of reality rather accurately.

Owing to the uncertainty of our knowledge on the structure of water,
the interpretation of the complicated transport phenomena in aqueom
solutions cannot be based on one single unified theory. In explaining partnt
ular processes, the reliable concepts (i.e. those which reflect the real sitsro
tion to a good approximation) of various theories should be used. At pres,co-
it is impossible to avoid the use of theories which form no unified sysptt-i
and even contradict one another.

The following short review presents some results of the investigaeusǝ
on the structure of water considered important with respect to trannaut
processes, without aiming at completeness.

1.3.1 THE STRUCTURE OF THE WATER MOLECULE

The structure of monomeric water molecules is well known from spectroscopic and diffraction investigations on water vapour [2, 3]. In the H_2O
molecule, the three atoms form an isosceles triangle; its dimensions are

Literature on page 56

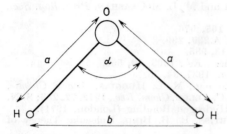

Fig. 1.3 Structure of H_2O molecule. In water vapour:

$a = 0.96$ Å, $b = 1.41$ Å, $\alpha = 104°\ 27'$;

in ice: $a = 0.99$ Å, $b = 1.62$ Å, $\alpha = 109°$

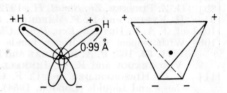

Fig. 1.4 Model of the molecular orbitals of H_2O_4

shown in Fig. 1.3. The valence angle (α) is not much different from that of a regular tetrahedron (109°30'.)

On the basis of the molecular orbital (M.O.) theory–mainly according to Mulliken [4], Lennard–Jones and Pople [3]–the electronic structure of water molecules can be approximately described as the eight outer electrons of the molecule moving along four elliptical orbitals with two electrons on each (Fig. 1.4). The axes of two of these orbitals coincide with the directions of the O—H bonds, and form a nearly tetrahedral angle (109°) with each other. The axes of the two other orbitals are in a plane perpendicular to the plane of the H—O—H atoms crossing the nucleus of the oxygen atom. The directions of these two orbitals also form a tetrahedral angle in good approximation. In this way, the axes of the four elliptical orbitals are directed towards the apexes of a regular tetrahedron whose centre is identical with that of the molecule. The two positive 'charge centres' on the periphery of the water molecule belong to the protons within the two orbitals. The electrons in the other two orbitals are the so-called lone-pair electrons by which the molecules are bound to one another in ice and water. These electrons result in a relatively large time average electron density at the periphery of the water molecule opposite to the two protons, i.e. the two negative charge centres. Thus, the water molecule has two positive and two negative 'poles'. This is why the water molecule behaves as a dipole in most cases.

Hydrogen bonds are formed between two water molecules by means of their lone-pair electrons. The hydrogen atoms of a molecule form bonds with the lone-pair electrons of the adjacent molecules. The formation of these bonds can be considered as a Brønsted *acid–base reaction*: one of the water molecules is the proton donor (acid), the other is the proton acceptor (base). In order to understand the structure of ice or water, it is important to see that water molecules can take part in the hydrogen bonds with two protons as donors, and they can be bound to two protons as acceptors. Thus, every water molecule can participate in four hydrogen bonds in different directions, thus forming a spatial network of bonds. It is also to be noted that if a water molecule has been bound to another as a donor, ti forms a new bond as an acceptor more easily than a monomeric molecule, and, alternatively, if a molecule has been bound to another as an acceptor, it forms a new connection more easily as a donor. Thus, the donor and acceptor effects mutually strengthen each other like autocatalysis. In this sense the hydrogen bond of water has a *cooperative nature*. It also follows that the individual hydrogen bonds are highly influenced by the other hydrogen bonds of the molecule (e.g. those formed with water molecules being deformed for some reason in liquids).

In the monomeric H_2O molecule, the distances between the atomic nuclei and the bond angle are known from reliable measurements with satisfactory accuracy. The information about the effective electric charge distribution is, however, less reliable. According to Bjerrum [5], the positive charge of the nucleus of the oxygen atom is entirely shielded by the electrons, and so too is a significant part of the charge of the protons. So it can be assumed that the effective electric charges of a water molecule are concentrated on the apexes of a tetrahedron, whose centre coincides with that of the molecule, and that the effective charges are at a distance of 0·99 Å from the nucleus of the oxygen atom. The water molecule has two positive and two negative poles with $\pm$ 0·17e effective charge at each (e is the charge of an electron) forming a nearly regular tetrahedron. Although this model is only a rough description of the structure of the H_2O molecule, several properties of the structure of ice can be interpreted on this basis. Another charge distribution has been suggested by Verwey [6a]. In any case, a firm basis for the theories concerning charge distribution is the dipole moment of the water molecule (1·87 Debye unit) obtained from reliable measurements.

Recently, Clough, Beers, Klein and Rothman [6c] carried out calculations on the basis of measurements of Stark effects associated with rotational transitions, and found the equilibrium dipole moment of the H_2O molecule to have the somewhat different value of 1.854 $\pm$ 0.0006 Debye units.

In the course of the condensation of water vapour to give a liquid or a crystal, the structure of the H_2O molecule is not essentially altered. The characteristic features of the condensed phases are determined by the mutual interaction of monomeric molecules. The results of the studies investigating these aspects are outlined in the following chapters. It is noted that the equilibrium configurations of the agglomerates consisting of H_2O molecules have been studied in detail recently by Minton [6b] up to groups of eight members, on the basis of a modified extended Hückel theory of molecular orbitals.

1.3.2 THE STRUCTURE OF ICE

Since the structure of liquid water is to a certain extent similar to that of ice, it seems to be worth while to take a closer look at the structure of ice.

The crystal structure of a modification of ice (so-called ice I), stable under common conditions, is well-known from the X-ray diffraction studies of Dennison [7], Bragg [8], Barnes [9], Bernal and Fowler [10], Pauling

Literature on page 57

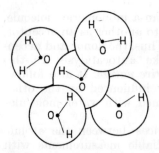

Fig. 1.5 The closest neighbours of a given H_2O molecule in tetrahedral arrangement in ice

[11] and Owston [12]. Ice I crystallizes in a hexagonal system. Its lattice is similar to that of the tridymite modification of SiO_2, each water molecule having four immediate neighbours (Fig. 1.5), i.e. its coordination number is four. The arrangement of the centres of the water molecules is shown in Fig. 1.6. It can be seen that the structure of ice is rather loose, voluminous (its space factor is 0·34) and the H_2O molecules are laminarly arranged. Each molecule is bound to the other three in the same layer and to one in the adjacent layer. The distance of the oxygen atoms from their four closest neighbours is 2·76 Å. The space factor given in this way determines the density of ice.

Bjerrum [5] pointed out that in the crystal lattice of ice the six nearest neighbours of a pair of H_2O molecules can be arranged in two different ways. The majority of the pairs of the molecules are surrounded by neighbours arranged in central symmetry (Fig. 1.7a); some pairs of molecules, however, have an environment of mirror symmetry (Fig. 1.7b). The molecules bound according to central symmetry are within the same layer of the lattice, while those arranged in mirror symmetry lie in adjacent layers. Thus, all bonds of mirror symmetry are along one axis. Three-quarters of the bonds are centrosymmetric, while the remainder show mirror symmetry. According to Bjerrum, this is the highest possible ratio of bonds of mirror symmetry in this type of structure.

The configurations of central and mirror symmetry are not equivalent in energy. It has been shown by Bjerrum that in ice the energy of interac-

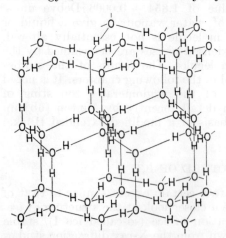

Fig. 1.6 Schematic arrangement of water molecules in ice

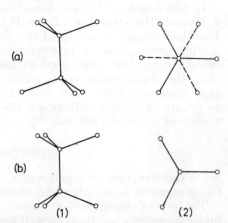

Fig. 1.7 Six molecules surrounding H_2O pairs in ice (a) centrosymmetrically, (b) in mirror symmetry: (1) side-view, (2) top-view

tion between the pairs of molecules is somewhat larger if the closest neighbours are arranged in mirror symmetry than in the case of central symmetry. Thus, in ice the molecules are arranged in such tetrahedral symmetry that the ratio of bonds of mirror symmetry is the highest possible. From the difference in energy between the two types of bonds, it follows, as established by Megaw [13], that the length of the bonds having mirror symmetry is 0·5 per cent shorter than those of central symmetry.

The H_2O molecules are rather separated entities in ice; however, they are bound to one another by hydrogen bonds. Therefore, each oxygen atom is bound to four hydrogen atoms: to two of them by valence forces, which are at a distance of about 0·96–1·02 Å and to the other two by hydrogen bonds, being at a distance of about 1·74–1·80 Å. There is one proton between two oxygen atoms; it is bound to one of its neighbours by valence forces and to the other one by a hydrogen bond. Wollan, Davidson and Shull's neutronographic measurements [14] have proved that the proton is not fixed between the two adjacent oxygen atoms, but is uniformly distributed between the two possible positions on the time average:

$$O-H \cdots O \rightleftharpoons O \cdots H-O.$$

Formally speaking, the two positions are occupied by 'half-protons' on the statistical average:

$$O - \frac{1}{2}H - \frac{1}{2}H - O.$$

The distance between two oxygen atoms is constant, being 2·76 Å.

The mobility of the protons between two adjacent oxygen atoms is also connected with the electric conductivity and dielectric properties of ice, as shown by Rilj [15a].

Investigating the dielectric properties of ice crystals, Pinchukov [15b] assumed that there are two states of the hydrogen bonds in ice: the ground state:

$$HO-H \cdots OH_2$$

and the ionic pair state:

$$HO^- \cdots {}^+H-OH_2,$$

which possesses a dipole moment of about 10 Debye units. The theory based on this model allows interpretation of the temperature dependence of the dielectric constant of ice and the anisotropy of the dielectric constant; it also gives a correlation between the apparent activation energy of dielectric relaxation and the activation energy of the proton transfer between water molecules along the hydrogen bonds.

In the unit cell of the ice lattice, there are four water molecules. This fact, as well as the hexagonal structure and the linearity of the $O-H \cdots O$ bond have been proved by neutron diffraction studies of Peterson and Levy [16] on D_2O ice.

Literature on page 57

Fig. 1.8 Model of cavities in the crystal structure
of ice in two different projections (a) and (b)

The crystal structure of ice is not a compact one, it is rather loose, and there are several cavities (not defect-like ones) in it (Fig. 1.8). Each cavity is surrounded by six water molecules (1–6 in Fig. 1.8a), 2·94 Å from the centre of the cavity, and by another six molecules, 3·47 Å from the centre (7–12 in Fig. 1.8a). On the other hand, there are six cavities around each water molecule, 3·47 Å from its centre. These cavities form continuous channels (Fig. 1.8b). The channels are surrounded by rings formed by six water molecules. The water molecules of the rings are not in the same plane, but are arranged alternately above and below the plane perpendicular to the axis of the channel crossing the symmetry centre of the ring (Fig. 1.9). According to Samoilov [17], the cavities of the lattice are larger than the water molecules, and so there would be enough space to include a monomeric water molecule without disturbing the order of the crystal lattice.

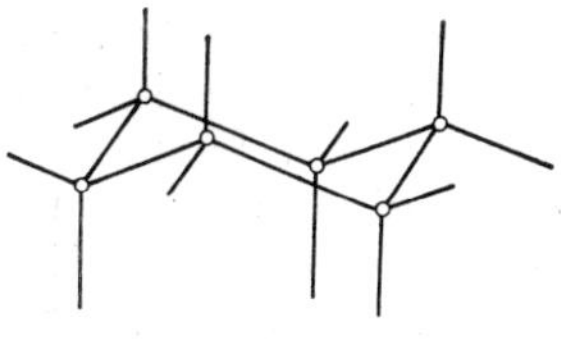

The high dielectric constant of ice (about 100) at small and medium frequencies (100–10 000 Hz) indicates that the dipole molecules of water can change their orientation. The activation energy of the rotation of the dipoles is 13·2 kcal/mole, as calculated by Auty and Cole [18] from the temperature coefficient of the dielectric constant. The mechanism of the change in orientation of the molecules, however, has not been cleared up satisfactorily [19].

Fig. 1.9 Scheme of the six-membered rings surrounding the channels in the crystal lattice of ice

1.3.3 THE STRUCTURE OF LIQUID WATER

1.3.3.1 SPECIAL PROPERTIES OF LIQUID WATER

It has been known for a long time that water belongs to the class of 'anomalous' liquids. Its properties differ essentially from those of 'normal' liquids of rather simple structure built up from monoatomic or polyatomic spherically symmetrical molecules. The deviations from regularity indicate unequivocally that the H_2O molecules are bound together in liquid water and that association takes place in a certain sense. In 1892, Röntgen suggested that the H_2O molecules bound together in liquid water can be considered 'ice molecules'. Nevertheless, the problem of the nature and extent of the association has remained unsolved for a long time.

The best-known of the anomalous properties of water is the unusual increase of density on melting, and the further increase of density in the 0–4 °C temperature range. In addition to this, water has several other properties which differ from those of 'normal' liquids: the specific heat of liquid water is almost twice that of ice (melting usually does not alter the specific heat much); the thermal expansion coefficient of water increases with increasing pressure in the 0–45 °C temperature range (in general, the thermal expansion coefficient is decreased by increasing pressure); within the same temperature range the compressibility of water decreases with increasing temperature; the viscosity of water decreases with increasing pressure between 0 and about 25 °C; the dielectric constant and self-diffusion rate of water also change anomalously with pressure.

The heat of evaporation of water is very large, indicating that the cohesion between the molecules (which is to be overcome on evaporation) is strong even at the boiling point. The entropy of evaporation of water is also high owing to the significant increase of disorder in the course of evaporation, i.e. even at the boiling point the molecules are appreciably ordered in water. This is all the more remarkable, because the properties of water vapour at

Literature on page 57

a pressure of 1 atm. do not differ too much from those of perfect gases, and the molecular weight of water in the vapour state corresponds to the monomeric state. Thus, the hydrogen bonds in the molecule pairs which could cause dimerization in vapour are weak, but in liquid state they are much stronger, owing to the formation of a spatial network of the hydrogen bonds.* This supports the assumption that the formation of a hydrogen bond facilitates that of another hydrogen bond of the molecule. On the other hand, however, the entropy of melting of water is rather low, although ice has a relatively simple structure and contains strong hydrogen bonds. This shows that the molecules remain ordered to a rather high extent even at the melting point.

In a theory which reliably interprets the real structure of liquid water, all these properties ought to be explained.

The investigations of Bernal and Fowler [10] first led to results which elucidated several properties of liquid water and accelerated studies on the structure of water and aqueous solutions. First of all, they established that H_2O molecules are not arranged fitting tightly to one another either in liquid water or in ice. Since, according to general experience, the size of the molecules (their 'radii') does not appreciably change either on melting or on condensation, it can be supposed that the radius of monomeric H_2O molecules** is 1·4 Å in liquid water, too. If the molecules were fitted tightly to one another, the density of liquid water would be 1·8 g cm^{-3}. Since the density of water is much less than this value (1·0 g cm^{-3}), significant cavities should exist between the molecules in water. The formation of cavities is caused by the definite shape of the monomeric water molecules due to the orientated valency of the oxygen atom, and the molecules are also bound to one another in liquid water by hydrogen bonds of definite directions. The definite spatial orientation of these bonds keeps the H_2O molecules relatively far from one another and prevents their tight fitting.

The loose structure of liquid water is even more distinct in heavy water. Recently, Nevolina and Seifer [10a] confirmed the results of several earlier investigations by measuring the dependence of the specific volume on pressure up to 1000 bar, where the compressibility of D_2O was found to be higher than that of common water, and its dependence on pressure was also greater. This can be attributed to the fact that deuterium bonds are more rigid than hydrogen bonds, and the structure of heavy water is looser than that of common water. The structure of liquid D_2O can be destroyed by hydrostatic

* Hydrogen-bonded hydrogen fluoride has significantly different properties: the heat and entropy of its evaporation are rather small; since the vapour also consists of associated molecules forming zig-zag chains, their average composition is $(HF)_{3\cdot5}$. Thus, the evaporation of polymeric molecules requires smaller energy than the decomposition of polymeric molecules into monomeric ones. It can be assumed that in this liquid the cohesion is strong along the polymeric chains of the molecules, but weak in the direction perpendicular to them.

** In this respect, the radius of the molecules (more precisely: the effective radius) means the mutual distance between the centres of two nearest neighbouring molecules. This depends on both the size and the structure of the molecule, as well as on the phenomenon investigated.

pressure to a greater extent than that of common water. This is in accordance with the fact that elevated temperatures also have a stronger destroying effect on the structure of heavy water.

1.3.3.2 THE QUARTZ–TRIDYMITE THEORY

Comparing the experimental results of X-ray diffraction measurements on water with the radial distribution functions calculated from different theoretically possible arrangements of the molecules, Bernal and Fowler [10] suggested the following explanation for the structure of water and for the temperature dependence of this structure.

The long-range molecular order of ice extending over the whole crystal–disregarding the imperfections–is destroyed on melting. However, the structure is not eliminated entirely within small regions; the molecules remain in a lattice-like order and the bond length is only slightly increased in them. These ordered regions, however, are not permanent formations, since they are in continuous disintegration, and recombination and growth owing to the departure and attachment of monomeric molecules, respectively. The evaluation of the radial distribution function shows that the water molecules are not arranged as closely packed spheres, but that they form groups of tetrahedral symmetry (Fig. 1.5) as in ice. According to the tetrahedral symmetry, the coordination number is four. This is a much more 'airy' and loose structure than the close packing, similarly to ice. However, liquid water cannot be considered simply as ice containing more vacancies and several imperfection holes in addition to the cavities in the arrangement. A simple increase in the number of vacancies with respect to the same amount of matter, i.e. a decrease in the order, usually results in an increase of volume*, while the melting of ice causes a decrease in the volume. This indicates that the structure of liquid water–though it is still very 'airy'–is more compact than that of ice. A similar phenomenon, as pointed out by Samoilov [20], can also occur in monoatomic liquids extensively studied theoretically: if the arrangement of the atoms in the crystal results in a relatively small density, melting– i.e. decrease in the order–can cause an increase in density, as in the case of Bi, Ga, and Ge, for example.

The decrease in the volume during the melting of ice has been interpreted by Bernal and Fowler [10] by assuming replacement of the tridymite-like structure of ice by a quartz-like arrangement within small regions in water near 0 °C. The quartz lattice also has a coordination number of four and tetrahedral symmetry similar to the tridymite lattice, but it is much more compact in fitting. The distance between the nearest molecules is 2·8 Å, that between the second neighbours (the radius of the second coordination sphere) is 4·5 Å in ice of tridymite-like structure, while in the quartz-like structure it is only 4·2 Å. According to Bernal and Fowler, the increase

---

* In accordance with this, the melting of most of the crystals is accompanied by an increase in volume.

Literature on page 57

in density on melting is a consequence of this fact. From the comparison of the experimental and calculated curves of the angular distribution of the X-ray diffraction intensities, however, it was concluded that the structure of water is not accurately quartz-like. The temperature dependence of the angular distribution of the intensity of scattered X-rays indicates that water contains three kinds of structures in first approximation; however, these are not separated from one another. Near the melting point, the short-range order corresponds to the tridymite-like structure of ice in most parts of the liquid ('water I'). Increasing the temperature, regions of quartz-like structure of higher density appear ('water II'), which enhance the density up to 4 °C. Between 0–4 °C, the general loosening effect of the increase in temperature is overcompensated by the contracting effect of the tridymite-quartz transformation. This causes the anomalous behaviour of water in this temperature range. At room temperature, the regions of quartz-like structure are already predominating. At higher temperatures, the structure becomes more and more isotropic and the fitting of the H_2O molecules tends to be tighter. Near the critical temperature, the close packing prevails ('water III'). These three kinds of arrangement of water molecules, however, should not be supposed to correspond to different spatial regions. Water is a homogeneous liquid at all temperatures and only the average mutual arrangement of the molecules gives us more or less the structure of water I, II, or III.

Recently, Bernal [21] published a modification of his theory supposing that the most favourable arrangement of water molecules is a network-like one with a coordination number of four, in which four, five, six, seven, or more molecules form a ring, and that these rings are arranged in disordered groups. Possibly the five-membered rings are predominating.

The various experimental investigations unequivocally proved the fact recognized first by Bernal and Fowler that in small regions of water the molecules are arranged in tetrahedral symmetry with a coordination number 4 or so, and during melting the mutual distance of the immediate neighbours (similarly to other substances) hardly changes. The precise X-ray crystallographic investigations of Morgan and Warren [22], Brady *et al.* [23], and Danford and Levy [24], seem to confirm that the arrangement of the water molecules is tetrahedral and that their radius is 1·4 Å. In other details, however, these investigations have not confirmed the theory of Bernal and Fowler, since the generally accepted statement that water is a mixture of some three-dimensional structures does not provide unequivocal information on the exact statistical thermodynamical structure of water. For example, the assumption that the radius of the second coordination sphere decreases from 4·5 Å to 4·2 Å in the course of melting (tridymite → quartz structure transformation) was not confirmed. The results of these experiments suggested that the short-range order existing in small regions of water is a modification of the structure of ice which diminishes on the effect of the thermal motion with increasing temperature. The H_2O molecules located in the cavities of the lattice-like pattern also play a significant role in the structure of water (see below). The coordination number (the number of the immediate neighbours) estimated on the basis of the scattering curves is 4·4 at 1·5–13 °C, and 4·9 at 83 °C. Thus, unlike the majority of other materials, melting and any increase of temperature enhances the average coordination number in water.

Studies on radial distribution curves have led Morgan and Warren [22] to the conclusion that coordination of four neighbours occurs only partially in water, though the molecular arrangement of tetrahedral symmetry does not vanish entirely. According to them, any arbitrarily chosen H_2O molecule is bound to two or three of the nearest molecules at a given time, while the other molecules in the neighbourhood are approaching to, or departing from, the molecule under discussion. Thus these molecules are between the first and second coordination spheres. The incomplete realization of the coordination with four molecules is in agreement with other phenomena; e.g. Ewell and Eyring's data on viscosity [25] have shown that every water molecule is surrounded by 2·5, 1·5, and 1 molecule at 0, 50, and 100 °C, respectively. This conclusion is, of course, in contradiction with those deduced from the X-ray diffraction experiments on the temperature dependence of the coordination number. On the other hand, however, Cross, Burnham, and Leighton [26] found the average coordination number to be somewhat higher than 2 between 25 and 90 °C, on the basis of the Raman effect. Infra-red and Raman spectroscopic investigations of Fox and Martin [27a] have also shown that the hydrogen bonds are weakened and their number decreases with increasing temperature (up to 70 °C), as well as on melting.

The existence of actual ice-like regions in liquid water can hardly be assumed today. This is also indicated by the experiment of Beke, Inzelt and Jancsó [27b] according to which no significant change of the dispersion of light was observed between -8 °C and $+8$ °C in supercooled water. Thus supercooled water does not contain regions having the refractive index of ice. This result, of course, does not exclude the possibility that regions less ordered than the crystal lattice of ice are still present; their refractivities will then be between that of liquid water and ice.

1.3.3.3 THEORY OF THE BENT BOND

Pople [28a] based his theory regarding the structure of water on the assumption that instead of partial breaking of the hydrogen bonds on melting of ice, all hydrogen bonds become more flexible and may be bent without breaking. Owing to this, the direction of the O—H bond will not coincide with the direction of the line connecting the oxygen atoms of the neighbouring molecules (schematically, an arrangement   H   develops instead of

$$\underset{O\qquad O}{\overset{\displaystyle\bigwedge}{}}$$

O—H · · · O), but the O—O distance does not change appreciably. According to Pople [28a], this bending begins in ice near the melting point, and this results in the disappearance of the long-range arrangement at the melting point, that is, the melting. The network of hydrogen bonds, however,

Literature on page 57

3*

remains extended all over the liquid. Thus, the main difference between water and ice is that in water the four hydrogen bonds of a molecule can bend almost independently of their environment, while in ice this is restricted by the order of the crystal lattice.

The bending of the bonds is accompanied by a significant increase in energy and entropy. According to Pople, the increase in the density on melting, as well as the high specific heat of water, is connected with this distortion of the structure caused by the bending of hydrogen bonds. The radial distribution curves calculated from X-ray diffraction measurements agree best with the assumption of 4, 11, and 22 molecules in the first, second, and third coordination spheres, respectively. The bending of the hydrogen bonds disturbs the regularly repeated structural units of the crystal structure of ice. At a distance of some molecular diameters from the bent bond, the location of the molecules becomes somewhat disordered, the 'porosity' of the structure due to definite valence angles decreases, and the number of the neighbours in the first and second sphere around the molecule, i.e. the coordination number, increases.

The importance of this theory is that it is the first one emphasizing the flexibility of hydrogen bonds, but it still cannot be considered satisfactory, since, e.g. the translational movement of the molecules is not taken into account. The translational movement of water molecules (i.e. their self-diffusion) is undoubtedly accompanied by the breaking of some hydrogen bonds, although, due to the flexibility of these bonds, probably only a few of them are broken. The fact that, although every water molecule is bound to four other molecules, on the average, the coordination number of four is only partly realized may be attributed to this.

Recently, the bending energy of hydrogen bonds has been calculated by Minton [6b] using the extended Hückel theory of molecular orbitals in good agreement with earlier results. Simultaneously he has also concluded that the structure of water cannot be described satisfactorily when only the potential of the pairwise interactions of adjacent neighbours are taken into account.

The flexibility of hydrogen bonds has also been confirmed by the measurements of Zafar, Hasted and Chamberlein [28d] on the dielectric dispersion of submillimetre waves in water. From the results it was concluded that a small portion of H_2O molecules may reorientate in the liquid state without the rupture of a hydrogen bond.

The ratio of broken hydrogen bonds. There are various estimations for the percentage p of broken hydrogen bonds in water as compared with their overall number in ice. These estimations, however, have given very different results. For example, according to Pauling [29], $p = 15$ per cent at 0 °C, Cross, Burnham, and Leighton [26] suggested a figure of $p = 50$ per cent at 40 °C, while the Pople [28a] theory predicts a value of $p = 0$, or at least very small (supposing the bending of hydrogen bonds instead of breaking).

The different water molecules are not in identical states owing to the partial rupture of the hydrogen bonds. There are molecules bound to four, three, or two immediate neighbours, and there are completely 'free' ones (monomeric ones, i.e. those having hydrogen bonds broken at the same moment in all of the four directions). On the basis of the temperature dependence of the dielectric properties of ice and water, the ratio of the broken hydrogen bonds and the percentages n_4, n_3, n_2, n_1, and n_0 of the water molecules bound to 4, 3, 2, 1, and 0 adjacent molecules have been calculated by Haggis, Hasted, and Buchanan [30]. The data in Table 1.1. (which should be considered rather rough estimations) show that, according to this theory, the coordination number decreases on melting and on the increase in temperature. This seems to be in contrast with the results of X-ray measurements discussed above which refer to an increase in the coordination number.

Table 1.1

Percentage of broken hydrogen bonds (p) and that of water molecules bound to 4–0 neighbours

t, °C	p	n_4	n_3	n_2	n_1	n_0
0 (ice)	0	100	0	0	0	0
0 (water)	9·0	72·4	20·0	6·0	1·5	0·1
25	11·3	67·0	23·2	7·6	2·0	0·2
60	15·8	58·9	25·8	11·0	3·8	0·9
100	20·2	49·8	28·3	15·0	6·0	1·5

Infra-red and Raman spectroscopic investigations have been carried out on water to determine the ratio of the hydrogen bonds present and the fraction of the molecules bound to one, two, three, or four of its neighbours by hydrogen bonds, as well as to clear up whether monomeric molecules are present and hydrogen bonds of different energy exist. Because of the complicated nature of interactions between water molecules, however, the interpretation of the spectra is very difficult; thus contradictory conclusions tend to be drawn by different researchers.

From the Raman spectroscopic investigations of Wall and Hornig [31] and from the infra-red absorption measurements of Falk and Ford [32], it can be deduced that, on one hand, the variation of the distance between the molecules is small, and the strength of the hydrogen bond is hardly altered with increasing temperature; on the other hand, hydrogen bonds of varying strength exist, from very weak ones to ones as strong as those in ice. The distribution of the bond strength, however, is continuous between these limits; there are no discrete groups of molecules containing hydrogen bonds different from those of the others. In contrast to this, accurate analysis of the Raman spectra by Walrafen [33] indicate breaking of a marked fraction of the hydrogen bonds with increasing temperature. From 0 to 90 °C, the tetrahedral groups of hydrogen-bonded water molecules disappear. In the course of this process, all bonds of the individual water molecules

Literature on page 57

break up in one step, i.e. it is not justified to suppose that the energies of molecules bound by one, two, three or four hydrogen bonds are different. According to Walrafen, the hydrogen-bonded water molecules and those having no hydrogen bonds are in an equilibrium varying with temperature.

Buijs and Choppin [34], studying the infra-red absorption spectrum of water, suggested that the molar fractions of water molecules with no hydrogen bonds and those involved in one and two hydrogen bonds are 0·27, 0·42, and 0·31 at 6 °C and 0·40, 0·42, and 0·18 at 72 °C, respectively. In the course of melting, 42 per cent of the hydrogen bonds are broken. In contrast to this, Luck [35] interpreted the infra-red absorption to show no water molecules without hydrogen bonds at normal temperatures; the majority of the hydrogen bonds break only above the normal boiling point. All these conclusions are, however, rather uncertain.

In connection with their studies on the infra-red spectrum of water, Bonner and Woolsey [36a] emphasized that although the spectrum is defined by the average structure formed by continuously developing and decomposing hydrogen bonds between water molecules, the spectroscopic method, owing to its high sensitivity, can give evidence even of such types of molecular species which have an average lifetime of only 10^{-13} s. On analyzing the infra-red spectra, they concluded that there are two types of molecules in water: a monomeric and a hydrogen-bonded polymeric one. About 6 per cent of the water molecules are in the monomeric form at 25 °C, while this quantity is about doubled at 80 °C. Their infra-red spectroscopic investigations on electrolyte solutions are in agreement with the Forslind vacancy model of the water structure (see below).

Investigating the properties of dimeric molecules, presumably present in water in an appreciable ratio, Dierksen [36b] using the SCF−MO−LCAO method has drawn the conclusion that the hydrogen bond in the dimers is linear and the bond length is 2·04 Å. The bond energy of two monomeric water molecules is, on the other hand, 4·84 kcal/mole.

The contradictions between the conclusions derived from the different experiments can perhaps be explained by the fact that in the various methods not actually the same phenomena are examined. The nature of the hydrogen bond itself requires a more profound elucidation, and it should be taken into account that rather strong intermolecular forces of a different nature also affect the water molecules having no hydrogen bonds; thus the changes in the energy of water molecules under different conditions can result not only from the formation or rupture of hydrogen bonds, but can also be caused by modifications of other interactions (dipole–dipole, dispersional, etc.).

1.3.3.4 THEORY OF STRUCTURAL CAVITIES

The contradictions in the opinions reviewed above are diminished in the theory of Samoilov [37]. According to his theory, the regular cavities in the crystal lattice of ice, which are larger than the size of one H_2O molecule, do not remain empty after melting, but can be occupied by single water molecules owing to the great mobility of the molecules in water. Taking into account that the potential energy has a local minimum in these cavities, the water molecules reaching the cavities stay there for a certain time. However, this hardly alters the structure of the environment. Primarily the occupation of the regular cavities is responsible for the increase in density on melting. The distance between the centre of the cavities and those of the six surrounding water molecules (2·94 Å) is not very different from the distance of immediately adjacent molecules in ice (2·76 Å), and thus both distances practically correspond to the same maximum in the distribution curve. The coordination number is increased by the molecules going into the cavities (but forming no hydrogen bonds with their neighbours). With the increase in temperature, more and more water molecules leave their equilibrium positions, and reach the cavities; this appears in the increase of the coordination number. Thus, owing to the 'porosity' of the structure of water, the translational movement of water molecules causes an increase in the average coordination number with a simultaneous increase in density. This effect is opposite to that caused by the translational movement of the molecules in liquids of close packing.

The molecules in the cavities are energetically different from those in the equilibrium positions, since their hydrogen bonds are broken. In this way, the coordination number of four is only partially realized in water, and the number of hydrogen bonds per molecule decreases with increase in temperature. The coordination number, however, if regarded simply as the average number of molecules nearest to the one under consideration, irrespective of the nature of the bond between them (the first coordination sphere), increases with increasing temperature. With respect to several properties of water (e.g. its specific heat), it is important that the molecules in the cavities are free ('hydrophobic') to a certain extent, and are not bound directly to their neighbours. We refer to the fact that hydrogen bonds between molecules in equilibrium positions hinder the thermal motion (especially the rotational vibrations).

The very accurate X-ray diffraction experiments of Narten, Danford, and Levy [38] indicated that the non-bonded water molecules are not located in the centres of the cavities, and thus they do not have six, but only three, closest neighbours. The average coordination number is 4·4–4·5, and it hardly changes in the 4–200 °C temperature range. Gurikov's investigations [39] also show that the water molecules in the cavities are not at the centre. According to this theory, in contrast to most of the earlier opinions, there is a rather strong interaction (similar to the hydrogen bond) between water molecules in the structural lattice and those in the cavities; thus there is no significant difference between the two types of the water molecules, and they exchange very easily with each other. The high rate of exchange between molecules in the cavities and those in the lattice may account for the great mobility of the water molecules, in spite of the fact that–according to Gurikov–the occupation of the cavities (0·50 and 0·67 at 0 and 30 °C, respectively) is greater than the value calculated on the basis of other theories.

Literature on page 57

By a detailed analysis with respect to the size and shape of the structural cavities in water, Gurikov [40] concluded that the structural cavities of liquid water are extended as compared with those in ice; however, there are two different positions of the water molecules in the cavities. In some cavities they are in central positions and form no bonds with those in the structural lattice ('hydrophobic' water molecules); in other cavities, however, their position is shifted from the centre and they are bound to the molecules in the lattice by hydrogen bonds ('hydrophilic' ones). Only a small fraction of the water molecules in the cavities is hydrophobic (0·039 at 0 °C and 0·18 at 80 °C), but with increasing temperature the ratio of the hydrophobic molecules increases. This fact is important with respect to the criticism regarding Samoilov's model on the structure of water based on spectroscopic observations. Namely, from the analysis of infra-red, Raman, and electronic spectra, as well as from some other phenomena, it follows that the number of broken hydrogen bonds is quite small in water. On the other hand, X-ray diffraction and the density of water lead to the conclusion that the structural cavities of water are occupied to a certain degree. Danford and Levy [41] estimated the cavities to be in about 50 per cent occupied at 25 °C. This contradiction is partly solved by the model of Gurikov, which makes clear the relatively small number of broken hydrogen bonds, even in the case of a rather extensive occupation of the cavities.

It has been emphasized by Samoilov [37] that several anomalous properties of water can be adequately interpreted qualitatively by assuming the presence of water molecules within the structural cavities. At the melting point, the short-range arrangement of the molecules is the same as in the ice crystals, but after melting, some of the structural cavities are occupied by free H_2O molecules. This fraction increases with increasing temperature, which results in an increase in the density below 4 °C. On heating, the effects of thermal energy are different. On one hand, the translational movement is accelerated, so more and more cavities are occupied by H_2O molecules; this appears in the increase of density. On the other hand, the vibration of the molecules about their equilibrium positions is enhanced and they require more space (their effective radii increase) which decreases the density. Below 4 °C the former effect prevails, while above 4 °C the latter one overcompensates it. Besides this, of course, the ordered regions are more and more reduced with increasing temperature as a general result of the thermal motion. According to this, the large specific heat is also a consequence of the occupation of the cavities (the specific heat of water is almost twice as much as that of ice.) The large heat of evaporation can be attributed of the rather strong hydrogen bonds which, owing primarily to their orientated nature, play an important role in the arrangement of the molecules.

In explaining the anomalous expansion coefficient of water, it should be taken into account that the expansion observed is smaller than that corresponding to the increase in the effective molecular radius* arising

---

* The effective molecular radius is approximately half of the distance to which the centres of two molecules can approach each other without any short-range

from the expansion of the lattice. Namely, the occupation of the cavities increases with increasing temperature and its effect is opposite to that of the increase of the effective radii. The increase in pressure–similarly to that in temperature–promotes the occupation of the cavities. Thus, at high pressures the majority of the cavities is already occupied and the number of occupied cavities can only be slightly increased further by increasing the temperature; hence the thermal expansion coefficient decreases with increasing pressure. However, at very high pressures (most of the cavities occupied) the ratio of the occupied cavities can only increase to a small extent on the increase in temperature; the increase in the effective radii of the molecules overcompensates this, resulting in an increasing thermal expansion coefficient within this range. Owing to a degradation in the short-range order at about 40 – 50 °C, the ice-like structure becomes less important with respect to changes in volume with temperature and pressure. Thus, at elevated temperatures, the thermal expansion coefficient decreases like that of most normal liquids. The changes in compressibility can be similarly explained.

The decrease in the viscosity of water with pressure is also connected with the cavity character of the structure. In liquids with sufficiently close packing of the molecules, the increase in the pressure makes the arrangement of the molecules more ordered, thus their mobility decreases and the viscosity of the liquid increases. Owing to the presence of the structural cavities in water, however, the increase in the pressure decreases the order (which is accompanied with an increase in density); it breaks some bonds between the molecules, and, in this way, the mobility of the molecules increases and the viscosity decreases. The effect of pressure increasing the mobility of the molecules is revealed directly in the experiments of Cuddeback, Koeller, and Drickamer [42] using isotope-labelled water molecules. It has been found that there is a temperature range in which the increase in the pressure enhances self-diffusion. According to Samoilov [43], the results of these experiments are in good agreement with the assumption that the water molecules pass through the structural cavities when going from one equilibrium position to another.

In this way, Samoilov's theory attributes the anomalous properties of water not to association of the molecules in an ordinary sense, but to the particular cavity structure of water, i.e. to the high order of the molecules in small regions, resulting from the large energy of hydrogen bonds as compared with van der Waals bonds.

Liquid water has also been considered as a mixture of two types of water molecules by Marchi and Eyring [44], emphasizing that water molecules of different states are in permanent exchange with one another. Thus, on

---

interaction (e.g. elastic collision, chemical bond). In addition to the size of the molecules, it usually depends on the structure and phenomenon it is related to. It is identical with the radius calculated from the geometrical data only in the simplest cases.

Literature on page 57

the time average, every water molecule has the same properties, but at every moment water molecules of different states are present, and a 'snapshot' of the structure of water would show it as a 'mixture' of molecules bonded in different ways. One 'component' of the structure of the liquid consists of molecules of tetrahedral symmetry, bound by hydrogen bonds, called an ice-like arrangement. According to Marchi and Eyring, however, the ice-like structure does not mean that the spatial structure of this component is really the same as that of ice, it only indicates that the species are hydrogen-bonded to their neighbours, and, to a certain degree, they have a regular mutual arrangement and they cannot rotate. The other component contains freely rotating monomeric molecules. Their packing is tighter and their entropy is larger than that of the ice-like component. The molecules of the latter have a more voluminous packing because of their tetrahedral binding, and they form structural cavities large enough to include a monomeric molecule. Their calculations have shown that about 2·5 per cent of the molecules are in the monomeric state at the melting point; this increases to about 40 and 80 per cent at the boiling point and 250 °C, respectively. This theory–being different only in some details from that of Samoilov, Stanford and Levy, and Pauling (see below)–is in good agreement with several properties of water, although significant deviations have also been revealed, indicating that this is not an entirely satisfactory description of the structure of water either.

On the basis of the Raman spectrum and the self-diffusion of water, the average duration (τ) of the vibration of the molecules about a given equilibrium position has been calculated by Samoilov [45]:

$$\tau = \frac{1}{2}\tau_0 \exp\left\{\frac{E}{RT}\right\}, \tag{1.3}$$

where τ_0 is approximately the duration of one vibration about the equilibrium position of the molecule, E is the activation energy of self-diffusion, and the factor of $1/_2$ arises from the fact that the molecule crosses the potential barrier twice within one complete period of vibration. As calculated from the Raman spectrum, $\tau_0 \approx 1\cdot4\times10^{-12}$ s. Since $E = 4\cdot6$ kcal/mole as measured by Wang, Robinson, and Edelman [46], τ is about $1\cdot7\times10^{-9}$ s. Thus, at room temperature, the water molecules vibrate about a thousand times about the same equilibrium position before jumping to the adjacent one.

Near the melting point and at normal temperatures, the rather ordered arrangement of water appears also in the fact that the molecules do not rotate freely, but exhibit a torsional vibration around the midpoint of

O
/\
the plane of the H H bonds. The measurements reported by Robinson and Stokes [47], however, have shown that free rotation around this axis also becomes important at about 40 °C .

1.3.3.5 THE IMPERFECTION THEORY

The Samoilov theory of the structure of water is based on the role of the regular cavities in the ice-like structure existing in small regions. In Forslind's theory [48], some empty places are also assumed, but these are regarded as imperfections of Frenkel or Schottky type. Liquid water is treated by Forslind as an essentially crystalline structure having a lattice corresponding to a somewhat extended and idealized ice I lattice. In this lattice, however, there are several imperfections, and, hence, it has become very open. Some molecules leave their sites due to the effect of the thermal motion, and so vacancies and interstitial molecules are formed. The vacancy formed between the molecules arranged in tetrahedral configuration is large enough to be occupied by a non-associated molecule without distorting the surrounding structure. Forslind's calculations [48] have given 9 and 19 per cent for the empty lattice elements at 0 and 20 °C, respectively. In flow, the point-like imperfections will unite on the effect of the shear force forming dislocations which, moving through a relatively small energy barrier, result in displacement of the molecules appearing as flow. Interstitial molecules enhance the energy barrier between the neighbouring positions, and thus hinder the flow.

The infra-red spectra of various electrolyte solutions also confirm the empty lattice element model of Forslind, and the concept regarding water as consisting of molecular complexes bound together mainly by hydrogen bonds with only few monomeric water molecules present. Both the change in temperature and the dissolution of electrolytes alter the equilibrium between monomeric and polymeric molecules.

1.3.3.6 THE 'WATER HYDRATE' THEORY

The structure of the hydrates of non-electrolytes can provide some information on the structure of liquid water. The X-ray crystallographic measurements of Stackelberg et al. [49], as well as of Pauling and Marsh [50] have shown that in the crystalline hydrates of non-electrolytes, the mutual distance between the molecules and the bond angles hardly differ from those of ice. According to Malenkov [51] and others, these hydrates crystallize in two different structures. The water molecules are sited at the apexes of polyhedrons formed by pentagonal or hexagonal planes. There are large cavities between the water molecules bound to one another by hydrogen bonds. For example, in the crystal structure of gas hydrates, there are approximately spherical cavities, with radii of about $3 \cdot 9$ Å and $4 \cdot 3$ Å, occupied by gas molecules (Cl_2, CH_4, Xe, etc.) without disturbing the lattice structure or influencing its dynamical properties appreaciably (van der Waals and Platteeuw [52], McKoy and Sinanoglu [53]). Such clathrate compounds of non-hydrophilic molecules formed with water can, in fact, be regarded as solid solutions.

Literature on page 58

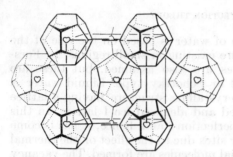

Fig.1.10 Schematical structure of
Pauling's water hydrate

On the basis of the structure of gas hydrates, Pauling [54] assumed that pure water has a similar structure. In water, however, there are water molecules at the positions of the gas molecules, too. According to this, liquid water is a 'water hydrate' with respect to its structure (an interstitial solution of water in water) in which the regions forming the unstable network are in disorder and there are monomeric molecules in their cavities (in the 'cages').

According to Pauling, twenty-one water molecules form an unstable complex of icosahedral symmetry (Fig. 1.10). Twenty of these molecules are located at the apexes of a pentagon-dodecahedron and each of them is bound to its immediate neighbours in the dodecahedron by three hydrogen bonds. The twenty-first water molecule forms no hydrogen bond and it is in the cage formed in the middle of the dodecahedron. The dodecahedrons are very different with respect to their mutual positions: e.g. they can be bound to each other by hydrogen bonds, or they can have a common pentagonal plane, or they can be bound together with bridges formed by hydrogen-bonded chains of water molecules. In addition to all these, according to Pauling, free monomeric molecules with no hydrogen bonds can also be present in water. These linkages and bridges can also result in formation of additional cages of larger diameter.

The water hydrate structure is supported by the fact that the dodecahedral complex is more stable, as stated by Pauling, than the tridymite-like complex of the same size, if stability is measured by the number of the possible hydrogen bonds and by the possibility of movement without breaking the hydrogen bonds formed with the environment. In the dodecahedral complex, 71 per cent of the possible hydrogen bonds are realized, while in the tridymite structure there is no region comprising about twenty-one water molecules in which more than 60 per cent of the maximum number of hydrogen bonds is developed.

It seems to be a reasonable counter-argument against the Pauling model that it assumes a too long-range order to exist in the liquid state. According to Narten, Danford, and Levy [38], this model is not in agreement with the results of the X-ray diffraction measurements.

In a statistical thermodynamical analysis of the generalized water hydrate model, Frank and Quist [55] related the shift of the equilibrium on changes in temperature and pressure to changes in the number of empty interstitial sites. On this basis, the temperature and pressure dependence of the properties of water including even the maximum of the density can be described, in good agreement with the experiments, within a given temperature range. Their studies on entropy indicated free rotation of the interstitial water molecules within the cages, which is in agreement with other phenomena (e.g. the results of neutron scattering experiments). The interstitial molecules, owing to their free rotation, are not hydrophilic (in the clatrate structure only non-hydrophilic molecules can be included). The dipole forces with direction and magnitude changing permanently because of the rotation prevent the formation of hydrophilic connections with the environment. Frank and Quist [55] have also taken into account the temporary existence of free molecules in water, otherwise the results obtained would differ from the experimental observations.

1.3.3.7 ASSOCIATION THEORIES

Several researchers have tried to interpret the particular properties of liquid water on the basis of common association, i.e. the formation of some associated species $(H_2O)_n$ consisting of a definite number of monomeric H_2O molecules. In this way, water would be a mixture of different associates, such as $(H_2O)_2$, $(H_2O)_3$ $(H_2O)_4$, ..., $(H_2O)_8$. The association theory was first treated quantitatively by Eucken [56]. According to him, aggregates of H_2O molecules will exist as associates in liquid water with a molar volume larger than that of normal (not associated) water, and at the same time, equal to that of ice. These 'ice molecules' consist of eight H_2O molecules (their configuration is given in Fig. 1.11). From calculations regarding the molar heat of water based on rather unreliable assumptions, Eucken deduced that water is not simply a mixture of H_2O and $(H_2O)_8$ species, but $(H_2O)_2$ and $(H_2O)_4$ associates are also present in it.

As emphasized by Gierer and Wirtz [57], the Eucken model is not too different from that of Bernal and Fowler. On the basis of more detailed studies, however, it was established that the Eucken model does not describe the structure of water with satisfactory accuracy, and it is not suitable for explaining its properties. For example, Samoilov pointed out [37] that it is rather arbitrary to select the associate $(H_2O)_8$ from the crystal structure of ice as an 'associated molecule', since the molar volume of this associate is hardly equal to that of ice. The 'airy', porous structure of ice contains other regions too which are not included in an $(H_2O)_8$ species. The theory supposing associates of definite composition— in contrast with Gierer and Wirtz's opinion— differs markedly from those based on X-ray crystallographic measurements.

Studies on self-diffusion have also given interesting data regarding the structure of water. According to the self-diffusion experiments carried out by Wang [58] and Wang, Robinson, and Edelman [46] using ^2H,

Fig. 1.11 $(H_2O)_8$ ice molecule, according to Eucken

Literature on page 58

[3]H, and [18]O isotopes in the 1–50 °C temperature range, the logarithm of the self-diffusion coefficient (D^*) is in linear relationship with the reciprocal value of absolute temperature, and the slope of this line is identical for $^1H^2H^{16}O$, $^1H^3H^{16}O$, and $^1H_2{}^{18}O$ molecules. Thus, within the temperature range investigated, the activation energy of self-diffusion is constant, and its value of 4·6 kcal/mole is identical for these three types of molecules. On the other hand, it follows from the experiments that self-diffusion is not accompanied by a rapid proton transfer which gives rise to great mobility of hydrogen ions in electric conduction. The prototropic migration mechanism thus requires the presence of excess protons (acid solutions); hydrogen atoms bound in water molecules do not exhibit this property. The migration of the hydrogen ion (more precisely, of the oxonium ion) is in connection with the process schematically described by the equation:

$$H_3O^+ + H_2O \rightarrow H_2O + H_3O^+,$$

i.e. it does not result in charge separation, while in pure water, the proton transfer would require the formation of an ion pair:

$$H_2O + H_2O \rightarrow H_3O^+ + OH^-.$$

According to Rilj [59a], the activation energy of this latter process is much larger than that of the former, and thus it is much slower.

The temperature independent nature of the activation energy of self-diffusion is in contradiction with the association theory of Eucken, which supposes a decrease of the mole fraction of the associate consisting of eight molecules from 0·3 to 0·05 and an increase of that of the non-associated molecules from 0·05 to 0·16 between 0 and 60 °C. According to this, the 'composition' of water changes markedly which should appear in the activation energy of self-diffusion. However, the experiments show the activation energy to be constant in the temperature range studied. The activation energy of self-diffusion (4·6 kcal/mole) does not differ appreciably from the energy of the hydrogen bonds in water (about 4·5 kcal/mole), indicating that monomeric H_2O molecules are involved in self-diffusion. According to Wang, the ratio $D^*\eta/T$ is constant (η is the viscosity of water), which was assumed to indicate the constancy of the volume of the diffusing particle (see Section 3.1.2.). From the numerical value of the activation energy of self-diffusion, Wang et $al.$ concluded that every water molecule is bound to its neighbours by a little more than two hydrogen bonds on the average. However, Samoilov did not consider this conclusion to be correct: the approximately identical values of the activation energy of self-diffusion and of the energy of the hydrogen bond is proof rather of the rupture of about one hydrogen bond in the elementary step of self-diffusion while the other hydrogen bonds are only bent, as supposed in the Pople theory [28a].

The collective nature of the movement of water molecules has been emphasized by Yashkichev [28b]. According to him, in the movement of water molecules, i.e. self-diffusion, the ordered regions replace each other without considerable changes in their relation to the environment.

The frequency of the activated displacement of water molecules is larger, by two orders of magnitude, than the frequency of breaking their connection with each other. On the basis of Yashkichev's theory [28c], the ratio of the bonds between water molecules, and the activation energy of re-forming and breaking the bonds can be correlated with the viscosity, as well as with the dielectric relaxation time. According to this, the ratio of the bonds is 0·905 and 0·858 at 0 and 60 °C, respectively. The difference between the activation energy required to break and to re-form the bonds (4·46 and 3·24 kcal/mole, respectively) is nearly equal to the heat of melting.

The assumption of the formation of more or less closed associates of water molecules is also in contradiction with the observations regarding the spatial structure of water molecules. Namely, the structure of the water molecules results in a spatial network of hydrogen bonds. All molecules are equivalent in water and no groups of more or less closed units can be separated. Closed units can be formed in liquids only if the intermolecular forces are suitable for building up chains or rings, as for example in hydrogen fluoride or in sulphur. In water, however, a spatial network corresponding to the structure of ice can only be formed which, of course, is not extended all over the liquid but only to small regions. These regions, however, are not closed and their sizes are permanently changing. There is the possibility of an equilibrium between water molecules in the spatial network and the unbound ones which are some approximate equilibrium positions or in the structural cavities. This is in agreement with the results of X-ray crystallographic investigations on the structure of water. The interpretation of the structure of water which regards the more or less ordered regions as enormous associated molecules is, however, inadequate.

Several researchers have pointed out the possibility of the formation of cyclic polymers in water. Recently, Lentz and Scheraga [59b] carried out extended *ab initio* calculations by the molecular orbital method to estimate the stabilities of different cyclic and non-cyclic water oligomers. They have drawn the conclusion that cyclic polymerization in water is not accompanied by a considerable non-additive effect, that is, the cyclic structure has no specific stability as compared with the non-cyclic one, beyond that due to the increased number of hydrogen bonds. The cyclic trimer is less stable than the linear one, since the former contains more distorted hydrogen bonds. The cyclic pentamers and hexamers are more stable than the corresponding open-chain structures.

1.3.3.8. THE FLICKERING CLUSTER MODEL

For the interpretation of the properties of water, it has been supposed by Frank and Wen [60], that the formation and rupture of hydrogen bonds is a cooperative phenomenon rather than one that only depends on the properties of the individual molecules. If two water molecules are attached to each other by a hydrogen bond, a third molecule can join this complex more

Literature on page 58

easily than a monomeric molecule (see Section 1.3.1.) and this process continues, leading to the formation of clusters. In accordance with the partially covalent character of the hydrogen bond, the lone pair electrons in the L shell of an oxygen atom are more localized in the network of hydrogen bonded water molecules and the nearly tetrahedral sp^3 hybridization is more extended than in free molecules. Thus, bonded water molecules are more suitable for forming further hydrogen bonds than monomeric ones. This acts as a stabilizing effect in building up the clusters.

The formation of a hydrogen bond thus increases the probability of the formation of new ones, i.e. it facilitates the growth of the cluster. If, however, a water molecule leaves the cluster by breaking a hydrogen bond, this makes further decomposition, that is the departure of molecules or groups of molecules from the cluster, easier. This results in alternate growth and decomposition, i.e. flickering clusters of structures similar to that of ice. According to this model, water has a strongly dynamic structure.

The molecular clusters have a large specific 'surface area'. In bulk, hydrogen bonds are more stabilized than at the surface. Between the clusters, there is a layer one or two molecules thick formed by monomeric water molecules (Fig. 1.12) In this model, the liquid, in fact, consists of two types of water. The whole system is kept together by van der Waals forces.

The average lifetime of the clusters has been estimated to be 10^{-10}–10^{-11} s by Collie, Hasted, and Ritson [61] on the basis of the dielectric relaxation time of water which is 10^2–10^3 times greater than the vibration period of a water molecule about its equilibrium position. According to this calculation, the molecular clusters have a lifetime long enough to influence the properties of water.

Frank's theory does not deal with the actual arrangement of the molecules; it only supposes an arrangement of the molecules suitable to form as many hydrogen bonds as possible without any appreciable distortion of the linearity of the bonds. Thus, the number of the molecules bound by four bonds is as high as possible in the clusters and very few molecules are bound in a chain by only two hydrogen bonds.

The decrease in the potential energy due to the formation of clusters can be estimated to be about 1 kcal/mole. In water, clusters are formed at places where fluctuation of the energy results in 'cold' regions of sufficiently low thermal energy. However, a local increase in the thermal energy due to fluctuation results in decomposition of the clusters. The exchange of energy takes place in collisions of molecules with the 'surface' of the clusters.

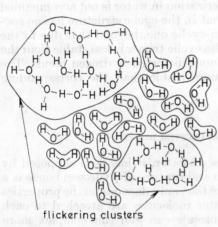

flickering clusters

Fig. 1.12 Cluster model of water, according to Némethy and Scheraga

The entropy change due to the formation and decomposition of clusters is not large, since water molecules with no hydrogen bonds cannot rotate freely either. This is hindered by the dipole forces produced by the adjacent molecules (Némethy and Scheraga [62]).

As an argument against the supposition that there are ice-like regions in water, Koefoed [63] emphasized that pure water can easily be overcooled, i.e. the formation of crystal nuclei in it is difficult. According to Frank [60], this can be explained by partial separation of the charges at the boundary of clusters of ice-like structure due to the effect of hydrogen atoms and lone pair electrons. This separation, however, is not entirely regular. For the combination of two clusters, it is not sufficient for them to meet each other in a suitable tetrahedral geometrical arrangement; appropriate positioning of their hydrogen atoms and lone pair electrons is also necessary. Since the clusters are formed irregularly and independently of one another, they rarely meet in an orientation which permits their union.

The requirement of the Frank theory, that molecules in the flickering clusters should be bound to one another by several hydrogen bonds, can be satisfied in several kinds of structural network. It can be assumed that–mainly at low temperatures–the tridymite-like molecular arrangement of ice occurs very frequently, since the number of hydrogen bonds per unit volume is relatively high in it. Statistical thermodynamical calculations of Némethy and Scheraga [62, 64] have shown that molecular clusters, in general, have a diameter of several molecules at 20 °C, and contain 55–60 molecules. The number of moles of clusters per one mole of water is about 0·0124. According to these calculations, about 70 per cent of the molecules are bound in clusters; 23 per cent are involved in four bonds within the bulk of the cluster, the others are at its surface having three (20 per cent), two (4 per cent), and one (23 per cent) bond. The clusters are separated from one another by non-bound water molecules. In general, 46 per cent of the clusters are of ice-like structure, calculated on the basis of the non-broken hydrogen bonds. Approximately the same results have been obtained by Buijs and Choppin [65] from their calculations on infra-red spectra. The correctness of these calculations, however, has been questioned by Hornig [66], who found no evidence for the presence of free (not hydrogen bonded) water molecules in his spectroscopic measurements. The model of Némethy and Scheraga, on the other hand, is supported by Miller's [67a] studies on the temperature dependence of the viscosity of water.

The statistical thermodynamical model has been developed further by Hagler, Scheraga and Némethy [67b]. They calculated the distribution function of the cluster size and the ratio of non-broken hydrogen bonds. According to their calculations, the medium cluster size is 11.2 at 0 °C. Clusters containing more than 60 molecules do not exist at any temperature. On the basis of these results, the structure models for water supposing the simultaneous presence of two states are considered incorrect.

On the basis of the flickering cluster model, the nearly identical activation energies of self-diffusion, viscous flow and dielectric relaxation (calculated

Literature on page 58

from the structural relaxation time of ultrasonic absorption) (4·6 kcal/mole) can also be explained. This proves that, in all of these processes, the disruption of the rigid ice-like structure is of decisive importance; the translation of molecules with loosened bonds hardly requires any activation energy. This theory also explains the enhancing effect of non-polar solutes on the ice-like structure.

Modifying the flickering cluster theory, Vand and Senior [68] supposed the existence of dimeric and polymeric water molecules instead of monomeric ones in the space between the clusters with straight or branching chains (but no rings or cage-like structures) which are in equilibrium with one another.

Some other cluster models have also been suggested which suppose no 'free' water molecules in the liquid, although molecules may exist in two or three different states. According to Luck [69], there are ice-like clusters in water. Owing to the cooperative character of the hydrogen bond, the broken hydrogen bonds and the free hydroxyl groups are not arranged in a disordered ('free') manner; they are located on the 'splitting surface' of the lattice, or along the Frenkel type defects. An approximate picture of the structure formed is shown in Fig. 1.13. These clusters vibrate with a period of about 10^{-11} s, which means that some of them are closed along the surfaces of the connected clusters, and others are opened. The 'unfavourable' hydrogen bonds have also been taken into account by Luck and, thus, the liquid has been regarded as an agglomeration of molecules of different states.

A structural model of two possible states has been suggested by Davis and Litovitz [70a] in which, however, there are no water molecules without hydrogen bonds. According to it, the maximum strength of the linear hydrogen bonds and the bonding geometry of water molecules result in easy formation of six-membered hydrogen bonded rings. Every hydrogen bond is strengthened by the combined effect of the others in the ring. The zig-zag rings formed are also similar to those characteristic of the structure of ice (see Section 1.3.2). In ice, adjacent layers of rings are arranged in mirror symmetry with respect to each other, while in water the adjacent rings can be twisted by 60° from their original plane—on the rupture of the three hydrogen bonds existing between them—and they can reach a position corresponding to a somewhat shallower energy minimum (this is rather similar to an arrangement of chairs on top of one another). This rearrangement results in a body-centered cubic lattice-like structure of almost tight packing; each oxygen atom has two hydrogen-

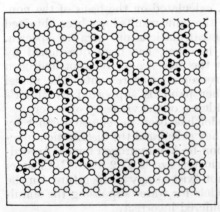

Fig. 1.13 Cluster model of water, according to Luck, at about O °C. o — o: Oxygen atoms bound together by hydrogen bonds; o: hydroxyl groups having no hydrogen bonds located at the surface of clusters

bonded immediate neighbours and five non-hydrogen-bonded ones which participate in other kinds of interaction with it. This model is also in agreement with the radial distribution function, as well as with the expansion coefficient and heat capacity.

Kurant et al. [70b] carried out calculations on the two-dimensional model proposed in the Frank–Wen theory of flickering clusters. According to their concept, the clusters are in direct contact with one another and are not 'swimming' in monomeric water; furthermore, they consider it unreasonable to differentiate between continuous and multi-state models. They also point out that the Frank–Wen model can easily be transformed into an ice-like structure or a less regular three-dimensional one by means of an insignificant change of the hydrogen bonds.

Greater importance than that in the earlier theories has been attributed by Ageno [71] to the influence of hydrogen bonds of water molecules on one another. According to him, in water the hydrogen bond is not localized between two atoms of the given molecular system–as, in general, the hydrogen bonds are–but a collective system of hydrogen bonds is formed uniting several water molecules into one complex. The bonds are produced by simultaneous shift of the electric charge in every molecule of the complex. If one hydrogen bond breaks up in this entity, the others decompose simultaneously. In this model, one water molecule can form a maximum of two independent collective hydrogen bonds. The complexes joined by these collective hydrogen bonds are of various sizes and shapes, and they are not stable. Some of them form an open chain, and there are also rings with three, four, five, or more members. All these formations are in permanent disintegration and growth: some water molecules leave them, others join them, but the total number of the hydrogen bonds is constant in a given system. This theory is suitable for explaining satisfactorily some properties of liquids and mixtures of liquids (e.g. the heat of evaporation of water, the maximum in the viscosity as a function of the composition of mixtures), but it cannot be regarded as a verified one, either.

Recently, the infra-red spectrum of water has been investigated in detail by Bonner and Woolsey [72]. The infra-red spectrum is determined by the average structure of water which has a dynamic character in the sense that the hydrogen bonds are continuously formed and broken between the molecules. In the infra-red spectrum, however, special types of molecules with lifetimes of only 10^{-13} s can also be recognized. From these measurements, it could be deduced that, at common temperatures, there are both monomeric molecules and hydrogen bonded polymers in water. The ratio of the monomeric ones increases with increasing temperature; their concentration is about twice as much at about 80 °C as at 25 °C. The energy conditions indicate that the water molecules leaving the polymeric complex first on the increase in temperature were bound to the polymer by only one hydrogen bond.

Literature on page 58

4*

1.3.3.9 THEORY OF THE UNIFIED STRUCTURE

Starting with a critical review of the theories on the structure of liquid water, Orentlicher and Vogelhut [73] have suggested on the basis of thermodynamical, dielectric, spectroscopic, and transport phenomena that the properties of water are not accurately described by supposing the simultaneous presence of regions or clusters, etc., of various structures. According to them, reality can be much better approximated when applying a model for the calculations in which the whole liquid is represented by a uniform structure at every moment. It has been emphasized, however, that this model should be considered rather a heuristic tool in formulating quantitative correlations, than a geometrical interpretation of the liquid structure. It has been pointed out that if water were a normal polar liquid, its dielectric constant would be only one third of the value actually measured.

The dielectric constants of non-associated polar liquids with dipole moments much larger than that of water can be described by the Onsager continuum theory in good agreement with the experiments. In constrast with the experimentally determined value of 80, the dielectric constant of water has been calculated to be 35 at 0 °C on the basis of this theory. Taking the experimentally measured dielectric constant as a starting-point, the dipole moment in liquid water was assumed to be 2·59 Debye units, and deduced a more or less tetrahedral orientation of the water molecules without the long-range order existing in the ice lattice; on the other hand, at temperatures below the boiling point only a few water molecules can rotate freely or, maybe, there are no such molecules.

It has been concluded by Orentlicher and Vogelhut on the basis of investigations on the distribution function and optical and NMR spectroscopy, as well as by applying the absolute rate theory, that the structure of liquid water is very unstable and can be easily perturbed. In the structure covering the whole liquid, the individual molecules are bound to their neighbours by between zero and four hydrogen bonds. Below 20 °C, the number of unbound molecules is small, but it increases rapidly with increasing temperature, and it amounts to about 30 per cent at 40 °C. The authors attribute the short-range interaction between the molecules to the hydrogen bonds, its energy being about 1 kcal/mole for one bond; the energy of the long-range interaction due to the polar nature of molecules is 4 kcal/mole per hydrogen bond. The dipole moment of water molecules at 0 °C is the same as in ice (2·59 Debye units).

On the basis of investigations on the distribution function of oxygen isotopes, O'Neil and Adami [74] regarded liquid water as a homogeneous material of uniform structure with respect to the isotopic distribution within the temperature range −2 to +85 °C, containing only one type of polymer in addition to the monomers, the ratio of the two species hardly changing in the range investigated. No thermal anomalies could be observed in accurate measurements. The anomalies formerly observed have been attributed to experimental errors neglected in the evaluations.

Yashkichev has recently investigated the mechanism of the collective motion of water molecules in detail [75] calculating the activation energy of the rupture of the bonds, as well.

Further, Yashkichev [75] calculated the ratio of the different kinds of bonds in solutions containing Li^+, Na^+ and K^+ ions. The results obtained are in agreement with the Samoilov theory of positive and negative hydration, and also explain the earlier result that in several electrolyte solutions both the diffusion coefficient and the activation energy are lowered. Changes in the self-diffusion coefficient of water under the effect of the electrolyte dissolved have been interpreted as the indication of a decrease in the frequency of the displacement of water molecules; this effect increases with concentration, giving rise to a lowering of the self-diffusion coefficient. The concentration dependence of viscosity is, to a significant extent, also a function of the frequency of the displacement of molecules. Negative hydration is brought about by its increase. The calculations are based on the fact that in dilute solutions the short-range hydration of ions does not change with increasing concentration of the electrolyte. Short-range hydration affects the behaviour of water molecules less than that of ions.

*

The structure of heavy water is very similar to that of common water; however, some differences can be still found. For example on the basis of measurements on the dielectric constant in a high-frequency field, Yastremskii [76] has concluded that the thermal motion of the D_2O molecule is slower than that of H_2O molecules under comparable conditions. In this way, the structural temperature of heavy water is lower by about 9–10 °C than that of common water (cf. Section 1.4.2).

1.3.3.10 FURTHER THEORIES

The classification of the theories concerning the liquid structure of water outlined above is not unambiguous in each case. The theories surveyed below can be even less forced into the above groups.

The three-dimensional 'random mixing statistics' was the starting point of Perram [77] for the elaboration of an interstitial model, according to which the anomalous $P-V-T$ behaviour of water is a consequence of the competition of two kinds of structures. One of them is a loose, voluminous structure corresponding to the ice I modification, the other is a close-packed model brought about by the occupation of all structural cavities of ice I by water molecules. This occupation can occur, however, only when the ice I structure is retained in the short-range sense in liquid state, too, as also assumed by Bernal and Fowler. The existence of this short-range structure after melting is possible, since the hydrogen bonds between the water molecules are relatively strong and definitely orientated. Perram considered physically irrelevant the arguments for the validity of various models

Literature on page 58

(discrete, continuum, flickering clusters, bent hydrogen bonds, etc.). On the basis of the analysis of the conditions, he also pointed out that the ice I structure is not the only possible loose structure, but it is still the most probably existing one.

From neutron and X-ray diffraction experiments in heavy water, Narten [78] calculated the correlation functions of molecule and atom pairs. The orientation correlation between molecule pairs was found to have a far shorter (≈ 5 Å) range of action than positional correlation operating between the centres of molecules (≈ 8 Å). The results of the neutron diffraction studies can be satisfactorily described by means of a model supposing an average preferred orientation between immediately adjacent molecules, with no orientation correlation between the second and farther neighbours. According to Narten, the structure of the D_2O molecule in the liquid state does not differ significantly from that of molecules in low-pressure gases. This model is in qualitative agreement with that derivable from molecular dynamical investigations.

Ztsepina [79] arrived at different conclusions. According to his hypothesis based on dynamical considerations, in the solid and liquid states intermolecular interactions produce such a deformation of the H_2O molecules that the atoms behave in a quasi-autonomous manner. Thus, ice should be regarded as an atom crystal rather than a molecule crystal. This theory assuming the presence of dynamically deformed H_2O molecules can explain the so-called anomalous properties of water well and simply. The theory makes questionable the correctness of those models which consider intermolecular interactions to be weaker than intramolecular ones also in the condensed states of water. Remarkable arguments have been forwarded for the assumption that discrete, independent water molecules can hardly be distinguished in the liquid state, owing to the strong deformation and loosening of H_2O molecules.

Weres and Rice [80] elaborated a relatively new model for liquid water utilizing the intermolecular potential function and a few simple assumptions regarding the structure. In fact, they proposed an advanced picture based on the earlier cell-theory of liquids, taking into consideration the contribution of structural entropy. In this model, the properties are expressed as the functions of the concentration of the different kinds of cells assumed, taking into account the fact that equilibrium concentrations correspond to minima in free energy. The calculated values of thermodynamical properties and density are in fair agreement with the experimental data; deviations can be attributed to the deficiencies of the cell model.

Summarizing the results of extensive investigations, Symons [81] arrived at the conclusion that liquid water has the structure expected on the basis of the electron structure of monomeric water. This is a normal electron structure, and the resulting chemical and physical characteristics should not be considered anomalous. Calculations are presented to show the case when water is dissolved in a hypothetic inert solvent. At low concentrations, water is dissolved in the monomeric state. With increasing concentration, the formation of dimers also starts. In the presence of dimers, the possibility of monomers getting attached to them to form trimers is higher than the possibility of combining with another monomer. Since the addition of each molecule produces two new suitable sites of attachment to the cluster, under unaltered conditions the total amount of water would be transformed into a giant polymer. However, this will not take place, since the new connection sites do not have equal reactivities, further, after the formation of tetramers cyclic oligomers also begin to appear (although only to a small extent at room temperature), their reactivities being not higher than that of the monomer. Symons has pointed out that tricentric hydrogen bonds can also be formed by the establishment of a weaker hydrogen bond in addition to the relatively strong one, through the formation of a weak bond between the non-bonded proton of water and a lone electron pair already participating in a hydrogen bond. This may be a transition state in the interchange of hydrogen bonds. Symons considers improbable all models supposing ice-like entities and monomers, or flickering clusters. In his opinion, the breaking of a hydrogen bond does not facilitate the rupture of further bonds, and the breaking of hydrogen bonds proceeding till the appearance of monomers is unlikely. In accordance with these, infrared spectroscopic measurements indicate that the monomer concentration is negligibly low at room temperature, but it increases with increasing temperature. The results of X-ray diffraction measurements have been explained by supposing that in liquid water the tetrahedral configuration of ice is retained in the immediate vicinity of the majority of H_2O molecules, however, this order is strongly disturbed already in the next layer. No long-range order was indicated by the results.

Hindman, Svirmickas and Wood [82] measured the proton spin-lattice relaxation time and they advanced an essentially identical mechanism for inter- and intramolecular relaxation in liquid water. The process predominating at elevated temperatures can be described as rotational diffusion, in which the amplitude of the angular motion increases with increasing temperature. Both the diffusional and non-diffusional processes contribute to relaxation.

Detailed theoretical calculations regarding the interaction of water molecules were carried out by Hankins, Moskovitz and Stillinger [83], further, by Sarkisov, Malenkov and Datevskii [84] applying the Monte Carlo method. The configurations calculated by the latter team show some tendency to the development of a three-dimensional skeleton formed by water molecules, in which the water molecules are attached to one another by means of bent hydrogen bonds in approximately tetrahedral orientation; however, these configurations do not correspond to those generally assumed for liquid water.

*

The above, in several respects contradictory, survey shows that extensive and intense experimental and theoretical research work has been in progress to elucidate the liquid structure of water in the early 1970's. Up to now, however, even the most recent scientific works have not led to gener-

Literature on page 58

ally accepted results satisfactory in all respects. In fact, these new theories are often in contradiction with one another not only in details, but even in the starting hypotheses. This clearly indicates that the experimental methods and theoretical treatments applied up to now are not suitable for the profound and detailed elucidation of the complex structure of this liquid.

In this respect, it should also be considered that for the decision whether an actual theory truly reflects reality or not, the agreement of its results with the measured data within the limits of the experimental errors unavoidable at the given period of time is not sufficient. Different theories can yield results identical within the limits of checking made possible by the experimental technique of the given period of time. Under such conditions, new experimental techniques should be developed to measure properties for which the different theories predict different results. Rigorously speaking, a theory can be accepted as really perfectly reflecting the state of the system investigated only when it can be proved that no other theory can achieve the same. The demand of this condition is, of course, in most cases on the verge of being unrealistic.

On the other hand, the requirement of developing new experimental techniques is certainly reasonable. The search for new methods should be continued which can detect new features or delicate differences in properties that have been unobservable. Application of the extremely sharp monochromatic laser radiation for excitation may also contribute to a very delicate analysis of the interacting atoms and molecules, allowing a deeper understanding of the liquid structure of water.

REFERENCES
to Section 1.3.

[1]　　The detailed survey of the structure of water can be found, e.g. J. L. KAVANAU, Water and Solute-Water Interactions (San Francisco, 1964); D. EISENBERG and W. KAUZMANN, Structure and Properties of Water (Oxford, 1969); N. E. DORSEY, Properties of Ordinary Water Substance in all its Phases (New York, 1969); J. Phys. Chem. 1970, 74, 3677–3819 (Symposium); E. U. FRANCK, Pure and Appl. Chem., 1970, 24, 13; O. YA. SAMOILOV, Struktura vodnykh rastvorov elektrolitov i gidratsiya ionov (Moskva, 1957); A. G. WILLIAMSON, An Introduction to Non-electrolyte Solutions (London, 1967); M. ORENTLICHER and P. O. VOGELHUT, J. Chem. Phys., 1966, 45, 4719; D. J. G. IVES and T. H. LEMON Roy. Inst. Chem. Rev., 1968, 1, 62; F. FRANS (ed.), Water, a Comprehensive Treatise, Vol. 1–4. (New York, 1972); R. A. HORNE (ed.), Water and Aqueous Solutions (Baffins Lane, 1971); N. H. FLETCHER, The Chemical Physics of Ice (Cambridge, 1970); J. F. COETZEE and C. D. RITCHIE (eds), Solute-Solvent Interactions, (Dekker, 1969).

[2]　　See, e.g. G. HERZBERG, Infrared and Raman Spectra of Polyatomic Molecules (New York, 1945).

[3]　　J. LENNARD-JONES and J. A. POPLE, Proc. Roy. Soc., 1950, A. 202, 166; J. A. POPLE, Proc. Roy. Soc., 1951, A. 205, 163.

[4]　　R. S. MULLIKEN, Phys. Rev., 1933, 43, 279.

[5]　　N. BJERRUM, Dan. Mat. Fys. Medd., 1951, 27, No.1.

[6a]　　E. J. W. VERWEY, Rec. Trav. Chim. Pays-Bas, 1941, 60, 837.

[6b]　　A. P. MINTON, Trans. Faraday Soc., 1971, 67, 1226.

[6c]　　S. A. CLOUGH, Y. BEERS, G. P. KLEIN and L. S. ROTHMAN, J. Chem. Phys., 1973, 59, 2254.

[7] D. M. DENNISON, *Phys. Rev.*, 1921, **17**, 20.
[8] W. H. BRAGG, *Proc. Roy. Soc.*, 1922, **34**, 103.
[9] W. H. BARNES, *Proc. Roy. Soc.*, 1929, **A. 125**, 670.
[10] J. D. BERNAL and R. H. FOWLER, *J. Chem. Phys.*, 1933, **1**, 515; I. G. MIKHAILOV and YU. P. SYRNIKOV, *J. Struct. Chem.*, 1960, **1**, 10.
[10a] N. A. NEVOLINA, A. L. SEIFER, *Zh. Strukt. Khim.*, 1973, **14**, 549
[11] L. PAULING, *Amer. Chem. Soc.*, 1935, **57**, 2680.
[12] P. G. OWSTON, *Adv. Phys.*, 1958, **7**, 171; *Quart. Rev.*, 1951, **5**, 344.
[13] H. D. MEGAW, *Nature*, 1934, **134**, 900.
[14] E. O. WOLLAN, W. L. DAVIDSON and C. G. SHULL, *Phys. Rev.*, 1949, **75**, 1348.
[15a] N. V. RILJ, *Zh. Fiz. Khim.*, 1955, **29**, 1372.
[15b] YU, E. PINCHUKOV, *Zhurn. Strukt. Khim.*, 1970, **11**, 415.
[16] S. PETERSON and H. A. LEVY, *Acta Cryst.*, 1957, **10**, 70.
[17] O. YA., SAMOILOV, *Zh. Fiz. Khim.*, 1946, **20**, 12.
[18] R. P. AUTY and R. H. COLE, *J. Chem. Phys.*, 1952, **20**, 1309.
[19] Cf. e.g. W. M. LATIMER, *Chem. Rev.*, 1949, **44**, 59; N. BJERRUM, *Science*, 1952, **115**, 385; H. GRÄNICHER, *Proc. Roy. Soc.*, 1958, **A. 247**, 453; C. T. O'KONSKI, *Rev. Mod. Phys.*, 1963, **35**, 732; J. D. DUNITZ, *Nature*, 1963, **197**, 860; D. EISENBERG and C. A. COULSON, *Nature*, 1963, **199**, 368; M. S. BERGQUIST and E. FORLIND, *Acta Chem. Scand.*, 1962, **16**, 2069.
[20] O. YA. SAMOILOV, *Zh. Fiz. Khim.*, 1946, **20**, 1411.
[21] J. D. BERNAL, *Proc. Roy. Soc.*, 1964, **A. 280**, 299.
[22] J. MORGAN and B. E. WARREN, *J. Chem. Phys.*, 1938, **6**, 666.
[23] G. W. BRADY and J. T. KRAUSE, *J. Chem. Phys.*, 1957, **27**, 304; G. W. BRADY, *J. Chem. Phys.*, 1958, **28**, 464; G. W. BRADY and W. J. ROMANOW, *J. Chem. Phys.*, 1960, **32**, 306.
[24] M. D. DANFORD and H. A. LEVY, *J. Amer. Chem. Soc.*, 1962, **84**, 3965; cf., A. H. NARTEN, M. D. DANFORD and H. A. LEVY, *Discuss. Faraday Soc.*, 1967, **43**, 97.
[25] R. H. EWELL and H. EYRING, *J. Chem. Phys.*, 1937, 5, 726.
[26] P. C. CROSS, S. BURNHAM and P. A. LEIGHTON, *J. Amer. Chem. Soc.*, 1937, **59**, 1134.
[27a] J. J. FOX and A. E. MARTIN, *Proc. Roy. Soc.*, 1940, **A. 174**, 234.
[27b] G. BEKE, G. INZELT and L. JANCSÓ, *Magyar Kémiai Folyóirat*, 1973, **79**, 70.
[28a] J. A. POPLE, *Proc. Roy. Soc.*, 1951, **A. 205**, 163.
[28b] V. I. YASHKICHEV, *Zh. Str. Khim.*, 1969, **10**, 780.
[28c] V. I. YASHKICHEV, *Zh. Str. Khim.*, 1970, **11**, 409.
[28d] M. S. ZAFAR, J. B. HASTED and J. CHAMBERLEIN, *Nature Phys. Sci.*, 1973, **243**, 106.
[29] L. PAULING, The Nature of the Chemical Bond (New York, 1966).
[30] G. H. HAGGIS, J. B. HASTED and T. J. BUCHANAN, *J. Chem. Phys.*, 1952, **20**, 1452.
[31] T. T. WALL and D. F. HORNIG, *J. Chem. Phys.*, 1965, **43**, 2079.
[32] M. FALK and T. A. FORD, *Can. J. Chem.*, 1966, **44**, 1699.
[33] G. E. WALRAFEN, *J. Chem. Phys.*, 1962, **36**, 1035; 1964, **40**, 3249; 1966, **44**, 1546; 1967, **47**, 114.
[34] K. BUIJS and G. R. CHOPPIN, *J. Chem. Phys.*, 1963, **39**, 2035.
[35] W. A. P. LUCK, *Ber. Bunsenges.*, 1965, **69**, 626.
[36a] O. D. BONNER and G. B. WOOLSEY, *J. Phys. Chem.*, 1968, **72**, 899.
[36b] G. H. F. DIERKSEN, *Theoret. Chim. Acta.*, 1971, **21**, 335.
[37] O. YA. SAMOILOV, Struktura vodnykh rastvorov elektrolitov i gidrotatsiya ionov, p. 60, 64. (Moskva 1957).
[38] A. H. NARTEN, M. D. DANFORD and H. A. LEVY, *Disc. Faraday Soc.*, 1967, **43**, 97.
[39] YU. V. GURIKOV, *Zh. Strukt. Khim.*, 1968, **9**, 599.
[40] YU. V. GURIKOV, *Zh. Strukt. Khim.*, 1968, **9**, 771.
[41] M. D. DANFORD and H. A. LEVY, *J. Amer. Chem. Soc.*, (1962) 84, 3965.
[42] R. B. CUDDEBACK, R. C. KOELLER and H. G. DRICKAMER, *J. Chem. Phys.*, 1953, **21**, 589.
[43] O. YA. SAMOILOV, *Dokl. AN SSSR*, 1955, **101**, 125.

[44]   R. P. MARCHI and H. EYRING, *J. Phys. Chem.*, 1964, **68**, 221.
[45]   O. YA. SAMOILOV, *Dokl. AN SSSR*, 1954, **95**, 587.
[46]   J. H. WANG, C. V. ROBINSON and I. S. EDELMAN, *J. Amer. Chem. Soc.*, 1953, **75**, 466.
[47]   R. A. ROBINSON and R. H. STOKES, Electrolyte Solutions (London, 1959).
[48]   E. FORSLIND, *Acta Polytechnica*, 1952, **115**, 9.
[49]   M. V. STACKELBERG, *Naturwiss.*, 1949, **36**, 327, 359; M. STACKELBERG and H. R. MÜLLER, *Z. Elektrochem.*, 1954, **58**, 25; M. STACKELBERG and W. JAHNS, *Z. Elektrochem.*, 1954, **58**, 162.
[50]   L. PAULING and R. E. MARSH, *Proc. Nat. Acad. Sci.*, 1952, **38**, 112.
[51]   G. G. MALENKOV, *Zh. Strukt. Khim.*, 1962, **3**, 206.
[52]   J. C. PLATTEEUW and J. H. van der WAALS, *Mol. Phys.*, 1958, **1**, 91.
[53]   V. McKOY and O. SINANOGLU, *J. Chem. Phys.*, 1963, **38**, 2946.
[54]   L. PAULING, Hydrogen Bonding (eds. D. Hadzi and H. W. Thompson) (London, 1959); The Nature of the Chemical Bond (New York, 1960).
[55]   H. S. FRANK and A. W. QUIST, *J. Chem. Phys.*, 1961, **34**, 604.
[56]   A. EUCKEN, *Z. Elektrochem.*, 1948, **52**, 255.
[57]   A. GIERER and K. WIRTZ, *Z. Naturforsch.*, 1950, **5a**, 577.
[58]   J. H. WANG, *J. Amer. Chem. Soc.*, 1951, **73**, 510, 4181; 1952, **74**, 1182, 1611, 6317; 1953, **75**, 1769, *J. Phys. Chem.*, 1954, **58**, 686.
[59a] N. V. RILJ, *Zh. Fiz. Khim.*, 1955, **29**, 1372.
[59b] B. R. LENTZ and H. A. SCHERAGA, *J. Chem. Phys.*, 1973, **58**, 5296.
[60]   H. S. FRANK and W. Y. WEN, *Disc. Faraday Soc.*, 1957, **24**, 133; H. S. FRANK, *Proc. Roy. Soc.*, 1958, **A. 247**, 481; *Nat. Acad. Sci.—Nat. Res. Council.Publ.*, 1963, **42**, 141.
[61]   C. H. COLLIE, J. B. HASTED and D. M. RITSON, *Proc. Phys. Soc.*, 1948, **60**, 145.
[62]   G. NÉMETHY and H. A. SCHERAGA, *J. Chem. Phys.*, 1962, **36**, 3382, 3401.
[63]   J. KOEFOED, *Disc. Faraday Soc.*, 1957, **24**, 216.
[64]   G. NÉMETHY and H. A. SCHERAGA, *J. Phys. Chem.*, 1962, **66**, 1773.
[65]   K. BUIJS and G. R. CHOPPIN, *J. Chem. Phys.*, 1963, **39**, 2035.
[66]   D. F. HORNIG, *J. Chem. Phys.*, 1964, **40**, 3119.
[67a] A. A. MILLER, *J. Chem. Phys.*, 1963, **38**, 1568.
[67b] A. T. HAGLER, H. A. SCHERAGA and G. NÉMETHY, *J. Phys. Chem.*, 1972, **76**, 3229.
[68]   V. VAND and W. A. SENIOR, *J. Chem. Phys.*, 1965, **43**, 1869, 1873, 1878.
[69]   W. A. P. LUCK, *Ber. Bunsenges.*, 1965, **69**, 626; *Disc. Faraday Soc.*, 1967, **43**, 115.
[70a] C. M. DAVIS and T. A. LITOVITZ, *J. Chem. Phys.*, 1965, **42**, 2563.
[70b] R. A. KURANT, B. D. REI and R. A. KHORN, *Elektrokhim.*, 1972, **8**, 581.
[71]   M. AGENO, *Proc. Nat. Acad. Sci. USA*, 1967, **57**, 856.
[72]   O. D. BONNER and G. B. WOOLSEY, *J. Phys. Chem.*, 1968, **72**, 899.
[73]   M. ORENTLICHER and P. O. VOGELHUT, *J. Chem. Phys.*, 1966, **45**, 4719.
[74]   J. R. O'NEIL and L. H. ADAMI, *J. Phys. Chem.*, 1969, **73**, 1553.
[75]   V. I. YASHKICHEV, *Zh. Strukt. Khim.*, 1970, **11**, 919; V. I. YASHKICHEV *Strukt. Khim.*, 1971, **12**, 163.
[76]   P. S. YASTREMSKII, *Zh. Strukt. Khim.*, 1971, **12**, 532.
[77]   J. W. PERRAM, *Molecular Physics*, 1971, **20**, 1077.
[78]   A. H. NARTEN, *J. Chem. Phys.*, 1972, **56**, 5681.
[79]   G. N. ZTSEPINA, *Zh. Fiz. Khim.*, 1973, **47**, 2005.
[80]   O. WERES and S. A. RICE, *J. Chem. Soc.*, 1972, **94**, 8983.
[81]   M. C. R. SYMONS, *Nature*, 1972, **239**, 257.
[82]   J. C. HINDMAN, A. SVIRMICKAS and M. WOOD, *J. Chem. Phys.*, 1973, **59**, 1517,
[83]   D. HANKINS, J. W. MOSKOVITZ and F. H. STILLINGER, *J. Chem. Phys.*, 1973. **53**, 45444.
[84]   G. N. SARKISOV, G. T. MALENKOV and V. T. DATEVSKII, *Zh. Strukt. Khim.*, 1973, **14**, 6.

1.4 EFFECT OF SOLUTES ON THE STRUCTURE OF WATER

The structure of liquid water is more or less altered by the solutes. The change depends on the type of interaction between the solute and the molecules of the solvent, and on how the solute molecules can be fitted into the cavities of the liquid structure or replace the water molecules. This latter is determined by the size of the solute molecules, in addition to the forces due to their interaction with water molecules. If the solute molecules do not appreciably distort the arrangement and the connections of the neighbouring solvent molecules, the conditions are rather simple, otherwise the liquid structure is significantly altered which results in changes of directly measurable properties. In this respect, there is an appreciable difference between uncharged neutral molecules and ions with positive or negative charges.

From the phenomena accompanying dissolution and from the properties of the solutions, several conclusions have been drawn regarding the structure of solutions, and various theories have been developed. All of them give some explanation for smaller or larger groups of the phenomena. However, none of the theories is successful in giving a unified and satisfactory interpretation of the structure of solutions, and of all phenomena of structural changes due to the solute, and, thus, none of them could be accepted generally.

Hydration phenomena and the effect of electrolytes on the structure of water will be discussed in detail in Section 5.2. In this section, only some general aspects of the interaction between the solute and the solvent are reviewed.

1.4.1 EFFECT OF SOLUTE MOLECULES ON THE STRUCTURE OF WATER

The concepts regarding the effect of solutes can be classified–rather schematically–into two groups: the 'iceberg' and 'soft ice' theories.

The iceberg theory originates from the investigations of Eley [1] based on the anomalous heat and entropy of dissolution of gases with non-polar molecules (noble gases, hydrocarbons, etc.) in water. The heat of dissolution (negative enthalpy) of these slightly soluble materials is much larger in water than in organic solvents; the entropy of dissolution, on the other hand, is more negative (by about 12 Clausius) than in 'normal' organic solvents (e.g. for methane, $\Delta H^0 = 3 \cdot 19$ kcal/mole, $\Delta S^0 = -31 \cdot 8$ Clausius/mole at 25 °C). According to Eley, no previous development of cavities is necessary for the dissolution of gas molecules (which would require a significant amount of energy) because the open worked structure of water gives a sufficient number of cavities suitable for the gas molecules. If the size of the gas molecules makes it necessary, the volume of the cavities can be increased at the expense of a relatively small energy. The

Literature on page 77

energy of the hydrogen bonds broken during this process is compensated by the new ones formed around the cavities. According to this theory, about 2 per cent of the interstitial cavities are occupied by non-polar gas molecules. Namiot [2] calculated the entropy change due to dissolution, and the results are in good agreement with the experimental data. The great decrease in entropy, as well as the small solubility, however, is not in accordance with the concentration-independent activity coefficient of dissolved gases.

According to the theory of Eley, the large non-polar molecules dissolved in water make the structure of water tighter, and they decrease the number of hydrogen bonds. On the other hand, the widely accepted theory of Frank and Evans [3] assumes the atoms or non-polar molecules of the gas to enhance the crystalline character of the water structure at normal temperatures, and to increase the number of hydrogen bonds; furthermore, according to Powell and Latimer [4a], they decrease the vibrational motion of the adjacent water molecules. The structure can be pictured as a submicroscopic 'iceberg' formed around the non-polar molecule, which results in a larger change in the enthalpy and entropy than would be expected. The term 'iceberg' does not really mean the development of a structure corresponding accurately to the lattice of ice due to the effect of solute molecules, it refers merely to the general strengthening of the structure. This phenomenon is also called *hydrophobic hydration*. Its existence is unquestionable, even if the mechanism of its formation has not been cleared up in detail.

Local structural changes caused by hydrophobic molecules in water have been concluded by Jolicoeur and Friedman [4b] on the basis of the investigation of EPR spectra, giving simultaneously a model for the dynamics of the interaction between hydrophobic particles and water.

According to Frank and Evans [3], the non-polar molecules are arranged along the borderlines of the water molecule complexes in the liquid, where the structure is more loose, and, in this way, there are more places to be occupied. Thus, the open structure is stabilized, and the iceberg grows. This process can be well described by the flickering cluster model (see, Section 1.3.3.8). Frank and Wen [5] considered the statistical extent of the ice-like character of the structure of water to be proportional to the average size of the clusters and to their half-lives. Cluster formation takes place if the development of hydrogen bonds overcompensates their decomposition in the negative phase of the fluctuation of energy. If, however, the decomposing forces prevail in the local regions, the ice-like clusters 'melt'. The non-polar molecules are relatively neutral in the sense that they do not cause or transmit any effects which decompose the structure (mainly because they are less polarizable and have no electrostatic effect). Thus, in the vicinity of non-polar molecules, ice-like clusters are formed more easily than elsewhere, and, if they have been formed, their lifetime is longer. As emphasized by Frank [6], this is supported by the fact that solutes of non-polar molecules increase the dielectric relaxation time of the solution.

The formation of icebergs has been treated by Némethy and Scheraga [7] on the basis of statistical thermodynamics. Within the bulk of the iceberg, water molecules forming four hydrogen bonds with the adjacent water molecules can also have a solute molecule as a neighbour. Thus the coordination number increases from four to five and, as a consequence, the energy of water molecules with four bonds becomes lower than that in the ground state corresponding to pure water. Permanent, induced, or transient dipole interactions result in a connection between the solute molecule and water. Around free water molecules this is altered in the sense that the coordination number of these molecules is larger than that of molecules forming four hydrogen bonds of definite orientation. Solute molecules can get into the vicinity of free water molecules only by replacing a water molecule at that place. Since the van der Waals dipole–dipole interactions between water molecules are much stronger than those between water and solute molecules, the energy of free water molecules is increased by solute molecules approaching their immediate neighbourhood. The energy of water molecules at the 'surface' of the icebergs forming one, two, or three hydrogen bonds also increases owing to the appearance of solute molecules in their vicinity. As a result of all these effects, the energy of water molecules in the first coordination sphere decreases with the effect of the non-polar solute. Thus, according to the Boltzmann distribution law, the solute increases the number of molecules in the ice-like clusters. For the sake of brevity, those molecules actually located in the first sphere around non-polar solute molecules, more definitely orientated and strongly involved in hydrogen bonds, are called 'iceberg'. For example, as calculated by Némethy and Scheraga, the percentage of the non-broken hydrogen bonds around dissolved hydrocarbon molecules is 59 per cent (in pure water, this calculation gives 46 per cent), that of molecules with four, three, two, and one hydrogen bonds is 43, 6, 18, and 12 per cent, respectively, while these figures in water are 23, 20, 4, and 23 per cent. The amount with no hydrogen bonds is 21 per cent which can be compared with a value of 29 per cent in water at room temperature.

Investigating infra-red spectra of aqueous solutions of organic substances, Bonner and Woolsey [8a] concluded that the structure of water is strengthened by saccharose, dextrose, and urea, not by the polarization of the solvent molecules, but rather by the increase in the number of hydrogen bonds per mole. The hydration number* of non-electrolyte molecules is large in comparison to that of monovalent electrolytes (21, 10, and 2·3 for saccharose, dextrose, and urea, while it is 9·0, 5·2, 2·3, and 4·6 for hydrogen, lithium, potassium, and cesium chlorides, respectively) as calculated from the spectra.

In contrast with the former results, Beauregard and Barrett [8b] assumed from studies on ultrasonic absorption and velocity, that urea solutions behave like pure water at elevated temperatures, i.e. the resultant effect of the presence of urea in water is the partial decomposition of the molecular clusters.

---

* For the non-unequivocal concept of the hydration number, see Section 5.2.

Literature on page 77

With respect to the effect of dissolved urea on the water structure, contradictory opinions can be found in the literature, such as those reviewed by Barone, Rizzo, and Vitagliano [8c]. On the basis of the infra-red absorption spectrum of urea dissolved in a mixture of H_2O and D_2O, the authors mentioned above have concluded that urea slightly destroys the water structure and the 'structural temperature' of the solution is higher than the temperature applied during the experiment. On the other hand, the derivatives of urea have a slight structure-making effect on water which is stronger, the higher the number of hydrophobic substituents in them.

Changes in the liquid structure of water due to the presence of dioxan were also investigated by Gorbunov and Naveruzhin [8d] by infra-red spectroscopy. They have concluded that the structure of water cannot be described satisfactorily by the mixed model assuming two states. In the description of the structure of non-electrolyte solutions, the presence of relatively stable structural formulations (globules, zones) should also be assumed. In fact, there are such complexes in pure water, too, but they are attached to one another by means of hydrogen bonds so that a liquid of uniform structure, having continuum properties is formed. The molecules of non-electrolyte solutes occupy the structural cavities of the globules, which thus become stabilized, and then the weak bonds between the globules are broken.

Kochnev [8e] arrived at rather similar conclusions by investigating the vibrational spectra of aqueous solutions containing alcohol and other non-electrolytes. According to him, the non-electrolyte will break a certain number of hydrogen bonds between the water molecules, depending on the dimensions of the molecule and on the degree of branching of its non-polar group, while the remaining hydrogen bonds will become stronger. This latter effect was, however, considered to be due not to complex formation by hydrogen bonds between the molecules of water, but mainly to van der Waals interactions arising between water and the non-polar part of the non-electrolyte. Further, it was proposed that the interaction between water and the non-polar groups is made possible primarily by the existence of certain structural zones or non-polar groupings corresponding to them, whose shapes and dimensions facilitate the van der Waals interaction. The extent of contact with the OH group of water is the greatest, when the van der Waals radius of the non-polar group is the same as that of the water zones. Under such conditions the hydrogen bonds of the OH groups of water molecules become linear. According to this picture, the stabilizing effect of non-electrolytes on the water structure can primarily be attributed to the fact that the interactions between molecules depend on their orientation. The hydrogen bonds in water become progressively deformed with increasing temperature, which results in an increasing coordination number, while the number of zones decreases. Its consequence will be the weakening of the effect of non-electrolyte solutes.

In addition to the increasing effect on the size and stability of the ice-like clusters, non-polar solutes also result in a change in the volume, during the formation or melting of the icebergs around them. Similarly to pure ice, there are significant structural cavities in the hydrogen-bonded flickering clusters, which are independent of the exact structure of these clusters.

Namely, the cavities are due to the nature of the three-dimensional tetrahedral hydrogen bonded network. At the border of the clusters, there are incomplete cages which can be partly occupied by the solute molecules. Thus, some places which are empty in pure water are occupied in solutions. This causes a decrease in volume on dissolution. For example, the partial molar volume of methane in n-hexane is 60 ml, while in aqueous solution it is 37 ml. (Gjaldbaek and Hildebrand [9], Masterton [10]).

The formation and melting of the icebergs or clusters is accompanied by a marked change in enthalpy and volume. To this can be attributed the fact that the structures of water and aqueous solutions are strongly influenced by temperature and pressure.

A thorough critical analysis has been given by Hertz [11] on the stabilization of the water structure by solute molecules and on the concept of icebergs. He suggested the term *secondary hydration* instead of that misleading name to denote the phenomenon that water molecules in the close vicinity of non-polar molecules behave in a certain sense as if their temperature were lower than that of the solution. As a measure of the 'temperature' of the structure, the sharpness of the maximum in the molecular distribution function, taking into account the orientation, has been considered. Both the experimental and theoretical determination procedures of the distribution function are, however, complicated, and no satisfactorily reliable methods are available for this purpose. Thus, no unequivocal conclusions can be drawn from the available experimental data for the structural changes due to the solute. Some properties indicate strengthening of the structure of the liquid, while others can be attributed to the decrease in the number of hydrogen bonds, i.e. the structure is loosened under certain conditions. The 'strengthening' of the structure is probably accompanied by rupture or deformation of hydrogen bonds too.

The structure-strengthening effect of the solute molecules in water is also connected with the problem of the neutral molecules in their immediate vicinity. On the basis of the quantum statistical theory of Golden and Guttman [12]–which is a new treatment of the phenomena in some respects–it can be supposed that the water molecules are orientated around neutral solute molecules too. On the basis of other assumptions, Luck [13a] obtained the same results, but he considered the orientation of water molecules contradictory to that given by quantum statistical theory.

The organic compounds containing proton acceptors (N, O, P, NO_2, CO, SO, PO, etc.) also influence the water structure by their hydrogen bonds in solution. According to Kariakin and Muradova [13b], a correlation can be established between the energy of the hydrogen bonds formed with water and the thermodynamic properties of the molecules involved in the interaction.

Recently, Naberukhin and Rogov reviewed the structure of aqueous solutions of non-electrolytes [13c].

Dahlberg [13d] endeavoured to elucidate the effect of solutes on the basis of thermodynamical properties, measuring the free energy and enthalpy

Literature on page 77

of the transfer of alcohols and ketones from H_2O into D_2O. The free energy of transfer is, in most cases, approximately zero; however, the value of enthalpy is between -514 cal/mole and $+70$ cal/mole. The results were interpreted as reflecting a destruction of the liquid structure of water by the polar groups of the solute, while a strengthening effect is exerted by the non-polar methylene groups. The overlapping of the action spheres of the groups producing the structural changes lowers the ability of these groups to alter the structure. Strikingly, the enthalpy of transfer of cyclic compounds is low, most probably because the water molecules cannot get into the inside of the rings.

1.4.2 EFFECT OF DISSOLVED IONS ON THE STRUCTURE OF WATER

The effect of dissolved ions on the structure of water is different from that of neutral molecules in several aspects. The changes in the structure of water caused by ions cannot be separated from the phenomena of ionic hydration that will be discussed in Section 5.2. Here, only some effects of the interaction between ions and water dipoles on the structure of water are reviewed [14]. The electric field of the ions alters the orientation of the dipole molecules of water, distorting the ice-like structure and also compressing them *(electrostriction effect)*. The electric field of ions in this way decomposes, and 'softens' the regions of ice-like structure to a certain extent, unlike the effect of neutral molecules which stabilizes the original water structure. Therefore, those regions exposed to such an ionic effect are called 'soft ice' (Davies and Rideal [15]).

The tetrahedral arrangement of water molecules is distorted by the electric field of dissolved ions, and a more or less altered order is developed, since the ion–water interaction is significantly different from the water–water interaction.

On one hand, ions deform the lattice-like arrangement of water molecules. This effect prevails in the case of anions and depends mainly on the size of the ions. Ions with crystallographic radii no larger than 1·3 Å (the size of the interstitial cavities) can occupy the structural cavities of water without altering their size appreciably (Benson and Copeland [16]). However, large ions reaching the cavities also deform them for merely geometrical reasons (Bergquist and Forslind [17]). The larger the ions, the larger the sterical deformation effect, so the deformation and rupture of hydrogen bonds between water molecules are extended to larger regions. Increase in temperature also results in a rupture of hydrogen bonds, however, these effects of different origin show little similarity.

In addition to the effect arising purely from the volume, the electric field of the ions also deforms the lattice structure, and alters the charge distribution in the dipole molecules; this is predominating mainly in the case of cations. The additional effects surpassing the steric ones have been divided into two groups by Samoilov [19]: those arising from incidental differences in the coordination numbers of the ion and the water molecule, and those reflecting the changes in the orientation of water molecules located immediately around the ions.

The electric field strength at the border ('surface') of ions not larger than Cs^+ and I^- is so large (10^5–10^6 volts/cm) that the immediately neighbouring water molecules are strongly orientated, polarized, and compressed by the ion-dipole forces ('immobile' layer). The immobility of the bond, however, has only a figurative meaning, since the water molecules in the close vicinity of ions are still in permanent exchange. The average residence time of water molecules in the first (nearest) sphere can be obtained from rates of exchange reactions taking place between these molecules and other ligands. According to Eigen's ultrasonic absorption measurements [18], the number of molecules exchanged per second is 0.5–9×10^8 for Li^+, 2–50×10^8 for Cs^+ (depending on the nature of the ligands), and 10^5 for Mg^{2+}. In the case of Ca^{2+} ions, however, the rate is about 1000 times higher. The exchange rate of water molecules between themselves around the ion is probably not too different from that with other ligands. The great difference in the rate of exchange of water molecules in the first hydrate spheres of Ca^{2+} and Mg^{2+} ions is surprising; however, this is also confirmed by some experiments on self-diffusion and the temperature coefficients of ionic mobilities. According to Samoilov [19], the activation energy of the exchange of water molecules in the close vicinity of ions is about 0.4 kcal/mole and 2.6 kcal/mole for Ca^{2+} and Mg^{2+} ions, respectively.

As calculated from NMR measurements in water enriched with ^{17}O isotope, the average dwelling time of water molecules in the first coordination spheres of Cu^{2+} and Ni^{2+} ions is 5×10^{-7} and 4×10^{-3} s, respectively, while the activation energy of the exchange is 5.0 and 11.6 kcal/mole, respectively (Swift and Connick [20]).

In solutions of transition metal ions, the average dwelling time of water molecules in the immediate vicinity of the ions can be much larger, as much as some hours. This is due to the fact that these ions form donor–acceptor bonds with water, and the structure of water is highly deformed in the vicinity of the ion. In solutions of transition metal ions the bonds between the ions and water molecules are strong enough to form complex ions of definite stoichiometric composition, as e.g. $Cr(H_2O)_6^{3+}$, which can be detected directly by different methods. In this complex, for example, the half-life of the dwelling time of a water molecule is about 40 hours (Hunt and Taube [21]).

The effect of ions on the structure of water strongly depends on their charges and sizes. It varies more or less linearly with the ratio of charge to radius (the so-called *polarizing power*). The water molecules bound in a definite orientation in the first coordination sphere of the ions move together with the ions. The separation between the hydrated ions and the solvent takes place generally at a distance corresponding to the diameter of one molecule. For example, according to the estimate given by Everett [22], the effect of alkyl ammonium and carboxylate ions on water molecules extends to about 5 Å.

Literature on page 77

The potential energy is increased, while the entropy, the volume and the heat of evaporation are very much decreased by adhesion of water molecules to ions, as well as by their polarization and compression (since their movement is hindered). This effect is opposite to the increase in the specific heat accompanying the formation of icebergs around neutral molecules.

Other effects of ions on the structure of water. Ions exhibit some other effects on the structure of water in addition to the electrostatic ones mentioned above (dielectric polarization, adhesion, and compression). Depending on the conditions, ions can break (loosen or distort), or strengthen the structure of water, i.e. they can increase or decrease the extent of ice-like structure appearing in small regions by shifting the equilibrium (Bernal and Fowler [79], Eucken [56b], Gurney [23]). The orientating effect on the water dipoles exerted by the electric field of ions also influence the formation and decomposition of the molecular clusters (Frank and Wen [5]).

The structure-breaking effect of dissolved ions on water has been explained by Vdovenko, Gurikov and Levin [82] by extending the two-structure model to ionic solutions. According to them, ions dissolve more easily in disordered structural regions which work on loosening the liquid structure.

Although several ions give rise to a decrease in the entropy similar to that produced by the formation of icebergs around non-polar molecules, there are some ions which decrease the loss in entropy in comparison with the effect of non-polar molecules, i.e. they increase the molecular disorder in water (Frank and Evans [3], Powell and Latimer [4a]). For example, the dissolution of one K^+ and one Cl^- ion causes significantly smaller loss in entropy than that of two argon atoms, although both ions have the same argon configuration. In connection with this, Bernal and Fowler [79] introduced the concept of *structural temperature*; this is the temperature at which the structure and the viscosity of pure water are identical with those of the given solution. Some ions increase the structural temperature, while others decrease it. According to the above authors, a hydrated ion can be formed if the energy of interaction between the ion and water is larger than the potential energy of water molecules in pure water (15·3 kcal/mole).

According to Frank and Evans [3], and Frank and Wen [5], dissolved ions are surrounded by water molecules in three concentric layers:

(i) in the first layer, there are 'immobile', polarized, and compressed water molecules,

(ii) in the second layer of a somewhat destroyed structure, water is less ice-like and more disordered than in pure state,

(iii) in the outer layer, water has a normal structure.

This model is suitable to explain some of the properties of aqueous solutions, however, it is evidently a very schematic interpretation.

The most favourable fitting of ions into the structure of water is the occupation of the centres of tetrahedra in the water clusters. In this way, ions have four water molecules as immediate neighbours, with their oxygen atoms at the apices of the tetrahedra (Bernal and Fowler [79], Gurney [23]). If the size of the ion is suitable to fit into the tetrahedron formed by water molecules, it does not geometrically distort the ice-like structure of water. Some deformation takes place even in this case owing to ion — dipole interaction: all the four water molecules are polarized towards the centre of the tetrahedron.

Perchlorate ions fit well into the structure of water with the four oxygen atoms arranged tetrahedrally around the chlorine atom. As calculated by Forslind [81], this tetrahedron has approximately the same size and shape as that of the water molecules. In addition, the oxygen atom of a perchlorate ion can form hydrogen bonds with the adjacent water molecules on the effect of the field of the environment without appreciably distorting the structure of water. Ammonium ions of similarly tetrahedral shape fit the structure of water as well, with almost no effect on the viscosity of water (Frank and Evans [3], Fajans and Johnson [24], Gillespie [25]). Of the ions, perhaps ammonium ions are the most similar to water, because this similarity covers their mass, molar volume, and interatomic distance, as well as the bond angles. Owing to the strong polarity of O–H bonds in water, as established by Vollmar [26], the electrostatic effect of water molecules on its environment does not differ considerably from that of the ammonium ion.

Recently Liashchenko [27a] investigated in detail how ions or hydrate complexes can be arranged in the structure of aqueous solutions. He assumed that the tetrahedral structural network of water is constant not only in diluted solutions, but up to high concentrations, i.e. the network of the hydrogen bonds is hardly altered in spite of the strong interaction between the ions. The ions and hydrate complexes are located partly in the cavities of the network, and partly they are built into the lattice. Ions which are not too different in size from water molecules (K^+, Rb^+, Ba^{2+}, Tl^+, NH_4^+, F^-), can occupy the cavities without any appreciable deformation of the structural network, since the cavities are large enough for ions of radii smaller than $1 \cdot 6$ Å. Anions of tetrahedral structure (ClO_4^-, SO_4^{2-}, PO_4^{3-}, CrO_4^{2-}, etc.) can be sited in the cavities without distorting appreciably the water structure; in these cases, two water molecules in the network are replaced by two oxygen atoms of the tetrahedron, while the two other oxygen atoms occupy two adjacent cavities. This is the probable arrangement of the complex ions of tetrahedral shape with ligands of nearly the same size as that of the water molecules (e.g. $Al(OH)_4^-$, $Be(F)_4^{2-}$). Ions of coplanar triangular form (NO_3^-, CO_3^{2-}, etc.) are arranged in the ice-like structure of water so as to occupy two positions in the network and one in a cavity, or one position in the lattice and two in adjacent cavities. Ions appearing as two tetrahedra with one common apex (e.g. $S_2O_7^{2-}$, $Cr_2O_7^{2-}$, etc.) occupy four cavities and replace four water molecules in the lattice. The octahedral complex ions (e.g. $Mg(H_2O)_6^{2+}$) replace three water molecules in the lattice and occupy three cavities, but to do this, they have to slightly deform the structure of water in their vicinity. According to these essentially geometrical studies of Liashchenko, ions partly replace water molecules in the lattice-like network of the liquid, and partly penetrate into the cavities of the

Literature on page 77

network. These two possibilities of insertion can influence the volume of the solution in various ways, which can also be approximately quantitatively described.

Shraiber and Tikhii [27b] investigated the alteration in the structure of heavy water on the effect of alkali metal chloride solutes, by means of precision density measurements. They conclude that in the temperature range between 10 °C and 40–50 °C, LiCl, NaCl and KCl destroy the structure of both heavy water and common water, but this effect is smaller in the former. This confirms the earlier observation that heavy water has a more definite structure than common water.

Ben-Naim's studies [28a] in which some calculations have been carried out regarding the chemical potentials of monomeric and cluster molecules on the basis of the two-state model give the strengthening effect of dissolved ions on the structure a new interpretation. For large enough clusters, the stabilizing effect on the structure is proportional to the number of molecules involved. The change in the chemical potential due to the cluster→ monomer transformation and the correlation between this change and the hydration energy of the solute show that the solute species form icebergs around themselves under certain conditions, and, in this way, their effect is rather passive.

On the basis of X-ray crystallographic investigations on concentrated hydrochloric acid solutions, Terekhova [28b] deduced that the coordination number of H_3O^+ ions in 1·85–5·59 mole/litre solutions is 3·0–3·3, and these ions destroy the structure of water. Chloride ions are sited in the structural cavities of water, and they exhibit a negative hydration (see Section 5.2.2). On the basis of the density, viscosity, diffusion coefficient and conductivity values of some very concentrated salt solutions, structural transformations of concentrated solutions have been discussed recently by Janz, Oliver, Lakshminarayanan and Mayer [28c].

On the basis of investigations on the ultrasonic absorption of electrolyte solutions, Breitschwerdt [28d] outlined a model for the structural relaxation of ionic solutions. On this basis some conclusions could be drawn regarding the number and kinetic properties of water molecules located at various distances from the ions in solutions of alkali halides. The results obtained for the relaxation parameters are in good agreement with the empirical experiments on compressibility, thermal expansion coefficient, molar volume, specific heat, and radial distribution.

Abraham and Hechler [28e] attempted to elucidate the structure of concentrated solutions of strong electrolytes on the basis of the structural model of salt melts containing vacancies.

The effect of organic ions on the structure of water is somewhat different from that of the inorganic ones. This is due to the fact that they are larger, and their hydrocarbon groups are hydrophobic and exert an overall structure-making effect on water, as shown by Némethy and Scheraga [29] and Schneider, Kreschek, and Scheraga [30]. This effect is related to changes in the energy levels of water molecules of different states, which modifies their statistical distribution in the immediate vicinity of solute ions. If there is a hydrophobic particle in the immediate neighbourhood of tet-rahedrally bound water molecules, the energy of the van der Waals bond, orientated towards the solute molecule, decreases (as compared with the case of a water molecule also at the site of the solute molecule), which stabilizes the tetrahedrally bound state; i.e. the water structure is strength-ened. This effect is significant, for example, in the thermodynamics of the hydration of tetra-alkyl ammonium ions, too, and it is also in connection with the thermodynamics of the dissociation of organic acids and bases, as pointed out by Conway, Verrall, and Desnoyers [31a].

From measurements on the heat of dilution Lindenbaum [31b] has concluded that large polymeric carboxylate (e.g. butyrate and valerate) ions have a structure-making effect on water, and they increase the number of hydrogen bonds.

The structure-forming effect on water can be accompanied by association of the ions. For example, according to Evans and Kay [32], a marked association must be assumed in solutions of tetramethyl ammonium bromide and iodide. Some attempts have been made by Diamond [33], and Linden-baum and Boyd [34a] to elucidate the correlation between the strengthen-ing effect on the structure of water and the association of ions.

Investigating the inelastic scattering of neutrons in solutions of $(CH_3)_4N^+$ ions, While [34b] has suggested a procedure by which the dynamics of the motion of organic ions with a large number of protons can be studied separately from that of the solvent molecules.

The difference between the effect of anions and cations has been studied extensively, but the processes could not be revealed unequivocally. Cations exhibit, in general, a stronger hydration than anions otherwise comparable with them. According to the conclusions of Latimer, Pitzer and Slansky [35], Buckingham [36], and Noyes [37], accepted generally in the liter-ature, the enthalpy and free enthalpy of the hydration of anions is somewhat larger than that of cations of the same charge and size, while their entropy of hydration is much larger.

The differences in the interactions between the cations and anions and water molecules are connected with the structure of water. One of the vertices of the triangle formed by the dipole molecule $-\,-\,O\!\!\begin{array}{c}{}^{\nearrow H^+}\\[-2pt]{}_{\searrow H^+}\end{array}$ is occupied by the oxygen atom which represents a negative pole by its attracting effect on the electrons of the covalent bond. At the other two vertices are hydrogen atoms and their partial charge centre is the positive pole. The two poles of the water dipole do not have the same structure. Since on hydration the water molecules turn their negative poles towards the cations and their positive ones towards the anions, the different effect of these two types of ion is understandable in this respect, too.

Around a *cation*, water molecules have the lowest energy when the oxygen atom representing the negative pole turns towards the ion and the two

Literature on page 77

OH bonds are orientated outwards. The bond type based on such an interaction is essentially different from the hydrogen bond. The cation, the
oxygen and the two hydrogen atoms are located in the same plane which
makes the free rotation of water molecules generally impossible ('irrotational
bond'), except, perhaps, the rotation about the dipole axis; however, this
has no influence on the orientation polarization. According to Noyes [37],
cations of different sizes develop a configuration during hydration which
permits a rather good fit into the surrounding water structure. Cations
interact with the two lone-pair electrons in the L shell of the oxygen atom
of the water molecule, therefore those in the immediate neighbourhood
of the cation can be bound to two water molecules only.

The protons representing the positive pole of a water molecule are orientated towards the *anions*. Noyes [37] concluded from this that the arrangement of the immediately neighbouring molecules is more sensitive to the
size of the ion than in the case of cations. It can be assumed that one of
the O—H bonds of a water molecule is perpendicular to the surface of the
ion, since the interaction between water molecules and anions is somewhat
similar to the hydrogen bonds (Stokes and Robinson [38], Moelwyn-Hughes
[39], Waldron [40], Sirnikov [41], Hindman [42]). The water molecule
bound to the anion in such an arrangement can rotate freely about this
bond which plays an important role in the orientation polarization; furthermore, it can form three hydrogen bonds with other water molecules.

These differences in the bonds formed by cations and anions with water
molecules, i.e. in the mechanism of *hydration*, have also been supported
by investigations on the dielectric constant (Haggis, Hasted and Buchanan
[43], Harris and O'Konski [44]).

The opposite polarization effect of the anions and cations can promote
the association in the case of small ions. At the other side of the water
molecule polarized by the cation, the binding of an anion is facilitated (Robinson and Harned [45], Diamond [46]; for some other effects of the ion-pair
formation see Kessler, Povarov, and Gorbanev [47]).

Another model of the interactions between water and ions has been
published by Buckingham [36], who assumed that the nearest water molecules are bound to the ions in a rigid tetrahedral or octahedral arrangement.
Around this primary hydration sphere, the liquid can be considered a
continuum having properties identical with those of pure water. It has
been pointed out that in the interaction of ions with water molecules,
quadrupole moment plays an important role. According to Vaslow [48],
the minimum of the potential energy in the vicinity of small cations does
not coincide with the dipole axis of water molecules because of the effect
of the quadrupole moments; it forms an angle with it. He assumed that,
owing to this effect, two or more water molecules around an actual small
cation are bound by bent hydrogen bonds.

The difference between the interactions of anions or cations with water
molecules can also arise from the fact that the centre of the positive charge
in the dipole molecule of water is nearer to the boundary of the molecule
than that of the negative charge. Consequently, the permanent dipole
can approach an anion to a smaller distance than a cation, and the energy

of the anion–dipole interaction is larger than that of the cation–dipole interaction (Latimer, Pitzer and Slansky [35], Buckingham [36], Noyes [37]). With respect to the deformation of the electronic shell of the water molecule (i.e. polarization), the effects of anions and cations are hardly different. The result of both these effects is that, on the average, a smaller number of molecules is located in the vicinity of the proton than otherwise. Near the oxygen atom turned towards the ion the average electron density of the water molecule is increased in the neighbourhood of cations, while anions attract the proton and repulse the electrons, which results also in an increase of the average electron density in the vicinity of the oxygen atom (Shoolery and Alder [49]).

The resultant of the structure-making and structure-breaking effects of ions depends mainly on the size and charge of the ion in question. A solution of higher viscosity than that of water is formed by the effect of relatively small ions (Li^+, Na^+, H_3O^+, F^-, OH^-) and polyvalent ions (e.g. Ca^{2+}, Ba^{2+}, Mg^{2+}, Al^{3+}, Er^{3+}), which leads to the conclusion that their resultant effect is the extension and strengthening of the water structure. The large electric field strength at the boundary of these ions polarizes the neighbouring water molecules; it makes them 'immobile', and compresses them by means of electrostriction *(primary hydration* or, according to Samoilov [50], *short-range hydration)*. On the other hand, it increases the order of water molecules beyond the first layer of the molecules and decreases the entropy (Bockris [51] suggested calling this phenomenon *secondary hydration* or, according to Samoilov's term, *long-range hydration*) since this action extends to the distorted structural regions, too. As it has been established by Samoilov [52], the activation energy of the exchange of water molecules in the immediate neighbourhood of these ions is positive, i.e. the mobility of water molecules in the primary hydrate sphere is smaller than in pure water or in other regions of the solution far from the ion *(positive hydration)*.

Large monovalent ions can also act as structure-making ones. For example, Frank [53] found that the water structure is stabilized appreciably by tetrabutyl ammonium ions which can be explained by the effect of the butyl group on water. As established by Diamond [46], a similar effect also results from the other tetra-alkyl ammonium ions and from all of the 'amphiphilic' ions which are large enough to have their charges embedded in their bulk.

The structure-making effect of Er^{3+} which is surrounded by water molecules in octahedral arrangement is particularly strong. Brady's [54], as well as Freed's, X-ray diffraction investigations [55] on $ErCl_3$ and ErI_3 solutions gave radial distribution curves with maxima exactly coinciding with those in ice, but these maxima cannot be found in the case of pure water. This shows that in a solution of Er^{3+} ions water molecules are ordered to an appreciably greater extent than in pure water.

Unlike small ions, large monovalent ions (e.g. K^+, NH_4^+, Cs^+, Cl^-,

Literature on page 78

I^-, Br^-, NO^-, BrO_3^-, IO_3^-) in low concentrations produce solutions of viscosities lower than that of water, which indicates that these ions destroy the structure of water and increase the entropy. Since the electric field strength is relatively small at the boundary of large ions, their polarizing, immobilizing and electrostriction effects on water molecules are efficient only in the first layer around the ions, i.e. only the molecules located in this layer are ordered to a certain extent. Beyond this layer, a strong structure-breaking effect prevails, probably increasing the mobility of water molecules (*negative hydration*, which has been studied in detail by means of thermodynamic and structural analysis in connection with NMR relaxation investigations of Engel and Hertz [56a]). On the other hand, the resultant effect on the structure of water of ions containing groups suitable to form hydrogen bonds ($-NH_2$, $-OH$) is small; since they fit into the ordered molecular clusters of water by hydrogen bonds, they do not deform their lattice-like order too much, and the formation and decomposition of the clusters is also slightly influenced by them (Hasted, Ritson and Collie [57], Frank and Wen [5], D'Orazio and Wood [58]).

The structure-breaking effect of anions and large monovalent cations is also supported by their effects on the infra-red spectra, the partial molar heat capacity, the change in the entropy on hydration and the relaxation time of the dielectric constant, as well as on self-diffusion.

With increasing temperature, as has been shown by the measurements of Kaminsky [59], the decrease in viscosity due to the effect of Cl^-, Br^-, I^-, and other similar anions becomes less significant as compared with that resulting from the increase in temperature. Evidently this can be attributed to the fact that the size of the ordered clusters of water molecules and the number of hydrogen bonds decreases with increasing temperature. In solutions of ions exerting a structure-making resultant effect on water, it can occur that at higher temperatures the conditions are more favourable to the binding of water molecules, because the fraction of water molecules with no hydrogen bonds will rise with temperature; in these cases the increase in temperature enhances the increasing effect of ions on viscosity (Hasted, Ritson, and Collie [57], Kaminsky [59]).

According to Kaminsky [59], the increasing effect of relatively small polyvalent ions (e.g. Be^{2+}, Mg^{2+}) on viscosity decreases with increasing temperature. To explain this, it can be assumed that the effect of these strongly hydrated ions is extended to several layers of water molecules, while in layers farther from the ions the arrangement of the molecules is more and more disturbed by the thermal motion.

Owing to the complicated nature of aqueous solutions of ions, the conclusions given above are not unequivocal, and the details of the interactions between ions and water are questioned even now. For example, the K^+ ion is considered to have a structure-breaking effect, while Bergquist and Forslind [17] concluded from NMR measurements that the potassium ion, in fact, does not decompose, but strengthens the water structure; it increases the energy of the hydrogen bonds, although it decreases the viscosity. It has also been shown that this ion decreases the density in a manner similar to that of dissolved noble gas atoms. This indicates that the potassium ion replaces an interstitial water molecule in the ordered molecular clusters (i.e. it does not occupy an additional place), which would have appeared there on the sterical deformation effect of the ions on the molecular lattice. In this way, the number of water molecules needed to maintain the thermal equilibrium of the distorted lattice is decreased by the presence of the large potassium ion. Forslind assumed that the combination of the stabilizing effect on the lattice (increasing the number of hydrogen bonds) with the disturbing one will decrease the viscosity of the system by hindering the increase in density. The viscosity of liquids is more sensitive to changes in density than to changes in the energy of the hydrogen bonds. According to this opinion, potassium ions decrease the viscosity of water because their effect on the density is larger than that on the energy of the hydrogen bond.

It still cannot be decided whether large monovalent ions form a complete hydrate sphere and whether they can immobilize water molecules at all, and have any electrostriction effect. According to Samoilov [52, 60] and Kreshtov [61], the activation energy of the exchange of water molecules in the immediate neighbourhood of some ions of relatively high hydration energy (K^+, Rb^+, Cs^+, Cl^-, Br^-, I^-) is negative, indicating that their mobility is larger in these solutions than in pure water (negative hydration). As established by Gurney [23], in the case of ions of radii larger than a certain critical value (this is about 1·6 Å for monovalent ions), the electric field strength is too small to order the water molecules by compensating their thermal motion. Thus, the electric field of the structure-breaking ions decreases the energy of water molecules arising from their ordered structure only by making this structure more sensitive and more easily destructable by the thermal motion. According to Mikhailov and Sirnikov [62], this is similar, in general, to the effect of the electric field of ions on the farther molecular layers around the ions. From the two-structure theory of water, it follows that the electric field shifts the equilibrium between the molecular clusters of the porous structure and the close-packed molecules towards the latter direction.

From NMR investigations, Hindman [42] concluded that there can also be thermally disordered molecules in the immediate vicinity of ions. In the first coordination sphere of Li^+ ion, there are four water molecules, and its orientation effect extends farther; however, the hydration number decreases with increasing radius of the cation. Of the anions, only the F^- ion is supposed to have a hydrate sphere (in accordance with the theoretical studies of Glueckauf [80]); the other halide ions (with radii larger than 1·7 Å) destroy the water structure in their vicinity, i.e. negative hydration takes place. On the other hand, Hindman's standpoint that the interaction energy of anions with water is lower than that of cations of similar size, is in contradiction with the generally accepted opinions based on the electrostatic model. It is, however, in agreement with some recent models of the structure of water (Lennard-Jones and Pople [63], Duncan and Pople [64], Burnelle and Coulson [65]).

Literature on page 78

On the basis of infra-red spectra of electrolyte solutions, Bonner and Woolsey [66], and Bonner [67] concluded from their studies on changes of the water structure caused by solutes that solute ions result in a similar change in the spectra to the formation of a hydrogen bond. Since the absorption bands are influenced by the interaction between the solute and the solvent, the hydration number of the ions can be calculated on the basis of proper hypotheses (see Section 5.2.4.5.). Depending on their size and the superficial charge density, molecules have two kinds of effects on the spectra: small ions or those of large charge density (e.g. Li^+, F^-, Mg^{2+}, La^{3+}) strongly orientate the water molecules in their vicinity, resulting in rupture of some hydrogen bonds, i.e. they loosen the water structure and hinder the formation of ionic pairs; large ions with small charge densities (e.g. Cs^+, organic ions), on the other hand, increase the number of hydrogen bonds, i.e. they strengthen the structure of water and promote the formation of ion pairs. The differences in the conclusions drawn from spectroscopic and other measurements have been attributed to the fact that these latter ones involve both the interaction of the ions with one another and the effect of ions on the solvent molecules, while in the infra-red spectra the ion-solvent interaction is only reflected.

Some cautiousness is necessary in connection with the conclusions drawn from the infra-red and Raman spectra regarding changes in water structure due to the ions, since the correlations between the molecular structure and the infra-red spectra are rather complicated [68], and oversimplified models are inadequate. In connection with this, Luck's finding [69] should be mentioned indicating that changes in the infra-red and Raman spectra of water due to an increase in temperature exhibit no discontinuity when passing the critical state. Taking into account the investigations carried out by Hornig [70], it does not seem to be proved that the alteration of the spectra due to the dissolution of ions is really a result of the shift in the equilibrium between the ice-like hydrogen bonded state and the monomeric state of water, and the role of this shift in the equilibrium in the spectroscopical phenomena has not been cleared up.

Dissolved ions alter the hydrogen bonds between the water molecules, but it would be an over-simplified picture of reality to suppose that this effect is no more than breaking and forming hydrogen bonds. This has been supported by the investigations of Lennard-Jones and Pople [71], Conway [72], Cohan et al. [73], and Bauer [74], and others who state that even the structure of pure water cannot be described in agreement with reality by supposing only two types of water molecules: hydrogen bonded and free ones (monomeric molecules). In order to interpret the various spectroscopic investigations, lengthening and bending of the hydrogen bonds due to various interactions with the environment should be assumed (Fig. 1.14) which result in the non-uniform energy of the hydrogen bonded water molecules, and the continuously changing energy levels which will correspond to the continuously changing deformation of the bonds. In this sense, free molecules would represent the limiting case of a continuous series of hydrogen bonded molecules with variously deformed bonds. It can be assumed that the statistical distribution of the differently deformed

hydrogen bonds is modified by the ions
to an extent depending on their size and
field strength, and the total rupture of
hydrogen bonds, or the bond formation
between free water molecules, or their
joining the molecular clusters, are only
limiting cases of this comprehensive effect.
With respect to this, experiments indicate
that the interaction between solvent molecules and solute ions is a very complicated
phenomenon (Section 5.2). Moreover, the
nature of the hydrogen bond itself is not
sufficiently elucidated. Widely diverging
opinions have been published on the covalent nature of this bond, the extent of its
orientation, and the magnitude of the
energy required to bend it. This explains
why the conclusions drawn from macroscopically observable properties of electrolyte solutions regarding the modifying
effect of dissolved ions on the water structure do not give a unified picture.

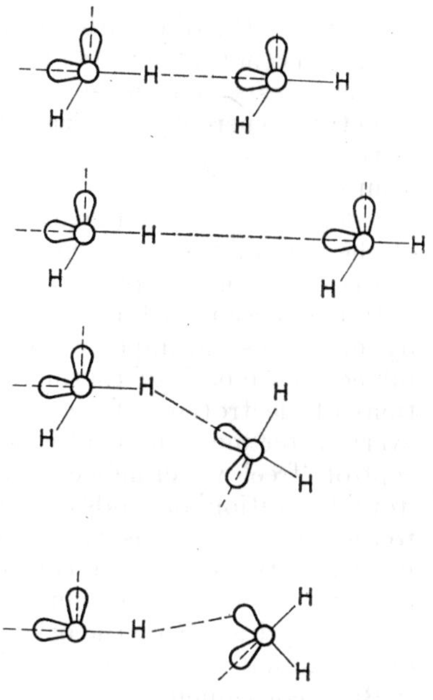

Fig. 1.14 Stretched and bent hydrogen bonds

*Insufficiencies in the electrostatic theory
of ion–water interaction.* A significant part
of the theories on the effect of ions in altering the water structure attributes the phenomena essentially to the electrostatic interaction of ions with
water dipoles. On this basis several phenomena can be interpreted, and
in certain respects the correlations obtained can be assumed to correspond
to reality. However, this interpretation involves several uncertainties and
assumptions and has not been proved satisfactorily. Moreover, the
conclusions obtained on this basis from different phenomena are often in
contradiction with one another. Thus, models taking into account only the
electrostatic interactions are of limited validity. In most cases, the covalent
elements of the interactions should also be considered.

The electrostatic concept is primarily limited by the fact that the experimental data always reflect the simultaneous effects of the cations and
anions, since hydration energies of anions or cations cannot be measured
separately. The data of the individual ions can only be calculated on the
basis of suitable assumptions, however, these assumptions are questionable.
This is shown by the considerable differences in the hydration energies calculated by the use of some apparently plausible assumptions in the Born
treatment–based essentially on electrostatic effects–and the respective experimental data. According to this, the Born model is a too rough picture of reality.

Magnusson [75] and Forslind [81] supposed on the basis of thermodynamical calculations and NMR measurements respectively, that the covalent
bond also has an important role in the ion–water interaction. According

Literature on page 78

to Forslind, the orientation of a free water molecule with respect to a polarizable ion is not influenced only by the resultant dipole moment of the water molecule, especially when the water molecule is a member of a lattice-like structure, where its polarization is determined essentially by its neighbours. If the interaction is strong enough to form a new type of molecule in the solution, the dipole orientation of the interacting particles will not be the decisive factor but a covalent bond will be formed on hybridization of the ion and the water molecule, irrespective of whether the formation of the new complex is based on hydrogen bonding or not.

In connection with the structural changes in the water structure caused by solute ions, the important results of Mishchenko and Dymarchuk [76], obtained on the basis of thermodynamical investigations, indicate that in solutions of electrolytes of high solubility there is a concentration above which every water molecule is already in the hydrate sphere of an ion, and the concept of 'free' water molecules cannot be applied. This concentration limit of 'total hydration' depends on the coordination number of the ions. According to these investigations, the structure of solutions more dilute than the limiting concentration can be considered–in a certain sense–as a water structure distorted by ions. The structure of solutions more concentrated than the limit of total hydration, on the other hand, is similar to that of the crystal hydrate of the actual electrolyte, or of the water-free crystals distorted by water.

By a consequent extension of the cluster theory of water to electrolyte solutions, Hümbelin [77] tried to calculate the thermodynamical data of these solutions from the various (vibrational, rotational, electronic, and mutual potential) energies of the particles.

The structure of electrolyte solutions can be approximated by the analysis of the empirical correlations between the densities and concentrations of solutions, as suggested by Lengyel [78]. According to this, in aqueous solutions of monovalent electrolytes there are 10–12 water molecules in the neighbourhood of each ion in a definite configuration; their arrangement is different from that in pure water. The structure of this molecular complex depends on the nature of the ion and on the temperature, but it is independent of the concentration of the solution and of the other ions present. If the solution is so concentrated that less than 10–12 water molecules are available for one ion, the outer sphere of water molecules around the ion is partly or completely absent, while the remaining inner sphere has the same configuration as in dilute solution. The density of the solution, and the partial volume of the components in the solution, are determined by the configuration of the water molecules surrounding the ions. The calculations based on this model are in agreement with the empirical equation containing three constants, but, with respect to the available information, it cannot be judged how truly this model reflects the real structure of solutions.

The structural characteristics of electrolyte solutions also affect the rate of formation of ice from the solutions. According to Rozental [83], the rate of growth of ice crystals from electrolyte solutions is the higher, the more randomly orientated the molecules and the lower the value of the hydration constant (ΔE). That is, structural disorder facilitates crystallization from solutions.

REFERENCES

to Section 1.4

[1]   D. D. ELEY, *Trans. Faraday Soc.*, 1930, **35**, 1281, 1421.
[2]   A. YU. NAMIOT, *Zh. Strukt. Khim.*, 1961, **2**, 408, 476.
[3]   H. S. FRANK and M. W. EVANS, *J. Chem. Phys.*, 1945, **13**, 507.
[4a]   R. E. POWELL and W. M. LATIMER, *J. Chem. Phys.*, 1951, **19**, 1139.
[4b]   C. JOLICOEUR and H. L. FRIEDMAN, *Ber. Bunsenges.*, 1971, **75**, 248.
[5]   H. S. FRANK and W. Y. WEN, *Disc. Faraday Soc.* 1957, **24**, 134.
[6]   H. S. FRANK, *Proc. Roy. Soc.*, 1958, **A. 247**, 481; *Nat. Acad. Sci.—Nat. Res. Council. Publ.*, 1963, **42**, 141.
[7]   G. NÉMETHY and H. A. SCHERAGA, *J. Chem. Phys.*, 1962, **36**, 3382, 3401.
[8a]   O. D. BONNER and G. B. WOOLSEY, *J. Phys. Chem.*, 1968, **72**, 899.
[8b]   D. V. BEAUREGARD and R. E. BARRETT, *J. Chem. Phys.*, 1968, **49**, 5241.
[8c]   G. BARONE, E. RIZZO, and B. VITAGLIANO, *J. Phys. Chem.*, 1970, **74**, 2230.
[8d]   B. Z. GORBUNOV and YU. I. NAVERUZHIN, *Zh. Strukt. Khim.*, 1972, **13**, 20.
[8e]   I. N. KOCHNEV, *Zh. Strukt. Khim.*,1973,**14**,362; I. N. KOCHNEV and A. I. KHA-LOMOV, *Zh. Strukt. Khim.*, 1973, **14**, 791.
[9]   J. C. GJALDBAEK and J. H. HILDEBRAND, *J. Amer. Chem. Soc.*, 1950, **72**, 1077.
[10]   W. L. MASTERTON, *J. Chem. Phys.*, 1954, **22**, 1830.
[11]   H. G. HERTZ, *Ber. Bunsenges.*, 1964, **68**, 907.
[12]   S. GOLDEN and C. GUTTMAN, *J. Chem. Phys.*, 1965, **43**, 1894.
[13a]   V. A. P. LUCK, *Ber. Bunsenges.*, 1965, **69**, 626.
[13b]   A. V. KARIAKIN and G. A. MURADOVA, *Zh. Fiz. Khim.*, 1971, **45**, 1054.
[13c]   YU. I. NABERUKHIN and V. A. ROGOV, *Usp. Khim.*, 1971, **40**, 369.
[13d]   D. B. DAHLBERG, *J. Phys. Chem.*, 1972, **76**, 2045.
[14]   K. P. MISHCHENKO and G. M. POLTORATSKII, Vopr. Termodinamikii stroeniya vodnykh i nevodnykh rastvorov elektrolitov (Leningrad, 1968).
[15]   J. T. DAVIES and E. K. RIDEAL, Interfacial Phenomena (New York, 1961).
[16]   S. W. BENSON and C. S. COPELAND, *J. Phys. Chem.*, 1963, **67**, 11 94.
[17]   M. S. BERGQUIST and E. FORSLIND, *Acta Chem. Scand.*, 1962, **16**, 2069.
[18]   M. EIGEN, *Pure Appl. Chem.*, 1963, **6**, 97.
[19]   O. YA. SAMOILOV, *Zh. Fiz. Khim.*, 1955, **29**, 1582; *Disc. Faraday Soc.*, 1957, **24**, 141, 216.
[20]   T. J. SWIFT and R. E. CONNICK, *J. Chem. Phys.*, 1967, **37**, 307.
[21]   J. P. HUNT and H. TAUBE, *J. Chem. Phys.* 1951, **19**, 602.
[22]   D. H. EVERETT, *Disc. Faraday Soc.*, 1957, **24**, 216, 220, 229.
[23]   R. W. GURNEY, Ionic Processes in Solutions (New York, 1953).
[24]   K. FAJANS and O. JOHNSON, *J. Amer. Chem. Soc.*, 1942, **64**, 668.
[25]   R. J. GILLESPIE, *Disc. Faraday Soc.*, 1957, **24**, 230.
[26]   P. M. VOLLMAR, *J. Chem. Phys.*, 1963, 39, 2236.
[27a]   A. K. LIASHCHENKO, *Zh. Strukt. Khim.*, 1968, **9**, 781.
[27b]   L. S. SHRAIBER and S. P. TIKHII, *Zh. Fiz. Khim.* 1973, **47**, 698.
[28a]   A. BEN-NAIM, *J. Phys. Chem.*, 1965, **69**, 1922; *J. Chem. Phys.*, 1965, **42**, 1512; *Trans. Faraday Soc.*, 1965, **61**, 821.
[28b]   D. S. TEREKHOVA, *Zh. Strukt. Khim.*, 1970, **11**, 530.
[28c]   G. J. JANZ, B. G. OLIVER, G. R. LAKSHMINARAYANAN and G. E. MAYER, *J. Phys. Chem.*, 1970, **74**, 1285.
[28d]   K. G. BREITSCHWERDT, *Ber. Bunsenges.*, 1971, **75**, 319.
[28e]   M. ABRAHAM and J. J. HECHLER, *Electrochim. Acta*, 1972, **17**, 1203
[29]   G. NÉMETHY and H. A. SCHERAGA, *J. Chem. Phys.*, 1962, **36**, 3382, 3401.
[30]   H. SCHNEIDER, G. C. KRESCHEK and H. A. SCHERAGA, *J. Phys. Chem.*, 1965, **69**, 1310.
[31a]   B. E. CONWAY, R. E. VERRALL and J. E. DESNOYERS, *Z. Phys. Chem.*, 1965, **230**, 157.
[31b]   S. LINDENBAUM, *J. Phys. Chem.*, 1971, **74**, 3027.
[32]   D. F. EVANS and R. L. KAY, *J. Phys. Chem.*, 1966, **70**, 366.
[33]   R. M. DIAMOND, *J. Phys. Chem.*, 1963, **67**, 2513.
[34a]   S. LINDENBAUM and G. E. BOYD, *J. Phys. Chem.*, 1964, **68**, 911.

[34b] J. W. WHILE, *Ber. Bunsenges.*, 1971, **75**, 379.
[35]   M. LATIMER, K. S. PITZER and C. M. SLANSKY, *J. Chem. Phys.*, 1939, **7**, 108.
[36]   A. D. BUCKINGHAM, *Disc. Faraday Soc.*, 1957, **24**, 151.
[37]   R. M. NOYES, *J. Amer. Chem. Soc.*, 1962, **84**, 513.
[38]   R. H. STOKES and R. A. ROBINSON, *J. Amer. Chem. Soc.*, 1948, **70**, 1870.
[39]   R. H. MOELWYN-HUGHES, *Proc. Cambridge Philos. Soc.*, 1948, **45**, 477.
[40]   R. D. WALDRON, *J. Chem. Phys.*, 1957, **26**, 809.
[41]   YU. P. SIRNIKOV, *Dokl. AN SSSR*, 1958, **118**, 760.
[42]   J. C. HINDMAN, *J. Chem. Phys.*, 1962, **36**, 1000.
[43]   G. H. HAGGIS, J B. HASTED and T. J. BUCHANAN, *J. Chem. Phys.*, 1952, **20**, 1452.
[44]   F. E. HARRIS and C. T. O'KONSKI, *J. Phys. Chem.*, 1957, **61**, 310.
[45]   R. A. ROBINSON and H. S. HARNED, *Chem. Rev.*, 1941, **28**, 419.
[46]   R. M. DIAMOND, *J. Amer. Chem. Soc.*, 1958, **80**, 4808.
[47]   YU. M. KESSLER, YU. M. POVAROV and A. I. GORBANEV, *J. Struct. Chem.*, 1962, **3**, 93.
[48]   F. VASLOW, *J. Phys. Chem.*, 1963, **67**, 2773.
[49]   J. N. SHOOLERY and B. J. ALDER, *J. Chem. Phys.*, 1955, **23**, 805.
[50]   O. YA. SAMOILOV, *J. Struct. Chem.*, 1962, **3**, 332.
[51]   J. O'M. BOCKRIS, *Quart. Rev.*, 1949, **3**, 173.
[52]   O. YA. SAMOILOV, *Disc. Faraday Soc.*, 1957, **24**, 141, 216.
[53]   H. S. FRANK, *Nat. Acad. Sci.—Nat. Res. Council. Publ.*, 1963, **42**, 141.
[54]   G. W. BRADY, *J. Chem. Phys.*, 1960, **33**, 1079.
[55]   S. FREED, *Rev. Modern Phys.*, 1942, **14**, 105.
[56a] G. ENGEL and H. G. HERTZ, *Ber. Bunsenges.*, 1968, **72**, 808.
[56b] A. EUCKEN, *Z. Elektrochemie*, 1948, **52**, 255.
[57]   J. B. HASTED, D. M. RITSON and C. H. COLLIE, *J. Chem. Phys.*, 1948, **16**, 1.
[58]   L. A. D'ORAZIO and R. H. WOOD, *J. Phys. Chem.*, 1963, **67**, 1435.
[59]   M. KAMINSKY, *Disc. Faraday Soc.*, 1957, **24**, 171.
[60]   O. YA. SAMOILOV, *Zh. Fiz. Khim.*, 1955, **29**, 1582.
[61]   G. A. KRESHTOV, *J. Struct. Chem.*, 1962, **3**, 137.
[62]   I. G. MIKHAILOV and YU. R. SIRNIKOV, *Zh. Strukt. Khim.*, 1960, **1**, 10.
[63]   J. LENNARD-JONES and J. A. POPLE, *Proc. Roy. Soc.*, 1951, **A. 205**, 155.
[64]   A. B. DUNCAN and J. A. POPLE, *Trans. Faraday Soc.*, 1953, **49**, 217.
[65]   L. BURNELLE and C. A. COULSON, *Trans. Faraday Soc.*, 1957, **53**, 403.
[66]   O. D. BONNER and G. B. WOOLSEY, *J. Phys. Chem.*, 1968, **72**, 899.
[67]   O. D. BONNER, *J. Phys. Chem.*, 1968, **72**, 2512.
[68]   See, e.g. the discussion of the infra-red spectrum of CO which is much simpler than water: H. FRIEDMANN and S. KIMEL, *J. Chem. Phys.*, 1965, **42**, 3327; cf. D. P. STEVENSON, *J. Phys. Chem.*, 1965, **69**, 2145.
[69]   W. A. P. LUCK, *Ber. Bunsenges.*, 1965, **69**, 626.
[70]   D. F. HORNIG, *J. Chem. Phys.*, 1964, **40**, 3119.
[71]   J. LENNARD-JONES and J. A. POPLE, *Proc. Roy. Soc.*, 1950, **A. 202**, 166, 323.
[72]   B. E. CONWAY, *Can. J. Chem.*, 1959, **37**, 178.
[73]   N. V. COHAN, M. COTTI, J. V. IRIBARNE and M. WEISSMANN, *Trans. Faraday Soc.*, 1962, **58**, 490.
[74]   W. H. BAUER, *Acta Cryst.*, 1965, **19**, 901.
[75]   L. B. MAGNUSSON, *J. Chem. Phys.*, 1963, **39**, 1953.
[76]   See, e.g., K. P. MISHCHENKO and N. P. DYMARCHUK, *Zh. Strukt. Khim.*, 1962, **3**, 411.
[77]   R. HÜMBELIN, *Electrochim. Acta*, 1964, **9**, 685.
[78]   S. LENGYEL, *Acta Chim. Acad. Sci. Hung.*, 1963, **37**, 87, 319.
[79]   J. D. BERNAL and R. H. FOWLER, *J. Chem. Phys.*, 1933, **1**, 515.
[80]   E. GLUECKAUF, *Trans. Faraday Soc.*, 1955, **51**, 1235.
[81]   Cf. in: J. L. KAVANAU, Water and Solute–Water Interactions, p. 60–66 (Linden 1964).
[82]   V. M. VDOVENKO, YU. V. GURIKOV and E. K. LEVIN, *Zh. Strukt. Khim.*, 1969, **10**, 576.
[83]   O. M. ROZENTAL, *Zh. Strukt. Khim.*, 1973, **14**, 796.

2. VISCOSITY

2.1 BASIC EMPIRICAL RELATIONSHIPS OF THE VISCOSITY OF LIQUIDS

Viscosity (internal friction) is the resistance against flow (deformation) caused by an external force in liquids (and gases, or solids). Its quantitative measure is the shear force which causes–in the case of laminar flow–a relative displacement of two layers of 1 cm² surface area at a distance of 1 cm from each other in the bulk of the liquid with a velocity of 1 cm s⁻¹.

Since an immobile liquid (or gas) layer is always adhered to the surface of wettable solids, viscosity plays an important role in the relative displacement of bodies and liquids (and gases).

The fundamental phenomenon of viscous flow under the simplest conditions is shown in Fig. 2.1. If there are two planes A and B in the liquid (gas) near to each other and the latter is displaced with respect to the former with velocity v in direction x, the liquid begins to move as well. Plane B carries along the liquid layer in immediate contact with it (located at position $y = 0$) as a result of adhesion, which gains an impulse mv (m is the mass of the liquid). This impulse is partly taken over by the next liquid layers which also move in direction x, but with velocity $v_x(y, t)$ decreasing with increasing distance y. Immediately after the beginning of the displacement of B, in the transient state, at a given point of the liquid the velocity depends on the time and at a given time on the position, i.e. the velocity distribution varies with time $[v_x = v_x(y, t)]$. In the case of laminar flow, a stationary velocity distribution is quickly attained. In the stationary state, the velocity of the liquid depends only on the position coordinate y, and it is independent of time $[v_x = v_x(y)]$.

Since the contiguous liquid layers transport impulse (and at the same time kinetic energy) continuously to one another, the movement of the neighbouring liquid layers with different velocities consumes energy and the preservation of the movement requires external work and force. After reaching the stationary state of laminar flow, the work of the external force needed to overcome viscosity will be transformed into thermal energy.

According to Newton's viscosity law, the shear force $f_{x,y}$ required to displace two

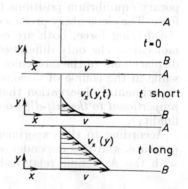

Fig. 2.1 Development of stationary and laminary velocity profiles in a liquid between planes A and B moving with velocity v with respect to each other

Literature on page 137

liquid layers of unit surface area located at a distance dy from each other in direction x with a relative velocity dv_x is proportional to the negative value of the velocity gradient at that place:

$$f_{x,y} = - \eta \frac{dv_x}{dy}, \tag{2.1}$$

were η [g cm^{-1} s^{-1}] is the viscosity coefficient of the liquid or, simply, the *viscosity*. The reciprocal value of viscosity is fluidity: $\varphi = 1/\eta$. For liquids of simple structure (not macromolecular) and for gases, η is independent of the velocity gradient *(Newtonian liquids)*. In addition to the thermal motion, viscosity depends on the mutual position, size, and shape of the liquid molecules, as well as on the forces arising from the interaction between them.

In equation (2.1), the impulse flow in direction y is expressed. In the course of the movement of the liquid, its impulse of direction x is transported to the adjacent liquid layers in direction y. In other words $f_{x,y}$ is the *viscous flux* of the *impulse* of direction x in direction y.* The direction of the viscous flux of impulse coincides with the direction of the negative velocity gradient according to equation (2.1), i.e. impulse flows from places of higher velocity towards those of lower one, similar to a sleigh slipping downwards on a slope or to the thermal energy flowing from warmer places to colder ones. The velocity gradient is the driving force of the transport of impulse. According to this, Newton's law in equation (2.1) can be discussed as a relationship regarding force which emphasizes the essentially mechanical nature of the phenomenon; on the other hand, it can be regarded as a correlation concerning the transport of impulse which points to the analogy with the energy and mass transport (heat conduction and diffusion).

In the course of viscous flow, molecules are transferred from their temporary equilibrium positions to adjacent ones in the direction of the shear force. The elementary process of self-diffusion is similar to this. Disregarding the driving force, both are essentially determined by the mobility of the molecules. The only difference is that in self-diffusion the direction of the displacement of the molecules is disordered (thermal wanderings at random) while in the course of viscous flow it is ordered. This analogy explains the experimental observation that *the viscosity coefficient of liquids is inversely proportional to the self-diffusion coefficient* (i.e. the latter is proportional to fluidity).

According to the experiments, viscosity decreases with increasing temperature, and this dependence is approximately exponential in accordance with the Arrhenius relationship:

$$\eta = A \exp \left[\frac{E_{\text{visc.}}}{RT} \right], \tag{2.2}$$

where A is a constant approximately independent of temperature and E_{visc}

---

* *Flux* is the velocity of a flow across unit area. The dimension of the flux of impulse, referred to area and time units, is [g cm s^{-1} cm^{-2} s^{-1}] = [g cm^{-1}s^{-2}], i.e. it is the same as that of the force related to unit area.

is *the activation energy of the viscous flow*. This is the energy required for local loosening of the liquid, resulting in the formation of a hole in the neighbourhood of a given molecule which will move into it, and the energy required for this action. Consequently, the activation energy also depends on the pressure and temperature.

Recently, some doubts have emerged as to whether the Arrhenius relationship is also valid for a broad temperature range. Eicher and Zwolinski [1b] described the temperature dependence of the viscosity of water and several other liquids by the following equation implying three parameters

$$\ln \frac{\eta_0}{\eta} = \frac{A(t-20) + B(t-20)^2}{C+t} \tag{2.2a}$$

in the -10 to $+150\ °C$ temperature range, where A, B and C are constants characteristic of the given liquid. The viscosity curve of supercooled water is a smooth continuation of that of the stable state.

The fact that the logarithm of viscosity varies approximately linearly with $1/T$ is interpreted, in general, on the analogy of the Arrhenius equation describing the rate of chemical reactions, and the slope of the straight line is used to calculate the activation energy. This is based on the concept that the flow of molecules is hindered by a barrier of some kind related to the quasi-lattice structure of the liquid. Such a barrier apparently exists in liquids with strong structure; however, Hildebrand *et al.* [1c] have recently demonstrated, also utilizing the earlier calculations of Dymond and Adler [1d], that in simple (non-associated) liquids (as well as in gases with a density higher than the critical value), the absolute values and the temperature dependence of viscosity and the diffusion coefficient can be calculated without assuming the existence of an activation barrier. The values calculated in this way agree with the experimental data within about 10%.

According to the theory elaborated by Hildebrand *et al.* on the basis of a thorough analysis of the knowledge available on simple liquids, every molecule participates equally in the thermal motion producing maximum random orientation. All characteristics distinguishing the crystal from the liquid are supposed to diminish on melting. The vacancies, if present in the crystal, are spread at random, to appear as intermolecular spaces in the liquid. Consequently, the theory of diffusion in solids assuming the presence of holes of molecular dimensions will have no physical meaning when extended to liquids. In non-polar liquids no orientating force (e.g., electrostatic field, gravitational field causing sedimentation) acts on the molecules. Their average displacement in time depends on the temperature and on the ratio of the intermolecular volume V to the volume V_0, in which the molecules approach one another so closely that bulk flow and diffusion become impossible. These authors regard it as verified by several examples that fluidity (φ) can be expressed by the following relationship over a broad range:

$$\varphi = \frac{1}{\eta} = B(V - V_0)/V_0.$$

Literature on page 137

The value of B is inversely proportional to the capacity of molecules to absorb the external moment of viscous flow due to their mass, flexibility, softness or the inertia of their rotation. The primary effects of temperature and pressure appear in the alteration of the value of V, but they also affect the value of B. When V exceeds the critical volume, the free path between the molecules increases, which allows velocities corresponding to free space. Under such conditions, the moments are proportional to $T^{1/2}$, while the fluidity and the specific volume are proportional to T^{-1}. The non-spherical shape of molecules lowers the freedom of motion, which is shown by a deviation from the straight line.

The Bachinskii rule often describes the correlation between viscosity and molar volume V of liquids in a good agreement with the experiments:

$$\eta = \frac{k}{V - V_s},\qquad (2.3)$$

where k and V_s are constants depending on the nature of the liquid, the latter being the molar volume of the substance in the solid state at the melting point, while k depends on the forces acting between the molecules. In a certain sense, k and V_s correspond to constants a and b in the van der Waals equation. V_s is the smallest volume to which 1 mole of liquid can be compressed; $V - V_s$ is the *free volume* in the liquid which makes possible the displacement of the molecules. It follows that fluidity depends linearly on the molar volume:

$$\varphi = \frac{1}{\eta} = \frac{V}{k} - \frac{V_s}{k}.\qquad (2.4)$$

Although, in the Bachinskii rule, temperature and pressure values do not appear, the dependence of viscosity on these parameters is involved, since the free volume increases with increasing temperature and decreases with increasing pressure.

In the case of *non-Newtonian liquids*, including mainly macromolecular substances and their solutions, the shear force (or the flux of impulse) causing the flow of the liquid is not proportional to the velocity gradient, i.e. viscosity η is not constant, but changes with the velocity gradient (or the shear force). Depending on the conditions, η can increase (pseudoplastic liquids) or decrease (dilatating liquids) with increasing value of $-\dfrac{dv_x}{dy}$.

The investigation of non-Newtonian liquids belongs to the topic of *rheology*. Here, we refer briefly to the several theories outlined [1a] for the correlation between the shear force in the equilibrium state and the velocity gradient. According to the theory of Bingham, we have:

$$f_{x,y} = -\eta_0 \frac{dv_x}{dy} \pm f_0, \quad \text{if } |f_{x,y}| > f_0 \qquad (2.5)$$

and

$$\frac{dv_x}{dy} = 0, \quad \text{if } |f_{x,y}| < f_0, \tag{2.6}$$

where $-f_0$ is a constant characteristic of the substance. The positive sign applies when $f_{x,y}$ is positive, and the negative sign should be used when $f_{x,y}$ is negative. Materials for which these correlations are valid are rigid if the shear force is smaller than f_0 and exhibit Newtonian properties in the case of shear forces higher than f_0 (the shear force varies linearly with the velocity gradient, see Fig. 2.2). This relationship applies to several suspensions and pastes.

According to the theory of Ostwald and De Waale

$$f_{x,y} = - \eta \left| \frac{dv_x}{dy} \right|^{n-1} \frac{dv_x}{dy}. \tag{2.7}$$

The limiting case of this correlation involving two parameters is the Newton law ($n = 1$), i.e. the value of n is the measure of deviation from the state of the Newtonian liquids. Pseudoplastic and dilatating behaviours correspond to $n < 1$ and $n > 1$, respectively.

According to the theory of Eyring, the following equation can be obtained:

$$f_{x,y} = A \operatorname{arc\,sinh} \left(- \frac{1}{B} \frac{dv_x}{dy} \right), \tag{2.8}$$

where A and B are constants. In the case of finite values of $f_{x,y}$, the liquid is pseudoplastic and its properties tend asymptotically to those of the Newtonian liquids as $f_{x,y}$ tends to zero.

Some equations containing three parameters are also used for the description of viscosity. It should be pointed out, however, that the equations describing the rheological properties of liquids are essentially empirical correlations based on experimental data, i.e. in fact, they are interpolation formulae. Their parameters depend on the temperature, pressure and composition, and mostly they are constant only in a rather narrow range of the velocity gradient.

Under non-stationary conditions, there are liquids with other properties also deviating from the Newtonian

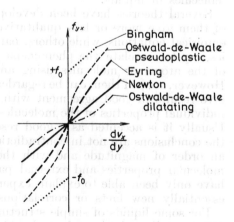

Fig. 2.2 Correlation between shearing force and velocity gradient in Newtonian and non-Newtonian liquids

Literature on page 137

6*

ones. These are, e.g. the thixotropic liquids, having viscosities decreasing
in time when exposed suddenly to a force $f_{x,y}$, and the rheopectic liquids
with viscosities increasing under such conditions. In addition, there are
liquids which partly regain their original shape when the shear force ceases
(viscoelastic liquids).

2.2 FUNDAMENTALS OF THE THEORIES OF VISCOSITY

2.2.1 GENERAL

The theory of viscosity of gases had already been elaborated by Maxwell
in the middle of the last century and his conclusions have also been proved
experimentally. On the basis of the theory further developed by Rayleigh,
Chapman and Enskog, the calculated values of the constants in the equation
of viscosity are in good agreement with the experimental data. The viscosity of gases can be calculated from the properties of their atoms or
molecules, and Lennard-Jones [2] deduced the law of forces acting between
different atoms from the measured values of the viscosity of gases. Our
theoretical knowledge on viscosity of liquids, however, is much more incomplete, which is a consequence of the more complicated and lesser known
structure of liquids as compared with that of gases.

In the viscous flow of gases, molecules transport impulse by their own
movement between the layers moving with respect to each other. The
viscous flow of liquids has another mechanism. Previously it was supposed
that molecules in liquids transport impulse not by their own movement,
but by means of interatomic forces, since owing to the density of liquids,
the mutual distance between the atoms or molecules in them is of the same
order of magnitude as the range of the interatomic forces. Only in about
the 1940's did it become evident that impulse is also carried by the displaced
molecules in liquids.

Several theories have been developed on the viscosity of liquids. Some
of them give more or less qualitative interpretation of the factors in the
empirical equations, while others, partly on the basis of quantum mechanics, try to correlate the phenomena of viscous flow with the properties
of the atoms or molecules using an extensive mathematical treatment.
However, none of them can be regarded as a satisfactory theory, which would
give results in good agreement with the experiments on the basis of the
individual properties of the molecules and the forces acting between them.
Usually it is accepted as a good result in the theories of viscosity when
the conclusions are not in contradiction with the experimental data within
an order of magnitude and with the main trends of the dependence on
molecular properties and external parameters. All theories known so far
have only been able to explain experimental facts without predicting any
essentially new facts or correlations valid in a broader range.

For some liquids of simple structure under definite conditions, Andrade
[3], as well as Kirkwood et al. [4], deduced some numerical values for
viscosity by using only molecular constants known from other phenomena,

but these results could not be generalized without introducing empirical parameters. Hence, they cannot serve as a basis for a general theory of the viscosity of liquids.

The viscosity of liquids plays a more or less important role in both scientific theory and practice. Therefore, it can be understood that the particular theories are evaluated from various points of view by the experts in different scientific fields. By means of a detailed analysis of the conditions, Brush [5] divided the theories into several groups with respect to their validity in the different applications.

Industrial experts (engineers, hydrologists, etc.) are interested primarily in a macroscopic description of the viscous properties of liquids; the correlations between these properties and the molecular structure of liquids or the actual molecular properties are of no importance for them.

On the other hand, physicists take interest in how the properties of viscous flow depend on the molecular structure of the liquid and on the properties of the individual molecules known from other phenomena or from other correlations. Physicists try to deduce the theory of viscosity from one or more general theorems or hypotheses. They try to obtain conclusions that can be checked experimentally and hence be either verified or excluded. They expect the theories to enable prediction of unknown phenomena. In this respect, it is only a necessary but not sufficient requirement of the correctness of a theory that its conclusions agree with the results of experiments made so far, since it cannot be judged in advance whether the same conclusions cannot also be achieved by a theory based on other hypotheses. On the other hand, a theory may be useful from the point of view of scientific development even when it leads to results contradictory to the experiments, because this gives evidence for being erroneous the starting hypotheses (or one of them). In this way, the hypotheses leading to contradictions with reality can be eliminated by means of the unsuccessful theories and the number of the hypotheses that may be regarded as correct is reduced.

The chemists' requirements for a theory are usually not so strict. For them, the theory of viscosity provides only a correlation between some measurable quantities involving some empirical constants. If the curve of the theoretical relationship fits the measured points when the constants are properly chosen, the theory is considered an applicable one. If not, the theory should not be rejected as a whole; it can be corrected by the introduction of further factors which take into account further assumed effects. This, of course, makes the theory more complicated. Now, however, chemists often require the theory to be based on molecular properties and the physical meaning of the empirical factors to be revealed (at least in a semi-quantitative manner).

Owing to the complicated nature of the mechanism of viscous flow, however, these interpretations are intuitive rather than unequivocal and logical consequences of some starting hypotheses.

The kinetic theory of the viscosity of gases satisfies the requirements of both the physicists and chemists mentioned above. Considering the

Literature on page 137

viscosity of liquids, the situation is not so unequivocal, since none of the theories known describes viscous flow satisfactorily. So far, only the theory of the viscosity of liquids of the simplest structure consisting of monatomic molecules with spherically symmetrical fields of very short range has been successfully elaborated (Born and Green [6], Kirkwood [7]). In fact, only the condensed noble gases are such liquids. The majority of liquids, however, particularly the chemically important ones, have far more complicated structures. The simple theory provides hardly any information on their properties of practical importance.

In the absence of an exact theory describing the important physico-chemical and chemical properties of the viscous flow in relation to other properties, one has to rely on less adequate theories. On the basis of various, apparently plausible suppositions, these theories endeavour to provide a picture of the mechanism and the elementary steps of viscous flow, as well as of the correlations between properties related to viscosity and other characteristics; they introduce constants with numerical values to be deter-mined empirically. Though each of these theories can be questioned in some respects and their conclusions are not unequivocal (owing to the empirical constants involved, since their numerical values should be deter-mined subsequently from the experiments), there are still some theories which appreciably promote an understanding of the properties of viscous flow and clarify their correlations with other phenomena.

In this respect, the judgement of the different theories is, of course, not uniform. A detailed discussion of the theories on the viscosity of liquids would lead too far (in connection with this we refer to, e.g., the review by Brush [5]). Here only some main characteristics of the corre-lations on viscosity will be reviewed.

The hydrodynamical, 'macroscopical' theories, treating the liquid as a continuum and neglecting its molecular structure, give results applicable for technical and certain physical purposes. However, physicists or physico-chemists who take an interest in the mechanism of viscous flow cannot utilize these theories satisfactorily; therefore, 'microscopical' theories based on the molecular structure of liquids are applied. Since, however, our knowl-edge on the structure of liquids is very limited, the theory regarded as the most applicable one to elucidate the phenomena is to a certain extent a matter of choice.

In the following sections, the absolute reaction rate theory of viscosity outlined by Eyring will be discussed in detail [8], because it seems effec-tively facilitate an understanding of the mechanism of the viscous flow of solutions and mixtures from physico-chemical points of view.

2.2.2 HOLES AND FREE VOLUME IN LIQUIDS

In liquids the molecules are rather tightly fitted to one another, although there are some gaps between them. Owing to this, the molecules can only vibrate around their equilibrium positions and translational displacement, i.e. transport of the molecules, can occur only when an empty place, i.e.

a hole (a vacancy) is formed in their vicinity. Thermal fluctuation results in continuous formation of holes in the liquid which are quickly occupied and new ones are formed.* The molecules move in the 'empty space' in gases; in a certain sense, holes 'move' similarly in liquids. The holes play the same role in liquids as molecules do in gases. The majority of the holes formed in the liquid, however, have probably a size smaller than the volume of one molecule.

The energy required for the formation of a hole *(energy of hole formation)* plays an important role in the theory of transport phenomena. In the course of the formation of a hole, the bonds between adjacent molecules are partly broken. Since the bonds between molecules are also broken in the course of evaporation, a correlation must exist between the heat of evaporation and the energy required for the formation of a hole. If ε is the bond energy of a molecule (taking into account all of its neighbours), the total bond energy in a liquid containing N molecules is $N\varepsilon/2$ (since each bond belongs to two molecules). The removal of one molecule from a liquid (i.e. its evaporation) requires an energy of $\varepsilon/2$, if the other molecules are rearranged in the liquid to eliminate the vacancy after the evaporation. If, however, a vacancy remains at the place of a molecule after evaporation, the energy required is ε, since the evaporated molecule must break its bonds with all of its neighbours and no energy is gained by closing the hole. Consequently, without evaporation, an energy of $\varepsilon - \varepsilon/2 = \varepsilon/2$ is needed to form a hole of molecular size. This means that the energy consumed by the formation of a hole of this size is identical with the internal heat of evaporation of a molecule. The formation energy of holes per mole is equal to the *internal molar heat of evaporation E_{evap}.***

Although the molecules in the liquid–disregarding the holes–are fitted rather tightly, this fitting is not without gaps. If the molecules of the liquid are considered in first approximation as rigid spheres of diameter d, these spheres are not in close contact with one another, and a *free volume v_f* is available for the movement of each molecule. The average volume v available for one molecule, calculated from the ratio of the macroscopical liquid volumes and the number of the molecules included, is larger than d^3 (i.e. the volume of the cube in which the molecule can just be placed, Fig. 2.3).

According to Eyring and Hirschfelder [9], liquids of simple structure can be supposed to consist of molecules which move in an average free volume v_f in a potential field produced by the neighbouring molecules. Each molecule is in a 'cage' separated from the adjacent one by an energy

---

* The holes formed on fluctuation are not necessarily identical with the regular cavities in the liquid structure.

** $E_{\text{evap}} = \Delta H_{\text{evap}} - RT$, if ΔH_{evap} is the heat of evaporation measured at constant pressure (i.e. the change in the enthalpy on evaporation) and the vapour behaves as an ideal gas.

Literature on page 138

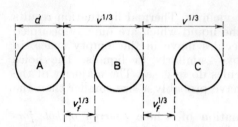

Fig. 2.3 Correlation between the diameter of liquid molecule (d) the average liquid volume available for one molecule (v) and free volume (v_f)

barrier. As given by Eyring and Hirschfelder, the distribution function of the molecules is:

$$F_l = \frac{(2\pi mkT)^{3/2}}{h^3} v_f b_l \exp\left[-\frac{E_0}{RT}\right], \qquad (2.9)$$

where m is the mass of the molecule, k is the Boltzmann constant, h is the Planck constant, b_i is the combined vibrational and rotational contribution, while E_0 is the energy difference of the molecules in the liquid and gaseous state at 0 °K related to 1 mole. As given above, E_0 is equal to the heat of evaporation to a good approximation.

Equation (2.9) can be considered to be the definition of free volume v_f. It can be shown that the following correlation applies for the external pressure p:

$$p = RT\left(\frac{\partial \ln F_l}{\partial v}\right)_T. \qquad (2.10)$$

Taking into account equation (2.9), one has:

$$p = RT\left(\frac{\partial \ln v_f}{\partial v}\right)_T - \left(\frac{\partial E_{\text{evap}}}{\partial v}\right)_T. \qquad (2.11)$$

The relationship between free volume v_f and volume v available for 1 molecule can be obtained as follows. For the sake of simplicity, let us suppose cubic packing of the molecules of the liquid. Any molecule chosen arbitrarily can be regarded as vibrating about its equilibrium position and its six immediate neighbours are fixed in their equilibrium positions in the three directions of space. Fig. 2.3 shows the situation along one axis: each molecule is at a distance $v^{1/3}$ from its neighbours. If the diameter* of the molecule which is considered incompressible is denoted with d, the central atom (in the above meaning) can be displaced to a distance $2v^{1/3}-2d$ in all the three directions. Thus the free volume can be expressed as:

$$v_f = (2v^{1/3} - 2d)^3 = 8(v^{1/3} - d)^3. \qquad (2.12)$$

* When compressing the liquid, the free volume decreases, but the size of the molecules is not altered.

In the case of another type of fitting, a similar expression is valid with a different numerical factor $c(\neq 2)$. Thus, in general,

$$v_f = c^3(v^{1/3} - d)^3. \tag{2.13}$$

Hence

$$\left(\frac{\partial \ln v_f}{\partial v}\right)_T = \frac{c}{v^{2/3} v_f^{1/3}}. \tag{2.14}$$

According to this, equation (2.11) gives:

$$\left\{p + \left(\frac{\partial E_{\text{evap}}}{\partial v}\right)_T\right\} v^{2/3} v_f^{1/3} = cRT. \tag{2.15}$$

It can be supposed that in good approximation:

$$\frac{\partial E_{\text{evap}}}{\partial v} = \frac{E_{\text{evap}}}{v}, \tag{2.16}$$

and furthermore, p can usually be neglected beside this value, so:

$$E_{\text{evap}} \frac{v_f^{1/3}}{v^{1/3}} = cRT. \tag{2.17}$$

If the molar volume of the liquid is V and N is the Avogadro number, it follows that:

$$v_f^{1/3} = \frac{cRTV^{1/3}}{N^{1/3}E_{\text{evap}}}. \tag{2.18}$$

This equation makes possible the calculation of the free volume of the molecules in liquid.

The free volume can also be calculated from the difference in the velocities of sound in liquids and gases. The velocity of sound is 5–10 times greater in most liquids than the average velocity of the molecules given by the kinetic theory. This can be explained as follows: let three molecules lie along a straight line (A, B, and C in Fig. 2.3) and suppose the front of the wave moves from the inner end of molecule A towards the adjacent end of molecule B with a velocity u_{gas} of sound which is equal to that measured in an ideal gas. According to the kinetic gas theory:

$$u_{\text{gas}} = \left(\frac{RT\varkappa}{M}\right)^{1/2}, \tag{2.19}$$

where $\varkappa$ is the specific heat quotient of the gas and M is its molecular weight. When A collides with molecule B, the sound signal passes almost instan-

Literature on page 138

taneously to the other end of B. In this way–although the front of the wave has passed apparently only a distance of $v_f^{1/3}$–it has reached the distance $v^{1/3}$ in reality. Thus, the following expression applies to the velocity of sound in liquid (u_{liq}):

$$\frac{u_{liq}}{u_{gas}} = \left(\frac{v}{v_f}\right)^{1/3},$$

(2.20)

from which, substituting equation (2.19), one obtains:

$$u_{liq} = \left(\frac{v}{v_f}\right)^{1/3} \left(\frac{RT\varkappa}{M}\right)^{1/2}.$$

(2.21)

So the free volume can be calculated from the velocity of the sound in liquids.

Both calculations on the free volume are only approximate ones, but that based on the velocity of the sound is probably more reliable.

2.2.3 THE EYRING THEORY OF VISCOSITY

In the theory of the viscosity of liquids, Eyring considered that molecules displaced relative to one another in the course of viscous flow [8, 10, 11] have to overcome the energy barrier between the adjacent positions. The rate of this process is determined by factors similar to those acting in chemical reactions, therefore the viscous flow can be treated on the basis of the absolute reaction rate theory with suitable modifications.

The viscosity coefficient is, by definition:

$$\eta = \frac{f\lambda_1}{\Delta u},$$

(2.22)

where λ_1 is the distance between two liquid layers of unit surface area, f is the shear force, and Δu is the relative velocity (Fig. 2.4 where λ_1, λ_2, λ_3 are the distances between adjacent molecules in the three directions of space respectively). In the mechanism of viscous flow, the displacement of a liquid layer with respect to the adjacent one can be described as the movement of the molecules of the layer from their equilibrium positions to the adjacent ones within the same layer.

However, a molecule can jump to the adjacent equilibrium position only when there is a hole to be occupied. Therefore, a hole must be formed in the liquid by pushing away the molecules, which requires a certain amount of energy. Thus, the jump of the molecule from its equilibrium position to an adjacent one can be regarded as a process accompanied by overcoming a potential energy barrier. Let λ be the distance between two adjacent equilibrium positions in the direction of the movement. This is not necessarily identical with λ_1, but its deviation from this is not too great. It can be assumed that–if no external shear force acts on the

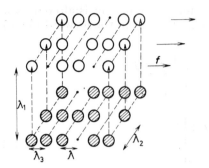

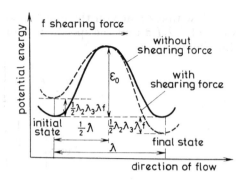

Fig. 2.4 Distances between molecules in a liquid, in the three directions of space $(\lambda_1, \lambda_2, \lambda_3)$; λ is the distance between two adjacent equilibrium positions of the molecule in viscous flow

Fig. 2.5 Potential barrier between adjacent equilibrium positions of a molecule, with and without shearing force resulting in viscous flow

liquid–the potential energy barrier is a symmetrical one (Fig. 2.5), i.e. the distance between the initial position and the peak of the barrier is $\lambda/2$. In this way, the activated state corresponding to the top of the energy barrier is halfway between the two adjacent equilibrium positions.

Since in the liquid an area of $\lambda_2\lambda_3$ is left for one molecule, a shear force $f\lambda_2\lambda_3$ acts on the molecule in the direction of the flow. During the jump of the molecule from its equilibrium position to the top of the barrier, it gains an energy of $f\lambda_2\lambda_3\lambda/2$. Thus the effect of the force resulting in flow decreases the energy barrier to a value of $f\lambda_2\lambda_3\lambda/2$ in the direction of the flow ('forwards', Fig. 2.5). In the opposite direction ('backwards'), it increases the height of the energy barrier by the same value. If the height of the energy barrier is ε_0 in a liquid at rest at 0 °K, this represents the activation energy of the displacement of the molecules.

The rate constant of the viscous flow *(specific rate, k)* gives the number of transitions of the molecule per second across the energy barrier. According to the absolute reaction rate theory, [12]:

$$k = \frac{kT}{h} \cdot \frac{F^{\ddagger}}{F} \exp\left[-\frac{\varepsilon_0}{kT}\right], \tag{2.23}$$

if the transmission coefficient* ≈ 1, and tunnelling is negligible. Here $F^{\ddagger}$ and F are partition functions of the molecules in the activated and initial states, respectively, related to unit volume (disregarding the translational degree of freedom perpendicular to the barrier).

---

* The transmission coefficient gives the ratio of the complexes decomposing at the top of the potential barrier (one portion vibrates back to the initial state).

Literature on page 138

If flow takes place on the effect of shear force, the barrier is altered by a value of $\lambda_2\lambda_3 f/2$ and the specific rate of the movement in the forward (force) direction will be:

$$k_f = \frac{kT}{h} \cdot \frac{F^{\ddagger}}{F} \exp \left[\frac{\varepsilon_0 - \frac{1}{2} f\lambda_2\lambda_3\lambda}{kT} \right] =$$

$$= k \exp \left[+ \frac{\frac{1}{2} f\lambda_2\lambda_3\lambda}{kT} \right], \tag{2.24}$$

while in the backward direction, it will be:

$$k_b = k \exp \left[- \frac{\frac{1}{2} f\lambda_2\lambda_3\lambda}{kT} \right]. \tag{2.25}$$

In a single transition across the potential barrier, the molecule is displaced to distance λ. Since k_f and k_b gives the number of transitions of a molecule per second across the energy barrier in the forward and backward directions, respectively, $k_f\lambda$ and $k_b\lambda$ are the distances the molecule travels in one second (forwards and backwards, respectively). This is the translational velocity of the layer in the two opposite directions. This makes possible the calculation of Δu, i.e. the velocity of the flow in the direction of force f:

$$\Delta u = (k_f - k_b)\lambda = \lambda k \left\{ \exp \left[\frac{\frac{1}{2} f\lambda_2\lambda_3\lambda}{kT} \right] - \exp \left[- \frac{\frac{1}{2} f\lambda_2\lambda_3\lambda}{kT} \right] \right\} =$$

$$= 2\lambda k \sinh \frac{f\lambda_2\lambda_3\lambda}{2kT}. \tag{2.26}$$

Inserting the viscosity coefficient into equation (2.22), we have:

$$\eta = \frac{\lambda_1 f}{2\lambda k \sinh (f\lambda_2\lambda_3\lambda/2kT)}. \tag{2.27}$$

Under normal conditions of viscous flow, f is small (in the order of magnitude of 1 dyn cm^{-2}), λ_2, λ_3 and λ are of molecular dimensions (10^{-8} cm). Thus $2kT \gg f\lambda_2\lambda_3\lambda$. This means that in a series expansion of the exponential terms, it is sufficient to retain the first member. Thus:

$$\eta = \frac{\lambda_1 kT}{\lambda_2\lambda_3\lambda^2 k}. \tag{2.28}$$

The value of k can be substituted from equation (2.23):

$$\eta = \frac{\lambda_1 h}{\lambda_2 \lambda_3 \lambda^2} \cdot \frac{F}{F^{\ddagger}} \exp\left[\frac{\varepsilon_0}{kT}\right].$$ (2.29)

Although λ is not necessarily equal to λ_1, the two quantities are still of the same order of magnitude, and, in this way, equation (2.29) can be transformed approximately as follows:

$$\eta = \frac{h}{\lambda_2 \lambda_3 \lambda_1} \frac{F}{F^{\ddagger}} \exp\left[\frac{\varepsilon_0}{kT}\right].$$ (2.30)

In the liquid, $\lambda_1 \lambda_2 \lambda_3$ is the volume available for one molecule, i.e.:

$$\lambda_2 \lambda_3 \lambda_1 = \frac{V}{N},$$ (2.31)

if V is the molar volume of the liquid. Thus:

$$\eta = \frac{hN}{V} \cdot \frac{F}{F^{\ddagger}} \exp\left[\frac{\varepsilon_0}{kT}\right].$$ (2.32)

The structural parameters of the Eyring theory for water were calculated by Pruppacher [12b] on the basis of his experiments on the self-diffusion of water. The values obtained for water are: $\lambda = 2.06 \times 10^{-8}$ cm, $\lambda_1 = 2.75 \times 10^{-8}$ cm, $\lambda_2 = \lambda_3 = 3.30 \times 10^{-8}$ cm.

The absolute reaction rate theory assumes that the activated complex is in equilibrium with the reactants and the law of mass action applies to this equilibrium. In analogy with this, the activated state of viscous flow can also be described by an equilibrium constant:

$$K^{\ddagger} = \frac{F^{\ddagger}}{F} \exp\left[-\frac{\varepsilon_0}{kT}\right].$$ (2.33)

If $\Delta G^{\ddagger}$ is the standard free enthalpy of activation per mole, the corresponding thermodynamical correlation is:

$$K^{\ddagger} = \exp\left[-\frac{\Delta G^{\ddagger}}{RT}\right].$$ (2.34)

Taking into account equation (2.31), we have:

$$\eta = \frac{hN}{V} \exp\left[\frac{\Delta G^{\ddagger}}{RT}\right],$$ (2.35)

and since

$$\Delta G^{\ddagger} = \Delta H^{\ddagger} - T \, \Delta S^{\ddagger},$$ (2.36)

Literature on page 138

($\Delta H^{\ddagger}$ is the enthalpy and $\Delta S^{\ddagger}$ is the entropy of activation):

$$\eta = \left\{ \frac{hN}{V} \exp\left[-\frac{\Delta S^{\ddagger}}{R} \right] \right\} \exp\left[\frac{\Delta H^{\ddagger}}{RT} \right]. \tag{2.37}$$

Since the molar volume of the liquid is little altered by the temperature and $\Delta S^{\ddagger}$ is approximately constant, the constant factors can be collected into constant B, and thus:

$$\eta = B \exp\left[\frac{\Delta H_{\text{visc}}}{RT} \right], \tag{2.38}$$

where ΔH_{visc} is the enthalpy of activation of the viscous flow. Both Arrhenius [13] and Andrade [3] obtained empirical relations of this form-equation (2.38) is in good agreement with the experiments.

According to the absolute reaction rate theory, approximately we have:

$$\frac{F}{F^{\ddagger}} = \frac{(2\pi \, mkT)^{1/2}}{h} v_f^{1/3}. \tag{2.39}$$

Taking this into account and expressing viscosity on the basis of equation (2.32), we obtain:

$$\eta = \frac{N}{V} (2\pi \, mkT)^{1/2} v_f^{1/3} \exp\left[\frac{\varepsilon_0}{kT} \right]. \tag{2.40}$$

Introducing the expression of the free volume from equation (2.18), $E_{\text{evap}} \approx \Delta H_{\text{evap}}$:

$$\eta = \left(\frac{N}{V} \right)^{2/3} \frac{cRT}{\Delta H_{\text{evap}}} (2\pi \, mkT)^{1/2} \exp\left[\frac{\varepsilon_0}{kT} \right], \tag{2.41}$$

where c is a factor characteristic of the fitting and packing of the liquid molecules. Inserting the values of N, k, R (the latter given in calorie units) and supposing cubic packing ($c = 2$), equation (2.41) is transformed into the following:

$$\eta = 1.09 \times 10^{-3} \frac{M^{1/2} T^{3/2}}{V^{2/3} \Delta H_{\text{evap}}} \exp\left[\frac{\Delta H_{\text{visc}}}{RT} \right], \tag{2.42}$$

where M is the molecular weight of the liquid, ΔH_{visc} is the activation energy of viscous flow (cal/mole).

This result of the Eyring theory can be compared with the empirical correlation of Andrade [13]

$$\eta = a V^{1/3} \exp\left[\frac{b}{T} \right], \tag{2.43}$$

where a and b are constants, and according to the experiments, $a \approx 0.4 \times \times 10^{-3}$ gives a good approximation for many liquids. If values $M = 121$,

$V = 125$ cm³/mole, and $\Delta H_{vap} = 5 \cdot 00$ cal/mole are substituted into equation (2.42), we obtain:

$$\eta = 0 \cdot 4 \times 10^{-3} \exp\left[\frac{\Delta H_{visc}}{RT}\right]. \tag{2.44}$$

When $V = 125$, the empirical equation (2.43) results in a value of one fifth that of the Eyring's formula.

Since in equation (2.42) the factor $T^{2/3}$ appears in the non-exponential term of the rate equation, constant B of equation (2.38) is not independent of the temperature, although its temperature dependence is much lower than that of the exponential term.

The Eyring theory is only an approximate description of the phenomena of viscous flow even for liquids of simple structure; its equations contain no data related to the molecular potential field and the structure of the liquid. In order to apply the theory in actual cases, the factors depending on these characteristics should be estimated empirically. A further deficiency of this theory is that no special term regarding the holes required for the displacement of molecules is involved. The theory of the viscous flow has been developed further by McLaughlin [14] to correct these deficiencies. Assuming a central symmetrical field of the liquid molecules and a lattice-like structure at temperatures far enough from the critical one, i.e. that the liquid is approximately a quasi-lattice, he considered that movement of the molecules accompanied by transport of moment or mass requires the existence of a hole in the quasi-lattice in their vicinity. If the potential of the forces $[\Phi(r)]$ acting between the molecules formulated according to the Lennard-Jones equation is:

$$\Phi(r) = 4\varepsilon\left\{\left(\frac{\sigma}{r}\right)^{12} - \left(\frac{\sigma}{r}\right)^{6}\right\} \tag{2.46}$$

where ε and σ are constants. If furthermore a is the distance of the nearest hole from the initial position of the displaced molecule, and χ is the fraction of the volume V/N available for the molecules, while V_0 is a constant related to the force function, then the equation takes the form:

$$\eta = \frac{(2\pi \, mkT)^{1/2}}{2\pi \, \chi a^2} v_f^{1/3} \exp\left[-\frac{\varepsilon\left(\dfrac{V_0}{V}\right)^4}{kT}\right] \exp\left[\frac{\varepsilon_0}{kT}\right]. \tag{2.47}$$

According to McLaughlin, this expression replaces equation (2.40).

2.2.4 ACTIVATION ENERGY OF VISCOUS FLOW

The activation energy of viscous flow, according to Sections 2.2.2 and 2.2.3, is related to the heat of evaporation, since both of them are accompanied by the formation of a hole in the liquid. In the case of viscous flow,

Literature on page 138

the hole which makes possible the jump of a molecule from its initial equilibrium position to the adjacent one is not necessarily identical with the size of the molecule, it can be smaller than that. Since the energy required for the formation of a hole of molecular size is equal to the heat of evaporation, the free enthalpy of activation of viscosity can be a fraction of this. According to equation (2.35), we can write:

$$\exp\left[\frac{\Delta G^{\ddagger}}{RT}\right] = \frac{V\eta}{hN},$$
(2.48)

from which it follows:

$$\Delta G^{\ddagger} = RT \ln \frac{V\eta}{hN}.$$
(2.49)

Thus, the free enthalpy of activation of viscosity can be calculated from the Eyring theory, if the molar volume and viscosity of the liquid are known at the actual temperature.

From experimental data on several different liquids, Eyring concluded that the following equation is approximately valid:

$$\Delta G^{\ddagger} = \frac{\Delta H_{evap}}{2 \cdot 45}.$$
(2.50)

According to this (on the basis of experiments on about 100 different substances), the free enthalpy of the hole formation in viscous flow is a constant fraction $\left(\text{somewhat larger than } \frac{1}{3}\right)$ of the enthalpy needed for the evaporation of one molecule. It has been emphasized by Powell, Roseveare, and Eyring [15] that this approximate correlation is valid for both non-associated liquids (e.g. hydrocarbons) and water and other associated liquids, e.g. glycerol.

Equation (2.35) can be applied to calculate approximately the viscosity of liquids taking into account the results expressed in equation (2.50). In this way:

$$\eta = \frac{hN}{V} \exp\left[\frac{\Delta H_{evap}}{2 \cdot 45 \, RT}\right].$$
(2.51)

Since the free enthalpy of activation of viscous flow can be calculated with an error less than 6 per cent on the basis of equation (2.50), the error in viscosity calculated by equation (2.51) does not exceed 30 per cent. Although this is not a high accuracy, it is still a good result for the absolute reaction rate theory, or of the approximate validity of this theory of viscous flow, taking into account the deficiencies in our knowledge of liquids.

For activation energy ΔH_{visc}, an expression similar to that of the free enthalpy of activation is valid, as shown by Ewell and Eyring [16]; it differs from the change in free enthalpy according to the equation $\Delta G^{\ddagger} = H_{visc} - T\Delta S_{visc}$ in the entropy change ΔS_{visc} accompanying the activation of viscous flow. The activation energy of the viscous flow is also a fraction (n) of the heat of evaporation:

$$\Delta H_{\text{visc}} = \frac{\Delta H_{\text{evap}}}{n}. \tag{2.52}$$

Taking into account this correlation, the viscosity can be formulated on the basis of equation (2.42) as follows:

$$\eta = 1 \cdot 09 \times 10^{-3} \frac{M^{1/2} T^{3/2}}{V^{2/3} \Delta H_{\text{evap}}} \exp\left[\frac{\Delta H_{\text{evap}}}{n \mathrm{R} T}\right], \tag{2.53}$$

where n is the measure of the size of the hole required for viscous flow in liquids. In order to estimate its numerical value, Ewell and Eyring substituted various values of n into equation (2.53) and calculated the viscosity at several temperatures. Plotting lg η vs. $1/T$, the activation energy of viscous flow has been calculated for several liquids (it proved to be about 1 to 6 kcal/mole), and this was compared with the heat of evaporation of the given liquid. According to these investigations, the value of n lies between 2 and 3 for non-associated liquids, while it is nearer to 3 for liquids of approximately spherical and non-polar molecules. For liquids consisting of molecules essentially different in their shape from central symmetry (hydrocarbons with long carbon chains), as well as for liquids of polar molecules, $n \approx 4$. This difference is, however, rather uncertain, since Eirich and Simha [17] showed that n increases with increasing temperature. It is noteworthy, however, that in several non-associated liquids the formation of a hole in the viscous flow requires about one quarter of the energy of evaporation.

It should be taken into account in the theory of viscous flow that the real conditions can be reflected by the mechanism supposed in the Eyring theory only when the molecules, after arriving at the new equilibrium position from the initial one, remain and vibrate there for a sufficient time to dissipate the energy gained to overcome the energy barrier [9]. It is supposed in the theory that the molecules stay at the position corresponding to the energy minimum for a time sufficient for the reestablishment of the Maxwellian energy distribution. If this condition is not satisfied, another mechanism of viscous flow should be assumed. Under these conditions, the mechanism of the viscous flow of liquids is similar to that of gases which can be described as the transport of impulse between two layers.

If the activation energy $\varepsilon_0 - f\lambda_2\lambda_3\lambda/2$ as given in equation (2.24) is relatively large ($\gg 2kT$), the molecule under consideration collides with other molecules between two jumps frequently enough to re-form the statistical distribution of the energy corresponding to the equilibrium state. However, if the energy of activation is small, this condition is not satisfied and the impulse transfer theory, proved to be correct for gases, can also be applied for liquids; the only modification is that in liquids mainly vibrational energy is transferred instead of translational energy, as shown

Literature on page 138

7

by Andrade [3]. This theory is in good agreement with the experimental data in the case of liquid metals where $\varepsilon_0 - f\lambda_2\lambda_3\lambda/2$ is very small.

Regarding a limiting case, if $\varepsilon_0 = 0$ in a liquid (as it is, in general, in gases) and it is assumed that the free volume is $v_f = \lambda^3$, we have:

$$\frac{F}{F^\ddagger} = \frac{(2\pi \, mkT)^{1/2} \, \lambda}{h},\qquad(2.54)$$

then, under such conditions, viscosity can be expressed according to equation (2.29):

$$\eta \approx \frac{\lambda(2\pi \, mkT)^{1/2}}{\lambda_2\lambda_3\lambda_1}.\qquad(2.55)$$

Furthermore, if $\lambda_1 = \lambda_2 = \lambda_3 = \lambda$, we obtain:

$$\eta \approx \frac{(2\pi \, mkT)^{1/2}}{\lambda^2} \approx BT^{1/2}\qquad(2.56)$$

(all the temperature independent factors are collected into constant B). According to this, the viscosity of liquids under such conditions is proportional to the square root of temperature, as in the case of gases.

The temperature dependence of the viscosity of liquids is significantly influenced by large pressures. Dudziak and Franck [18] concluded from their measurements up to 3500 kbar and 560 °C that the temperature coefficient of the viscosity of water is negative when its density is larger than 0·8 g/cm³; however, it becomes positive when the density falls below this value on changes in the pressure. The experimental data deviate by about 25 per cent from those calculated on the basis of the statistical theory of rigid spheres.

The viscosity of water, as shown by the studies of Horne and Johnson [19], decreases with increasing pressure between 2 and 20 °C, then it increases after passing a minimum at a pressure of about 1000 kg/cm², but the change is not too much up to 200 kg/cm². The higher the temperature, the smoother the minimum. This change is attributed to the fact that the density of the structural clusters in water is smaller than that of monomeric ('free') water. Pressure acts towards the destruction of the clusters which alters the conditions in two respects. On one hand, the ratio of the monomeric water molecules which have a higher mobility than the clusters increases, which results in decreasing viscosity; on the other hand, the structure of water becomes tighter which increases viscosity. Under relatively lower pressures the former effect is predominating, while after passing the minimum the latter overcompensates this.

The fact that the viscosity of water decreases with increasing pressure at temperatures lower than about 33 °C, unlike other liquids (except for very high pressures), can be attributed to the destruction of the liquid structure by pressure, as shown by Nevolina, Samoilov, and Seifer [20]. With increasing pressure, the structural cavities in the tetrahedral lattice of water become more occupied by monomeric water molecules, and this is accompanied by the rupture of some of the hydrogen bonds. So the general 'binding' of the water molecules decreases with increasing pressure and their mobility is enhanced. At very high pressures and elevated temperatures, however, the viscosity of water also increases with increasing pressure.

Water has a large viscosity as compared with non-associated liquids. However, taking into account its strong ice-like structure, it can be considered as a small one. This has been explained by Ageno and Frontall [21, 22] with the existence of linear polymers in water which easily lose a water molecule from one of their ends and, due to the suitability of the protons to form hydrogen bonds at the ends of the chain, a monomeric molecule can also very easily join the polymer. The permanent exchange between polymeric and monomeric water molecules decreases viscosity (see Section 2.3).

(Deryagin et al. [23] carried out experiments in which the formation of liquid modifications having viscosities far higher than that of common water (so-called anomalous water or polywater) was claimed resulting from the condensation of water vapour on a quartz surface in vacuum. This suggestion has been disproved; Gingold [24a], as well as Deryagin and Churaev [24b] established that the anomalous properties of this liquid were due to the presence of quartz exhibiting unusually high solubility under the special experimental conditions.)

2.2.5 THE ROLE OF HOLES IN THE LIQUID WITH RESPECT TO VISCOSITY

The fundamental assumptions in the theories of viscosity according to which viscous flow is made possible by the presence of holes in the liquid are in agreement with the temperature dependence of viscosity observed. At *constant pressure*, viscosity decreases appreciably with increasing temperature, while at *constant volume*, on the other hand, the temperature coefficient is *very small* in comparison to the former. If the displacement of the molecules requires holes, it can be assumed that viscosity – under comparable conditions – is proportional to the number of holes in unit volume. However, it can be concluded on the basis of some other experiments that the volume of the molecules is hardly altered when compressing the liquid or increasing its temperature, while both the volume and the number of the holes decreases. Thus, at constant volume the number of holes is almost unaltered with changing temperature which explains the very small temperature coefficient of viscosity at constant volume.

The Bachinskii rule [25]:

$$\eta = \frac{k}{V - V_s} \tag{2.57}$$

(where k is a constant determined by the forces between the molecules, V_s is the molar volume of the given substance in solid state at the melting

Literature on page 138

7*

point) also supports this hole theory expressed for non-associated liquids. Since, in this respect, the main difference between the two phases is the negligibly small number of holes in solid state in comparison with the liquid phase, $V - V_s$ is approximately proportional to the number of holes in the liquid, and, at the same time, it is inversely proportional to viscosity.

On the basis of the Eyring theory, the experimental fact that the entropy of melting of solids containing molecules which can rotate in solid state (these mainly consist of monoatomic molecules, but some others consisting of molecules of high symmetry also belong to this group) is $\Delta S_{melt} \approx 2$, can also be interpreted. If it is supposed that liquid is a disordered mixture of molecules and holes and that the ratio of the molecules and holes is approximately the same in different liquids at the melting point, the melting of one mole of a substance can be regarded as the mixing of N molecules and N_l holes. The increase in the entropy in the course of this process is:

$$\Delta S = - N \text{k} \ln \frac{N}{N + N_l} - N_l \text{k} \ln \frac{N_i}{N + N_l}. \tag{2.58}$$

This is, at the same time, the entropy of melting as well. Since, according to experiment, $\Delta S_{melt} \approx 2$, thus $N_i \approx 0.54 \, N$ in equation (2.58). In this way, the melting of substances of simple structure (for which $\Delta S_{melt} \approx 2$) is accompanied by the introduction of a certain number of vacancies into the liquid. If the viscosity depends only on the number of the holes in unit volume, the viscosities of each of these substances would be the same at their melting points, according to this theory. Experiments show that, for example, the viscosity of Hg, Cd, Pb, Bi, CS_2, CCl_4, and some paraffin hydrocarbons with small molecules is $\eta \approx 0.02$ poise at their melting point.

The activation energy of the hole formation and of the transition into the hole. The elementary step of the viscous flow is accompanied partly by *hole formation* (empty equilibrium position) and partly by the *transition* of the liquid molecule *into the hole*. Both processes require activation energies, but the ratio of the two activation energies depends on the structure of the liquid. The sum of these two activation energies is the *total activation energy of viscous flow*. According to Powell, Roseveare, and Eyring [15], the ratio can be estimated as follows: if an activation energy $\Delta \varepsilon_f$ is required by a molecule to be transferred into an adjacent hole already present, and k_1 is a constant also involving the number of holes, the fluidity can be given as:

$$\frac{1}{\eta} = \frac{\lambda k_1}{f \lambda_1} \exp\left[- \frac{\Delta \varepsilon_f}{\text{k}T} \right] \times 2 \sin \text{h} \frac{f \lambda_2 \lambda_3 \lambda}{2\text{k}T}. \tag{2.59}$$

If the approximation $2\text{k}T \gg f \lambda_2 \lambda_3 \lambda$ is considered valid in this case too, and the temperature independent factors are reduced to k_2, then:

$$\frac{1}{\eta} = k_2 T^{-1/2} \exp\left[- \frac{\Delta \varepsilon_f}{\text{k}T} \right]. \tag{2.60}$$

According to the Eyring theory, this is the relation between the activation energy of viscosity and that of the transition into a preformed hole. Consequently, if η_1 and η_2 are viscosities corresponding to temperatures T_1 and T_2, respectively, the activation energy of the transfer of molecules into preformed holes is ΔE_f per mole, and we have from equation (2.60):

$$\mathrm{R}\ln\frac{\eta_2}{\eta_1} = -\frac{1}{2}\ln\frac{T_1}{T_2} - \Delta E_f\left(\frac{1}{T_1} - \frac{1}{T_2}\right). \qquad (2.61)$$

The activation energy of the transition into a vacancy can be calculated from this equation knowing the viscosity measured at constant volume and at two different temperatures.

According to measurements on several metals, hydrocarbons, phenyl halides, carbon disulphide and carbon tetrachloride, $\Delta E_f = 0\cdot5$ kcal/mole (approximately) which is about a fifth to a tenth of the total activation energy of the viscous flow. In these liquids, the main part of the total activation energy of viscous flow is consumed by the *hole formation*. The transfer into a preformed hole requires a relatively smaller activation energy. On the contrary, ΔE_f is much larger for some associated liquids containing hydroxyl groups, e.g. in alcohols it is about 3–4 kcal/mole, approximately equal to the energy of the hydrogen bond. Thus in associated liquids, in addition to the formation of holes, the transition process also requires a significant activation energy. On the other hand, ΔE_f is negative in the case of water which shows that the change in temperature at constant volume results in a change of structure, i.e. of the coordination number.

The properties of *associated liquids*, mainly those containing hydroxyl groups, show an appreciable deviation from those of non-associated ones (they are 'anomalous liquids'). Their viscosity is much higher than that of similar non-associated ones under identical conditions (e.g. at 20 °C the viscosity of water is $\eta = 1\cdot00$ cP, that of glycol $\eta = 19\cdot9$ cP, while that of ethyl ether and pentane is $\eta = 0\cdot23$ and $\eta = 0\cdot24$ cP, respectively). The viscosity decreases rapidly with increasing temperature (for water, see Table 2.1) and the relation $\ln\eta$ vs. $1/T$ is not linear. Thus the activation energy of viscosity depends on the temperature. It can be seen in Table 2.1, that the ratio $\Delta H_{evap}/\Delta H_{visc}$ increases with increasing temperature in the case of water.

As established by Eyring, the great viscosity of water can be attributed to the fact that, in addition to the energy required for the formation of holes, the hydrogen bonds between the water molecules have to be broken to achieve an activated state for viscous flow. In addition to the 'normal' activation energy, a '*structural activation energy*' is needed in water and in other associated liquids. The number of hydrogen bonds which should be broken to make the flow possible decreases with increasing temperature which results in a decrease in the activation energy. It can be assumed that at 150 °C structural changes do not appreciably affect the activation

Literature on page 138

Table 2.1

Temperature dependence of the viscosity and heat of evaporation of water,
as well as of the activation enthalpy of its viscous flow and the ratio of the heat
of evaporation and activation energy

Temperature °C	η, millipoise	ΔH_{visc}, kcal · mole^{-1}	ΔH_{evap}, kcal · mole^{-1}	$\dfrac{\Delta H_{evap}}{\Delta H_{visc}}$
0	17·95	5·1	10·2	2·0
50	5·49	3·4	9·6	2·8
100	2·84	2·8	9·0	3·2
150	1·84	2·1	8·3	3·9

energy. The relatively large viscosity of liquids containing hydroxyl groups
is a consequence of the high activation energy.

Studying the activation energy of the transport phenomena as a function
of temperature, Horne, Courant, and Johnson [27] established that the
activation energy of the viscous flow of water shows no maximum at the
temperature of the maximum density (4 °C), unlike the activation energy
of electric conduction which has a maximum at this temperature. They
concluded from this that, in the mechanism of the viscous flow, the vacancy
formation is not so predominating as the rotational 'stumbling' of the
molecular clusters between the monomeric water molecules. However,
this opinion should be supported by further studies.

Recently the temperature dependence of the viscosity of water has been
studied by Korson, Drost-Hansen, and Millero [28a]. Under high pressures
(up to 1406 kg/cm²) and in the 2–30 °C temperature range, Stanley and
Batten [28b] found the activation energy of the viscous flow of water
to decrease smoothly with increasing temperature and pressure.

When the atoms of water are replaced by corresponding isotopes, the
viscosity becomes altered. This change depends on the molecular weight,
molar volume and heat of evaporation of the replaced species, as well as on
the activation energy of the viscous flow. According to Kudish, Wolf and
Steckel [28c], the ratio of the viscosities of $H_2^{18}O$ and $H_2^{16}O$ corresponds to
the ratio of the molecular weights; nevertheless, the isotope effect of $D_2^{16}O$
has been found to be far higher, which can be attributed to the higher activation energy of the viscous flow of heavy water.

The effect of the rotation of molecules on viscosity. The theories of viscosity
mentioned above do not deal with the effect of the rotation of molecules
on viscosity. It has been emphasized by Davies and Matheson [29] that,
in some cases, the rotational state and its changes influence the viscosity.
The change in the viscosity of liquids with temperature can be described
with good accuracy by the equation of Andrade which is similar to the
Arrhenius equation:

$$\ln \eta = A + \frac{\Delta H_{visc}}{RT}.$$

There are some liquids (e.g. neon, methane, neopentane) for which this correlation is valid in the whole temperature range. For other liquids (e.g. propane, toluene, chloroform) the Andrade equation holds at temperatures far from the melting point, while nearer to it a deviation occurs.

Analysing the conditions, Davies and Matheson [29] concluded that in the Arrhenius–Andrade range the viscosity is determined by the probability of the transition of liquid molecules from one equilibrium position to the adjacent one. Between two jumps these molecules exhibit rotation about at least two different rotational axes. Approximately spherically symmetrical molecules can rotate freely at every temperature of the liquid; for such liquids the Andrade equation is valid in the whole temperature range of their existence. However, for liquids following the Andrade equation only at higher temperatures, the possibility of free rotation diminishes on approaching the freezing point. In this temperature range, the molecules can rotate only about one axis in the time interval between two translational jumps. Thus, it may happen that the liquid molecules have energy enough to jump to the adjacent equilibrium position due to the fluctuation, but this may be prevented by unfavourable orientation of the molecule, and its neighbours can only rotate about one axis in the translational period. In some liquids (e.g. isopropylbenzene) there is a discontinuity in the non-Arrhenius–Andrade range as well, which can be attributed to the fact that in the vicinity of the melting point the rotation about even one axis is restricted or impossible.

The studies of Dexter and Matheson [30] also supported this interpretation of the unusual behaviour by establishing that there is a structural factor in the molar heat capacity of such liquids which cannot be observed in that of liquids exhibiting the Arrhenius–Andrade behaviour. This structural part of the molar heat capacity correlates with the changes in the rotational degree of freedom.

2.2.6 ACTIVATION ENTROPY OF VISCOUS FLOW

The activation energy of the viscous flow of associated liquids is large compared with the non-associated ones, while in the free enthalpy of activation there are no such differences (see Section 2.2.4). This is due to the fact that the entropy of activation of viscous flow of associated liquids has a large positive value.
Since

$$\Delta G^{\ddagger} = \Delta H^{\ddagger} - T\Delta S^{\ddagger}, \tag{2.62}$$

the high enthalpy of activation $\Delta H_{\mathrm{visc}} = \Delta H^{\ddagger}$ is compensated by the large change in the entropy, which results in a normal $\Delta G^{\ddagger}$ value (i.e. corresponding to that of non-associated liquids). If it is supposed, in accordance with Eyring, that the viscous flow of associated liquids involves the elementary step of displacement of individual molecules, and breaking of several

Literature on page 138

hydrogen bonds is necessary to form the activated complex, it can be understood that the entropy of the activated state is significantly higher than that of the initial one.

For liquids of spherically symmetrical molecules or approximately such ones, $\Delta H_{evap}/\Delta H_{visc} \approx 3$, while, for non-spherically symmetrical molecules, it is about 4. It follows that the activation energy of the viscous flow of liquids of spherically symmetrical molecules is relatively large and, owing to this, the entropy of activation of the flow of associated liquids is also high. This is in agreement with the experiments, and it reveals that—as expected—the symmetrical molecules are closely packed in the normal liquid state.

2.2.7 VOLUME VISCOSITY

The 'common' ('dynamic') viscosity discussed above is the measure of the friction or the energy dissipation which takes place when liquid molecules are displaced relatively to each other, or the liquid layers slip upon each other on the effect of an external force. The dissipation of mechanical energy as heat can also be caused by external forces, changing only the volume of the liquid without any change in its shape and any slipping-like displacement. The measure of the friction accompanying this change in volume is the *volume viscosity* (η'); it is sometimes called secondary viscosity [31].

Volume viscosity is observed when the compression or dilatation of the liquid is so fast that the rearrangement of the disturbed thermodynamic equilibrium in the liquid lasts for a longer time than the process of the change in the volume. The equilibrium is disturbed, for example, by changes in the energy distribution of the molecules or by the modification of the quasi-lattice structure of the liquid. A certain time is required to attain the new equilibrium and this can be characterized by the relaxation time τ. If the change in volume takes place in a time shorter than the relaxation time, the mechanical energy is partially transformed into heat and the measure of this dissipation is the volume viscosity. If, on the other hand, the change in the volume proceeds in a longer time than that of the relaxation, there is time enough for equilibration at each stage of the process, so no appreciable energy is dissipated and the volume viscosity is $\eta' = 0$.

In the majority of phenomena of viscous flow, the volume of the liquid does not change appreciably (i.e. the liquid can be considered incompressible) and no volume viscosity appears. The Stokes equation of viscosity involves the assumption $\eta' = 0$. However, fast processes, as, for example ultrasonic absorption, are significantly influenced by the volume viscosity. If c_0 denotes the velocity of the common sound in a liquid of density ϱ, and the velocity of high-frequency ultrasound is c_u [$c_u \gg 1/\tau$], the volume viscosity is [31]:

$$\eta' = \tau \varrho (c_u^2 - c_0^2). \tag{2.63}$$

The volume viscosity can be calculated from measurements on the absorption of normal (low-frequency) sound and ultrasound. For moderately high-frequency ultrasonic absorption, η' is constant, while for very high frequencies it changes, i.e. volume viscosity also has a dispersion.

The volume viscosity of liquids is, in general, many times larger than the dynamical one, and its magnitude is determined by the interaction between the liquid molecules. In water and aqueous solutions, the volume viscosity arises mainly from structural relaxation processes, i.e. it is in connection with the transformation of the quasi-crystalline structure into a more tight one on the effect of pressure.

It follows from the correlation between the volume viscosity and the structure of the liquid that η' is altered by the hydration of ions in electrolyte solutions. According to the theoretical study of Fisher and Zaitseva [32], the hydration of ions influences the volume viscosity three or four times as much as the dynamical one. This can be attributed to the fact that hydration appears essentially in the compression of water by the electric field of the ions. The molecules within the hydrate spheres are also displaced relative to one another by compression or dilatation. In common viscous flow, the hydrate sphere participates more or less as a unit and the displacement of the molecules within the hydrate sphere plays no significant role in this.

Some aspects of the liquid structure could be discussed by analyzing the ratio η'/η, however, the experimental facts available do not cover a sufficient range. For several liquids, e.g. for water and ethanol, as established by Pinkerton [33], η'/η is independent of temperature, thus the correlation between the liquid structure and volume viscosity cannot be revealed by measuring this ratio. For liquid mixtures, however, η'/η changes with temperature which permits some conclusions, as pointed out by Akhmetsianov and Petrea [34]. In 26 per cent and 46 per cent aqueous ethanol solutions, η'/η increases with decreasing temperature from $+20\ ^\circ C$; it reaches a maximum between 0 and $-5\ ^\circ C$, and afterwards it begins to decrease. It is assumed that in these mixtures relaxation is connected with the formation and disappearance of inhomogeneities of unknown nature.

2.3 VISCOSITY OF LIQUID MIXTURES AND NON-ELECTROLYTE SOLUTIONS

The correlation between the viscosity of liquid mixtures and that of the pure components is very complicated, and no satisfactory theories taking into account the deviations from the properties of ideal mixtures have been developed so far. The conditions are relatively simple when the rate-determining elementary step of viscous flow is the transition of a molecule from an equilibrium position to the adjacent one, and when the free enthalpy of activation (which differs from the energy of hole formation and transition into a preformed hole in the entropy term) is the same in mixtures as in the pure components and furthermore, when the mixture is ideal (i.e. the

Literature on page 138

interaction between the different molecules of the components is not appreciably different from the interaction of identical molecules). Under such conditions, it can be deduced on the basis of equations (2.23)–(2.28) that the reciprocal value of the viscosity of a two-component mixture, i.e. its *fluidity*, can be obtained *additively* from the reciprocal viscosities of the components:

$$\frac{1}{\eta} = \frac{x_1}{\eta_1} + \frac{x_2}{\eta_2}, \tag{2.64}$$

where η_1 and η_2 are the viscosities of the two components, while x_1 and x_2 are their mole fractions.

The rigorous validity of this correlation cannot in general even be expected for ideal mixtures, since it can hardly be assumed that the energy required for the hole formation and for the transfer into the hole (on which the activation energy of viscous flow is strongly dependent) is the same in the mixture and in the pure components. According to Eyring [26], the real conditions can be better approximated when the average value $x_1 \Delta G_1^{\ddagger} + x_2 \Delta G_2^{\ddagger}$ of the free enthalpies of activation of the viscous flows of the two components is taken into account. Since equation (2.35) results in the following correlation for pure liquids:

$$\frac{1}{\eta} = \frac{V}{hN} \exp\left[-\frac{\Delta G^{\ddagger}}{RT}\right], \tag{2.65}$$

the case of a liquid mixture can be described by a similar equation. If V_{12} is the average value of the molar volumes of the two components, we have:

$$\frac{1}{\eta} = \frac{V_{12}}{hN} \exp\left[-\frac{x_1 \Delta G_1^{\ddagger} + x_2 \Delta G_2^{\ddagger}}{RT}\right]. \tag{2.66}$$

If V_1 and V_2 are not too different from each other, $V_{12} \approx V_1 \approx V_2$. In this way, taking into account equations (2.65) and (2.66), we can write:

$$\ln\frac{1}{\eta} = x_1 \ln\frac{1}{\eta_1} + x_2 \ln\frac{1}{\eta_2}. \tag{2.67}$$

This relationship corresponds to that obtained by Kendall and Monroe [35].

Equation (2.66) is in good agreement with the experiments on chemically similar liquids. However, mixtures of liquids of different nature deviate significantly from the additive correlation given in equation (2.66). For example, the viscosities of mixtures of benzene and ethanol are smaller than the calculated one, while in the case of compound formation (e.g. in mixtures of ether and chloroform), the viscosity of the mixtures is higher than that expected on the basis of additivity. The highest deviation from the values calculated on the basis of the additivity of fluidity can mostly be observed in mixtures of about 50 mole per cent concentration, as in the case of deviation from the Raoult law. As established by Powell, Roseveare, and Eyring [15], the deviations from the ideal behaviour in fluidity and vapour pressure are proportional to each other in several liquids. The deviation in the free enthalpy of activation of viscosity from that of the ideal behaviour ($\Delta G^{\ddagger}_{real}$) and the free enthalpy of the mixture ΔG_{mixt} (which is the measure of the deviation from the Raoult law) are correlated by the following equation:

$$\Delta G^{\ddagger}_{real} = \frac{\Delta G_{mixt}}{2 \cdot 45} \tag{2.68}$$

which is valid for several liquids. It should be pointed out that the value 2·45 in the equation is approximately equal to the ratio of the free enthalpy of viscous flow and the heat of evaporation (i.e. the energy of vacancy formation, see Section 2.2.4). Taking into account the correlation between the deviation from the relationship given in equation (2.66) and the free enthalpy of mixing, this equation can be written as follows:

$$\frac{1}{\eta} = \frac{V_{12}}{hN} \exp\left[- \frac{x_1\,\Delta G^{\ddagger}_1 + x_2\,\Delta G^{\ddagger}_2 - \dfrac{\Delta G_{mixt}}{2 \cdot 45}}{RT} \right], \tag{2.69}$$

or, since in approximation $\Delta G^{\ddagger} = \Delta H_{evap}/2 \cdot 45$, one has:

$$\frac{1}{\eta} = \frac{V_{12}}{hN} \exp\left[- \frac{x_1 \Delta H_{evap,1} + x_2 \Delta H_{evap,2} - \Delta G_{mixt}}{2 \cdot 45\,RT} \right]. \tag{2.70}$$

Knowing the vapour pressure of the mixture, ΔG_{mixt} can be calculated, and the two calculations can be checked. The viscosity values calculated on the basis of equation (2.70) are in good agreement with the experimental data even for mixtures of hydrocarbons with associated liquids (e.g. benzene and phenol). This can partly be attributed to the presence of an empirically determined factor in the equation.

With respect to these facts it can be established that the changes in the viscosity of binary mixtures with composition cannot be unequivocally described. Fialkov [36], reviewing the viscosity conditions of about 100 binary liquid mixtures, showed that the following isotherm:

$$\eta = (x_1\eta_1^{1/3} + x_2\eta_2^{1/3})^3 \tag{2.71}$$

is the upper limit of the family of curves representing the relationship between the viscosities of the mixtures and those of the components, while a lower limit is expressed by:

$$\ln \frac{1}{\eta} = \frac{1}{\dfrac{x_1}{\eta_1} + \dfrac{x_2}{\eta_2}} \tag{2.72}$$

Literature on page 138

According to him, the $\ln(1/\eta)$ term is additive only for a given group of mixtures of liquids for which $\eta_1/\eta_2 \leq 6$.

As for the viscosity of liquid mixtures consisting of dissimilar molecules, Fort and Moore [37] investigated the 'excess viscosity' (η^E) arising from the difference between the viscosity of the mixture η and those of the components:

$$\eta^E = \eta - x_1\eta_1 - x_2\eta_2$$

as a function of the composition of the mixture. In several cases, they observed that $\eta^E < 0$ if a weak bond is formed between the components (e.g. benzene–chloroform, benzene–carbon tetrachloride). If, however, a hydrogen bond or a stronger connection is formed between the components, $\eta^E > 0$ in most cases. Although the larger η^E, the stronger the interaction between the molecules of the components, it still cannot be regarded as a measure of the strength of the interaction. The difference in the size of the molecules can also play an important role in this respect.

Changes in viscosity due to the interaction between the components. Non-electrolytes alter the viscosity of the solvent even in dilute solutions. If there is no specific interaction between the solvent and the solute, this effect can often be attributed to changes in the streamlines, i.e. they give rise to an Einstein effect for which we refer to Section 2.4.2. Interaction between the solute and the solvent (i.e. between the components, in general), however, occurs frequently, and owing to this, the viscosity does not change monotonically with concentration but in some cases has an extreme value.

The mixtures of substances of chemically similar nature can show various behaviours in this respect, too. For example, as it has been shown by Erdey-Grúz, Kugler and Reich [38] and others for mixtures of methanol and water, and by Erdey-Grúz, Kugler and Hidvégi [39] for ethanol–water and propanol–water mixtures, the viscosity has a maximum as a function of the alcohol content at about 20–25 mole per cent alcohol concentration. It is much sharper at 5 °C and appears at lower concentrations than at 25 °C (Fig. 2.6). The longer the carbon chains in the alcohols, the larger the maximum viscosity. On the other hand, Erdey-Grúz and Kugler [40, 41a] for example found that in ethylene glycol–water and glycerol–water mixtures the viscosity increases monotonically with increasing alcohol concentration. The maximum viscosity is undoubtedly a consequence of a specific interaction between water and alcohol in which hydrogen bonds formed between the alcohol and water molecules play an important role. However, it is evident from the above facts that the lack of a viscosity maximum does not indicate the absence of interactions (e.g. hydrogen bonds) between the components, since there is no reason to suppose that only the hydroxyl groups of monoalcohols can form hydrogen bonds with the water molecules and that those of polyalcohols are not suitable for this.

The maximum of viscosity in mixtures of water and monovalent alcohols and the absence of this viscosity maximum in mixtures of polyvalent alcohols and water have been interpreted by Ageno and Frontall [22] on the basis

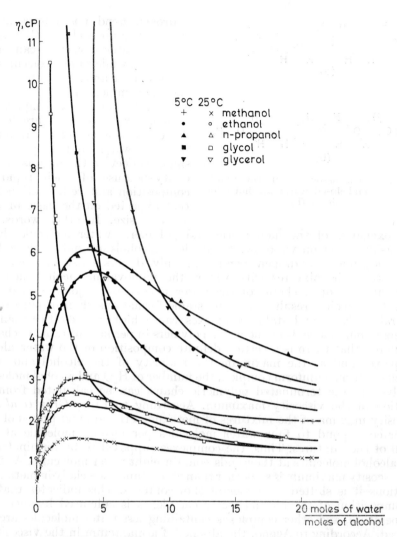

Fig. 2.6 Variation of viscosities of some mixtures of mono- and polyvalent alcohols and water with concentration at 5 and 25 °C

of the theory supposing collective hydrogen bonds in the water structure
(see Sections 1.3.3.8 and 2.2.4). According to this theory, the H_2O molecules
bound together by hydrogen bonds form chains and rings of varying
length and size in water. An individual water molecule is suitable for
forming two independent hydrogen bonds because there are unoccupied

Literature on page 138

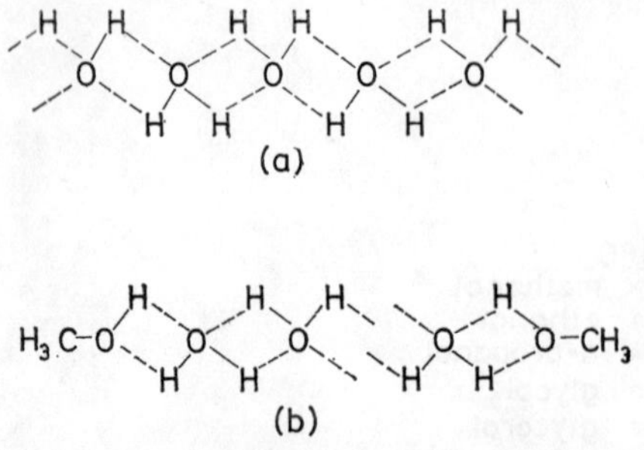

(a)

(b)

Fig. 2.7 'Open' associates of water molecules (a) and closed water–alcohol associates (b)

hydrogen bond possibilities at the ends of the open chains (Fig. 2.7a). Further water molecules can easily join these, while water molecules can easily depart and form a bond with another complex with the same bond energy. In this way, complexes of various size and shape are in permanent decomposition and re-formation in water and they have a smaller effect on the increase in viscosity–because of their rapid decomposition and re-formation–than that expected on the basis of their average size. In other words, the easy exchange of the monomeric and polymeric water molecules has a decreasing effect on viscosity. Monovalent alcohols are only able to form one independent hydrogen bond and only dimers exist in their pure state. In their mixtures with water, the monovalent alcohol molecules terminate the open chains of the water molecules by joining their ends (Fig. 2.7b) which results in more stable formations than in pure water. The rather large and stable complexes cause higher viscosities than the dimers in pure alcohols or the water polymers in permanent rapid exchange. Assuming that there is a certain definite composition of the water–alcohol complexes and at the maximum the majority of the alcohol and water molecules form a uniform complex, the number of H_2O and alcohol molecules involved in the terminated polymeric chains can be calculated from the position of the viscosity maximum. According to Erdey-Grúz *et al.*, the viscosity maxima of the mixtures of methanol, ethanol, and propanol with water correspond to four water molecules per alcohol molecule at 5 °C in all of the mixtures. Thus, the complexes supposed to be terminated by two alcohol molecules at their ends contain eight water molecules. At 25 °C the viscosity maximum is smoother than at 5 °C and–especially in methanolic solutions–it is shifted to higher alcohol contents. This indicates that the decomposition of the alcohol–water complexes is enhanced by increasing temperature, and other complexes containing less water molecules are also formed. According to Ageno, the absence of a maximum in the viscosity of glycol–water and glycerol–water mixtures as a function of concentration can be attributed to the fact that these molecules are suitable for forming more then one independent hydrogen bond. Thus, if polyvalent alcohol molecules join the water complexes, these are not terminated by the alcohol molecules, but the exchange of the water molecules at the ends of the chains can continue. It should be noted, however, that the theory of the collective hydrogen bonds is not proved by the viscosity maxima in water-monovalent alcohol mixtures; it is only in agreement with it.

In order to explain the viscosity maxima in alcohol–water mixtures, it is not necessary to assume the formation of actual complexes, and the phenomenon should not be attributed entirely to this. The fact that on the isotherms of various properties of water–alcohol mixtures an extreme value appears at an alcohol content of about 0·2 mole fraction has already been explained by Mitchell and Wynne-Jones [41b] and Mikhailov [41c] with the occupation of the structural cavities in water. At lower concentrations, the molecules of the dissolved non-electrolyte occupy the structural cavities in water, which deforms the water structure but does not destroy it. At higher concentrations, however, there is not enough room for all of the solute molecules in the structural cavities of water, and this results in the formation of a new type of liquid structure which alters the properties in another way than the deformation of cavities. According to this theory, these points of inflection and extreme values decrease with increasing temperature because of the gradual disappearance of the water structure with this effect. The position of the points of inflection or the extreme values on the isotherms, however, hardly depends on the temperature. Of course, the structural cavities in water can be occupied by the solute molecules without destroying the structure only if the size of the molecules is suitable for fitting into the cavities and deforming but not destroying them.

The formation of complexes bound by hydrogen bonds and the occupation of the structural cavities of water do not exclude each other. In fact these two types of effects act together resulting in changes of the properties of water–alcohol mixtures. It can hardly be assumed that the alcohol molecules do not form hydrogen bonds with the neighbouring water molecules or that they do not modify the hydrogen bonds between them.

2.4 VISCOSITY OF ELECTROLYTE SOLUTIONS

The viscosity of water is altered by dissolved electrolytes, namely they increase or decrease viscosity depending on the properties of the ions, mainly their size and charge. It is easily seen that this effect varies approximately linearly with concentration in dilute (about 0·002 to 0·1 M) solutions. At very low concentrations, however (more dilute than 0·002 M), an appreciable deviation from linearity appears, as observed by Grüneisen [42], and this reveals an effect which always increases the viscosity.

In the most dilute electrolyte solutions, viscosity increases with increasing concentration, but not linearly. The linear effect that increases or decreases viscosity, depending on the nature of the solute ions, prevails only at a definite (but unsharp) concentration limit. On the basis of investigations on several electrolyte solutions, Jones and Dole [43] found the following empirical correlation valid at constant temperature, regarding the relative viscosity of the solution η_r and the concentration of the electrolyte (c):

$$\eta_r = \frac{\eta}{\eta_0} = 1 + A\sqrt{c} + Bc \qquad (2.73)$$

Literature on page 139

(*Jones–Dole equation*: η is the viscosity of the electrolyte solution, η_0 is the viscosity of the pure solvent at the same temperature, A and B are empirical constants independent of the concentration of the given system). Here A is always positive, but B can be either positive or negative depending on the nature of the solute and the solvent. The temperature coefficient of B is positive in most cases. The typical change in the relative viscosity against the electrolyte concentration of KCl solutions is shown in Fig. 2.8, on the basis of Gurney's [44a] experiments.

According to Werblan, Rotowska, and Minc [44b], the Jones–Dole equation is also valid in solutions of $LiClO_4$, $NaClO_4$, and NaCl in mixtures of water and methanol.

Equation (2.73) is in agreement with the experiments up to about 0·8 mole/l concentration. Including a term containing the square of the concentration, its validity can be extended to higher concentrations as well:

$$\eta_r = 1 + A\sqrt{c} + Bc + Dc^2. \qquad (2.74)$$

The constants in the Jones–Dole equation can be determined empirically by measuring the viscosity of the solutions. According to equation (2.63), we have:

$$\left(\frac{\eta}{\eta_0} - 1\right)\frac{1}{\sqrt{c}} = A + B\sqrt{c}. \qquad (2.75)$$

By plotting the left-hand side term of this equation as a function of $\sqrt{c}$ (Fig. 2.9), a straight line is obtained with a slope B and an intercept A on the ordinate.

From investigations on the concentration dependence of viscosity–as compared with other properties of solutions as well–it can be concluded that factor A is related to the electrostatic interaction of the solute ions with one another, while constant B corresponds to the interaction between the solvent and the solute ions. The prevailing effect is, in general, the latter one. The mutual interaction of the ions becomes predominating only in very dilute solutions.

The dependence of the viscosity of an electrolyte solution on concentration can be described by the following relationship (Andrussow [45]):

$$\frac{\eta_{c+dc}}{\eta_c} = \left(\frac{c + dc}{c}\right)^{\psi^*}, \qquad (2.76)$$

where η_c and η_{c+dc} are the viscosities of solutions of concentration c and $c + dc$, respectively and ψ^* is a properly defined 'intrinsic concentration exponent' depending on the concentration. It has been pointed out by Kaminsky [46] that this equation gives a less satisfactory agreement with the experiments, for example, Li_2SO_4 solutions, than the Jones–Dole equation.

Dissolved electrolytes alter not only the dynamic viscosity, but also the *volume viscosity* (Section 2.2.7). According to Fisher and Zaitseva [47], the volume viscosity increases proportionally to the concentration

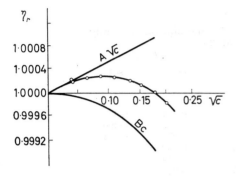

Fig. 2.8 Relative viscosities of aqueous KCl solutions at 18 °C, ∘: measured values

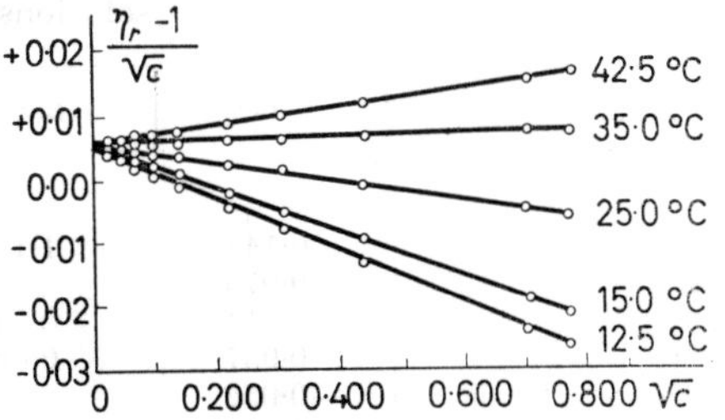

Fig. 2.9 Correlation $(\eta - 1)\, c^{-2/1}$ vs. $c^{1/2}$ for aqueous ammonium chloride solutions, according to Kaminsky

because of the hydration of ions. On the basis of their theory, the change in the difference between the volume viscosity of the solution η_1' and that of the pure solvent η_0' with concentration is:

$$\eta_1' - \eta_0' = A\sqrt{c} + Bc. \qquad (2.77)$$

The first term of this relationship corresponds to the electrostatic interaction between ions, while the second one is attributed mainly to the ion–solvent interaction, that is the hydration.

2.4.1 EFFECT OF THE INTERACTION OF DISSOLVED IONS AND SOLVENT ON THE VISCOSITY OF SOLUTIONS

The change in the viscosity due to the interaction between the solvent and the solute is formally taken into account in coefficient B of the Jones–Dole equation for dilute solutions, and has been investigated extensively by Kaminsky [48]. In aqueous solutions, B is negative for many electrolytes (Table 2.2), while in less associated solvents, the value of B is generally a positive one, and for a given electrolyte its magnitude depends on the nature of the solvent.

The additivity of coefficient B. It is characteristic of the values of B that they can be obtained additively from terms corresponding to the anion and cation. It has been established by Cox and Wolfenden [49] that, according to their investigations in binary electrolyte solutions, the differences in the B values of solutions of salts of one cation and various anions (both ions monovalent) are independent of the common cation. Similarly, the differences in the B values of solutions containing the same anion and various cations are independent of the anion. Although the additivity

Literature on page 139

Table 2.2

B coefficients (litre×mole^{-1}) of some aqueous electrolyte
solutions at 25 °C

	Cl$^-$	Br$^-$	I$^-$	SO$_4^{2-}$	Picrate
Li$^+$	0·143	1·106	0·081	0·508	0·451
Na$^+$	0·0793		0·018	0·390	
K$^+$	−0·0140	−0·048	−0·075	0·194	
Rb$^+$	−0·037	−0·061	−0·099		
Cs$^+$	−0·052		−0·114		
NH$_4^+$	−0·0144	−0·037	−0·08		
Mg^{2+}	0·371			0·5937	
La^{3+}	0·5672				
N(C$_2$H$_5$)$_4^+$		0·343			0·743

can be concluded unequivocally from this behaviour, the resolution to
B coefficients corresponding to the individual ions requires an assumption
that cannot be proved directly.*

Gurney [50] and Kaminsky [48] assumed that the effect of K$^+$ and Cl$^-$
ions on coefficient B at 25 °C and 14–45 °C is the same, i.e. one half of the
B value in a KCl solution corresponds to the K$^+$ ions and the other half
to the Cl$^-$ ions. It makes reasonable the assumption that B is very small
in KCl solutions and the mobilities of the K$^+$ and Cl$^-$ ions hardly differ
from each other in the above temperature range, which shows that these
two types of ions influence the water structure similarly. The coefficient
B for some ions, calculated on this basis, can be seen in Table 2.3 (the
dimension of B is reciprocal concentration, in most cases litre/mole).

Although Nightingale [51a] supposed identical values of B for the cation
and anion in RbCl or CsCl solutions, with respect to the division of coefficient
B into two equal parts, the former assumption still seems to be more reliable.
This is so not only because of the small value of B in KCl solutions and the
identical mobilities of the cation and anion, but because this case can also
be correlated with the thermodynamic properties. Nightingale obtained
B = −0·05 for the perchlorate ion which he explained by the suitability
of the hydrated perchlorate ion to loosen or destroy the water structure,
decreasing the viscosity in the vicinity of the ion. Such an effect can be
observed, in general, only in water and in connection with ions which have
small effective ionic radii in the hydrated state.

The soundness of the method suggested by Cox and Wolfenden and further,
by Gurney and Kaminsky for the division of coefficient B into parts characteristic of the cation and the anion was questioned recently by Krumgalz
[51d]. The above methods are essentially based on the assumption that

---

* The equivalent conductivity of electrolyte solutions can be divided into the
ionic mobilities by means of the directly measurable transference numbers; for viscosity, however, there is no such quantity that would correspond to the transference
number.

Table 2.3

B coefficients of ions (litre×mole^{-1}) in aqueous solutions

°C	Li$^+$	Na$^+$	K$^+$	Cs$^+$	Mg^{2+}	La^{3+}	NH$_4^+$
15	0·1615	0·0860	−0·0200		0·4091		−0·0137
25	0·1495	0·0863	−0·0863	−0·045	0·3852	0·588	−0·0074
35	0·1385	0·0851	+0·0049		0·3625		−0·0027
42·5	0·1210	0·0861	+0·0121		0·3472		+0·0018

	Cl$^-$	Br$^-$	I$^-$	SO$_4^{2-}$	Picrate		
15	−0·0200			0·1899			
25	−0·0070	−0·032	−0·0685	0·2085	0·434		
35	+0·0049		−0·0536	0·2277			
42·5	+0·0121			0·2399			

	N(CH$_3$)$_4^+$	N(C$_2$H$_5$)$_4^+$	N(C$_3$H$_5$)$_4^+$	N(C$_4$H$_9$)$_4^+$			
20	0·1223	0·3999					
25	0·1175	0·3807	1·092	1·396			
30	0·1158	0·3636					
42·5	0·1143	0·3520					

B_+ and B_- are proportional to the volume of the solvated ion and inversely proportional to the third power of their mobilities. The inconsistency of the above procedure is also indicated by the fact that different B values are obtained for a given ion, when its salts formed with different counter-ions are considered. Krumgalc suggests that in the division of B according to ions, a more correct starting point is offered by the fact that the R$_4$N$^+$ (tetra-alkylammonium) ions (when R $\geq$ Bu) and other complex organic ions do not become solvated, their size remaining the same in different solvents. Assuming that the B coefficients of such ions are proportional to their volumes, the B values calculated for some solvents are, in good approximation, independent of the counter-ion.

The viscosity of aqueous solutions of alkali halides has been recently measured by Lengyel, Tamás, Giber and Holderith [52a] between 10 and 40 °C, covering the whole concentration range. On the basis of these measurements and some earlier literature data, Lengyel [52b] tried to explain the dependence of viscosity on the concentration by means of a model in which the solution is regarded as consisting of point-like charges distributed in a continuum; the dielectric constant of the medium depends on the electric field strength.

For the interaction between the ions and the solvent (i.e. for coefficient B) no quantitative theory has been outlined so far. The theory developed

Literature on page 139

by Finkelstein [53], proved to be unsatisfactory, since it does not take into account the effect distorting the liquid structure. Some valuable qualitative correlations could still be established. If the electrolyte in the solution is entirely dissociated and the association of the ions can be disregarded,* the interaction between the ions and solvent can appear mainly in two effects: in the solvation of ions and in the alteration of the liquid structure by the ions.

The effective size of the ions is increased on solvation which helps to increase the viscosity. The extent of solvation depends on the charge and other properties (mainly the radius) of the ions. In general–under comparable conditions–the larger the charge of the ions and the smaller their radii, the higher the extent of the solvation.

Ions partly alter the solvent structure by means of their charges ordering solvent molecules to a certain extent (see Sections 1.4.2 and 5.2). This orientation polarization increases viscosity. On the other hand, the field of ions causes depolymerization, as shown by Eucken [54], or, according to Frank and Evans [55a], it partially destroys the short-range ice-like structure. This effect decreases viscosity.

Depending on whether the common effect of solvation and orientation polarization is larger or smaller than that destroying the structure, coefficient B is positive or negative respectively.

The facts mentioned above explain that the temperature coefficient of B is positive in the most cases. Since the more or less ice-like structure of pure water decomposes more and more with increasing temperature, the structure-breaking effect of the ions decreases. On the other hand, hydration is probably favoured by the increase in temperature, since the number of monomeric water molecules increases owing to the partial decomposition of the water structure and they can join ions more easily than the associated ones. This can increase the radius of the hydrated ion which causes an increase in viscosity. Of course, this effect can be decreased by the loosened binding of the water molecules in the hydrate sphere to the bulk of the liquid due to the partial decomposition of the hydrogen bonds.

The increase in the relative viscosity of the solution and coefficient B with increasing temperature can be interpreted according to this qualitative picture of the effect of ions. Due to the structure-breaking effect, the viscosity of the solvent decreases more rapidly than that of the solution, since in the latter the water structure has already been destroyed to a certain extent by the ions.

The altering effect of solute ions on the liquid structure is also responsible for the fact that alkali chlorides result in a higher relative decrease in viscosity in D_2O than in H_2O solutions, as established by Ostroff, Snowden, and Woessner [55b]. This has been attributed to the assumption that, owing to the higher ratio of the hydrogen bonds in D_2O, the structural order is higher in it than in H_2O, and the disordering effect of the solute

---

* Dissociation and association are not precisely opposite processes: dissociation is the decomposition of covalent molecules into ions, while association reflects mainly coulombic forces keeping together the ions in pairs or in some larger complexes.

ions is also favoured in the former solvent. The effect of the salts investigated increases with the increasing radius of the cation. In respect to the effect exerted on the liquid structure and in respect to solvation, the electrolytes behave in heavy water to a certain extent similarly to that in common water. According to the viscosity measurements of MacDonald [55c], the 9 electrolytes examined by him in heavy water solutions can be divided into three groups: (1) in the environment of both ions of KCl, KBr, KI and $KMnO_4$ the solvent molecules are less ordered than in the bulk of the liquid (negative hydration); (2) in LiF and NaF solutions, the molecules are slightly more ordered around the cations (weak positive hydration) and less ordered around the anions (weak

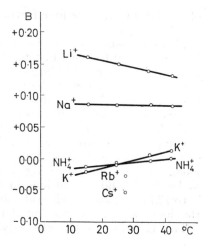

Fig. 2.10 Variations of B coefficients of some ions with temperature

positive hydration), as compared with the bulk of the liquid; (3) in solutions of Li_2SO_4, Na_2CO_3 and $MnSO_4$, both the cations and anions enhance the order of solvent molecules.

According to Horne and Young [55d], the coefficient B is also related to the extent of the elimination of ions from the boundary layer of the solid phase being in contact with the solution. An increasingly strengthened liquid structure is developed in water as the interface of phases is approached, thus the ions that would alter the structure of water are forced to leave the surface layer.

The dependence of coefficient B on the size of the ions and the temperature. In a column of the periodic table, the value for B was found to change regularly with the crystallographic radii of the ions (see Table 2.3). The ions of smaller atom increase viscosity to a higher extent than the larger ones. Moreover, at adequate temperatures the large atom ions decrease viscosity ($B < 0$). But in solutions of large molecule ions (e.g. tetra-alkyl ammonium ions, picrate ion) $B > 0$, while for the ammonium ion $B \approx 0$. The value of B for polyvalent ions (with elementary charges greater than 1) is much larger than that for the monovalent ones.

The temperature coefficient of B ($\partial B/\partial T$) changes regularly in the periodic system. Of the alkali metals, the temperature coefficient of B for Li^+ is negative, for Na^+ it is almost zero, and for K^+ it is positive (Fig. 2.10).

The five components of viscosity. The regularities outlined above appear as a result of various effects. Also taking into account the results of Kaminsky [46], mentioned above, as well as those of Stokes and Mills [56], the viscosity

Literature on page 139

of a solution involves four further components in addition to the viscosity η_0 of the solvent:

(i) An increase in the viscosity due to the size and shape of the ions (corresponding to the Einstein effect, η_E, see Section 2.4.2); this is always positive and higher the larger the size of the ions.

(ii) The orientation of the polar molecules of the solvent by the electric field of ions (η_{or}) which also increases viscosity, since the mobility of these orientated molecules decreases.

(iii) The structural component (η_{str}) corresponding to the structural decomposition of the solvent which decreases viscosity; this is a resultant of forces causing structural changes in the bulk of the liquid (far from the ions), and the effects arising from the field of ions and from the solvent molecules bound to the ions in a certain orientation.

(iv) The increase in viscosity due to the electrostatic interaction between the ions ($\eta_{el.\,st.}$, see Section 2.4.3).

The viscosity of the electrolyte solutions consists of these five terms:

$$\eta = \eta_0 + \eta_E + \eta_{or} + \eta_{str} + \eta_{el.\,st}. \tag{2.78}$$

Taking into account the Jones–Dole equation:

$$\eta_E + \eta_{or} + \eta_{str} + \eta_{el.\,st} = \eta_0(A\sqrt{c} + Bc). \tag{2.79}$$

Since coefficient A corresponds to the electrostatic interaction between the ions, it can be written (see Section 2.4.3):

$$\eta_{el.\,st} = \eta_0 A\sqrt{c}. \tag{2.80}$$

This term can be eliminated from both sides of the equation, thus we finally have:

$$\eta_E + \eta_{or} + \eta_{str} = \eta_0 Bc. \tag{2.81}$$

According to this equation, coefficient B is related to the size of the solvated ion, the orientation of the solvent molecules, and the change in the solvent structure; more precisely: it is related to the effects of these phenomena on viscosity.

Atom ions of relatively small size or large charge (e.g. Li^+, Mg^{2+}) strongly orientate the water molecules. It can be assumed that these ions are surrounded by a tight hydrate sphere bound to them, carried along in their movement as a kinetic entity. Consequently, η_E is relatively large, since the hydrated ion forming a kinetic entity has a rather extended size. In addition to this, at room temperature the ion and the field of the molecules within the inner sphere probably orientate the water molecules near to them although out of their immediate vicinity (the inner sphere), i.e. η_{or} is also positive. Under such conditions, the increase in viscosity due to $\eta_E + \eta_{or}$ overcompensates the decrease due to η_{str}. It can be assumed, under such conditions, that the structure-breaking effect of the ions is relatively small, since the hydrogen atoms orientated outwards from the hydrate sphere fit rather well into the tetrahedral arrangement of the molecules in water. Therefore, $\eta_E + \eta_{or} \gg \eta_{str}$

applies to this group of ions, and B has a relatively large positive value.

The charge density at the 'boundary' of large atom ions, e.g. Cs^+, I^-, is relatively small, thus the orientation effect acting on water molecules is weak even in the first coordination sphere, and no strongly bound hydrate sphere is formed. Under such conditions, η_E corresponds to the size of the 'bare' ion, its value is small and η_{or} is not large either. However, on the other hand, the structure of water is considerably distorted in the vicinity of the large ion, thus η_{str} is high. For these ions $\eta_E + \eta_{or} < \eta_{str}$, and B is negative.

In solutions of ions of medium size (e.g. K^+ and Cl^-), the effects influencing viscous flow compensate one another, i.e. $\eta_E + \eta_{or} \approx \eta_{str}$ and coefficient B does not differ much from zero. Ammonium ions also belong to this group; their presence hardly deforms the water structure owing to its tetrahedral structure. Since their polarization effect is small, their coefficient B is approximately zero. The mobility of ammonium ions is also similar to that of K^+ ions.

In solutions of relatively large molecule ions (e.g tetra-alkyl ammonium ions), η_E is high, while the structure-breaking and orientation effect is small: $\eta_E + \eta_{or} \gg \eta_{str}$, which results in a large positive value for B. However, according to Frank [57], another effect may act in addition to these: it can be assumed that the ice-like structure is strengthened around large molecule ions (growth of the 'icebergs' in water, see Section 1.4.1). This effect also increases the viscosity of the solutions, where it plays a really significant role.

The negative B values become positive with increasing temperature which can be interpreted as follows. The structure-breaking effect of the ions with negative B value (e.g. Cs^+, I^-) is strong at low temperatures, that is, η_{str} is significant. However, with increasing temperature the water structure is more and more broken, so the structure-breaking effect of the field of the ions is restricted. η_E does not markedly depend on the temperature, and η_{or} slightly decreases with increasing temperature. Thus, it can occur that $\eta_E + \eta_{or} < \eta_{str}$ and, in this way, B becomes positive. According to Nightingale [51a], hydration can also increase with increasing temperature, since the ratio of monomeric water molecules increases; however, the thermodynamic data do not indicate this.

As established by Kaminsky [46], the negative temperature coefficient of the values of B for relatively small ions of large charge can be explained by a loosened binding of the water molecules in the second sphere of the hydrated ions; since, owing to the thermal motion, they are less orientated than those in the first sphere, this causes a decrease in η_{or} with increasing temperature. Taking into account the facts outlined above, we have: $\eta_E + \eta_{or} > \eta_{str}$ even in this case.

One of the factors affecting the change in microviscosity is the alteration of the activation energy required for the rotation of water molecules in the vicinity of ions. Starting with the theory of Samoilov, Uedaira [51b] has calculated this change of the activation energy. According to this

Literature on page 139

deduction, the relative microviscosity is higher than 1, if the ion exhibits positive hydration, while under the effect of negatively hydrated ions the relative microviscosity is lower than unity.

For simple atom ions, the coefficient B decreases with the increase in ionic radius, while it increases with it in solutions of the relatively large tetra-alkyl ammonium ions. Thus the curve of B vs. r_{ion} passes a minimum (at about $r_{ion} \approx 2 \cdot 5$ Å). This effect is attributed, in general, to the structure-breaking effect and the Einstein effect (which predicts a monotonic increase of the value of B). According to Criss and Mastroianni [51c], this variation of the values of B can hardly be ascribed unequivocally to the structure-breaking effect of ions, since B exhibits a similar variation in solutions having a structure much less pronounced than that of water (methanol, acetonitrile). In liquids of loose structure or in structureless ones, ions can exert a structure-making effect on the liquid–presumably due to electrostriction. This affects viscosity similarly to a decrease in the structure-breaking effect in liquids of appreciable structure.

The correlation between coefficient B, the mobility of the ions and the entropy of dissolution. Not only the temperature coefficient of B changes regularly with the crystallographic radii of the ions, but also the temperature coefficient of their mobility. The mobility of large ions (e.g. Cs^+, I^-) shows a smaller increase with temperature than that of the ions of smaller radii (e.g. Li^+). Gurney [50] observed that there is a linear relationship between the ratio of the products of the ionic mobility λ and the viscosity values measured at two different temperatures (e.g. at 0 and 18 °C) and coefficient B of a given ion (Fig. 2.11).

In agreement with the observations discussed above, all these can evidently be attributed to the marked destructive effect of non-hydrated ions (or those having a very small hydrate sphere) on the water structure in their vicinity; this decreases viscosity, while in the vicinity of small ions with a large hydrate sphere, this effect is small. Ionic mobilities mainly depend on the viscosity of the solvent in the vicinity of the ion (microviscosity), and the increase in temperature favours the structure-breaking effect in water (and in this way, the decrease in viscosity). In solutions of large ions (which decrease the structural order of the water molecules simply by their presence), the decreasing effect of increasing temperature on viscosity is relatively less than in solutions of small ions which hardly deform the liquid structure. As a consequence of this, the temperature

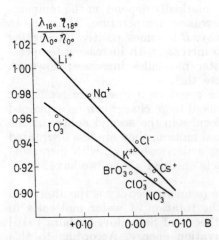

Fig. 2.11 Correlation between the temperature coefficient of the product $\lambda\eta$ and the B coefficient

coefficient of the mobility of large ions is smaller than that of the small ones. Considering that coefficient B is related to similar effects, its correlation with the ionic mobility can be explained.

The correlation between coefficient B and the *change in the entropy* of ionic hydration can also be attributed to similar phenomena. A part of the entropy of the dissolved ions is a measure of the disorder around them, and this disorder also depends on the interaction between the ions and the solvent. According to Gurney [50], a linear relationship exists between the partial molar entropy of atom ions and their B value. This correlation is represented by one single straight line, if a value of $S_{h^+}^0 = -5 \cdot 5$ cal mole^{-1} degree^{-1} is taken for the hydrogen ion in calculating the entropies of the ions. Asmus [58] and Nightingale [51a] found a more simple relationship between the hydration entropy of ions and coefficient B. The partial entropy of hydration S_h^0 can be calculated from the entropy of the solute ions (S_{aq}^0) by subtracting the sum of the entropy of translation and rotation (S_g^0) calculated for the ion in gaseous state:

$$S_h^0 = S_{aq}^0 - S_g^0, \qquad (2.82)$$

S_{aq}^0 can be calculated from the entropy of dissolution (ΔS^0, i.e. the difference between the partial molar entropies of the electrolyte in 1 M ideal solution and that in the crystal) and from the conventional reference entropy S_{ref}^0 of a given ion, one has:

$$S_{aq}^0 = \Delta S^0 + S_{ref}^0. \qquad (2.83)$$

The hydration entropy calculated in this way is in linear correlation with the B values for both atom and molecule ions (Fig. 2.12).

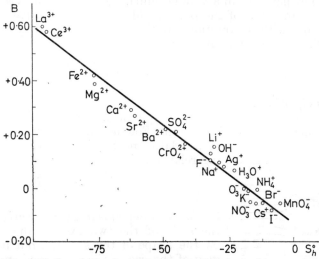

Fig. 2.12 Correlation between the B coefficient of ions and the partial molar entropy of hydration

Literature on page 139

Investigating some solutions of monovalent electrolytes, Burns [59] also concluded that the B values depend linearly on the partial molar entropy of the ions.

2.4.2 EFFECT OF THE SIZE AND SHAPE OF THE DISSOLVED PARTICLES ON VISCOSITY (THE EINSTEIN EFFECT)

The solvent structure is influenced by the size and shape of the dissolved ions, and viscosity is altered by this effect as well. This is similar to the effect observed in suspensions: the viscosity of the liquid is altered by suspended solid particles of microscopic size, resulting in higher viscosity of the suspensions than that of the pure solvent. With respect to the size of the suspended particles, the liquid can be considered a continuum, and thus the effect can be interpreted by changes in the streamlines due to presence of the suspended particles (Fig. 2.13). A similar effect can appear in electrolyte solutions in which the hydrated ion forming a kinetic entity is large in comparison with the solvent molecules, i.e. the solvent can be considered to be a continuum with respect to the movement of the ions in first approximation. Such ions are, for example Li^+, Mg^{2+}, La^{3+} ions with their large hydrate spheres. This effect is similar in several respects to changes in viscosity due to dissolved non-electrolytes.

The viscosity of non-electrolyte solutions. The theory of the hydrodynamic movement of rigid spheres suspended in a continuum has been developed by Einstein [60]. If the treatment is limited to dilute suspensions (containing only a few particles in a volume unit), the ratio of the viscosity of the suspension η and that of the pure liquid η_0, i.e. the relative viscosity, increases linearly with the fraction Φ of the volume occupied by the suspended particles (i.e. the volume fraction):

$$\frac{\eta}{\eta_0} = \eta_r = 1 + 2{\cdot}5\,\Phi \tag{2.84}$$

(Einstein's equation).

If the suspended species is not spherical, the proportionality factor of the volume fraction Φ is different. For suspended particles of rotational ellipsoidal shape, Simha [61] deduced the following relation:

$$\frac{\eta}{\eta_0} = 1 + a_1\Phi \tag{2.85}$$

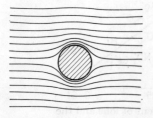

Fig. 2.13 Streamlines in a liquid around a solid particle

in which a_1 depends in a complicated manner on the ratio of the two axes of the ellipsoid (i.e. on the shape of the species), but it is independent of the size of the suspended species. In this equation, a_1 is greater when the ellipsoid is more eccentric; for spherical particles, $a_1 = 2{\cdot}5$. It has been established by Simha that under normal conditions shear forces do not result in any marked order with respect to the direction of the translation of the species because of the counter-effect of the thermal motion. The effect of long-shaped species, resulting in a significant increase in viscosity, is important in solutions of polyelectrolytes and other macromolecules.

The Einstein correlation (2.84) is valid for dilute suspensions in which the streamlines around the adjacent particles do not interfere with one another. However, in more concentrated suspensions, the streamlines around the adjacent species interact significantly. Under such conditions, the following equation holds as shown by Vand [62a]:

$$\ln \frac{\eta}{\eta_0} = \frac{2 \cdot 5 \varPhi}{1 - Q \varPhi} , \qquad (2.86)$$

where interaction factor Q can be calculated only approximately. Equation (2.86) is reduced to equation (2.84) in the case of small volume fractions. Thus, introducing the concentration c of the suspended material, one has:

$$\ln \frac{\eta}{\eta_0} = kc \qquad (2.87)$$

which corresponds to the empirical relation of Arrhenius.

Recently, Eagland and Pilling [62b] has found the Vand equation to hold for concentrated solutions of tetraalkylammonium bromides, too. They also established a correlation between $\varPhi$ and coefficient B in the Jones–Dole equation.

In equations (2.84)–(2.86), the volume fraction $\varPhi$ involves also the solvent molecules bound to the surface of the suspended particles and moving together with it.

The Einstein equation (2.84) has been found to hold for latex spheres of polystyrene up to a radius of 1300 Å in the experiments of Cheng and Schachmann [63a]. In these suspension, the volume of water bound to the surface of the species can be neglected. In solutions of small species, however, the volume of the bound water should be taken into account as well.

The effect on viscosity of alcohols, urea and other amides dissolved in water has been analyzed by Herskovits and Kelly [63b], on the basis of the variation of coefficients B and C in the following empirical expression:

$$\frac{\eta}{\eta_0} = Bc + Cc^2 .$$

They found the slope of the line representing the above expression to be steeper in these solutions than expected from the Einstein and Simha equations. The higher increment of viscosity has been attributed to the strengthening and destruction of the liquid structure of water, caused by the presence of non-polar hydrocarbon groups and the polar OH and $CONH_2$ groups, respectively.

Literature on page 139

The effect of the hydration and the shape of the particles. According to Stokes and Mills [64], in the case of appreciable binding of water molecules at the boundary of the suspended particles, the volume fraction is:

$$\Phi = cV_h, \tag{2.88}$$

if c is the concentration, (mole/litre) and V_h (litre/mole) is the molar volume of the solute together with the volume of the water molecules bound to it. It has been assumed that a correlation similar to equation (2.86) is also valid for solutions containing non-spherical species in higher concentrations, but other values of constants a and b should be applied:

$$\ln \frac{\eta}{\eta_0} = \frac{ac}{1 - bc}, \tag{2.89}$$

where, taking into account equation (2.88), $a = a_1 V_h$ and $b = QV_h$. On the basis of the experimental data, this equation gives reasonable values for solutions of non-electrolytes with medium sized rigid molecules. For example, the saccharose molecule can be considered to be a rotational ellipsoid in good approximation, with an axis ratio of about 0·5. For such a shape, $a_1 = 2·85$, as shown by Simha. However, according to the empirical equation describing the viscosity of aqueous solutions of saccharose [64], one has:

$$\ln \frac{\eta}{\eta_0} = \frac{0·875\,c}{1 - 0·231\,c}, \tag{2.90}$$

from which $V_h = 0·875/2·85 = 0·307$ litre/mole. The apparent molar volume of saccharose calculated on the basis of its molecular weight is 0·212 litre/mole. Thus, $V_h - 0·212 = 0·307 - 0·212 = 0·095$ litre/mole which is the volume of the water bound in the hydrate sphere corresponding to about five water molecules per saccharose molecules. These data are in agreement with the thermodynamical properties of saccharose solutions at 25 °C.

In a similar calculation on penta-erythritol molecules $C(CH_2OH)_4$, which can be considered spherically symmetrical and somewhat smaller, a value of $V_h = 0·136$ litre/mole can be obtained at 25 °C instead of 0·102 litre/mole calculated without the volume of water molecules, which corresponds to about two water molecules for one penta-erythritol molecule. According to a similar calculation, glucose and mannitol have two and four water molecules, respectively.

It has also been emphasized by Moulik [65], investigating the viscosity of suspensions of higher concentration within the validity range of the Einstein equation, that the volume fraction Φ in equation (2.84) involves the volume of solvent molecules located in an immobilized state at the surface of the suspended species. Taking into account equation (2.88), the Einstein equation can be transformed into the following one:

$$\frac{\eta}{\eta_0} = 1 + 2·5\,cV_h = 1 + Kc, \tag{2.91}$$

where $K = 2 \cdot 5 \, V_h$. This equation, which is valid for dilute solutions only, neglects the interaction between the solvent and solute, although this means that the microviscosity is not identical with the macroscopic viscosity. In more concentrated solutions, the streamlines around the adjacent particles are not independent of one another. According to Moulik [65], the following form of the Einstein equation should be applied to investigate the concentration dependence:

$$\frac{\partial \left(\dfrac{\eta}{\eta_0} \right)}{\partial c} = K \, . \tag{2.92}$$

For solutions or suspensions of higher concentrations, for which this correlation does not hold, the following equation seems to be valid, as given by Moulik's experiments:

$$\frac{\partial \left(\dfrac{\eta}{\eta_0} \right)}{\partial c} \cdot \frac{\dfrac{\eta}{\eta_0}}{c} = K' \, , \tag{2.93}$$

or, after integration:

$$\left(\frac{\eta}{\eta_0} \right)^2 = K' c^2 + I \, , \tag{2.94}$$

where I is an integration constant.

The effect of ionic charges. As can be expected on the basis of the reasonable values obtained for non-electrolyte solutions, some conclusions can also be drawn on the hydration of ions from the viscosity of electrolyte solutions. Taking into account equation (2.84), the Jones–Dole equation for electrolyte solutions transforms into:

$$\frac{\eta}{\eta_0} = 1 + A \sqrt{c} + Bc = 1 + 2 \cdot 5 \Phi \, , \tag{2.95}$$

from which

$$2 \cdot 5 \Phi = A \sqrt{c} + Bc \, . \tag{2.96}$$

According to the experiments, $A \sqrt{c} \ll Bc$, and

$$\Phi \approx c V_i \tag{2.97}$$

where V_i is the molar volume of the hydrated ion in the solution. If it is assumed that the viscosity-increasing effect of large ions (considered together with their hydrate spheres) can be attributed to the Einstein effect in

Literature on page 139

aqueous solutions, i.e. the effect arises from them bending the streamlines, from equations (2.96) and (2.97) one has:

$$2 \cdot 5 \, V_i \approx B. \tag{2.98}$$

Under such conditions, the factor B in the Jones–Dole equation is *approximately proportional to the volume of the dissolved ions*.

From the corresponding coefficient B at 25 °C, 0·154 litre/mole can be calculated for the strongly hydrated Mg^{2+} ion, if it is attributed entirely to the Einstein effect. For the La^{3+} ion, this value is 0·235 litre/mole. From these molar values the radii of the atom ions are 3·94 and 4·53 Å, respectively. These values are not in contradiction with those calculated from the ionic mobilities using Stokes' Law (3·46 and 3·95 Å) allowing for the fact that the Stokes law usually results in smaller values than the real ones in the case of small species.

The correlation between coefficient B and the size of the ions can be investigated in solutions of large molecule ions as well. For example, the average radius of the $N(CH_3)_4^+$ ion calculated from the crystal lattice is 2·67 Å. According to equation (2.95), a value of $B = 0·120$ is obtained which is in good agreement with that determined by Hückel and Schaaf [66] ($B = 0·123$). The radius of the $N(C_2H_5)_4^+$ ion is 3·5 Å in the crystal lattice, but, if the ethyl groups are stretched, the radius is nearer to 4 Å. The ionic radius obtained by Hückel and Schaaf on the basis of viscosity measurements is about 3·8–4·0 Å, supposing spherical particles. In the case of $N(C_3H_7)_4^+$ and $N(C_4H_9)_4^+$ ions, the B coefficients indicate that the alkyl groups are stretched in solution and some water molecules are embedded between them in a more or less immobile state. In this way, these ions have no spherical shape, so they are not suitable for quantitative test of the validity of equation (2.95).

Investigating the viscosity of solutions of chloroamine platinum(IV) salts in the concentration range 0·005–0·1 M, Nightingale and Kuecker [67] concluded that coefficient B is determined mainly by the size of the ions in solutions of species of hydrophobic surface and the charge of the ions has almost no effect. Polyatomic ions containing hydrophilic groups ($-Cl$, $-NH_2$, $-OH$) in which the charge of the ion is more effectively delocalized, decrease the value of B. On the other hand, in solutions of small monoatomic hydrated ions, B also depends on the electric field around the ion. It increases with increasing ionic charge and with the decrease of the crystallographic ion radius. The limiting value of coefficient B is determined by the electric field strength which polarizes the water molecules so that ion Me^{Z+} transforms hydrolytically into $Me(OH)_n^{(Z-n)+}$ ion.

The Jones–Dole equation proved to be valid for the viscosity of solutions of tetra-alkyl ammonium halides in H_2O, D_2O, CH_3OH, and CH_3CN within the temperature range of 0 to 60 °C as published by Kay, Vituccio, Zawoyski and Evans [68]. Structure-making effects have been shown in the case of aqueous solutions of $N(C_3H_7)_4^+$ and $N(C_4H_9)_4^+$ ions, while the $N(CH_3)_4^+$ ions have a structure-breaking effect. In solutions of $N(C_2H_5)_4^+$ ions, the two opposite effects compensate each other. In solutions of structure-making ions, B and the corresponding product $\lambda^0 \eta$ (λ^0 is the limiting equivalent conductivity) change in the same way, while for structure-breaking ions, when B is small, $\lambda^0\eta$ is large, and if B increases with temperature, $\lambda^0\eta$ decreases with it.

It has been found by Isono and Tamamushi [69a] in their studies on the correlation between the coefficient B of viscosity and the molar volume of the solute, that this can be established unequivocally only for the whole electrolyte, and its division into some parts corresponding to individual ions is rather arbitrary. It has also been emphasized by them that the B coefficients can be expressed as linear functions of the molar volume in agreement with the facts mentioned above only in the case when the molar volume of the hydrated ion is taken. A more precise relationship between the B value of the electrolyte as a whole and the molar volume of the hydrated salt $V^0_{s,h}$ can be given as:

$$B = aV^0_{s,h} - b \qquad (2.99a)$$

(at 25 °C, $a = 2\cdot8$, $b = 0\cdot19$, while at 5 °C, $a = 2\cdot6$ and $b = 0\cdot20$). From this empirical correlation $V^0_{s,h}$ can be calculated using the $B = 0$ value; at 25 and 5 °C, it is 67 and 79 ml, respectively. This can be considered to be the volume of the water molecules associated with the ions. The equation above is valid mainly for macromolecular substances, but the deviations are not too large for solutions of small molecules either. In solutions of ions found to be structure-breaking ones on the basis of other phenomena, $V^0_{s,h}$ is smaller than for solutions of ions without this effect.

The viscosity of concentrated aqueous salt solutions has been investigated recently by Breslau and Miller [69b]. They found the Einstein correlation in equation (2.84) to be valid in the form given by Thomas as follows:

$$\frac{\eta}{\eta_0} = 1 + 2\cdot5\Phi + 10\cdot05\Phi^2. \qquad (2.99b)$$

The 'effective rigid volume' calculated on the basis of the equation:

$$\overline{V}_e = \frac{\Phi}{c}, \qquad (2.99c)$$

taking into account concentration c of the solution and the volume fraction Φ of the solute particles ($\Phi < 0\cdot25$), proved to be independent of concentration. Analysing numerous data on the viscosity of concentrated solutions of several salts, a relationship has beeen found between V_e and coefficient B in the Jones–Dole equation. According to this, for monovalent ions, one has:

$$B = 2\cdot90\,\overline{V}_e - 0\cdot018, \qquad (2.99d)$$

while for polyvalent ions it is:

$$B = 6\cdot06\,\overline{V}_e - 0\cdot041. \qquad (2.99e)$$

The difference between these two relationships has been attributed partly to hydrodynamic effects and partly to interactions between the ions and the solvent molecules.

Literature on page 139

2.4.3 EFFECT OF ELECTROSTATIC INTERACTION OF
DISSOLVED IONS ON VISCOSITY

The term $A\sqrt{c}$ in the Jones–Dole relationship in equation (2.73) takes into account the retardation force arising from the electrostatic interaction of the dissolved ions that can be calculated on the basis of the Debye–Hückel theory [70] of strong electrolytes (Section 5.1). Owing to the simultaneous effects of the electrostatic attraction and repulsion and the thermal motion, each ion is surrounded by an excess of ions of opposite charge in solution according to this theory. The ionic atmosphere, surrounding each ion and having an opposite charge from that under consideration, is spherically symmetrical, on the statistical average, around the 'resting' ion in equilibrium (see Fig. 2.14a); thus the resultant force acting on the ion in the middle of the ionic atmosphere is zero. If, however, the liquid flows and the velocity gradient is, for example, linear, the ion cloud is deformed (Fig. 2.14b,c), which can be attributed to the fact that the development of the ion cloud requires a certain time (relaxation time). If the velocity of the liquid flow increases upwards in Fig. 2.14, the upper right and lower left quadrants of the ionic atmosphere of a positive ion in question would contain excess negative ions in comparison with the spherically symmetrical case. In the other two quadrants, however, the amount of the negative charges is less than in the spherically symmetrical state. This leads to a shear force acting against the flow and increasing viscosity. The range of action of the electrostatic interaction of the ions is larger than that between the ions and the solvent molecules.

According to the theory of Falkenhagen *et al.* [71a], coefficient A can be calculated, in the limiting case of very dilute solutions, on the basis of the theory of strong electrolytes from the dielectric constant ε_0 of the solvent, the valence z_1 and z_2 of the ions, and their mobility $(\lambda_1^0, \lambda_2^0)$ in infinitely dilute solutions:

$$A = \frac{1 \cdot 461}{\eta_0 \sqrt{\varepsilon_0 T}} \sqrt{\left(\frac{v_1 |z_1|}{|z_1| + |z_2|} \right)} \cdot \frac{1}{\lambda_1^0 \lambda_2^0} \psi, \qquad (2.100)$$

where v_1 is the number of the types of the ions formed from a 'molecule and

$$\psi = \frac{\lambda_1^0 z_2^2 + \lambda_2^0 z_1^2}{4} - \frac{(|z_2| \lambda_1^0 - |z_1| \lambda_2^0)^2}{\left\{ \sqrt{(\lambda_1^0 + \lambda_2^0)} + \sqrt{(|z_2| \lambda_1^0 + |z_1| \lambda_2^0)} \sqrt{\left(\frac{|z_1| + |z_2|}{|z_1 z_2|} \right)} \right\}^2} \qquad (2.101)$$

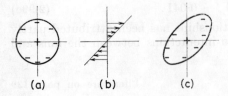

Fig. 2.14 (a) Scheme of the ion atmosphere around an immobile ion; (b) shearing forces in a flow of linear velocity distribution; (c) ion atmosphere deformed by shearing forces

(a)   (b)   (c)

Table 2.4

Measured and calculated A values for some electrolytes
in aqueous solutions at 25 °C

Electrolyte	A_{found}	A_{calc}	Electrolyte	A_{found}	A_{calc}
NaCl	0·0067	0·0060	$BaCl_2$	0·0201	0·0147
KCl	0·0052	0·0050	$MgSO_4$	0·0225	0·0218
KBr	0·0047	0·0049	$MnSO_4$	0·0231	0·0227
CsI	0·0039	0·0048	$CuSO_4$	0·0230	0·0230
$KClO_3$	0·0050	0·0054	$CdSO_4$	0·0232	0·0225
$KBrO_3$	0·0058	0·0058	$Cr_2(SO_4)_3$	0·0495	0·0507
KNO_3	0·0050	0·0052	$K_4]Fe(CN)·]$	0·0369	0·0369
NH_4Cl	0·0057	0·0050	$Ca_2[Fe(CN)·]$	0·0495	0·0490
$KMnO_4$	0·0058	0·0056			

The A values calculated according to this equation show good agreement with the experimental ones in aqueous and several non-aqueous solutions (Table 2.4).

On the basis of measuring the viscosities of aqueous solutions of tetrabutylammonium nitrate in the 10–50 °C temperature range, Kudryavtseva, Krumgalz and Mishchenko [71b] stated that the A values in the Jones–Dole equation calculated theoretically are in good agreement with the experimentally determined ones. The high positive value of B in these solutions indicates that the ions of this salt exert a significant orientating effect on the water molecules, which, however, decreases with increasing temperature.

The correlations described above are valid only for not too high flow rates. If the flow rate is large, the ionic atmosphere can be only partially formed, owing to the finite relaxation time, while in the case of a very high velocity gradient, as shown by Deubner [72], the term containing $\sqrt{c}$ vanishes (the Wien effect in viscosity, see Section 4.2.3.6). As established by Asmus [73], viscosity is determined by ionic strength I in solutions containing several types of electrolytes:

$$\eta = \eta_0 (1 + a\sqrt{I} + bI),\qquad(2.102)$$

where a and b are constants and a can be calculated using the Onsager–Fuoss theory [74].

Davies and Malpass [75] studied extensively the effect of ion association (the formation of ion pairs) on the viscosity of dilute electrolyte solutions. They found the Jones–Dole equation to be in good agreement with experiments. It has been established that viscosity is decreased by ionic pair formation; however, in the case of some electrolytes with small or negative coefficients B, the influence of the ionic pair formation is negligible.

Literature on page 140

2.4.4 VISCOSITY OF CONCENTRATED SOLUTIONS

The dependence of viscosity on the concentration of electrolyte solutions is described by the Jones–Dole equation (equation (2.73)), in good agreement with the experiments up to a concentration of about 0.1 M. Studies on more concentrated solutions ($c > 1$ M) have dealt mainly with the temperature dependence of viscosity. In this concentration range, the empirical equation of Andrade [3] is valid in several cases:

$$\eta = A \exp\left[\frac{b}{T}\right], \tag{2.103}$$

where A and b are constants (see Section 2.2.4). Regarding the viscous flow as a rate process, this equation corresponds to the Arrhenius–Andrade type correlation: if ΔH_{visc} is the enthalpy of activation of the viscous flow, one has:

$$\eta = A \exp\left[\frac{\Delta H_{\text{visc}}}{RT}\right]. \tag{2.104}$$

From this equation

$$\Delta H_{\text{visc}} = R \frac{d \ln \eta}{d \frac{1}{T}}, \tag{2.105}$$

i.e. the enthalpy of activation can be calculated from the slope of the straight line defined by the $\ln \eta \sim 1/T$ function.

The Eyring theory enables a detailed description of the temperature dependence of viscosity. From equation (2.48) one obtains:

$$\eta = \frac{hN}{V} \exp\left[\frac{\Delta G^{\ddagger}}{RT}\right], \tag{2.106}$$

where $\Delta G^{\ddagger}$ is the free enthalpy of activation. The relationship between this and the enthalpy of activation ($\Delta H^{\ddagger} = \Delta H_{\text{visc}}$) can be given as follows

$$\Delta G^{\ddagger} = \Delta H^{\ddagger} - T \, \Delta S^{\ddagger} \tag{2.107}$$

and for the entropy of activation we have:

$$\eta = \frac{hN}{V} \exp\left[\frac{\Delta H^{\ddagger}}{RT} - \frac{\Delta S^{\ddagger}}{R}\right]. \tag{2.108}$$

If V and $\Delta S^{\ddagger}$ are constant in good approximation, the Eyring theory also leads to

$$\eta = A \exp\left[\frac{\Delta H^{\ddagger}}{RT}\right] = A \exp\left[\frac{\Delta H_{\text{visc}}}{RT}\right] \tag{2.109}$$

where A involves the constant factors.

These relationships are related to pure non-associated liquids, and agree with the experiments in several such cases. In associated liquids (water, alcohols, etc.), the viscous flow is accompanied by a deformation of the structure which contributes to the normal molecular friction. The flow of a liquid with internal structure requires an excess energy, i.e. a *structural activation energy*.

The internal structure of associated liquids, i.e. the order of the molecules is altered by the solute, in addition to the effect of temperature. With increasing temperature, the order begins to disappear and the structure is loosened owing to the enhanced thermal motion. The liquid structure is altered by ion–solvent interactions, its extent depending on the concentration and nature of the dissolved electrolytes. In a given temperature and concentration range, the changes in the entropy of both effects depend on the liquid structure, i.e. on the degree of the order of molecules. $\Delta S^{\ddagger}$ is not necessarily constant, thus equation (2.108) is not reduced to equation (2.104) in every case. With respect to the effects on the structure, it cannot be expected that simple rules should hold for the viscosity of associated liquids containing an electrolyte in higher concentrations. The correlation lg η vs. $1/T$ is still linear for several electrolytes of this type; in other cases, however, there are some deviations from it even in dilute aqueous solutions [76d].

The available experimental data on concentrated solutions are in contradiction to one another. According to Suryanarayana and Venkatesan [76b], the linear behaviour seems to be valid for solutions of KCl, NaCl, KNO_3, $CuSO_4$, and $ZnSO_4$ from 1 M to saturation, which can be explained perhaps by the model suggested by Satoh [77] for the structure of aqueous electrolyte solutions. According to this, in solutions more concentrated than 1 M, the normal water structure is destroyed by the interaction between the hydrate spheres of the ions. Thus, in concentrated solutions the ions, in fact, are embedded in a solution having properties other than those of the dilute solution. The simple Andrade equation is supposed to be valid for concentrated solutions. However, this assumption cannot be regarded as proved, since the experimental data are not unequivocal.

On the other hand, the correlation lg η vs. $1/T$ is not linear in solutions of NaSCN, $NaClO_4$, NaI, and LiCl in higher concentration, as shown by Miller and Doran [78]. The approximate value of the activation energy of the viscous flow can still be evaluated from the measurements carried out in a narrow temperature range. Calculating the free enthalpy of activation from equation (2.106), the entropy of activation can be obtained from equation (2.108). It can be seen in Table 2.5 that $\Delta S^{\ddagger}$ is smaller in 2 M solutions than in water, which indicates that 'initially' a structure-breaking effect is exhibited by the dissolved salts (up to medium concentrations). The entropy of activation increases rapidly with increasing salt concentration which shows that, under such conditions, a local order starts to be built up around the ions. From the partial molar entropy of concentrated aqueous solutions of various salts, Miller and Doran concluded that the local order is extended to a distance of some ionic diameters in NaSCN, $NaClO_4$, and

Literature on page 140

9*

Table 2.5

The enthalpy $\Delta H^{\ddagger}$, free enthalpy $\Delta G^{\ddagger}$, and entropy $\Delta S^{\ddagger}$ of activation of viscous flow in concentrated solutions of some electrolytes at room temperature

Electrolyte	conc mole · 1 $^{-1}$	kcal/$\Delta H^{\ddagger}$mole	kcal/$\Delta G^{\ddagger}$mole	cal/$\Delta S^{\ddagger}$degree · · mole
— (water)	—	3·89	2·17	5·7
NaSCN	2	3·63	2·27	4·9
	7	4·44	2·88	5·2
	9	5·70	3·30	7·9
	10	8·18	3·55	15·3
NaClO$_4$	2	3·54	2·26	4·2
	7	4·68	2·97	5·6
	9	8·70	8·70	17·3
	9·5	9·85	3·59	20·7
NaI	2	3·40	2·22	3·9
	6	3·90	2·59	4·3
	8	4·42	2·92	5·0
LiCl	10·75	4·68	3·38	4·3
	13	8·50	3·72	15·7

NaI solutions, while in LiCl solutions the high values of $\Delta S^{\ddagger}$ are probably due to extensive hydration of the Li$^+$ ion.

In order to calculate the values of $\Delta S^{\ddagger}$ for some ions, Nightingale and Benck [79a] started from the Jones–Dole equation given in equation (2.73), neglecting the small term $A\sqrt{c}$ arising from the electrostatic interaction between the ions. The value of $\Delta G^{\ddagger}$ had been calculated from (2.106), while they substituted the value given by the Jones–Dole equation for η in equation (2.105) to determine the energy of activation:

$$\Delta H_{\text{visc}} = R \frac{\mathrm{d} \ln \eta_0}{\mathrm{d} \dfrac{1}{T}} + \frac{1}{1 + Bc} \frac{\mathrm{d}(1 + Bc)}{\mathrm{d} \dfrac{1}{T}}. \qquad (2.110)$$

Using the B values obtained by other authors, $\Delta H_{\text{visc}}(= \Delta H^{\ddagger})$ and, after that, $\Delta S^{\ddagger}$ can be calculated. The values of $\Delta S^{\ddagger}$ have been divided into parts resulting from the individual ions similarly to the division of the B values (see Section 2.4.1). According to this approach, hydrated structure-making ions with a positive coefficient B have a positive entropy of activation, while the structure-breaking ions with negative B coefficients have negative entropy of activation. This result, of course, is based partly on the value of the B coefficients used.

On the basis of a detailed analysis of the results obtained in experiments on concentrated solutions of several electrolytes (in the concentration range of 1–8 mol/litre), Ebert and Wendorff [79b] have concluded–also taking into account the viscosity-increasing effect of ions–that the structure-breaking effect of ions on liquids is greater than that supposed earlier. They have calculated the average time interval between two displacements of the water molecules located around ions from the activation energy of viscosity. Their results, which are otherwise in agreement with some earlier concepts regarding this effect, indicate that the water molecules of the primary hydrate sphere are strongly bound to the ions, but the mobility of the molecules outside the primary hydrate sphere shows a decrease only in solutions of certain electrolytes (e. g. Li_2SO_4, $LiOOCCH_3$). Thus the structure-making effect in water seems to be rare.

For the change of viscosity with concentration in concentrated solutions of strong electrolytes, a new theory has been developed by Suryanarayana and Venkatesan [80]. Starting with the stability of the saturated solution as the highest possible one at a constant temperature, the concentration (mole fraction) and viscosity of the solution were related to the concentration and viscosity of the saturated solution. The ratio of the mole fraction of the given solution and that of the saturated one has been called *concentration potential* (C_p). The ratio of the viscosity η_p of the given solution and that of the saturated one was found to be:

$$\eta_p = A \exp [BC_p], \qquad (2.111)$$

where A and B are empirical constants, their numerical values depending on the temperature. A is the value of η_p when $C_p = 0$ (infinitely dilute solution), i.e. the ratio of the viscosity of pure water and that of the saturated solution. On the basis of a comparison with the Arrhenius equation of viscosity, the constant B is determined by the activation energy (ΔH_{visc}) of viscous flow:

$$B = \frac{\Delta H_{visc}}{RT}. \qquad (2.112)$$

According to the experiments of the above authors, $\ln \eta_p$ changes linearly with C_p within the experimental errors in concentrated solutions of strong electrolytes, but there is a discontinuity in the straight lines at one or two points. Thus, in fact, the relationship between $\ln \eta_p$ and C_p consists of two or three straight lines crossing each other and that corresponding to higher concentrations has the larger slope. The 'knees' on the isotherms have been attributed to changes in the polymerization of water, or to ion pair formation or hydration. Similar correlations have been observed for the viscosity of $LiNO_3$ in ethanol–water mixtures too, and, furthermore, for non-electrolyte solutions. This latter condition indicates that the hydrodynamic forces in the solution influencing the viscosity are not just of electrostatic nature, but some other interactions are also involved. On the basis of the analysis of the experimental data obtained for concentrated solutions of several strong electrolytes, Suryanarayana and Govindaswamy [81a] concluded that the change in the viscosity of concentrated solutions with the internal pressure π of the liquid is proportional to the viscosity:

$$\frac{d\eta}{d\pi} = K\eta \qquad (2.113)$$

where K is constant.

Literature on page 140

The empirical relationships found by Suryanarayana *et al.* are note-worthy, but their theoretical fundamentals are not adequate and their validity range should be elucidated.

Angell and Bressel [81b] have treated the viscosity and conductivity of aqueous electrolyte solutions unconventionally. They do not distinguish the role of the solvent and the solute, thus their theory remains applicable to concentrated solutions, too, where extended regions of bulk solvent are no longer present. In generalizing the results obtained by measuring the viscosities and conductivities of $0-26$ mole-% $Ca(NO_3)_2$ solutions between $+80$ and -60 °C, they rely upon the concepts used in the description of the relaxation processes in liquids forming holes. The following quasi-empirical isotherm is proposed to describe the dependence of viscosity on concentra-tion, over the whole concentration range from the most dilute solutions to the most concentrated ones:

$$\frac{1}{\eta} = A \exp\left[-\frac{B}{x_0 - x}\right],$$

where x is the concentration of the salt in mole-%, x_0, A and B are constants. According to these authors, the properties of this type of electrolyte solu-tions can be understood by assuming three regions of composition: (1) a relatively high concentration range, where no bulk water is present, and all motions of the particles present are strongly coupled; (2) the middle range of concentration with very complex properties, the structure probably containing microscopic inhomogeneities and worm-hole-like regions, where particles rich in water and those rich in hydrated ions are fluctuating, their size being greater with decreasing temperature; (3) the range of dilute solu-tions, where, in agreement with the conventional picture, the ions are dis-persed together with their hydrate spheres in the bulk water continuum.

2.4.5 EFFECT OF ELECTROLYTES ON THE VISCOSITY OF LIQUIDS AT HIGH PRESSURES

Unlike other liquids, the viscosity of water does not show a monotonic increase with increasing pressure, but initially it decreases, then, after passing a minimum, it increases at high pressures. This can be interpreted by the destructive effect of the hydrostatic pressure on the short-range lattice-like structure, distorting the binding of the water molecules to their neighbours and hence increasing their mobility. According to Horne and Johnson [82], the structure-breaking electrolytes which decrease the viscosity in water diminish the anomalous effect of pressure, since the fur-ther loosening of the structure partly destroyed by the electrolytes due to the effect of pressure is less significant than in the case of the original water structure. The effect of such electrolytes is, in this respect, similar to that of temperature. On the other hand, the structure-making electrolytes enhance viscosity and the effect of pressure. For example, NaCl increases viscosity, i.e. it is structure-making. However, this is in contradiction with the fact that the activation energy of the viscous flow is decreased with increasing electrolyte concentration, similarly to the effect of increasing pressure. If the decrease in the activation energy is attributed to the loosening and destroying effects on the liquid structure, one has to assume that NaCl is a structure-breaking electrolyte. The interpretation of this contradiction requires further studies.

2.4.6 VISCOSITY OF TERNARY ELECTROLYTE SOLUTIONS

For the viscosity of solutions containing more than one solute, only a few data are available. However, these indicate conditions essentially similar to those mentioned above to exist in ternary systems as well, although some specific deviations can be revealed in some cases.

Galinker, Tyagai, and Fenerli [83] measured the viscosity in the systems $CdCl_2-HCl-H_2O$, $CdCl_2-LiCl-H_2O$, $CdCl_2-NaCl-H_2O$, $CdCl_2-KCl-H_2O$, and $CdCl_2-NH_4Cl-H_2O$ in solutions of identical molarity within the range of 25 to 35 °C. In all cases the viscosity proved to be lower than that calculated on the basis of additivity; however, the composition of the complexes formed in the mixture cannot be determined reliably from the deviations. As an interpolation formula, the following relationship seems to be valid in good approximation:

$$\eta = x_1\eta_1 + x_2\eta_2 - kx_1x_2, \qquad (2.114)$$

where k is a constant independent of the composition of the system, and the subscripts correspond to the two dissolved electrolytes (x is the mole fraction).

Investigating the viscosity of the systems $LiCl-NaCl-H_2O$, $LiCl-KCl-H_2O$, and $NaCl-MgCl_2-H_2O$, Wu [84] checked the Jones–Dole law in solutions of ionic strength $I = 0.01-1$ M at 25 °C:

$$\frac{\eta}{\eta_0} = 1 + a\sqrt{I} + bI, \qquad (2.115)$$

in which a is the slope of the straight line corresponding to the Falkenhagen limiting law (see Section 5.1.3). He proved the statement of Harned and Owen that, in ternary systems, coefficient b is additively composed of the corresponding terms of the individual electrolytes b_1 and b_2, in proportion to the ionic strength fraction y^*:

$$b = y_1b_1 + y_2b_2. \qquad (2.116)$$

On the basis of measurements on the viscosity of aqueous solutions of metal chlorides and HCl, Berecz and Vértes [85] made a critical survey to decide which equation, suggested in the literature to describe the viscosity

* The ionic strength fraction is a quantity defined similarly to the mole fraction.

Literature on page 140

of binary systems as a function of the composition, can be extended to ternary systems in the best agreement with the experimental data. The most suitable one for this extension was the equation, suggested by Zdanovskii [86], correlating the viscosity of a ternary system with those of two suitable binary systems. This is essentially based on the assumption that the kinematic fluidity (ϱ/η, where ϱ is the density of the mixture) is an additive quantity obtained from the kinematic fluidities of the components in proportion to the volume fractions. If the viscosity and density of a ternary system are η' and ϱ', and if the viscosities and densities of the binary systems, in which the concentration of the electrolyte is identical with the total concentration of the ternary system, are η'_1, ϱ'_1 and η'_2, ϱ'_2, respectively, while ϑ'_1 is the volume fraction of one of the components,* one has:

$$\frac{\varrho'}{\eta'} = \frac{\varrho'_2}{\eta'_2} + \left(\frac{\varrho'_1}{\eta'_1} - \frac{\varrho'_2}{\eta'_2}\right)\vartheta'_1. \qquad (2.117)$$

The experimental data can be described well by this equation.

Dissolving a second electrolyte in a binary electrolyte solution in some cases makes the viscosity change in a complicated way. For example, Berecz and Horányi [87a] observed that, adding hydrogen chloride to LiCl solutions of various concentrations, the change of viscosity caused by HCl ($\Delta\eta/\Delta c_{\text{HCl}}$) depends on the LiCl concentration, as shown in Fig. 2.15 for low hydrogen chloride concentrations. The viscosity of dilute LiCl solutions hardly changes on the addition of small amounts of HCl. The effect of hydrogen chloride reaches a maximum at about 10 M concentration of LiCl, while a further increase in the LiCl concentration decreases this effect rapidly. It should be noted that the viscosity curve of the binary system LiCl—H$_2$O is also broken at this concentration.

On the basis of viscosity measurements with the systems CsCl–HCl–H$_2$O and MgCl$_2$–HCl–H$_2$O at various concentrations, Berecz and Bader [87b] studied the dependence of the viscosity of binary systems on the addition of a third component. The experimental results have led to conclusions regarding the isothermal structural changes occurring in solutions, as well as the variation of the dissociation of HCl in ternary systems.

The viscosity of non-electrolyte solutions is also altered by an electrolyte dissolved as a third component. For example, according to the measurements of Emmerich and Rosskopf [88], the viscosity of aqueous solutions of saccharose changes linearly with concentration, while the viscosity

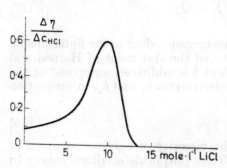

Fig. 2.15 Changes in viscosity of aqueous LiCl solution on the addition of HCl

* The volume fraction can be expressed by the equation $\vartheta'_1 = \dfrac{\varrho_2 - \varrho'}{\varrho_2 - \varrho_1}$.

increases by three orders of magnitude. The viscosity of sugar solutions is increased by LiCl and NaCl, on the other hand, it is decreased by CsCl and HCl. The viscosity of sugar solutions is altered by salts approximately to the same degree as in the case of pure water, although the viscosities themselves are very different. The order of the effects of salts corresponds to the order in their ionic radii, and it can be supposed that it is related to their hydration or solvation. According to this, the decrease in viscosity is due to the effect of the field of the ions destroying the water-saccharose complexes to a degree depending on the conditions, and, therefore, the mobility of the particles is increased. The increase in viscosity due to the strongly hydrated Li^+ and Na^+ ions is probably caused partly by the fact that the ratio of the water molecules in the solvate sphere of the ions is larger than that of the saccharose molecules, and so the solution behaves as if the concentration of the sugar has increased. On the other hand, the solvate complexes are also somewhat ordered around the ions, and these complexes have a relatively large size due to the presence of built in saccharose molecules and thus their mobility is low.

On the basis of the investigations on changes in the viscosity of $BaCl_2$ solutions caused by non-electrolytes (urea, glycine, glycerol, mannitol, saccharose), Lakshmanan and Rao [89] found that the viscosity of the solution increases rapidly with increasing ionic strength of the $BaCl_2$ solution in the presence of non-electrolytes, and the extent of this increase depends on the size of the non-electrolyte molecules. The activation energy of the viscous flow of the liquid is decreased by the non-electrolyte, which can be attributed to the decreasing effect of the non-electrolyte molecules on the average distance between ions; at the same time, they decrease the potential at the position of the ion produced by the neighbouring ions. The decrease in the forces between ions consequently facilitates viscous flow.

REFERENCES

to Chapter 2

[1a]   For a more detailed treatment see, e.g. R. B. BIRD, W. E. STEWART and E. N. LIGHTFOOT, Transport Phenomena (New York, 1960); R. M. MAZO, Statistical Mechanical Theories of Transport Processes (London, 1967).

[1b]   L. D. EICHER and B. J. ZWOLINSKI, J. Phys. Chem., 1971, **75**, 2016.

[1c]   J. H. HILDEBRAND, Science, 1971, **174**, 490; J. H. HILDEBRAND, J. M. PRAUSNITZ and R. L. SCOTT, Regular and Related Solutions, Chapter 3 (New York, 1970); J. H. HILDEBRAND and R. H. LAMOREAUX, Proc. Nat. Acad. Sci. U. S., 1972, **69**, 3248.

[1d]   J. H. DYMOND and B. J. ADLER, J. Chem. Phys. 1966, **45**, 2061.

[2]   J. E. LENNARD-JONES, Proc. Roy. Soc., 1924, **A. 106**, 441.

[3]   E. N. DA C. ANDRADE, Phil. Mag., 1934, (Ser. 7) **17**, 497, 698.

[4]   J. G. KIRKWOOD, F. P. BUFF, and M. S. GREEN, J. Chem. Phys., 1949, **17**, 988; R. ZWANZIG, J. Chem. Phys., 1961, **34**, 1931.

[5]   S. G. BRUSH, Chemical Rev., 1962, **62**, 513.

[6]   M. BORN and H. S. GREEN, Proc. Roy. Soc. (London), 1946, **A. 188**, 10; 1947 **A. 190**, 455; H. S. GREEN, The Molecular Theory of Fluids (Amsterdam, 1952)

[7]   J. G. KIRKWOOD, J. Chem. Phys., 1946, **14**, 180, 347; J. H. IRVING and J. G KIRKWOOD, J. Chem. Phys., 1950, **18**, 817; R. ZWANZIG, J. Chem. Phys., 1961,

34, 1931.
[8]   S. GLASSTONE, K. J. LAIDLER and H. EYRING, The Theory of Rate Processes,
        p. 477 (New York, 1941).
[9]   H. EYRING and J. HIRSCHFELDER, *J. Phys. Chem.*, 1937, **41,** 249; J. O. HIRSCH-
        FELDER, *J. Chem. Ed.*, 1940, **16,** 540.
[10]   J. F. KINCAID, H. EYRING and A. E. STEARN, *Chem. Rev.*, 1941, **28,** 301.
[11]   H. EYRING, *J. Chem. Phys.*, 1936, **4,** 283; R. H. EWELL and H. EYRING, *J. Chem.
        Phys.*, 1937, **5,** 726; J. HIRSCHFELDER, D. STEVENSON and H. EYRING, *J. Chem.
        Phys.*, 1937, **5,** 896; R. H. EWELL, *J. Appl. Phys.*, 1938, **9,** 252.
[12a]   As a review, see e.g. S. GLASSTONE, K. J. LAIDLER and H. EYRING, The Theory
        of Rate Processes, p. 153 (New York, 1941); ERDEY-GRÚZ T. and SCHAY G.,
        Elméleti fizikai kémia vol. II, p. 501 (Budapest, 1962).
[12b]   H. R. PRUPPACHER, *J. Chem. Phys.*, 1966, **45,** 2061.
[13]   S. ARRHENIUS, *Z. Phys. Chem.*, 1887, **1,** 285.
[14]   E. MCLAUGHLIN, *Trans. Faraday Soc.*, 1959, **55,** 29.
[15]   E. R. POWELL, W. E. ROSEVEARE and H. EYRING, *Ind. Eng. Chem.*, 1941,
        33, 430.
[16]   R. H. EWELL and H. EYRING, *J. Chem. Phys.*, 1937, **5,** 726.
[17]   F. EIRICH and R. SIMHA, *J. Chem. Phys.*, 1939, **7,** 116.
[18]   K. H. DUDZIAK and E. U. FRANCK, *Ber. Bunsenges.*, 1966, **70,** 1120.
[19]   R. A. HORNE and D. S. JOHNSON, *J. Phys. Chem.*, 1966, **70,** 2182.
[20]   N. A. NEVOLINA, O. YA. SAMOILOV and A. L. SEIFER, *Zh. Strukt. Khim.*,
        1969, **10,** 203.
[21]   M. AGENO and C. FRONTALL, *Bull. Soc. Ital. Phys.*, 1966, **50,** 19.
[22]   M. AGENO and C. FRONTALL, *Proc. Nat. Acad. Sci. USA*, 1967, **57,** 856.
[23]   B. V. DERYAGIN, N. V. CHURAEV, N. N. FEDYAKIN, M. V. TALAEV and I. G.
        ERSHOVA, *IZV. AN SSSR, Ser. Khim.*, 1967, **10,** 2178.
[24a]   M. P. GINGOLD, *Nature Phys. Sci.*, 1972, **75,** 235.
[24b]   B. V. DERYAGIN and N. V. CHURAEV, *Nature*, 1973, **244,** 431.
[25]   A. J. BATSCHINSKII, *Z. Phys. Chem.*, 1913, **84,** 643.
[26]   S. GLASSTONE, K. J. LAIDLER and H. EYRING, The Theory of Rate Processes,
        p. 488 (New York, 1941), cf. T. M. REED and T. E. TAYLOR, *J. Phys. Chem.*,
        1959, **63,** 58.
[27]   R. A. HORNE, R. A. COURANT and D. S. JOHNSON, *Electrochim. Acta*, 1966,
        11, 987.
[28a]   L. KORSON, W. DROST-HANSEN and F. J. MILLERO, *J. Phys. Chem.*, 1969,
        73, 34.
[28b]   E. M. STANLEY and R. C. BATTEN, *J. Phys. Chem.*, 1969, **73,** 1187.
[28c]   A. I. KUDISH, D. WOLF and F. STECKEL, Faraday Transact. I. **1972,** 2041.
[29]   D. B. DAVIES and A. J. MATHESON, *J. Chem. Phys.*, 1966, **45,** 1000; *Trans.
        Faraday Soc.*, 1967, **63,** 596.
[30]   A. R. DEXTER and A. J. MATHESON, *Trans. Faraday Soc.*, 1968, **64,** 2632.
[31]   As a review, see e.g. S. M. KARIM and L. ROSENHEAD, *Rev. Modern Phys.*,
        1952, **24,** 108; L. N. LIEBERMANN, *Phys. Rev.*, 1949, **75,** 1415.
[32]   I. Z. FISCHER and A. M. ZAITSEVA, *Dokl. AN SSSR*, 1964, **154,** 1175; *Zh.
        Strukt. Khim.*, 1963, **4,** 261.
[33]   J. M. M. PINKERTON, *Nature*, 1947, **160,** 128; *Proc. Phys. Soc.*, (B) I., 1949,
        62, 129.
[34]   K. G. AKHMETSIANOV and I. C. PETREA, *Revue Roumaine de Chimie*, 1965,
        10, 939.
[35]   J. KENDALL and K. P. MONROE, *J. Amer. Chem. Soc.*, 1917, **39,** 1787.
[36]   YU. YA. FIALKOV, *Zh. Fiz. Khim.*, 1963, **37,** 1745, 1938, 2139.
[37]   R. J. FORT and W. R. MOORE, *Trans. Faraday Soc.*, 1966, **62,** 1112.
[38]   T. ERDEY-GRÚZ, E. KUGLER and A. REICH, *Acta Chim. Acad. Sci. Hung.*,
        1958, **13,** 429.
[39]   T. ERDEY-GRÚZ, E. KUGLER and J. HIDVÉGI, *Acta Chim. Acad. Sci. Hung.*,
        1959, **19,** 89; T. ERDEY-GRÚZ, E. KUGLER and L. MAJTHÉNYI, *Electrochim.
        Acta*, 1968, **13,** 947.
[40]   T. ERDEY-GRÚZ, E. KUGLER and J. HIDVÉGI, *Acta Chim. Acad. Sci. Hung.*,
        1959, **19,** 363.

[41a] T. Erdey-Grúz and E. Kugler, *Acta Chim. Acad. Sci. Hung.*, 1968, **57**, 301.
[41b] A. G. Mitchell and W. F. K. Wynne-Jones, *Disc. Faraday Soc.*, 1953, **15**, 16ᴸ.
[41c] V. A. Mikhailov, *Zh. Strukt. Khim.*, 1961, **2**, 677.
[42] E. Grüneisen, *Wiss. Abhandl. phys.-techn. Reichsanstalt*, 1909, **4**, 239.
[43] G. Jones and M. Dole, *J. Amer. Chem. Soc.*, 1929, **51**, 2950; G. Jones and S. K. Talley, *J. Amer. Chem. Soc.*, 1933, **55**, 624, 4124.
[44a] R. W. Gurney, Ionic Processes in Solutions p. 163, New York, 1953.
[44b] L. Werblan, A. Rotowska and S. Minc, *Electrochim. Acta*, 1971, **16**, 41.
[45] L. Andrussow, *Z. Elektrochem.*, 1958, **62**, 608.
[46] M. Kaminsky, *Z. Elektrochem.*, 1960, **64**, 867; L. Andrussow, *Z. Elektrochem.* 1960, **64**, 1244.
[47] I. Z. Fisher and A. M. Zaitseva, *Dokl. AN SSSR*, 1964, **154**, 1175.
[48] M. Kaminsky, *Z. phys. Chem. NF.*, 1955, **5**, 154; 1956, **8**, 173; 1957, **12**, 206; *Disc. Faraday Soc.*, 1957, **24**, 171.
[49] W. M. Cox and J. H. Wolfenden, *Proc. Roy. Soc.*, 1934, **A. 145**, 475.
[50] R. W. Gurney, Ionic Processes in Solutions, p. 168 (London, 1953).
[51a] E. R. Nightingale, *J. Phys. Chem.*, 1959, **63**, 742, 1381; 1962, **66**, 894.
[51b] H. Uedaira, *Zh. Fiz. Khim.*, 1971, **45**, 2550.
[51c] C. M. Criss and M. J. Mastroianni, *J. Phys. Chem.*, 1971, **75**, 2532.
[51d] B. S. Krumgalz, *Zh. Fiz. Khim.*, 1973, **47**, 1691.
[52a] S. Lengyel, J. Tamás, J. Giber and J. Holderith, *Acta Chim. Acad. Sci. Hung.*, 1964, **40**, 125.
[52b] S Lengyel, *J. de chimie physique*, 1969, 28.
[53] B. N. Finkelstein, *Phys. Z.*, 1930, **31**, 130, 165.
[54] A. Eucken, *Göttinger Nachr. Math. Phys. Kl.* **1946**, 38; **1947**, 20.
[55a] H. S. Frank and M. W. Evans, *J. Chem. Phys.*, 1945, **13**, 507.
[55b] A. G. Ostroff, B. S. Snowden and D. E. Woessner, *J. Phys. Chem.*, 1969, **73**, 2784.
[55c] J. C. MacDonald, *Electrochim. Acta*, 1972, **17**, 1965.
[55d] R. A. Horne and R. P. Young, *Electrochim. Acta*, 1972, **17**, 763.
[56] R. H. Stokes and R. Mills, Viscosity of Electrolytes and Related Properties, p. 59 (Oxford, 1965), (Oxford, 1959).
[57] H. S. Frank: *Proc. Roy. Soc.*, 1958, **A. 247**, 481; H. S. Frank and W. Y. Wen *Disc. Faraday Soc.*, 1957, **24**, 133.
[58] E. Asmus, *Z. Naturf.*, 1949, **4. A.** 589.
[59] D. T. Burns, *Electrochim. Acta*, 1965, **10**, 985.
[60] A. Einstein, *Ann. Phys.*, 1906, **19**, 289; 1911, **34**, 591.
[61] R. Simha, *J. Phys. Chem.*, 1940, **44**, 25; *J. Appl. Phys.*, 1942, **13**, 147.
[62a] V. Vand, *J. Phys. Chem.*, 1948, **52**, 277.
[62b] D. Eagland and G. Pilling, *J. Phys. Chem.*, 1972, **76**, 1902.
[63a] P. Y. Cheng and H. K. Schachmann, *J. Polym. Sci.*, 1955, **16**, 19.
[63b] T. T. Herskovits and T. M. Kelly, *J. Phys. Chem.*, 1973, **77**, 381.
[64] R. H. Stokes and R. Mills, Viscosity of Electrolytes and Related Properties, p. 48 (Oxford, 1965).
[65] S. P. Moulik, *J. Phys. Chem.*, 1968, **72**, 4682.
[66] E. Hückel and H. Schaaf, *Z. Phys. Chem. NF.*, 1959, **21**, 326.
[67] E. R. Nightingale and J. F. Kuecker, *J. Phys. Chem.*, 1965, **69**, 2197.
[68] R. L. Kay, T. Vituccio, C. Zawoyski and D. F. Evans, *J. Phys. Chem.*, 1966, **70**, 2336.
[69a] T. Isono and R. Tamamushi, *Electrochim. Acta*, 1967, **12**, 1479.
[69b] B. R. Breslau and I. F. Miller, *J. Phys. Chem.*, 1970, **74**, 1056.
[70] P. Debye and E. Hückel, *Physik. Z.*, 1923, **24**, 185, 305; L. Onsager, *Physik. Z.*, 1926, **27**, 388; 1927, **28**, 277; P. Debye and H. Falkenhagen, *Physik. Z.*, 1928, **29**, 121; As a review, see e.g. H. Falkenhagen, Elektrolyte p. 197 (Leipzig, 1953); R. H. Stokes and R. Mills, Viscosity of Electrolytes and Related Properties p. 22 (Oxford, 1965).
[71a] H. Falkenhagen and M. Dole *Physik. Z.*, 1929, **30**, 611; H. Falkenhagen, *Physik. Z.*, 1931, **32**, 745; H. Falkenhagen and E. L. Vernon, *Physik. Z.*, 1932, **33**, 140.

[71b] I. V. KUDRYAVTSEVA, B. S. KRUMGALZ, K. P. MISHCHENKO, *Zh. Strukt. Khim.*, 1972, **13**, 217.
[72]   A. DEUBNER, *Physik. Z.*, 1940, **41**, 110.
[73]   E. ASMUS, *Ann. Phys.*, 1939, **35**, 1; *Z. Phys. Chem.*, 1940, **B. 47**, 357, 365.
[74]   L. ONSAGER and R. M. FUOSS, *J. Phys. Chem.*, 1932, **36**, 2689.
[75]   C. W. DAVIES and V. E. MALPASS, *Trans. Faraday Soc.*, 1964, **60**, 2075, 2078.
[76a] For detailed treatment see, R. H. STOKES and R. MILLS, Viscosity of Electrolytes and Related Properties, p. 63 (Oxford, 1965).
[76b] C. V. SURYANARAYANA and V. K. VENKATESAN, *Monatsh. für Chem.*, 1958, **89**, 824.
[77]   T. SATOH, *J. Phys. Soc. Japan.*, 1960, **15**, 1134.
[78]   M. L. MILLER and M. DORAN, *J. Phys. Chem.*, 1956, **60**, 186; R. GOPAL, *J. Indian Chem. Soc.*, 1953, **30**, 708.
[79a] E. R. NIGHTINGALE and R. F. BENCK, *J. Phys. Chem.*, 1959, **63**, 1777.
[79b] G. EBERT and J. WENDORFF, *Ber. Bunsenges.*, 1971, **75**, 82.
[80]   C. V. SURYANARAYANA and V. K. VENKATESAN, *Nature*, 1956, **178**, 1461; *Acta Chim. Acad. Sci. Hung.*, 1958, **16**, 149, 339, 345; *Monatsh. für Chemie*, 1959, **90**, 36.
[81a] C. V. SURYANARAYANA and S. GOVINDASWAMY, *Acta Chim. Acad. Sci. Hung.*, 1962, **31**, 373.
[81b] C. A. ANGELL and R. D. BRESSEL, *J. Phys. Chem.*, 1972, **76**, 3244.
[82]   R. A. HORNE and D. S. JOHNSON, *J. Phys. Chem.*, 1967, **71**, 1147.
[83]   V. S. GALINKER, V. A. TYAGAI and G. N. FENERLI, *Zh. Fiz. Khim.*, 1962, **36**, 2638.
[84]   Y. C. WU, *J. Phys. Chem.*, 1968, **72**, 2663.
[85]   E. BERECZ and G. VÉRTES, *Acta Chim. Acad. Sci. Hung.*, 1963, **39**, 437.
[86]   A. B. ZDANOVSKII, *Zh. Fiz. Khim.*, 1955, **29**, 209.
[87a] E. BERECZ and GY. HORÁNYI, *Acta Chim. Acad. Sci. Hung.*, 1961, **29**, 157.
[87b] E. BERECZ and I. BADER, *Acta Chim. Hung. Acad. Sci.*, 1972, **74**, 213.
[88]   A. EMMERICH and F. ROSSKOPF, *Z. Elektrochem.*, 1958, **62**, 1115.
[89]   G. LAKSHMANAN and K. NAGARAJA RAO, *Electrochim. Acta.*, 1969, **14**, 1173.

3. DIFFUSION

3.1 DIFFUSION OF NON-ELECTROLYTES

Diffusion is the transport of a chemical substance in a material system consisting of two or more components, from places where its concentration is higher in the given phase towards those of lower concentration or–in non-ideal mixtures–of lower activity. The driving force of diffusion is the difference in the chemical potential of the diffusing substance which has, in general, the same sign as the difference in its concentrations, in the same phase and at uniform and constant temperature all over the system.*

Rigorously speaking, in a system of different local concentrations (an inhomogeneous but not heterogeneous system), all components participate in the diffusion process; the rates of the transport of the individual components, however, are in general different, which should be accounted for in describing diffusion [1].

Under real conditions, substances usually diffuse in all directions. However, the essential features of the correlations concerning mass transport due to concentration differences can be revealed satisfactorily by focusing our interest on mass transport taking place in one direction (one-dimensional or *linear diffusion*). In this way, the treatment of the phenomena becomes simpler without loss in understanding, even less so since, in the apparatuses designed for experimental determination of the rate of diffusion, one-dimensional change of concentration is aimed at and the concentration is kept constant in the directions perpendicular to this.

3.1.1 FUNDAMENTAL RELATIONSHIPS OF DIFFUSION

For a description of mass flow due to diffusion, one can start from the basic laws of transport processes, according to which the flux in direction y is proportional, within certain limits, to the driving force. For the linear diffusion of any ith component present in the mixture (solution), we have:

$$J_i = - D_i \frac{\partial c_i}{\partial y} \qquad (3.1.1)$$

---

* The diffusion defined above is usually quoted as *common* or *isothermal diffusion*, in order to distinguish this from *thermal diffusion* which is due to a difference in temperature, or *pressional diffusion* (actually, flow) due to a difference in pression, or from *forced diffusion* caused by some other external force. Besides common diffusion, this book deals only with forced diffusion due to an electric field, i.e. with ionic migration (conductivity).

Literature on page 247

('*Fick's law*'). Here J_i is the *mass flux (mass flow)*, i.e. the amount of the material (in moles, grams, or other units), diffusing across a unit cross-section perpendicular to direction y of the current in unit time. y is taken to be positive in the direction of the mass flow. $\partial c_i/\partial y$ is the *concentration gradient* in the direction of the flow, i.e. the change in concentration per unit distance. In general, the concentration c_i should be expressed in the same unit of quantity as the flux (in moles, grams, etc.) and the unit volume applied should be the cube of the unit of y (if, for example, y is measured in cm and the concentration unit is mole $\times$ cm^{-3}, the unit of flux is mole $\times$ cm$^{-2} \times$ $\times$ s^{-1}). D_i is the *diffusion coefficient* * of the ith component; its value is positive and its dimension is (length)^2/time; in CGS units it is [cm^2 s^{-1}]. The negative sign indicates that the material is transported in the direction of decreasing concentration. If the concentration and flux, as given above, are expressed in identical units, the numerical value of D_i will be independent of the unit of mass used.

The diffusion flow can also be described by the mean velocity of the diffusing entity ($\bar{v}_i$).** If the whole liquid is at rest macroscopically, the flux of the ith component is:

$$J_i = c_i \bar{v}_i.$$ (3.1.2)

Substituting this relation into equation (3.1.1), the rate of diffusion will be:

$$\bar{v}_i = -D_i \frac{1}{c_i} \frac{\partial c_i}{\partial y} = -D_i \frac{\partial \ln c_i}{\partial y}.$$ (3.1.3)

Diffusion can also be expressed in terms of the mole fractions of the components. The mole fraction of a particular component is:

$$x_i = \frac{c_i}{\Sigma c_i},$$ (3.1.4)

where c_i is the number of moles of the ith component in unit volume.

If diffusion is not accompanied by a significant change in volume, and a dilute solution is considered in which diffusion does not cause any appreciable change in the total number of moles, the molar mass flux is:

$$J_i = -D_i \Sigma c_i \frac{\partial x_i}{\partial y}.$$ (3.1.5)

Thus, each of the components in the mixture diffuses in the direction of its decreasing mole fraction, similarly to the case of viscous flow, where impulse is transported towards decreasing 'concentration of impulse', and

---

* Earlier, D_i was called *diffusion constant*, however, as it depends not only on the nature of the material and on the temperature, but also on the concentration, the term diffusion coefficient is more correct.

** $\bar{v}_i$ is not the velocity of the individual molecules or other entities of the ith material, but it can be interpreted as the sum of the rates of transport of all molecules of component i present in a given small volume, divided by the number of these molecules. $\bar{v}_i$ is, thus, the mean velocity of the ith kind of particles.

to the case of heat conduction, where energy is carried in the direction of decreasing 'concentration of energy'.

In solutions containing one single solute, the flux of the solute with respect to the solvent supposed to be at rest (see later) can be expressed (counting with molality (m_2) instead of mole fraction):

$$J_2 = -\frac{c_1 M_1}{1000} D \frac{\partial m_2}{\partial y} \qquad (3.1.6)$$

(M_1 and c_1 are the molecular weight and concentration of the solvent respectively).

The diffusion coefficient generally depends not only on the nature of the system and on the temperature, but, to a small extent, on the concentration as well. In several cases, it is just the investigation of the concentration dependence of D_i that gives information on the mechanism of the process. In order to obtain easily interpretable results, however, the conditions of the experiments should possibly be chosen so as to get a D_i value approximately constant during the measurement.

The fundamental equation of diffusion, equation (3.1.1), can also be given in a vectorial form independent of the actual coordinate system:

$$J_i = -D_i \operatorname{grad} c_i, \qquad (3.1.7)$$

implying that the vector of the diffusional mass flux is of opposite direction with respect to the concentration gradient, while being proportional to it within certain limits.

If the ith component diffuses in all directions of space and not only in one direction, the concentration gradient expressed in Cartesian coordinates (x, y, z), is:

$$\operatorname{grad} c_i = \frac{\partial c_i}{\partial x} + \frac{\partial c_i}{\partial y} + \frac{\partial c_i}{\partial z}. \qquad (3.1.8)$$

Diffusion is usually accompanied by a change in the density of the solution due to the change in concentration; this often results in flow of the whole liquid, i.e. it gives rise to *convection*. This convection accelerates the mass transport in addition to diffusion. The treatment of convective diffusion is rather complicated, and it will not be discussed here, since it would not help appreciably in understanding the mechanism of mass transport. (See for example Neumann's work [2] on convective diffusion between coaxial cylinders.)

In order to make the fundamental equation of diffusion, equation (3.1.1), —which also defines the diffusion coefficient—unequivocal, one has to elucidate what the mass flux is referred to, i.e. the location of that unit cross-section should be determined at which the amount of the transported material is to be regarded as the flux. In the case of the basic equation of heat and electric conductance formally similar to that of diffusion, as well as for the diffusion in non-swelling solids, a reference plane with unchanged

Literature on page 247

position during the process can be determined easily and unequivocally. For diffusion in liquid mixtures and solutions, however, this is not so simple. Namely, in liquids the change in concentration due to diffusion results in changes in density and volume. Consequently, a plane of constant position with respect to the coordinate system fixed to the vessel containing the liquid (that could be regarded as the starting plane of diffusion), usually cannot be regarded as a reference plane characterized by a constant physical parameter. The reference plane for the mass flux may be defined in different ways. These possibilities essentially correspond to various definitions of the diffusion coefficient, thus the numerical values of the diffusion coefficient calculated for different reference systems are also different, although the differences between them are usually small.

In detailed investigations of the conditions of diffusion it should be taken into account that rigorously, in mixtures and solutions, all the components are diffusing simultaneously. In fact, it is only in ideal solutions extrapolated to infinite dilution that diffusion can be correctly defined as the transport of the solute in a solvent at rest. Even in solutions (not only in concentrated mixtures) the transport of the solute in a given direction is inseparably accompanied by the transport of the solvent in the opposite direction, since there is also a difference in the chemical potentials of the solvent in the solution portions of different concentrations.

Limiting the discussion to two-component solutions, linear diffusion should be described by two simultaneously valid equations, regarding the flow of the two components in opposite directions:

$$J_1 = -D_1 \frac{\partial c_1}{\partial y}$$

and   (3.1.9)

$$J_2 = -D_2 \frac{\partial c_2}{\partial y},$$

where the subscripts 1 and 2 in the flux, concentration, and diffusion coefficient refer to the first and second component, respectively.

The discussion of diffusion is simplest when the reference plane is defined so as to make the two diffusion coefficients identical with each other. This can be achieved by referring the diffusion flow to that plane where the changes in volume due to the two-directional mass flow crossing the plane compensate each other, i.e. across this plane there is no 'transport of volume'. If the partial molar volumes of the two components are V_1 and V_2 and J_1^V and J_2^V are the fluxes for the volumes fixed in the above sense, we have:

$$J_1^V V_1 + J_1^V V_2 = 0.$$ (3.1.10)

Introducing the fluxes from equation (3.1.9), and denoting the diffusion coefficients related to this reference plane by D_1^V and D_2^V, we obtain:

$$D_1^V V_1 \frac{\partial c_1}{\partial y} + D_2^V V_2 \frac{\partial c_2}{\partial y} = 0.$$ (3.1.11)

Since, in accordance with the definition of the molar volume:

$$V_1 c_1 + V_2 c_2 = 1, \qquad (3.1.12)$$

and

$$c_1 dV_1 + c_2 dV_2 = 0,$$

it follows that:

$$V_1 dc_1 + V_2 dc_2 = 0. \qquad (3.1.13)$$

Taking this into account, it follows from equation (3.1.11), that the two diffusion coefficients must be equal:

$$D_1^V = D_2^V = D_{12}^V \qquad (3.1.14)$$

where the common diffusion coefficient *(interdiffusion coefficient)* is denoted by D_{12}^V. This is often practically equal to the diffusion coefficient determined experimentally by performing a diffusion process in which two dilute solutions of different concentration diffuse toward each other through an initially sharp horizontal boundary.

The reference plane of the diffusion flow can be fixed in another way. For dilute solutions, the reference plane can be suitably fixed to the solvent, since the diffusion of the solvent in the direction opposite to that of the solute is negligible. The reference plane can be chosen, for example, so as to obtain zero net mass transport through it. It can be shown that under such conditions the mass flux given in g $m^{-2} s^{-1}$ units ($J_1 + J_2 = 0$), is:

$$J_1 = - D_1 \varrho \frac{\partial w_1}{\partial y},$$

and

$$\qquad (3.1.15)$$

$$J_2 = - D_2 \varrho \frac{\partial w_2}{\partial y},$$

where ϱ is the density of the mixture, and w_1 and w_2 are the mole fractions of the two components. The two diffusion coefficients defined in this way are also equal to each other:

$$D_1 = D_2 = D_{12}. \qquad (3.1.16)$$

The relationship between the diffusion coefficients defined with respect to different reference planes has been discussed in detail by Hartley and Crank [3], and for non-isothermal conditions by Alexander [4].

Diffusion against concentration gradient. The 'driving force' of diffusion is the chemical potential gradient. However, the changes in chemical potential are determined unequivocally by the changes in concentration in ideal mixtures only:

$$d\mu_i = RT \, d \ln c_i.$$

Literature on page 247

In real mixtures, activity a_i should be applied:

$$d\mu_i = RT \ln a_i = RT \ln \gamma_i c_i, \tag{3.1.17}$$

where γ_i is the activity coefficient of the ith component at a given point in the mixture. Since in real mixtures the activity coefficient depends on the nature and concentration of all of the components, in multicomponent systems it can occur that the direction of the activity gradient of a component does not coincide with that of its concentration gradient. This component diffuses from places of lower concentration towards those of a higher one ('upwards' along the concentration gradient).

This often occurs in the course of diffusion through an interface of different phases in heterogeneous systems, since the requirement of the equilibrium between different phases containing a given material is its identical activity in them. However, from experiments on the partition ratio, it is well known that this is rarely accompanied by the identity of concentrations. Thus, solutes often diffuse through interfaces toward points of higher concentration (e.g. at the interface of chloroform and water, iodine diffuse into the chloroform solution of higher concentration).

Apparently, there is an essential difference between diffusion and heat conduction in this respect, although they are rather similar in phenomenological sense. Thermal energy is always transported in the direction of decreasing temperature, and this applies to its transport through interfaces too (with a discontinuity only in the temperature gradient). In each case the requirement of the equilibrium in heat conduction is the identical temperature at every point of the system. As for diffusion, however, the concentrations are usually different at the two sides of the interface even after reaching the equilibrium state. The requirement of the diffusional equilibrium is the identity of chemical potentials at every point of the system, and widely different concentrations can correspond to identical chemical potentials in the different phases. In diffusion, chemical potential and not concentration is the analogue of temperature in heat conduction. The reason why concentration is still taken into consideration is that there is no instrument which directly measures the chemical potential or the absolute value of activity.*

Development of the steady state and the change of concentration in time. In electric conduction, a steady state is generally considered. A constant potential gradient can easily be realized (by applying potential difference between two electrodes located at a given distance), and the electric current reaches a steady state within a fraction of a second. Thus, the rate of development of the steady state would be studied in exceptional cases only.

---

* The description of heat conduction would be more closely similar to that of diffusion if the heat flow were described not by the temperature gradient $\left(-\lambda \dfrac{\partial T}{\partial y} \right)$, but by the gradient $\left(-\lambda \dfrac{\partial h}{\partial y} \right)$ of specific enthalpy h. After reaching the isothermal state, the specific enthalpies are usually not identical at the two sides of the interface.

In diffusion or heat conduction, however, it is often important to study how the phenomenon takes place in time, since the gradient is often not constant and so no steady state can be attained; even in cases when a steady state exists, its development takes a long time.

The description of changes in concentration with time at a given point of the system due to diffusion can be based on equation (3.1.1), taking into account that the concentration gradient itself also changes with the position coordinate. If flux $(J_i)_y$ corresponds to a concentration gradient $\left(\dfrac{\partial c_i}{\partial y}\right)_y$ at a given point y, the flux will be $(J_i)_y + \left(\dfrac{\partial J_i}{\partial y}\right) \mathrm{d}y$ at point $y + \mathrm{d}y$. In a layer of thickness $\mathrm{d}y$ perpendicular to the flux and representing 1 cm^2 surface area, the amount of material accumulated during a time interval $\mathrm{d}t$ is:

$$\left\{(J_i)_y - \left((J_i)_y + \frac{\partial J_i}{\partial y}\,\mathrm{d}t\right)\right\}\mathrm{d}t = -D_i\,\frac{\partial c_i}{\partial y}\,\mathrm{d}t - \left\{D_i\left(\frac{\partial c_i}{\partial y} + \frac{\partial^2 c_i}{\partial y^2}\,\mathrm{d}y\right)\right\}\mathrm{d}t. \quad (3.1.18)$$

The left-hand side of this equation corresponds to the change in the amount of substance i in a layer of volume $\mathrm{d}y$ cm^3 during time $\mathrm{d}t$, and the change in its concentration ∂c_i is obtained on dividing by ∂y. Thus, the rate of change in concentration at point y is:

$$\left(\frac{\partial c_i}{\partial t}\right)_y = D_i\left(\frac{\partial^2 c_i}{\partial y^2}\right)_t, \quad (3.1.19)$$

if the diffusion coefficient is independent of concentration. This is the so-called *Fick's second law**, and its integration with respect to boundary conditions determined by the experimental conditions leads to the diffusion coefficient based on directly measurable data [5].

When diffusion is not one-dimensional, but takes place in three dimensions, the change in concentration with time at a given point is (omitting the obvious subscripts of the partial differential quotients):

$$\frac{\partial c_i}{\partial t} = D_i\left(\frac{\partial^2 c_i}{\partial x^2} + \frac{\partial^2 c_i}{\partial y^2} + \frac{\partial^2 c_i}{\partial z^2}\right) = D\varDelta^2 c_i = D\ \mathrm{div\ grad}\ c_i, \quad (3.1.20)$$

if the diffusion coefficient is independent of concentration and $\varDelta^2$ represents the Laplace operator.

---

* It was this equation which Fick considered the law of diffusion originally, while equation (3.1.1) was regarded as the starting point required to obtain it. The statement that in a given system and at constant temperature the diffusion coefficient is independent of concentration can also be regarded as the Fick law, but this is valid only approximately.

Literature on page 247

The diffusion relation can sometimes be more suitably expressed by means of polar coordinates instead of Cartesian ones. In the special case of spherically symmetrical diffusion it is:

$$\frac{\partial c}{\partial t} = D \left(\frac{\partial^2 c}{\partial r^2} + \frac{2}{r} \frac{\partial c}{\partial r} \right), \tag{3.1.21}$$

where r denotes the distance measured from the centre of the sphere. For cylindrically symmetrical diffusion, we have:

$$\frac{\partial c}{\partial t} = D \left(\frac{\partial^2 c}{\partial r^2} + \frac{1}{r} \frac{\partial c}{\partial r} \right), \tag{3.1.22}$$

where r measures the distance from the axis of the cylinder.

Dependence of the diffusion coefficient on concentration. The assumption introduced above, namely that the diffusion coefficient is independent of concentration, is seldom correct rigorously (e.g. in ideal solutions and in self-diffusion). In real systems the diffusion coefficient usually depends on the concentration of the diffusing substance, and thus, on the position too. In several cases (e.g. in electrolyte solutions), just the dependence of the diffusion coefficient on concentration can be utilized to get some information on the mechanism of the process.

If the diffusion coefficient depends on concentration, it should be taken into account in equation (3.1.19) that D_i also depends on y. For one-dimensional diffusion:

$$\frac{\partial c_i}{\partial t} = \frac{\partial}{\partial y} \left(D_i \frac{\partial c_i}{\partial y} \right). \tag{3.1.23}$$

If the diffusion coefficient does not depend explicitly on the position coordinate, this equation can be transformed into the following one:

$$\frac{\partial c_i}{\partial t} = D_i \frac{\partial^2 c_i}{\partial y^2} + \frac{\partial D_i}{\partial c_i} \left(\frac{\partial c_i}{\partial y} \right)^2. \tag{3.1.24}$$

For three-dimensional diffusion:

$$\frac{\partial c_i}{\partial t} = \mathrm{div}\ (D_i\ \mathrm{grad}\ c_i). \tag{3.1.25}$$

In experimental investigations, it is difficult to consider the changes in the diffusion coefficient with concentration. It is much simpler to measure the rate of diffusion produced by such a small concentration difference that the change of the diffusion coefficient should be negligible in the concentration range involved. The *average diffusion coefficient* calculated in this way is approximately equal to the *differential diffusion coefficient* corresponding to average concentration $\frac{1}{2} (c_i' + c_i'')$. By performing measurements in solutions of various concentrations, the dependence of the diffusion coefficient on concentration can be established.

The actual calculations on diffusion are rendered difficult by the dependence of the diffusion coefficient on concentration, particularly if non-linear diffusion is discussed. The equations of non-linear diffusion in systems with concentration-dependent diffusion coefficients have been deduced by Rozenstok [6].

* * *

The laws of diffusion discussed above are valid, if the mass transport is due to the concentration gradient only. It often happens, however, that there is a temperature gradient or an electric potential gradient in the solution, in addition to the concentration gradient, which contributes to the mass transport too. Under such conditions, the cross-effects can also become important. For example diffusion of dissolved electrolytes–owing to the different mobilities of ions–results in an electric potential difference in the solution which affects the rate of diffusion. For the details of electrolyte diffusion, important in both theoretical and practical aspects, see Section 3.2.

3.1.2 FUNDAMENTALS OF THE THEORY OF DIFFUSION IN LIQUIDS

Since there is no satisfactory theory of the structure of liquids, there is no theory of diffusion in liquids describing the phenomena of this transport process precisely. Nevertheless, the approximate theories outlined so far on diffusion are suitable for providing valuable information on several details of this process and on some aspects of its mechanism, furthermore they give some correlations between the diffusion coefficient and other properties of a given system.

The theories on diffusion can, by and large, be classified into three groups: hydrodynamic theories; the theories based on the kinetic theory of liquids, and those based on the thermodynamics of irreversible processes.

3.1.2.1 HYDRODYNAMIC THEORY OF DIFFUSION

The hydrodynamic theory of diffusion [5] considers the liquid as a continuum in which the diffusion flow of the components is determined by a balance between the driving force and the frictional resistance force. In the theory of Sutherland [7] and Einstein [8] the gradient of osmotic pressure was regarded as the driving force, while in the up-to-date theories the gradient of chemical potential is taken to be responsible for this. If substance 2 gives

Literature on page 247

an ideal mixture on dissolving in liquid 1, the gradient of the chemical potential of the solute ($\mu_2 = \mu_2^0 + RT \ln c_2$) in direction y, i.e. the measure of the driving force of diffusion, is:

$$\frac{\partial \mu_2}{\partial y} = RT \frac{\partial \ln c_2}{\partial y} = \frac{RT}{c_2} \frac{\partial c_2}{\partial y}. \tag{3.1.26}$$

If f_s is the *frictional resistance* of the solute per molecule (i.e. Nf_s per mole), the average velocity of one-dimensional diffusion is*:

$$\bar{v}_2 = -\frac{RT}{Nf_s} \frac{\partial \ln c_2}{\partial y} = -\frac{kT}{f_s} \frac{\partial \ln c_2}{\partial y}. \tag{3.1.27}$$

According to hydrodynamic considerations, and taking into account equation (3.1.3), the diffusion coefficient is:

$$D_{12} = \frac{kT}{f_s}. \tag{3.1.28}$$

In physical chemistry, *mobility* is often used instead of frictional resistance, which is the reciprocal value of the former. If U_2 is the mobility of the solute particles, i.e. their velocity gained as a result of unit force, the diffusion coefficient is:

$$D_{12} = kTU_2. \tag{3.1.29}$$

This is the so-called *Nernst–Einstein equation.*

[The most difficult and still not in every respect reliably solved problem of the hydrodynamic theory of diffusion is the calculation of the frictional resistance from other data on solutions. This deficiency can mainly be attributed to our inadequate knowledge of the liquid structure. Thus, in the calculation of the diffusion coefficient from other properties of a solution, the assumptions applied regarding the frictional resistance are approximate and of limited validity]

An early assumption, that can still be applied in several cases, was suggested by Stokes (1850); he calculated the frictional resistance acting on a spherical body moving in a continuum on the basis of classical hydrodynamics. This frictional resistance depends on radius r_2 of the sphere, on viscosity η_1 of the solvent, and on the sliding frictional coefficient β between the sphere and the liquid. According to Stokes, the hydrodynamic frictional resistance is:

$$f_s = 6\pi\eta_1 r_2 \frac{1 + \dfrac{2\eta_1}{\beta r_2}}{1 + \dfrac{3\eta_1}{\beta r_2}}. \tag{3.1.30}$$

* Frictional resistance f_s represents the force necessary to make the particles move with unit velocity.

Sliding friction thus varies between two limiting cases. In one of them, the liquid entirely wets the sphere moving in it, and $\beta = \infty$, so the frictional resistance is:

$$f_s = 6\pi\eta_1 r_2 . \quad \text{— large molecule} \tag{3.1.31}$$

If, however, the liquid does not wet the sphere at all–the other limiting case–, i.e. it does not adhere to it, $\beta = 0$ and the frictional resistance is:

$$f_s = 4\pi\eta_1 r_2 . \quad \text{— small molecule} \tag{3.1.32}$$

These hydrodynamic relationships are, in principle, only of very limited validity for molecules diffusing in a solution, since the liquid is not a continuum but consists of discrete molecules with sizes approximately of the same order of magnitude as those of the diffusing molecules. Furthermore, the molecules are not spherical (except monoatomic ones); even the most symmetrical molecules can be considered spheres only in approximation. The extent of sliding friction is also uncertain. It can still be established, in general, that the approximation given by the Stokes' frictional law for diffusing molecules improves with increasing size of the solute molecules as compared with those of the solvent and with their increasingly centrosymmetrical character.

For the diffusion of large spherically symmetrical molecules in a solvent with small molecules, Sutherland [7] supposed that sliding hardly occurs in a hydrodynamic sense, i.e. practically $\beta = \infty$. Under such conditions, equation (3.1.31) holds for the frictional resistance accompanying diffusion, and the diffusion coefficient is:

$$D_{12} = \frac{kT}{6\pi\eta_1 r_2} \tag{3.1.33}$$

(Stokes–Einstein equation). In fact, this relation is often valid in a rather good approximation even for solutions in which this would not be expected considering the ratio of the sizes of the molecules.

Recently, Loflin and McLaughlin pointed out [9] that several relationships found earlier for the interdiffusion coefficient of binary liquids can be deduced in a unified treatment from the theory of Bearman [10], if suitable assumptions are applied for the frictional coefficients.

Diffusion coefficients of glycol–water mixtures have been investigated by Byers and King [11], and they stated that the ratio $D\eta/T$ is independent of temperature, but varies linearly with the mole fraction of glycol. This leads to the conclusion that the activation energies of diffusion and viscous flow are equal, while the energy barrier between adjacent diffusional equilibrium positions is increased by the glycol molecules which are larger than those of water.

Longsworth [12] determined the limiting mobility λ^0 of uncharged species from measurements on the diffusion coefficients of different solutes in methanol–water mixtures. The change in the product $\lambda^0\eta$ in the diffusion of

Literature on page 247

non-electrolytes is smaller than in the case of monoatomic ions of similar mobility, and it decreases with increasing size of the particles. Starting with the assumption that large tetra-alkyl ammonium ions are not hydrated in water, and are not solvated in methanol, he concluded that the solvation number of dissolved non-electrolytes is somewhat lower in methanol than their hydration number. This indicates that the low value of $\lambda^0\eta$ in solutions of small particles can be attributed primarily to the relatively large size of the methanol molecules in the solvation sphere. The solvation number calculated on the basis of diffusion is not greater than the number of sites on the dissolved molecule available for forming hydrogen bonds.

Effect of the shape of molecules on diffusion. In aqueous solutions of macromolecular substances, the above requirement regarding the validity of Stokes' law is approximately satisfied by the size relations of the solvent and solute molecules; however, macromolecules are usually not spherically symmetrical, even approximately. In connection with the diffusion of non-spherical molecules, Perrin [13] has taken into account three frictional coefficients f_1, f_2, and f_3. Under such conditions the diffusion coefficient is:

$$D_{12} = \frac{kT}{3}\left(\frac{1}{f_1} + \frac{1}{f_2} + \frac{1}{f_3}\right). \qquad (3.1.34)$$

If the particles are rotational ellipsoids, f_2 is equal to f_3:

Considering the dependence of diffusion on the shape of the diffusing entity, Perrin [13] and Herzog, Illig and Kudar [14] have shown that, in certain cases, some information can be obtained on the *shape of the solute molecules* from the diffusion coefficient. Determining empirically the diffusion coefficient of non-solvated large molecules of rotational ellipsoidal shape, the empirical frictional coefficient $f_{\rm emp}$ related to the actual conditions can be calculated from equation (3.1.28):

$$\frac{1}{f_{\rm emp}} = \frac{1}{3}\left(\frac{1}{f_1} + \frac{2}{f_2}\right). \qquad (3.1.35)$$

On the other hand, it can be calculated approximately how large the frictional resistance would be, if the molecules were spherical. Namely, if the molar volume of the solute is V_2 and spherical molecules are assumed, their hypothetical radius r_f can be obtained from the molar volume by means of the following equation:

$$\chi V_2 = \frac{4}{3}\pi r_f^3, \qquad (3.1.36)$$

where $\chi(\le 1)$ is a packing factor depending on the arrangement of molecules in the liquid. In calculations regarding diffusion of large molecules, no appreciable error occurs when taking $\chi = 1$. According to this, the frictional resistance of a molecule assumed to have a spherical shape would be:

$$f_f = 6\pi\eta r_f. \qquad (3.1.37)$$

Comparing this with the value calculated from the experimentally determined diffusion coefficient, the shape can be estimated. If the ratio f_{emp}/f_s deviates from unity, the molecule is either hydrated or has a non-spherical shape. According to Wang [15], the effect of hydration can be taken into account in a correction factor and the ratio of the axes can be computed from the term $(f_{emp}/f_s)_{corr}$.

The alteration of the viscosity of a liquid by a solute is non-uniform, being greater in the vicinity of the solute species than at larger distances from them. This also makes the detailed discussion of the phenomena of diffusion difficult.

For the diffusion of small molecules, the conclusions are more uncertain. Under such conditions, the estimation of the best numerical factor corresponding to the real frictional conditions is still less reliable than in the case of solutions of large molecules, and the assumption of $\chi = 1$ cannot be accepted either, since this would mean that the molecules are fitted together with no gaps, which is impossible. If the liquid is built up from tightly packed spheres, $\chi = 0 \cdot 74$; this leads to a radius less by about 10 per cent than that obtained on the basis of $\chi = 1$.

In the course of diffusion of small molecules in a solvent consisting of molecules large or commensurable in comparison with them, solute molecules pass through the cavities existing between the solvent molecules, as suggested by Sutherland. Under such conditions, β may be nearly equal to zero, and, according to equation (3.1.32), one has:

$$D_{12} = \frac{kT}{4\pi\eta_1 r_2}. \tag{3.1.38}$$

It can be assumed that, in first approximation, this relationship applies to self-diffusion; from this a correlation between the self-diffusion coefficient and the molar volume can be obtained. If it is assumed that spherical liquid molecules are arranged in a cubic lattice in contact with one another:

$$2r_2 = \left(\frac{V}{N}\right)^{1/3} \tag{3.1.39}$$

(where V is the molar volume of the liquid) and:

$$D = \frac{kT}{2\pi\eta}\left(\frac{N}{V}\right)^{1/3}. \tag{3.1.40}$$

This equation is valid for several liquids of polar and non-polar molecules with an error of about ± 12 per cent. Some calculations of Collins and Raffel [16] on self-diffusion based on the kinetic theory resulted in a correlation corresponding to equation (3.1.38), but their numerical factor was lower than 4.

Although various methods regarding diffusion coefficients lead to different numerical factors in the equation of frictional resistance, it can be

Literature on page 247

established that the frictional resistance is approximately proportional to viscosity:

$$f_s = \alpha\eta, \tag{3.1.41}$$

where α is a parameter with a dimension of length, approximately independent of concentration. Thus:

$$D_{12} = \frac{RT}{N\alpha\eta} = \frac{kT}{\alpha\eta}. \tag{3.1.42}$$

This relationship is consistent with the early general observation that the product $D\eta$ changes much less with the composition of the system and with the other state parameters than D itself. In the case of solutions, however, it has not been properly cleared up, whether the viscosity of the pure solvent or that of the solution in bulk should be used in the calculations. It can be supposed that in very dilute solution ($c_2 \rightarrow 0$) the viscosity of the solvent is predominating in the diffusion coefficient D_{12}^0:

$$D_{12}^0 = \frac{kT}{\alpha\eta_1}. \tag{3.1.43}$$

Furthermore, it has been known for a long time that, under comparable conditions, the higher the molecular weight of the diffusing species, the lower the diffusion coefficient. According to Walden [17], in solutions of various substances of molecular weight M_2, the following equation holds:

$$D_{12}\eta_1\sqrt{M_2} \approx \text{const.} \tag{3.1.44}$$

However, this correlation is of limited validity even in solutions of large, approximately spherically symmetrical molecules. In solutions of organic substances of nearly identical specific volume v_2, Polson [18] found the diffusion coefficient to be proportional to $M_2^{1/3}$, since the radius of the molecule is, in approximation:

$$r_2 = \left(\frac{3M_2v_2}{4\pi N}\right)^{1/3}. \tag{3.1.45}$$

From measurements on the diffusion of polypeptides, proteins, and carbohydrates with shapes essentially different from spherical, Longsworth [19] concluded that:

$$D_{12} = \frac{A}{M_2^{1/3} - B} = \frac{A'}{v_2^{1/3} - B'}, \tag{3.1.46}$$

where A, B, A', and B' are empirical constants.

Diffusion in non-ideal mixtures. The equations discussed above are valid–even approximately–only for ideal mixtures and dilute solutions. In real solutions and mixtures, it should be taken into account that chemical potential, regarded as the driving force of diffusion, is determined not by concentration, but in fact by activity ($a_i = \gamma_i c_i$), and the activity coefficient γ_i depends on the composition of the mixture. Since the composition changes in the course of diffusion, the activity coefficient–i.e. the correlation between activity and concentration–also changes and this should be accounted for in the equations describing the dependence of the diffusion flow on concentration.

On the basis of the studies of Hartley [20] and Onsager and Fuoss [21], the gradient of the chemical potential of the solute (i.e. the driving force of diffusion per mole) is:

$$(\text{grad } \mu_2)_{T,P} = \frac{RT}{c_2}\left\{1 + \left(\frac{\partial \ln \gamma_2}{\partial \ln c_2}\right)_{T,P}\right\}\text{grad } c_2, \qquad (3.1.47)$$

if the activity coefficient of the solute is denoted by γ_2. Introducing the following abbreviation:

$$B_2^c = \left\{1 + \left(\frac{\partial \ln \gamma_2}{\partial \ln c_2}\right)_{T,P}\right\} = \left\{1 + c_2\left(\frac{\partial \ln \gamma_2'}{\partial c_2}\right)_{T,P}\right\}, \qquad (3.1.48)$$

the average velocity of the diffusing species in non-ideal solutions is:

$$\bar{v}_2 = \frac{kT}{f_s}B_2^c\frac{d \ln c_2}{dy}, \qquad (3.1.49)$$

if equation (3.1.27) is also taken into consideration. On the other hand, the diffusion coefficient, according to equation (3.1.28), is:

$$D_{12} = \frac{kT}{f_s}B_2^c = D_{12}^0 B_2^c, \qquad (3.1.50)$$

if D_{12}^0 is the diffusion coefficient calculated for ideal behaviour of the solution.

According to Gordon [22], the correlation between the diffusion coefficient and the viscosity of the solution in the non-ideal case is:

$$D_{12}\,\eta_{12} = D_{12}^0\,\eta_1\,B_2^c \qquad (3.1.51)$$

considering equations (3.1.41)–(3.1.43). This relation has proved to be valid in good approximation for dilute solutions of several electrolytes, and non-electrolytes as well [23]. However–as emphasized by Gordon [24]–it can hold only for dilute solutions, since it neglects the diffusion flow of the solvent opposite to that of the solute.

The Stokes–Einstein law has been supplemented by Broersma [25] taking into account the changes in viscosity with distance, measured from the solute species. The complicated relationships deduced in this way make the calculation of the diffusion coefficient of suspended particles possible.

In more concentrated non-ideal solutions, the movement of the solvent opposite to the diffusion of the solute must also be considered, as well as the

Literature on page 247

fact that parameter α in equation (3.1.41) describing the frictional resistance
is different for the two components. On this basis, Hartley and Crank [26]
obtained the following correlation between the diffusion coefficient and mole
fractions x_1 and x_2 of the two components:

$$D_{12} = \frac{kT}{\eta} \left(\frac{\partial \ln \gamma_2 x_2}{\partial \ln x_2} \right)_{T,P} \left(\frac{x_1}{\alpha_2} + \frac{x_2}{\alpha_1} \right). \qquad (3.1.52)$$

If α_1 and α_2 are approximately independent of concentration, this equation
becomes identical with that deduced by Carman and Stein [27], according
to which:

$$D_{12}\, \eta_{12} = \left(\frac{\partial \ln \gamma_2 x_2}{\partial \ln x_2} \right)_{T,P} (x_1 \eta_1 D_1 + x_2 \eta_2 D_2), \qquad (3.1.53)$$

if η in equation (3.1.52) is identified with the viscosity η_{12} of the solution.
Here, D_1 and D_2 are the limiting values of the interdiffusion coefficient
when c_2 and c_1 tend to zero, respectively; η_1 and η_2, on the other hand,
are the viscosities of the two components. If α_1 and α_2 are not independent
of concentration, the relationship will be modified [28].

3.1.2.2 KINETIC THEORIES OF DIFFUSION

Hydrodynamic theories of diffusion have revealed certain relationships,
but cannot provide a deeper understanding of the molecular mechanism
of the phenomenon, because they neglect the molecular structure of the
liquid, regarding it as a continuum. The kinetic statistical theories tend to
elucidate the molecular mechanism of diffusion. The most general of them
is the theory published by Eyring *et al.*, and its advanced modification gives
a deeper explanation of the elementary processes of diffusion, although its
results cannot be considered satisfactory in all respects [29].

Fundamentals of the Eyring theory. Eyrings' theory [29] discusses the diffusion
coefficient on the basis of a relatively simple model of the liquid state
by applying the theory of absolute reaction rates. It is assumed that diffusion can be described similarly to the rate processes of monomolecular
reactions, involving the temporary development of such a configuration of
the species and its environment which can be considered an activated state.

Diffusion–with respect to its mechanism–is in many aspects similar to
viscous flow, with the difference that unlike molecules are involved in
the former process. In solution, diffusion requires the slipping of the solute
and solvent molecules past one another. In the course of their movement,
the solute molecules should cross the potential barrier (free enthalpy barrier)
separating two adjacent equilibrium positions in the liquid structure.
If two adjacent equilibrium positions are at a distance λ from each other,
the molecules cover a distance λ in each jump in the direction of the negative concentration gradient (Fig. 3.1). In ideal solutions, the standard
free enthalpy is the same in all equilibrium positions of the diffusing molecule, irrespective of concentration.

If the standard free enthalpy is identical in the successive equilibrium positions occupied by the diffusing molecule, and a symmetrical energy barrier is assumed, the free enthalpy of activation of the process in the direction of decreasing concentration ('forward') is identical with that in the opposite direction ('backward'). Consequently, the rate constants (specific rates k_f and k_b) of the process in the two directions are equal to each other ($k_f = k_b = k$).

For the sake of simplicity, let us

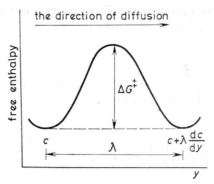

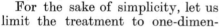

Fig. 3.1 Free-enthalpy barrier of diffusion in ideal solutions

limit the treatment to one-dimensional diffusion, i.e. let us investigate the case where the concentration difference occurs only in direction y. If the concentration of the solute is c mole cm^{-3} at a given point, it will be $c + \lambda \dfrac{dc}{dy}$ at the adjacent one, where the molecules arrive by means of diffusion (see Fig. 3.1.). The solute diffuses towards decreasing concentration, $dc < 0$, and thus, $c + \lambda \dfrac{dc}{dy} < c$. Under such conditions, the number of molecules moving 'forward' through a cross-section of 1 cm^2 perpendicular to direction y per second (the rate of transport in the forward direction) is:

$$\bar{v}_f = Nc\lambda^2 k \text{ molecule cm}^{-1}\text{s}^{-1} \qquad (3.1.54)$$

where k is the specific rate of diffusion (rate constant), i.e. the number of passages of one molecule across the energy barrier per second, or the number of jumps to an adjacent position.* Due to the thermal motion, molecules move in the opposite direction, too. The rate of movement in the backward direction (from the right- to the left-hand side in the figure) is:

$$\bar{v}_b = N \left(c + \lambda \frac{dc}{dy} \right) \lambda^2 k \text{ molecule cm}^{-1}\text{s}^{-1}. \qquad (3.1.55)$$

The resultant velocity of the solute molecules (i.e. the rate of diffusion) is:

$$\bar{v} = \bar{v}_f - \bar{v}_b = -N \lambda^2 k \frac{dc}{dy}. \qquad (3.1.56)$$

* The number of molecules is $Nc\lambda$ in a layer of cross-section 1 cm^2 and thickness λ measured from the energy barrier, and the number of such layers crossing the barrier in unit time is λ.

Literature on page 247

According to the Fick law, this quantity (expressed by the number of diffusing molecules) is:

$$\bar{v} = -\mathrm{D}N\frac{dc}{dy}.\tag{3.1.57}$$

From the comparison of the latter two equations, it can be seen that the diffusion coefficient in ideal solutions is:

$$D = \lambda^2 k.\tag{3.1.58}$$

If λ and k are the same as the corresponding quantities in the theory of viscosity, (e.g. in the case of self-diffusion) equation (2.28) gives:

$$\mathrm{D} = \frac{\lambda_1 \mathrm{k}T}{\lambda_2 \lambda_3 \eta} \approx \frac{\mathrm{k}T}{\lambda \eta}\tag{3.1.59}$$

since $\lambda_1/\lambda_2\lambda_3$ does not differ in order of magnitude from $1/\lambda$.

The expression for the diffusion coefficient given by Eyring does not differ from that of Stokes (equation 3.1.33), except for using the term $\lambda_2\lambda_3/\lambda_1$ instead of $6\pi r_2$. Both quantities are approximately in the order of magnitude of 10^{-8} cm, but within this, they can differ from each other even by a factor of ten. It should be noted, however, that the Eyring and Stokes relationships are not, in fact, comparable, since the latter is based on the assumption that the diffusing molecules move in a continuum (i.e. the solvent molecules are small in comparison with those of the solute) and hence, the laws of classical hydrodynamics are applicable at least in approximation. On the other hand, the Eyring theory is based on the assumption that the sizes of the solute and solvent molecules are the same order of magnitude, and the movement of both the solvent and solute molecules has been taken into account during diffusion.

The mechanism of diffusion given above can hardly be valid for the diffusion of large species in a solvent consisting of small molecules, since it cannot be supposed that, under such conditions, the rate-determining step will be the jump of the solute molecule from one equilibrium position to the adjacent one, because the formation of the space required for this would require a large amount of energy. More probably, the rate-determining step is the jump of the small solvent molecules (Fig. 3.2). In the course of this, the large molecules of solute are displaced in the opposite direction, and they occupy the spaces left empty by the solvent molecules. Diffusion takes place in this way.

Recently, these two types of mechanisms of diffusion have been investigated in detail by Nir and Stein [30b]. On the basis of the assumption that the activation energy of diffusion depends on the size of the diffusing molecule, they have calculated semi-empirically the diffusion coefficients of small and medium size molecules dissolved in water.

The fact that the temperature coefficient of diffusion of large molecules is equal to the temperature coefficient of viscosity of the solvent is consistent with the statement that the rate of diffusion is determined by the movement of the solvent species under such conditions. The small solvent molecules, of course, go past the large solute species in several steps in the direction shown by the arrow in Fig. 3.2, taking a series of jumps between the successive equilibrium positions. A solvent particle has to cover a minimum distance πr_2 (right-hand side in the figure) while the solute molecule moves a distance of about λ (in the left-hand side direction in the figure; λ is the distance between two adjacent equilibrium positions of the solvent molecule). If the solute molecule were small (e.g. 2' in the figure), this would be the one moving a distance of λ in each jump, from the right to the left-hand side.

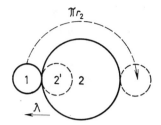

Fig. 3.2 Diffusion of a large molecule (2) due to movement of small solvent molecules (1)

Thus, in order to have a large solute molecule diffused a distance λ, the pathlength of the movement making this displacement possible should be πr_2; when the radius r_1 of the solute molecule does not differ too much from that of the solvent species, this requires only a pathlength of about λ. So, according to this theory, the diffusion coefficient D_l of a large molecule is smaller than that of small molecules (D_s) by the ratio:

$$D_l = D_s \frac{\lambda}{a\pi r_2},$$ (3.1.60)

where a is a factor not too different from unity, which takes into account that the solvent molecule does not necessarily move past the large one along the shortest path possible.

Since the value of D_s can be obtained from equation (3.1.59), we have:

$$D_l = \frac{kT}{a\pi r\eta}.$$ (3.1.61)

According to Stokes' equation, $a = 6$; however, this is valid only if the shape of the diffusing molecules is spherical and they are so large that the solvent can be considered a continuous medium in comparison with them, and if the solvent is adhering strongly to the 'surface' of the diffusing molecules. It is still supported by the fact that, in general, a does not differ too much from six in the diffusion of large species in a solvent consisting of small molecules.

The assumption that λ and k are the same in diffusion and in viscous flow can be rigorously correct in the case of self-diffusion only. In the diffusion of a solute, some deviations can be expected; this is partly because the sizes of the solute and solvent molecules are different, and partly because the free enthalpy of activation of the rate-determining process (which influences the value of k) depends on the properties and concentrations of both components. The product $D\eta$ in different solvents is related to the ratio $\lambda_1/\lambda_2\lambda_3$, even if in the expressions for η and D the term $\lambda^2 k$ is not

Literature on page 247

Table 3.1

Product $D\eta$ for the diffusion of I_2 and the molar volume of the solvent

Solvent	$D\eta$	Molar volume cm³
Dibromoethane	1·23	87
Tetrabromoethane	1·54	117
Chloroform	1·06	81
Carbon tetrachloride	1·13	97
Ethyl acetate	0·85	98
Iso-amyl acetate	0·93	149

eliminated. Thus, a correlation between the product $D\eta$ and the molar volume of the solvent can be expected as pointed out already by Taylor [31]. For example, in Table 3.1, the product $D\eta$ calculated for the diffusion of iodine increases with increasing molar volume of the solvent, when solvents of similar nature are compared.

Absolute value of the diffusion rate. The rate constant term k in equation (3.1.58) of the diffusion coefficient can be calculated on the basis of the theory of absolute reaction rates [32]:

$$k = \frac{kT}{h} \cdot \frac{F^{\ddagger}}{F} \exp\left[-\frac{\varepsilon_0}{kT}\right], \qquad (3.1.62)$$

where $F^{\ddagger}$ and F are the partition functions of the activated and ground states respectively, and ε_0 is the energy of activation per molecule at 0 °K. In order to calculate the partition function approximately, let us assume that the degree of freedom corresponding to diffusional movement is of translational character, similarly to the case of viscosity; we obtain, according to the statistical theory:

$$\frac{F^{\ddagger}}{F} = \frac{h}{(2\pi\, mkT)^{1/2}\, v_f^{1/3}}, \qquad (3.1.63)$$

where m is the mass of a molecule and v_f is the free volume in the liquid as shown in Fig. 2.3.

Furthermore, it should be taken into account that the elementary process of diffusion, i.e. the transfer of a molecule from one equilibrium position to the adjacent one, requires the formation of an empty place, a hole in the liquid; its size, however, is not necessarily identical to that of the molecule. The energy required for hole formation is related to the heat of evaporation (cf. Sections 2.2.2. and 2.2.5) since, after the evaporation of a molecule from the liquid surface an empty place is left there for a moment; its formation requires an activation energy of evaporation ΔH_{evap} per mole. Since the formation of an empty place is involved in the activation energy of both diffusion and evaporation, it can be supposed that the activation energy of diffusion per mole, ΔH_{diff}, is related by a factor of $(1/n)$ to the activation energy of evaporation:

$$\Delta H_{\text{diff}} = \frac{\Delta H_{\text{evap}}}{n}. \tag{3.1.64}$$

Considering equations (3.1.58), (3.1.62), and (3.1.63), the diffusion coefficient can be calculated, as given by Stearn, Irish and Eyring [33], since:

$$\frac{\varepsilon_0}{kT} = \frac{\Delta H_{\text{diff}}}{RT}$$

and

$$D = \lambda^2 \frac{kT}{h} \cdot \frac{F^{\ddagger}}{F} \exp\left[-\frac{\varepsilon_0}{kT}\right] =$$

$$= \lambda^2 \frac{kT}{h} \frac{h}{(2\pi mkT)^{1/2} v_f^{1/3}} \exp\left[-\frac{\Delta H_{\text{evap}}}{nRT}\right] =$$

$$= \frac{\lambda^2}{v_f^{1/3}} \left(\frac{kT}{2\pi m}\right)^{1/2} \exp\left[-\frac{\Delta H_{\text{evap}}}{nRT}\right]. \tag{3.1.65}$$

The distance λ can be calculated approximately from the molar volume:

$$\lambda \approx \left(\frac{V}{N}\right)^{1/3}, \tag{3.1.66}$$

and the free volume v_f can be estimated as given in Section 2.2.5. Eyring assumed that $n \approx 3$. However, in calculating the diffusion coefficient by means of equation (3.1.65) using the activation energy of evaporation, it should be considered that, in general, at least two kinds of molecules participate in diffusion (those of the solute and the solvent). Since our present knowledge cannot take these details into account, in first approximation properly calculated mean values should be employed for λ, v_f, and ΔH_{evap}. λ and v_f can be the weighed arithmetic means of the values for the two pure components:

$$\lambda = x_1 \lambda_1 + x_2 \lambda_2, \tag{3.1.67}$$

where x_1 and x_2 are the mole fractions of the two components, and:

$$v_f^{1/3} = x_1 v_{f,1}^{1/3} + x_2 v_{f,2}^{1/3}. \tag{3.1.68}$$

The mean value of ΔH_{evap} has been calculated as follows:

$$\Delta H_{\text{evap}}^{1/2} = x_1 \Delta H_{\text{evap},1}^{1/2} + x_2 \Delta H_{\text{evap},2}^{1/3}. \tag{3.1.69}$$

Literature on page 247

Table 3.2

The diffusion of tetrabromoethane in tetrachloroethane

Temperature, °K	$\lambda^2 \cdot 10^{16}$	$v_l^{1/3} \cdot 10^9$	E, kcal	$D \cdot 10^5 \, \text{cm}^2 \, \text{s}^{-1}$	
				calculated	found
273·4	3·12	6·45	9·85	0·64	0·35
288·0	3·15	6·92	9·72	0·92	0·50
308·6	3·19	7·59	9·52	1·42	0·74
324·1	3·23	8·13	9·38	1·89	0·99

For the mass of the molecule, the reduced mass of the two types of molecules can be taken:

$$m = \frac{m_1 m_2}{m_1 + m_2}. \tag{3.1.70}$$

This means essentially that the moving entity in solution is a complex formed from one solute and one solvent molecule which, of course, can hardly be correct, but the error caused by this choice is probably not large.

Table 3.2 shows an example of the agreement between measured and calculated diffusion coefficients of tetrabromoethane dissolved in tetra-chloroethane. It can be seen that the calculated values are about twice the measured ones. This result of the theory is still remarkable, since the diffusion coefficient is calculated from data related not to the diffusion process itself.

Similar deviations can be observed for data obtained in other organic solvents as well, which can evidently be attributed to the simplifying assumptions. Such an assumption is, for example, that the contribution of rotation and vibration to the ratio of the partition functions is identical in the activated and the ground states. On the contrary, in structureless liquids, rotation has probably more degrees of freedom in the ground state than in the activated one, thus $F^{\ddagger}/F$ is lower than unity. This effect decreases the calculated value of diffusion. In solvents of considerable internal structure (e.g. water), the effect of rotational components is opposite, and this leads to a calculated diffusion coefficient lower than the measured one.

In water and other liquids containing hydroxyl groups, monomeric molecules are bound together by a network of hydrogen bonds rather extended as compared with the molecular dimensions. Taking into account the experiments on viscosity, diffusion is accompanied by a certain distortion of the internal structure of the liquid and by partial rupture of the hydrogen bonds in the activated state. This decreases the theoretically calculated values. In Table 3.3 the diffusion coefficients of some substances, calculated theoretically from equation (3.1.65), are compared with those measured in dilute aqueous solutions. The data used in the calcula-

Table 3.3

Diffusion coefficients of some substances in water

Solute	Temperature, °K	$D \cdot 10^5 \, cm^2 \, s^{-1}$	
		calculated	found
Methyl alcohol,	291	0·39	1·37
Amyl alcohol	291	0·34	0·88
Glycerol	293	0·36	0·83
Mannitol	296	0·43	0·61
Glucose	291	0·32	0·57
Saccharose	293	0·34	0·57
Phenol	291	0·34	0·80
Urea	293	0·38	1·18

tion and the relationships applied are: $\Delta H_{evap} = 9 \cdot 7$ kcal/mole, $n = 2 \cdot 4$, while v_f is, according to Eyring and Hirschfelder [34]

$$v_f^{1/3} = \frac{cRT V^{1/3}}{N^{1/3} \Delta H_{evap}}.$$ (3.1.71)

It can be seen that, under such conditions, the calculated values are smaller than the measured ones. However, the calculations give proper values with respect to the orders of magnitude in this case, too.

Diffusion in non-ideal mixtures. For solutions or mixtures of higher concentrations, deviations from the properties of ideal mixtures should also be taken into account in the kinetic theory of diffusion, as in the hydrodynamic one (see Section 3.1.2.1). For diffusion in non-ideal mixtures, the chemical potential, i.e. the free enthalpy, should also be different at different points of the given material system. However, the free enthalpy of the diffusing component is determined unequivocally by its concentration only in ideal mixtures. In non-ideal mixtures, activity $a = \gamma c$ is the relevant quantity. In concentrated solutions or mixtures, the activity coefficient γ of the diffusing component is not unity and it depends not only on the qualitative composition of the mixture, but also on the concentrations of all of the substances present. This implies that the free enthalpy of the diffusing molecules is not indentical in the activated and initial states of the individual elementary steps.

The non-ideal behaviour of concentrated solutions and mixtures can be taken into consideration by calculating the free enthalpy change ΔG due to the deviation from the ideal case. Assuming that the energy barrier in the elementary process of diffusion is symmetrical, the free enthalpy arising from non-ideal behaviour is divided equally between the free enthalpies of activation of the two processes in opposite directions (Fig. 3.3).

Literature on page 248

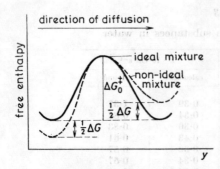

Fig. 3.3 Variation of the free enthalpy of activation in the course of diffusion

On this basis, Stearn, Irish and Eyring [33] calculated the diffusion coefficient in non-ideal solutions and mixtures in the following way. If the activity coefficient of the diffusing species is γ_1, the change $\gamma\left(\dfrac{\partial \ln \gamma}{\partial y}\right)$ of γ_1 corresponding to one elementary step (jump) of the diffusing molecule will be the determining factor with respect to changes in free enthalpy arising from non-ideal behaviour of the solution:

$$\Delta G = \mathrm{R}T\,\lambda\,\frac{\partial \ln \gamma_1}{\partial y} = \mathrm{R}T\,\lambda\,\frac{\partial \ln \gamma_1}{\partial x_1}\cdot\frac{\partial x_1}{\partial y} = \lambda\,\frac{\partial x_1}{\partial y}\,\Gamma_1, \qquad (3.1.72)$$

where x_1 is the mole fraction of the diffusing component and Γ_1 is, by definition:

$$\Gamma_1 = \mathrm{R}T\,\frac{\partial \ln \gamma_1}{\partial x_1}. \qquad (3.1.73)$$

If the free enthalpy of activation in ideal solutions is $\Delta G_0^{\ddagger}$, in non-ideal solutions the free enthalpy of activation in the 'forward' direction will be:

$$\Delta G_f^{\ddagger} = \Delta G_0^{\ddagger} + \frac{1}{2}\,\Delta G = \Delta G_0^{\ddagger} + \frac{1}{2}\,\lambda\,\frac{\partial x_1}{\partial y}\,\Gamma_1. \qquad (3.1.74)$$

In the 'backward' direction, on the other hand, the free enthalpy of activation is:

$$\Delta G_b^{\ddagger} = \Delta G_0^{\ddagger} - \frac{1}{2}\,\lambda\,\frac{\partial x_1}{\partial y}\,\Gamma_1. \qquad (3.1.75)$$

The net rate of diffusion, i.e. the modified form of equation (3.1.56) representing this case, is:

$$\bar{v} = N_0 c_1 \lambda k_f - N_0\left(c_1 + \lambda\,\frac{\partial c_1}{\partial y}\right)\lambda\,k_b, \qquad (3.1.76)$$

where k_f and k_b are the rate constants for diffusion in the forward and backward directions, respectively. According to the theory of absolute reaction rates:

$$k_f = \frac{\mathrm{k}T}{\mathrm{h}}\,\exp\left[-\frac{\Delta G_f^{\ddagger}}{\mathrm{R}T}\right] \qquad (3.1.77)$$

and

$$k_b = \frac{\mathrm{k}T}{\mathrm{h}}\,\exp\left[-\frac{\Delta G_b^{\ddagger}}{\mathrm{R}T}\right]. \qquad (3.1.78)$$

Taking into account equation (3.1.74), we have:

$$k_e = \frac{kT}{h} \exp\left[-\frac{\Delta G_0^{\ddagger}}{RT}\right] \exp\left[-\frac{1}{2} \lambda \frac{\partial x_1}{\partial y} \cdot \frac{\Gamma_1}{RT}\right] =$$

$$= \frac{kT}{h} \exp\left[-\frac{\Delta G_0^{\ddagger}}{RT}\right] \exp\left(-\alpha\right), \qquad (3.1.79)$$

where

$$\alpha = \frac{1}{2} \lambda \frac{\partial x_1}{\partial y} \cdot \frac{\Gamma_1}{RT}. \qquad (3.1.80)$$

In ideal solutions, the rate constant in both the forward and backward directions is:

$$k = \frac{kT}{h} \exp\left[-\frac{\Delta G_0^{\ddagger}}{RT}\right]. \qquad (3.1.81)$$

Consequently, equation (3.1.79) can be transformed into

$$k_f = k \exp[-\alpha] = k e^{-\alpha}. \qquad (3.1.82)$$

Similarly

$$k_b = k \exp[\alpha] = k e^{\alpha}. \qquad (3.1.83)$$

Inserting these expressions into equation (3.1.76) we obtain:

$$\bar{v} = N c_1 k (e^{-\alpha} - e^{\alpha}) - N \lambda^2 k \frac{\partial c_1}{\partial y} e^{\alpha} \text{ molecule cm}^{-1}\text{s}^{-1}. \qquad (3.1.84)$$

Since $2RT \gg \lambda \dfrac{\partial x_1}{\partial y} \Gamma_1$ and $\alpha \ll 1$, the exponential terms can be expanded into series and only the first term retained. With this restriction we get:

$$\bar{v} = - N c_1 \lambda k (2\alpha) - N \lambda^2 k \frac{\partial c_1}{\partial y} (1 + \alpha) =$$

$$= - N c_1 \lambda k (2\alpha) - N \lambda^2 k \frac{\partial c_1}{\partial y}. \qquad (3.1.85)$$

Substituting the value of α from equation (3.1.80) into equation (3.1.85), it becomes:

$$\bar{v} = - N c_1 \lambda^2 k \frac{\partial x_1}{\partial y} \cdot \frac{\Gamma_1}{RT} - \lambda^2 k \frac{\partial c_1}{\partial y}. \qquad (3.1.86)$$

Literature on page 248

Instead of the concentration gradient, the gradient of mole fraction can be taken in first approximation*:

$$c_1 \frac{\partial x_1}{\partial y} = x_1 \frac{\partial c_1}{\partial y}. \qquad (3.1.87)$$

Substituting this into equation (3.1.86), we have:

$$\bar{v} = - N\lambda^2 k \frac{\partial c_1}{\partial x} \left(\frac{x_1 \Gamma_1}{RT} + 1 \right). \qquad (3.1.88)$$

Since, according to the Fick equation, the rate of diffusion of component 1 related to unit cross section is:

$$\bar{v} = - D_1 N \frac{\partial c_1}{\partial y},$$

it follows that:

$$D_1 = \lambda^2 k \left(\frac{x_1 \Gamma_1}{RT} + 1 \right). \qquad (3.1.89)$$

According to equation (3.1.58), D is equal to $\lambda^2 k$ in ideal solutions. If this value is denoted by D_1^0, and Γ_1 is taken from equation (3.1.73), equation (3.1.89) gives:

$$D_1 = D_1^0 \left(1 + x_1 \frac{\partial \ln \gamma_1}{\partial x_1} \right) = D_1^0 \left(1 + \frac{\partial \ln \gamma_1}{\partial \ln x_1} \right) = D_1^0 \frac{\partial \ln a_1}{\partial \ln x_1}, \qquad (3.1.90)$$

where $a_1 = \gamma_1 x_1$ is the activity of solute 1. Essentially, this equation is identical with that given by Onsager and Fuoss [35] for dilute electrolyte solutions (cf. Section 3.2.1).

In the deduction given above it has been assumed that the rate-determining step of the diffusion process is the jump of a molecule from an actual equilibrium position to the adjacent one. In reality, of course, not only the solute molecules are displaced in the course of diffusion, but–in the opposite direction–those of the solvent as well. Taking this into account in the expression of ΔG, a similar term with an opposite sign would be required in equation (3.1.72) to describe the change of the activity coefficient of the solvent. However, from a comparison of equation (3.1.90) with the experimental data, it can be concluded that, although both types of molecules are displaced in the course of diffusion, the rate-determining process is in practice, the activated displacement of the solute molecules only.

The comparison of equation (3.1.90) with the experimental data is helped by investigation of the product $D_1 \eta$, where η is the viscosity of the solution:

$$D_1 \eta = D_1^0 \eta \frac{\partial \ln a_1}{\partial \ln x_1}. \qquad (3.1.91)$$

* Rigorously speaking, a small correction factor is also involved in this transformation, but it can be incorporated in the value of λ^2.

Since, according to equation (3.1.59):

$$D_1^0 \eta = \frac{\lambda_1 kT}{\lambda_2 \lambda_3}, \qquad (3.1.92)$$

consequently:

$$D_1 \eta = \frac{\lambda_1 kT}{\lambda_2 \lambda_3} \frac{\partial \ln \alpha_1}{\partial \ln x_1}. \qquad (3.1.93)$$

Since the activity of a component of appreciable vapour pressure is proportional to the partial vapour pressure p_1 above the solution, equation (3.1.93) can be formulated as follows:

$$D_1 \eta = \frac{\lambda_1 kT}{\lambda_2 \lambda_3} \frac{\partial \ln p_1}{\partial \ln x_1}, \qquad (3.1.94)$$

and:

$$\frac{D_1 \eta}{\dfrac{\partial \ln p_1}{\partial \ln x_1}} = \frac{\lambda_1}{\lambda_2 \lambda_3} kT, \qquad (3.1.95)$$

where λ_1, λ_2, and λ_3 are the mean values corresponding to the components of the mixture. Assuming that these are linear functions of the mole fraction, the left-hand side in equation (3.1.95) will be a linear function of the mole fraction of the two-component mixture, which is in agreement with the observations of Powell, Roseveare and Eyring [36] in the case of chloroform-ether mixtures (Fig. 3.4). Thus, according to this equation, the product $D\eta$ has a maximum, in agreement with the experimental data. Investigation of some other mixtures has led to similar results when the activity of the components could be obtained by measuring the partial vapour pressures.

According to the present state of our knowledge, the Eyring theory reflects the mechanism of diffusion correctly on the whole. In its details, however, it is based on very simple assumptions, so it is not surprising that marked deviations can be observed between the calculated and measured numerical values of some properties. In this way, more or less reliable conclusions can be drawn from this theory only for the main features of diffusion. Some studies have been published about the extension of this

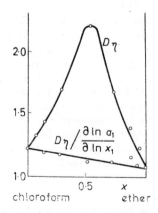

Fig. 3.4 Variation of $D\eta$ and $D\eta \Big/ \dfrac{\partial \ln a_1}{\partial \ln x_1}$ with concentration in chloroform–ether mixtures

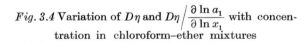

Literature on page 248

theory and the elimination of its deficiencies (e.g. in the theory of Li and Pin Chang [37], it has been taken into account that the molecules are not in an immobile environment, but they diffuse past adjacent moving molecules as well), however, development of a theory, which is satisfactory on the details of diffusion has not been successful so far.

The Panchenkov theory [38a] gives a treatment of the phenomena of viscosity and diffusion different from that of the Eyring theory; however, the essential conclusions are similar. According to this theory, two requirements must be met for the displacement of a molecule from its equilibrium position to an adjacent one: first, a sufficient kinetic energy must be gained by the particle to break the van der Waals bonds with its neighbours; on the other hand, the adjacent molecules must be temporarily far enough apart to enable the diffusing molecule to pass between them. The former requirement yields an energy term, while the latter gives an entropy term: the two terms together are equivalent to the free enthalpy of activation. Assuming a Maxwellian distribution of energy and a packing factor $\chi = 1$, the self-diffusion ceofficient is–according to Panchenkov–:

$$D_1^* = 4\left(\frac{3V_1^0}{4\pi N}\right)^{1/3} \frac{E^W + 2RT}{(2\pi M_1 RT)^{1/2}} \exp\left[-\frac{\bar{\gamma}E^W + 4U_{i,T}}{\bar{\gamma}RT}\right], \quad (3.1.96)$$

where E^W is the binding energy per mole due to van der Waals forces, $U_{i,T}$ is the molar internal heat of evaporation of the liquid at temperature T; $\bar{\gamma}$ is the average number of neighbours around the diffusing species (the coordination number of the liquid). This equation–although different in details–is similar to equation (3.1.65) in its form.

The theory of transfer diffusion. During the diffusion of solute molecules or other entities in multicomponent systems, in addition to the migration of the whole entities in the direction of the concentration gradient, exchange reactions with another component of the solution may also contribute to the apparent translation. Ruff and Friedrich [38b] called attention to this mechanism. In a system, where an AX component diffuses in the presence of component A, the effective (apparent) diffusion coefficient (D') of X is higher than the diffusion coefficient (D) corresponding only to the migration of AX, since in this case, X can be directly transferred from AX to a free, A. In this way, AX appears to be displaced without the transport of A. The apparent diffusion coefficient is thus the sum of the classical diffusion coefficient representing the translation of the whole entity, and that corresponding to the transfer. The contribution of this process, which can be called *transfer diffusion*, to the effective diffusion is significant, if the rate of the exchange reaction

$$AX + A^* \rightarrow A + A^*X$$

is high. For example, according to Ruff, Friedrich, Demeter and Csillag [38c], in solutions containing ferrocene and ferricinium ion, the latter need not really travel the distance between the iron atom of the activated complex and the iron atom in each case; when it collides with a ferrocene molecule in the solution, apparent diffusion may simply occur by exchange of the electron determining the oxidation state. After this transfer, the ferricinium ion formed will continue its translation motion from the position that belonged previously to the ferrocene molecule. Ruff *et al.* [38d] also assumed the possibility of transfer diffusion of tri-iodide and tribromide ions in solutions containing iodide and bromide ions, respectively [38e].

Ruff *et al.* elaborated the kinetic theory of transfer diffusion in detail, permitting also the calculation of the rate constant of rapid, second-order exchange reactions of the above type. Although Lengyel [38f] pointed out that Ruff's theory assumes erroneously a planar uniform distribution related to the axis of the activated complex during diffusion, whereas in reality a uniform steric distribution should be reasonably assumed, the theoretical statements regarding the transfer diffusion are correct. According to the correction made by Lengyel the factor $\pi/4$ should be replaced by $1/3$ in the flux term taking into account the transfer, and this results in numerical changes also in the other equations; this, however, does not affect the principles of the theory.

Other molecular theories. On the basis of measurements on the scattering of neutrons in normal and heavy water solutions between -5 and $+95$ °C, Blanckenhagen [38g] calculated the diffusion coefficient, also taking into consideration the frequency distribution of intermolecular vibrations. The experienced quasi-elastic scattering of neutrons indicates that near the melting point the molecules become activated individually, and the diffusion can be described by a jump model. At about the boiling point, however, it is globule diffusion which provides a reasonable interpretation of the phenomena. According to the latter, molecule groups (globules) having relatively long lifetimes migrate in a medium consisting of individually moving ('monomeric') molecules. A molecular dynamical study does not, however, support this conclusion; according to this, neither jumps occur, nor clusters are formed, but the diffusion is based on a vibrational mechanism; the molecules are displaced by means of an intermolecular translational vibration.

In the theory proposed by Frenkel, Wegdam and van der Elsken [38h], the rotational motion due to collisions in a liquid containing dipole molecules is regarded as the starting point. They arrive at the conclusion that in a liquid with strong interactions, the distribution of the collisions of the diffusing species determining rotational diffusion is a function of the frequency distribution of the medium, and, in addition to rotation, vibration also has an effect.

A theory independent of the kinetic model of diffusion has been elaborated by Ahn, Jensen and Kivelson [38i]. They correlated the diffusion coefficient with viscosity through time-dependent correlation functions.

Literature on page 248

3.1.2.3 THERMODYNAMIC THEORY OF DIFFUSION

The relationships for transport phenomena cannot be deduced from classical thermodynamics, since this science deals with equilibria and reversible processes only, (this is why it is also called thermostatics) while transport phenomena are irreversible. From the thirties, however, the thermodynamics of irreversible processes has shown a marked development initiated by the works of Onsager, de Groot, Prigogine and Meixner [39].

One of the starting points of the extension of thermodynamics to irreversible processes is the fact that, according to the phenomenological laws of Fourier, Ohm and Fick regarding transport processes (conduction of heat and electricity, diffusion), the correlation between the fluxes of heat, electricity or mass and the forces giving rise to them is linear. In general, the relation between the rates of flow (fluxes, J) and the thermodynamic driving forces (denoted by X) responsible for them, can be described by a power series of the force. However, in a state close to equilibrium, it is sufficient to retain the term containing the first power, as shown by the experiments. This is the basis of the *Onsager linear law* stating that in a thermodynamic system, the phenomenological correlation between force X_i acting on component i and the flux J_i produced by this force is:

$$J_i = L_i X_i, \qquad (3.1.97)$$

if this is the only force influencing component i. Here L_i is the *conduction coefficient* (or *phenomenological coefficient*) of flow.

According to the thermodynamics of irreversible processes in isothermal systems, the gradient of chemical potential is the thermodynamic force producing diffusion*. But it is not obvious how the chemical potential corresponding to equilibrium states (a quantity of free enthalpy character) can be applied to the description of the rate of irreversible processes–and among them, of diffusion. It is well known, for example, that the rate of chemical reactions is not proportional, in general, to the change of free enthalpy in the process. However, in a detailed analysis of the problem, Onsager and de Groot [39] found that in diffusion the change in free enthalpy due to mixing of solutions of various concentrations (which is measured by the chemical potential gradient) can be identified with the work of the diffusing species against the friction of the medium. Since diffusion is a slow process, deviations from the equilibrium state are much smaller in the course of this process than in most of the chemical reactions. Under such conditions, the total change in free enthalpy is almost equal to the energy dissipated in friction. Thus, in an isothermal system, the thermodynamic force resulting in diffusion of the ith component is:

$$X_i = - \operatorname{grad} \mu_i. \qquad (3.1.98)$$

---

* The first theory of diffusion developed by Nernst considered the osmotic pressure of the solute to be the driving force. In our present knowledge, however, osmotic pressure cannot be regarded as a pressure in the narrow sense, therefore it is more correct to use chemical potential in the calculations. In the limiting case of infinitely dilute solutions, a correct correlation can be obtained from the original Nernst's theory as well.

In reality, usually it is not one single force which acts in transport processes, but *several forces are responsible for the mass flow*, and the different flows are not independent of one another. If, for example, not only the concentration but also the temperature is different in different parts of a solution, not only the chemical potential gradient gives rise to mass transport but it is also influenced by the temperature gradient. Similarly, in a solution containing electrically charged ions, the chemical potential gradient results in not only mass transport, but electric charge transport too, and the electric force produced in this way affects mass transport. The *cross-effects* arising in this way are significant under certain conditions (e.g. in the diffusion of dissolved electrolytes).

In general, if thermodynamic forces $X_1, X_2, \ldots, X_i, \ldots, X_n$ simultaneously influence the ith component of a thermodynamic system, the flux of this component (within the validity range of the linear law) is:

$$J_i = \sum_{k=1}^{n} L_{ik} X_k. \tag{3.1.99}$$

The conduction (phenomenological) coefficients of the cross-effects are 'symmetrical' as shown by Onsager, i.e. the following correlation holds:

$$L_{ik} = L_{ki}. \tag{3.1.100}$$

This is the *Onsager reciprocity relationship* which, according to investigations on diffusion, proved to be in agreement with the experiments within the limits of experimental error. The validity of the reciprocity relationship within the experimental error has been proved, for example, by Ellerton and Dunlop [40] in aqueous solutions of mannitol and sugar, and by Miller [41] in aqueous NaCl and KCl solutions.

In two-component solutions, the flux J_a of mass flow is determined by the gradient of the chemical potential X_a and the temperature gradient X_q simultaneously, if the concentration and the temperature are not constant:

$$J_a = L_{aa} X_a + L_{aq} X_q. \tag{3.1.101}$$

The effect of temperature gradient on transport processes has recently been analysed in detail by Ćukrowski and Baranowski [42], who deduced detailed phenomenological equations of the phenomenon and applied them also to the diffusion of tracer ions.

Similarly, the flux J_q of heat flow is determined by two thermodynamic forces:

$$J_q = L_{qa} X_a + L_{qq} X_q.$$

The conduction coefficient L_{aa} and the coefficient L_{qq} correspond to the effect of the chemical potential gradient on mass transport in diffusion and to that of the temperature gradient on heat transport, respectively. Of the cross-effects, L_{aq} is the conduction coefficient corresponding to the effect to the temperature gradient on mass transport, while L_{qa} stands

Literature on page 248

for that of the chemical potential gradient on heat transfer. According to the Onsager reciprocity relationship:

$$L_{aq} = L_{qa}. \tag{3.1.102}$$

Similarly, if the gradients of chemical potential X_a and electric potential X_e act simultaneously in the solution, the flux of mass flow is:

$$J_a = L_{aa}X_a + L_{ae}X_e, \tag{3.1.103}$$

while that of electric current is:

$$J_e = L_{ea}X_a + L_{ee}X_e. \tag{3.1.104}$$

The conduction coefficient L_{ae} of the cross-effects refers to the effect of the gradient of electric potential on mass flow, while L_{ea} stands for the effect of the chemical potential gradient on electric current. According to the reciprocity relationship, we have:

$$L_{ae} = L_{ea}. \tag{3.1.105}$$

The cross-effects are related equally to effects of gradients caused by some external conditions in the given thermodynamic system (preparation method, heating or cooling, etc.), and to effects produced by the cross-effects themselves. For example, if a temperature difference is maintained between the two ends of a tube containing a solution or a gas of uniform concentration, a concentration gradient develops in the solution which will remain after reaching the steady state *(Soret effect)*. On the other hand, if a concentration gradient is maintained in the steady state of a solution, it will result in a temperature gradient *(Dufour effect)*. Similarly, if there is a concentration gradient in the solution of an electrolyte consisting of ions of different mobilities, an electric potential gradient will appear ('diffusion potential') due to the different diffusion velocities of the ions and it will affect the mass flux. This phenomenon plays an important role in the diffusion of electrolytes (cf. Section 3.2).

The conduction (phenomenological) coefficients introduced in the thermodynamics of irreversible processes are related to the coefficients applied in the empirical and hydrodynamic equations of transport phenomena. If there are no cross-effects, the relationships are relatively simple.

Such a process is, for example, the *diffusion of non-electrolytes in two-component mixtures at constant temperature*. Both components diffuse in the direction of the negative gradient of chemical potential (opposite to each other) and the thermodynamic driving force is determined by the combined effect of the two potential gradients. At constant temperature and pressure, expressing the mass flux in g cm^{-2} sec^{-1} units and relating it to that plane in the diffusing system for which the resultant of diffusions of opposite directions is zero, i.e. through which there is no net mass transfer, and applying weight fractions w_1 and w_2 in the chemical potential, we obtain:

$$J_1 = -L_{11}(\text{grad } \mu_1 - \text{grad } \mu_2), \tag{3.1.106}$$

and:

$$J_2 = -L_{22}(\text{grad } \mu_2 - \text{grad } \mu_1). \tag{3.1.107}$$

The correlation between chemical potentials of the two components is given by the Gibbs–Duhem equation*:

$$w_1 \operatorname{grad} \mu_1 + w_2 \operatorname{grad} \mu_2 = 0. \tag{3.1.108}$$

With respect to this:

$$J_1 = -L_{11} \left(\operatorname{grad} \mu_1 + \frac{w_1}{w_2} \operatorname{grad} \mu_1 \right) =$$

$$= -\frac{L_{11}}{w_2} \operatorname{grad} \mu_1 \tag{3.1.109}$$

and similarly:

$$J_2 = -\frac{L_{22}}{w_1} \operatorname{grad} \mu_2.$$

Since, according to the definition of the reference plane:

$$J_1 + J_2 = 0, \tag{3.1.110}$$

consequently:

$$L_{11} = L_{22}. \tag{3.1.111}$$

The phenomenological equations of diffusion in multicomponent systems have recently been deduced in detail by Pamfilov, Lopushanskaia and Tsvetkova [43] on the basis of the general equation of mass transport (cf. Section 3.3.2). The concentration dependence of phenomenological coefficients has been analysed by Schönert [44] who expanded the concentration function into a Taylor series. It has also been shown by Schönert [45a] that the transport processes of hydrated components are coupled to one another by means of hydration even if there is no impulse exchange between them. Recently, Kett and Anderson [45b] discussed the phenomena of diffusion in non-associated multicomponent liquid systems on the basis of the hydrodynamic theory. They have deduced a general correlation for the flux of each component and correlated the diffusion coefficients and the phenomenological ones. The Onsager reciprocity relationship can also be obtained from their theory, and they showed that the phenomenological coefficients are independent of activity data.

Correlations between phenomenological and diffusion coefficients. Comparing equation (3.1.109) of diffusion obtained from the thermodynamics of irreversible processes with equation (3.1.15) in the case of one-dimensional diffusion, the correlation between the conventional diffusion coefficient and the phenomenological one is the following:

$$J_2 = -D_2 \varrho \frac{\partial w_2}{\partial y} = -\frac{L_{22}}{w_1 M_2} \frac{\partial \mu_2}{\partial y}. \tag{3.1.112}$$

* It should be noted, however, that the applicability of the Gibbs–Duhem equation under these non-equilibrium conditions is questionable.

Literature on page 248

Expressing the chemical potential gradient with mole fractions:

$$\frac{\partial \mu_2}{\partial y} = \frac{\partial \mu_2}{\partial x_2} \cdot \frac{\partial x_2}{\partial w_2} \cdot \frac{\partial w_2}{\partial y}. \tag{3.1.113}$$

If the mixture is ideal, the following equation holds:

$$\frac{\partial \mu_2}{\partial x_2} = \frac{RT}{x_2}. \tag{3.1.114}$$

And, in accordance with the definition of mole fractions and weight fractions:

$$\frac{\partial x_2}{\partial w_2} = \frac{(x_1 M_1 + x_2 M_2)^2}{M_1 M_2}, \tag{3.1.115}$$

the diffusion coefficient is, under such conditions:

$$D_{12} = \frac{RT L_{22}}{\varrho w_1 x_2} \frac{(x_1 M_1 + x_2 M_2)^2}{M_1 M_2^2}. \tag{3.1.116}$$

If the mixture is non-ideal, i.e. the activity coefficient depends on the mole fraction, we have:

$$D_{12} = \frac{RT L_{22}}{\varrho w_1 x_2} \frac{(x_1 M_1 + x_2 M_2)^2}{M_1 M_2^2} \left(1 + \frac{\partial \ln \gamma_2}{\partial \ln x_2}\right). \tag{3.1.117}$$

The theory of diffusion based on the thermodynamics of irreversible processes does not have many advantages in the discussion of simple diffusion. In the case of cross-effects, however, it is important.

* * *

Surveying the main lines of investigation of diffusion, it can be established that the up-to-date theoretical discussion of this transport phenomenon starts partly with the Fick law or its generalization; these equations are fundamental in the interpretation of the experimental data. The proportionality factor between mass flux and concentration gradient is, by definition, the conventional diffusion coefficient D. On the other hand, theoretical considerations based on the thermodynamics of irreversible processes lead to correlations which express the proportionality between flux and chemical potential gradient, this latter being the driving force of diffusion. The proportionality factors are the phenomenological coefficients L_{ij}. If flux is related to the same reference system in these two methods, then the factor belonging to the gradient $\partial c/\partial y$, separated from the theoretical expression, is identical with the diffusion coefficient defined by Fick. Thus, D can be related to the macroscopical properties of the solution, for example, to the activity coefficients of the components. The conventional diffusion coefficients D_{ij} implied in the Fick law and the phenomenological diffusion coefficients L_{ij} have been analysed by Wendt [46] in multicomponent systems. The diffusion taking place in an isothermal system containing n neutral components can be described by two sets of equations: partly on the basis of the thermodynamics of irreversible processes, and partly by the Fick correlations. Relating the mass flux to a solvent at rest (this is denoted by superscript 0), we have, on one hand:

$$J = - \sum_{j=1}^{n-1} L_{ij}^0 \, \text{grad} \, \mu_j, \qquad (i = 1, 2, \ldots, n-1), \qquad (3.1.118)$$

on the other hand:

$$J_i^0 = - \sum_{j=1}^{n-1} D_{ij}^0 \, \text{grad} \, c_j, \qquad (i = 1, 2 \ldots, n-1) \qquad (3.1.119)$$

J_i^0 means the mass flux with respect to the solvent-fixed frame of reference, or–in other words–these data give the mass flux of solutes with respect to the local velocity of the solvent. According to Wendt, the relationship between the diffusion and the phenomenological coefficients is:

$$D_{ij}^0 = \sum_{k=1}^{n-1} L_{ik}^0 \frac{\partial \mu_k}{\partial c_j} \qquad (i, j, k = 1, 2, \ldots, n-1). \qquad (3.1.120)$$

The coefficients measured in diffusion experiments correspond to a coordinate system fixed with respect to the diffusion cell, and they are equal, in good approximation, to the diffusion coefficient D_{ij}^V belonging to fixed volume. If V_k is the partial molar volume (cm^3 mole^{-1}) of the kth component, the relationship between the two types of diffusion coefficient is, according to Wendt:

$$D_{ij}^0 = D_{ij}^V + \frac{c_i}{c_0 V_0} \sum_{k=1}^{n-1} V_k D_{kj}^V \qquad (i, j = 1, 2, \ldots, n-1). \qquad (3.1.121)$$

In the limiting case of infinitely dilute solutions $D_{ij}^0 = D_{ij}^V$, while in solutions with a total molarity of 1·0, the difference between them is only 1–5 per cent. With respect to further complicated relationships we have to refer to the original paper.

3.1.2.4 DIFFUSION IN CONCENTRATED SOLUTIONS

In order to describe diffusion in concentrated solutions, it is not just sufficient to take into account the change in the diffusion coefficient with the activity coefficient of the solute, but the participation of the solvent in the process should be considered as well. First of all, diffusion of the solvent in the direction opposite to that of the solute cannot be neglected since, under such conditions, the concentration of the solvent also changes appreciably with the position coordinate. Diffusion is also influenced by solvation of the solute molecules, and they diffuse together with the solvate sphere

Literature on page 248

bound strongly to them (with respect to this, see the paper of Schönert [45a]). Furthermore, solutes can alter the solvent structure too, which mainly modifies the viscosity conditions. Of these effects, counter-diffusion of the solvent can be discussed unequivocally.

Essentially, on the basis of the theory of Hartley and Crank [47], diffusion in concentrated solutions can be described as follows. In concentrated solutions, it is unwise to refer the diffusion flux to a plane fixed with respect to the solvent, since the flux of the solvent cannot be neglected. The conditions can be best handled if plane P is chosen as the plane of fluxes, characterized by unaltered volumes at both sides during diffusion. This essentially means a reference plane fixed with respect to the apparatus. The mass fluxes of the two components across unit surface area of the plane are given by equation (3.1.1) thus:

$$J_1^V = - D_1^V \frac{\partial c_1}{\partial y}, \quad \text{and} \quad J_2^V = - D_2^V \frac{\partial c_2}{\partial y}, \qquad (3.1.122)$$

where diffusion coefficients D_1^V and D_2^V are referred to constant volumes of the two components, and c_1 and c_2 represent their concentrations (the number of moles per unit volume).

Since in concentrated solutions diffusion is accompanied by changes in volume, there is a 'volume flow' through plane P in both directions. If $\overline{V}_1$ and $\overline{V}_2$ denote partial molar volumes* of the two components in the solution–considered independent of concentration–the volume fluxes in the two directions are:

$$- D_1^V \overline{V}_1 \frac{\partial c_1}{\partial y} \quad \text{and} \quad - D_2^V \overline{V}_2 \frac{\partial c_2}{\partial y}. \qquad (3.1.123)$$

Since plane P has been chosen so as to have a net volume transfer equal to zero:

$$D_1^V \overline{V}_1 \frac{\partial c_1}{\partial y} + D_2^V \overline{V}_2 \frac{\partial c_2}{\partial y} = 0. \qquad (3.1.124)$$

Also since c_1 and c_2 are the number of moles of the two components in unit volume, we can write:

$$\overline{V}_1 c_1 + \overline{V}_2 c_2 = 1. \qquad (3.1.125)$$

Differentiating with respect to y, we obtain:

$$\overline{V}_1 \frac{\partial c_1}{\partial y} + \overline{V}_2 \frac{\partial c_2}{\partial y} = 0 \qquad (3.1.126)$$

* Partial molar volume is independent of concentration in good approximation if diffusion takes place between solutions of not too different concentration, i.e. when the differential diffusion coefficient is considered.

(since $\overline{V}_1$ and $\overline{V}_2$ are independent of the position coordinate). Equations (3.1.124) and (3.1.125) are valid simultaneously, consequently:

$$D_1^V = D_2^V = D^V. \tag{3.1.127}$$

That is, in binary systems with a reference plane fixed with respect to constant volume, the diffusion coefficients of the two components are identical if the partial molar volumes do not change in the course of diffusion. Under such conditions, the phenomenon of diffusion can be described by one single mutual diffusion coefficient D^V *(interdiffusion coefficient)* even in concentrated solutions, calculated from concentration gradient of either component 1 or 2.

The flux defined above is, however, due not only to diffusion in a narrower sense, since through volume-fixed reference plane P a certain amount of component 1 is, so to speak, pressed in one direction, and the same volume of component 2 will cross it in the opposite direction; in fact, not by diffusion but in consequence of diffusion. This is why Hartley and Crank introduced *intrinsic diffusion coefficients* D_1' and D_2' for the two components. The total flux of both components consists of two parts: *intrinsic diffusion flux* and *volume flux* which arises from the difference in the partial molar volumes of the two components. In principle, we can choose a plane Q through which no volume flow occurs; in this case, the intrinsic diffusion coefficients D_1' and D_2' refer to the mass flow across a unit surface area of this plane. Since the former assumption implies that the partial molar volumes are constant, $\dfrac{\partial c_1}{\partial y}$ and $\dfrac{\partial c_2}{\partial y}$ are of opposite sign. When, for example in Fig. 3.5, c_1 increases upwards, while c_2 decreases in that direction, the velocity of the increase in volume above plane Q caused by the penetration of component 1 will be $-\overline{V}_1 D_1' \left(\dfrac{\partial c_1}{\partial y}\right)$, and the velocity of the decrease in volume arising from the departure of component 2 will be $+\overline{V}_2 D_2' \left(\dfrac{\partial c_2}{\partial y}\right)$. The net rate of the increase in volume (V') is:

$$\frac{\partial V'}{\partial t} = -\left(\overline{V}_1 D_1' \frac{\partial c_1}{\partial y} + \overline{V}_2 D_2' \frac{\partial c_2}{\partial y}\right). \tag{3.1.128}$$

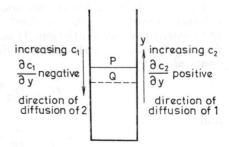

increasing c_1

$\dfrac{\partial c_1}{\partial y}$ negative

direction of diffusion of 2

y

increasing c_2

$\dfrac{\partial c_2}{\partial y}$ positive

direction of diffusion of 1

Fig. 3.5 P is the plane at the two sides of which the volume remains constant in the course of diffusion. Q is the plane across which no liquid volume flows

Literature on page 248

Since no volume-transfer takes place across Q, the displacement of Q with respect to P is equal to the volume flow across plane P, i.e. equation (3.1.128) represents the downward volume flow through P. The volume flow is, thus, accompanied by the transport of component 1 upwards, i.e. in the direction of the diffusion of component 1 whose volume flux is:

$$J_{1, \text{vol}} = -c_1 \frac{\partial V'}{\partial t}. \qquad (3.1.129)$$

In addition to this, there is a transport of component 1 across P by means of intrinsic diffusion:

$$J_{1, \text{i, diff}} = -D_1' \frac{\partial c_1}{\partial y}. \qquad (3.1.130)$$

The total flux of component 1 across plane P is:

$$J_{1, \text{total}} = -D_1' \frac{\partial c_1}{\partial y} + c_1 \left(\overline{V}_1 D_1' \frac{\partial c_1}{\partial y} + \overline{V}_2 D_2' \frac{\partial c_2}{\partial y} \right). \qquad (3.1.131)$$

The total flux across the fixed plane P defines the mutual diffusion coefficient D^V determined experimentally:

$$J_{1, \text{total}} = -D^V \frac{\partial c_1}{\partial y}. \qquad (3.1.132)$$

Comparing this equation with equations (3.1.131) and (3.1.126), it can be seen that:

$$D^V = D_1' + \overline{V}_1 c_1 (D_2' - D_1'). \qquad (3.1.133)$$

The diffusion coefficient D_1' of concentrated solutions can be expressed by the limiting value of D_1^0 extrapolated to infinite dilution, if the deviation of the solution from ideal mixtures is taken into account by the factor $\frac{\partial \ln a_1}{\partial \ln c_1}$. Probably the relative viscosity η/η_2^0 of the mixture should be taken instead of the viscosity η_2^0 of one of the pure components.

The activity in the above thermodynamic correction can be expressed in arbitrary units, since the constant transformation factor is eliminated in the logarithmic differentiation. If activity is expressed by means of mole fractions $(x_1\gamma_1, x_2\gamma_2)$, the intrinsic diffusion coefficients of the two components are:

$$D_1' = D_{12}^0 \frac{\eta_2^0}{\eta} \frac{\partial \ln x_1 \gamma_1}{\partial \ln c_1}, \qquad (3.1.134)$$

and

$$D_2' = D_{22}^0 \frac{\eta_2^0}{\eta} \frac{\partial \ln x_2 \gamma_2}{\partial \ln c_2}, \qquad (3.1.135)$$

where D_{12}^0 denotes the diffusion coefficient of component 1 related to component 2 in infinite dilution, and D_{22}^0 is the self-diffusion coefficient of component 2 in the pure state. Since:

$$c_1 = \frac{x_1}{x_1 \overline{V}_1 + x_2 \overline{V}_2}, \tag{3.1.136}$$

we obtain:

$$\frac{\partial \ln c_1}{\partial \ln x_1} = 1 - \frac{x_1(\overline{V}_1 - \overline{V}_2)}{x_1 \overline{V}_1 + x_2 \overline{V}_2} = \frac{\overline{V}_2 c_1}{x_1} \tag{3.1.137}$$

(considering that $x_1 + x_2 = 1$). Similarly:

$$\frac{\partial \ln c_2}{\partial \ln x_2} = \frac{\overline{V}_1 c_2}{x_2}. \tag{3.1.138}$$

Inserting these equations into equations (3.1.134) and (3.1.135):

$$D_1' = D_{12}^0 \frac{x_1}{\overline{V}_2 c_1} \frac{\eta_2^0}{\eta} \frac{\partial \ln x_1 \gamma_1}{\partial \ln x_1}, \tag{3.1.139}$$

$$D_2' = D_{22}^0 \frac{x_2}{\overline{V}_1 c_2} \frac{\eta_2^0}{\eta} \frac{\partial \ln x_2 \gamma_2}{\partial \ln x_2}. \tag{3.1.140}$$

Further, since the Gibbs–Duhem equation yields:

$$\frac{\partial \ln x_1 \gamma_1}{\partial \ln x_1} = \frac{\partial \ln x_2 \gamma_2}{\partial \ln x_2}, \tag{3.1.141}$$

the mutual diffusion coefficient (interdiffusion coefficient) D^V can be obtained from equations (3.1.133), (3.1.139) and (3.1.140):

$$D^V = \frac{\partial \ln x_1 \gamma_1}{\partial \ln x_1} \cdot \frac{\eta_2^0}{\eta} \left\{ D_{12}^0 x_1 \left(\frac{1}{\overline{V}_2 c_1} - \frac{\overline{V}_1}{\overline{V}_2} \right) + D_{22}^0 x_2 \frac{c_1}{c_2} \right\}. \tag{3.1.142}$$

Considering that:

$$\frac{x_1}{x_2} = \frac{c_1}{c_2} \quad \text{and} \quad c_1 \overline{V}_1 + c_2 \overline{V}_2 = 1,$$

then:

$$D^V = \frac{\partial \ln x_1 \gamma_1}{\partial \ln x_1} \frac{\eta_2^0}{\eta} (x_2 D_{12}^0 + x_1 D_{22}^0), \tag{3.1.143}$$

and, for symmetry reasons:

$$D^V = \frac{\partial \ln x_2 \gamma_2}{\partial \ln x_2} \frac{\eta_1^0}{\eta} (x_1 D_{21}^0 + x_2 D_{11}^0). \tag{3.1.144}$$

This is the so-called *Hartley–Crank equation*.

Literature on page 248

Several experimental investigations have been outlined in order to prove the validity of the Hartley–Crank equation. Wishaw and Stokes [48] observed good agreement in experiments in aqueous solutions of $LiNO_3$ and NH_4NO_3, although in solutions of this latter salt, there are definite effects of ion pair formation. Harned [49] has shown the agreement of this equation with the measured values for several other electrolyte solutions. On the other hand, Janz, Lakshiminarayanan, Klotzin and Mayer [50] observed a significant ion pair formation when analysing the experimental results in $AgNO_3$ solutions on the basis of the extended Hartley–Crank law within the concentration range $0 \cdot 1 - 9$ м.

$$* \ * \ *$$

In concentrated solutions and mixtures, diffusion can give rise to marked changes in volume. Generalizing the Fick law on the basis of the thermodynamics of irreversible processes, taking into account the changes in volume, Sauer and Freise [51] obtained equations which are independent of the origin of the reference coordinate system.

$$* \ * \ *$$

Lamm [52] introduced the frictional coefficient φ referred to unit volume and the molar friction Φ instead of common viscosity in his theory regarding the phenomenological coefficients and diffusion. In a two-component liquid, if $\Phi_1 c_1 = \Phi_2 c_2 = \varphi_{12}$ and the velocities of local movements of the individual components are $\bar{v}_1$ and $\bar{v}_2$, the frictional coefficient referred to unit volume is defined by the following equation:

$$- \operatorname{grad} \mu_1 = \frac{(\bar{v}_1 - \bar{v}_2)\varphi_{12}}{c_1}. \qquad (3.1.145)$$

On the other hand, the diffusion coefficient, according to this theory, is:

$$D_{12} = D_{21} = \frac{RTx_1}{\Phi_2} \cdot \frac{\partial \ln a_1}{\partial \ln x_1} = \frac{RTx_2}{\Phi_1} \cdot \frac{\partial \ln a_2}{\partial \ln x_2}. \qquad (3.1.146)$$

Starting with the Lamm theory, Dullien [53] concluded from investigations on mixtures containing associated liquids (ethanol + benzene, methanol + benzene), that in these cases the kinetic entities of diffusion are the individual molecules rather than the associated complexes. The anomalous concentration dependence of the diffusion coefficient in these liquids has been attributed to the fact that alcohol and benzene molecules are not in disorder. His studies support the opinion that the frictional coefficient is the same for diffusion and viscosity. Dunlop [54] investigated the frictional coefficient starting with a general set of equations, while Albright [55] studied the conditions in a ternary system in which chemical reaction (isomerization) also takes place.

Some complications in the thermodynamics of the theories of diffusion processes have been studied by Irani and Adamson [56a], emphasizing that the diffusion model depends on the liquid structure and its association conditions. From the data on diffusion, however, limited conclusions can only be drawn regarding the liquid structure and the association conditions. Owing to the special mechanism of the exchange between molecules in solutions, the binary diffusion coefficients are not necessarily identical.

Only a small number of detailed investigations have been performed on the diffusion of non-electrolytes in *solvent mixtures*. Recently, the diffusion of iodine has been studied by Nakanishi and Ozasa [56b] in mixtures of water and ethanol, and water and t-butanol. According to their measurements, the diffusion coefficient does not change linearly with composition. The ratio $D\eta/T$ has a maximum in solutions of $0 \cdot 05 - 0 \cdot 1$ mole fraction alcohol content. In accordance with other phenomena, they attribute this to the enlargement of the open structural cavities due to the structure-making effect of alcohols on water, which facilitates translational movement of I_2 molecules.

3.1.2.5 DIFFUSION WITHOUT ACTIVATION ENERGY

The theories of diffusion usually assume that during the migration of the diffusing entities some barrier has to be overcome, which requires an activation energy. In accordance with this concept, the experimentally obtained relationship between $\log D$ and $-1/T$, being nearly linear in several cases, has been interpreted on the analogy of the Arrhenius equation derived for chemical reactions, and an activation energy has been calculated from the slope of the line. In liquids with strong interaction between the entities this is, in the main, justified. However, Hildebrand et al. [56c] pointed out that in simple liquids, consisting of molecules which can be considered rigid spheres, and in gases with densities higher than the critical value, the diffusion phenomena can be understood without assuming a quasi-lattice structure producing an energy barrier. According to Hildebrand's theory, in such liquids all of the molecules participate in the thermal motion, which gives rise to a maximal disorder. In such liquids, the average displacement of the molecules depends partly on the temperature, partly on the ratio of the intermolecular volume V to volume V_0 of the compressed liquid in which the molecules approach one another so closely; that is, they are in such a close-packed arrangement that diffusion becomes impossible. Dymond [56d] also demonstrated on the basis of molecular dynamical considerations that the V/V_0 ratio can accurately be calculated from the self-diffusion coefficients of systems consisting of rigid spheres.

3.2 DIFFUSION OF ELECTROLYTES

Although the diffusion of dissolved electrolytes does not differ in principle from that of dissolved non-electrolytes, two related factors should be taken into account here which do not act in solutions of non-electrolytes. One

Literature on page 248

of them is the electric potential gradient, caused by ions of different mobilities, which markedly alters the rate of the diffusion of ions by means of cross-effects; the other one is due to the electrostatic interaction between ions which does not allow macroscopic separation of ions of opposite charge (i.e. in a macroscopic volume of the solution, no free charge can be agglomerated; this is the *law of electroneutrality*). All of these influence the mobility of ions.

The diffusion of dissolved electrolytes is similar in a certain sense to the conduction of electric current by electrolytes. In both phenomena, ions are displaced by the chemical or electric potential gradient. The main difference between these two phenomena is that, on one hand, anions and cations move in the same direction in the course of diffusion (in one-dimensional processes), while with an electric current, anions and cations move in opposite directions. On the other hand, in the limiting case of infinitely dilute solutions, anions and cations carrying electricity move independently of each other with different velocities, while in diffusion (disregarding the starting moment of diffusion) ions of opposite charge move together at least macroscopically, since otherwise charges would also be separated macroscopically in the solution.

This is connected with the fact that the association of ions (if not too extensive) only slightly influences the diffusion coefficient (mostly enhancing it) because ions diffuse 'pairwise'.* Electric conductivity, however, is decreased appreciably by ion association, since the number of particles transporting electricity decreases.

The conditions are rather simple when one (binary) electrolyte dissociating into two kinds of ions is dissolved in a solvent with no marked dissociation. Under such conditions, the two kinds of ions diffuse with identical velocity, i.e. the process can be described by one single diffusion coefficient. In the presence of more than two types of ions, different ions do not necessarily diffuse with the same velocity, thus the phenomenon can only be described by means of several diffusion coefficients.

Since the mobilities of ions (regarded as the measure of frictional resistance) can be different, in general, different ions move with different velocities under the effect of the same force. The diffusion of electrolytes is initiated by the chemical potential gradient. In the case of diffusion of *one single* dissolved electrolyte, this influences each ion equally, but it accelerates ions of higher mobility to a higher velocity than those of lower mobility.** This results in the development of a layer in the solution which contains an excess of ions of higher mobility, and this creates an electric potential gradient in that layer of the solution in which diffusion takes place. Owing to this, the diffusion of ions of lower mobility is accelerated while that of ions of higher mobility is slowed down so that they move further together with the same velocity. If this were not so, electric charges would

* Namely, the frictional resistance decreases on association of two (or more) ions forming one larger entity.

** The chemical potential gradient influences the associated ion pairs to the same extent as the ions, since the two states of the electrolyte are in equilibrium with each other.

be separated macroscopically in the solution, and, in some macroscopic regions of the solution, free electric charge would be agglomerated. However, this is impossible according to results summed up in the law of electroneutrality.

In electrolytic solutions, each ion is affected by two forces: the chemical potential gradient of ions, and–as a cross-effect–the electric potential gradient resulting from the different mobilities. This latter force is the same for all types of diffusing ions in every case, while the gradient in the chemical potential is identical for the two kinds of ions only in solutions containing *not more than one* electrolyte. In solutions containing several electrolytes, it can be different for the different kinds of ions. Thus, the flux of the diffusion mass current of the ith kind of ions is, at constant temperature,

$$J_i = c_i \bar{v}_i = - \sum_k L_{ik}(\text{grad } \mu_k + z_k \mathbf{F} \text{ grad } \psi), \qquad (3.2.1)$$

according to the thermodynamics of irreversible processes, where $\bar{v}_i$ is the average velocity of diffusion for the ith kind of ions [see equation (3.1.2)], z_k is the valence of the kth kind of ions, ψ is the electric potential resulting from diffusion, and the summation is taken over all the components present. In dilute solutions, the diffusion of the non-dissociating solvent is not significant. Under such conditions, the mass flux is related most conveniently to the solvent which is assumed to be at rest.

The phenomenological coefficient L_{ik} of diffusion depends on the mobility of the ith kind of ions. According to the common definition in electrochemistry, the absolute mobility U_i of ions is the velocity produced by unit electrical potential gradient, if the chemical potential and the temperature are constant.* Considering equation (3.2.1), one has:

$$U_i = \left| \frac{\bar{v}_i}{\text{grad } \psi} \right| = \frac{\mathbf{F}}{c_i} \Sigma_k L_{ik} z_k. \qquad (3.2.2)$$

In the limiting case of dilute ideal solutions, the interaction between ions can be neglected. Consequently, the coefficients corresponding to the interaction between different ions disappear: $L_{ik} = 0$ ($i \neq k$), and the flux of the ith ion is determined by the conduction coefficient L_{ii} only. In this way, equation (3.2.1) has the following simple form:

$$J_i = c_i \bar{v}_i = - L_{ii} (\text{grad } \mu_i + z_i \text{ grad } \psi), \qquad (3.2.3)$$

at constant temperature.

---

* In general, the absolute mobility of a species moving in a viscous medium is the velocity gained stationarily under the effect of unit force (after a sufficient time). In the c.g.s. system, U is the velocity in cm s^{-1} units produced by a force of 1 dyne. Since $1 \text{ V} = 1/299 \cdot 8$ e.s.u. and the charge of a monovalent ion is $4 \cdot 802 \times 10^{-10}$ e.s.u., an electric field strength of 1 V cm^{-1} exerts a force of $z \times 1 \cdot 602 \times 10^{-12}$ dyne on an ion of valence z.

Literature on page 248

3.2.1 DIFFUSION OF BINARY ELECTROLYTES IN DILUTE SOLUTIONS

The relationships regarding the diffusion of electrolytes are relatively simple in dilute solutions of binary electrolytes.*

According to the *thermodynamic theory*, under such conditions the diffusion flux of the cation is:

$$J_c = c_c \bar{v}_c =$$
$$= - L_{11} (\text{grad } \mu_c + z_c \, \mathbf{F} \, \text{grad } \psi) =$$
$$= - L_{11} \, \text{grad } \tilde{\mu}_c, \tag{3.2.4}$$

at constant temperature, as can be obtained from equation (3.2.3), while the flux of the anion is:

$$J_a = c_a \bar{v}_a =$$
$$= - L_{22} (\text{grad } \mu_a + z_a \mathbf{F} \, \text{grad } \psi) =$$
$$= - L_{22} \, \text{grad } \tilde{\mu}_a, \tag{3.2.5}$$

where c_c and c_a are the concentrations of the cations and anions, respectively (the number of moles in unit volume of the solution), z_c and z_a are their charge numbers (valences), μ_c and μ_a are their molar chemical potentials, and $\tilde{\mu}_c$ and $\tilde{\mu}_a$ are the molar *electrochemical potentials* defined as:

and
$$\tilde{\mu}_c = \mu_c + z_c F \psi,$$
$$\tilde{\mu}_a = \mu_a + z_a F \psi. \tag{3.2.6}$$

If electric potential ψ arises only from diffusion (i.e. there is no external potential difference), we have:

$$z_c J_c + z_a J_a = z_c c_c \bar{v}_c + z_a c_a \bar{v}_c = 0, \tag{3.2.7}$$

and since the electric charges cannot be macroscopically separated, i.e. the solution has no net electric charge.

$$z_c c_c + z_a c_a = 0. \tag{3.2.8}$$

Since:
$$J_c = c_c \bar{v}_c$$
and
$$J_a = c_a \bar{v}_a,$$

it follows that:

$$\bar{v}_c = \bar{v}_a = \frac{J_c}{c_c} = \frac{J_a}{c_a}. \tag{3.2.9}$$

---

* Here, the term binary electrolyte means an electrolyte containing two different ions of not necessarily the same valence.

In general, if the total dissociation of 1 mole of electrolyte results in ν_c moles of the cation and ν_a moles of the anion, the flux J_2 of the dissolved electrolyte has the following relationship with the flux of ions:

$$J_2 = \frac{J_c}{\nu_c} = \frac{J_a}{\nu_a}. \tag{3.2.10}$$

The concentration c_2 of the electrolyte, on the other hand, is:

$$c_2 = \frac{c_c}{\nu_c} = \frac{c_a}{\nu_a} = \frac{c_c + c_a}{\nu_c + \nu_a}. \tag{3.2.11}$$

Considering equation (3.2.9) we have:

$$\bar{v}_c = \bar{v}_a = \frac{J_2}{c_2}. \tag{3.2.12}$$

The *average* (or *mean*) *chemical potential* (μ_{ca}) of the electrolyte is defined as:

$$(\nu_c + \nu_a)\mu_{ca} = \nu_c\mu_c + \nu_a\mu_a. \tag{3.2.13}$$

Thus, taking into account equations (3.2.4) and (3.2.5), we can write:

$$(\nu_c + \nu_a) \operatorname{grad} \mu_{ca} = \nu_c \operatorname{grad} \mu_c + \nu_a \operatorname{grad} \mu_a =$$
$$= \nu_c \operatorname{grad} \tilde{\mu}_c + \nu_a \operatorname{grad} \tilde{\mu}_a =$$
$$= -\left(\frac{\nu_c J_c}{L_{11}} + \frac{\nu_a J_a}{L_{22}}\right). \tag{3.2.14}$$

The quantities L_{11} and L_{22} can be expressed by means of the mobilities on the basis of equation (3.2.2):

$$L_{11} = \frac{U_c c_c}{z_c \, \mathrm{F}}, \qquad L_{22} = \frac{U_a c_a}{z_a \, \mathrm{F}}. \tag{3.2.15}$$

In the case of one-dimensional diffusion, the flux of the electrolyte related to the solvent is:

$$J_2 = -\frac{U_c U_a}{U_c + U_a} \cdot \frac{(\nu_c + \nu_a)c_2}{\nu_c |z_c| \, \mathrm{F}} \frac{\partial \mu_{ca}}{\partial y} =$$
$$= -\frac{U_c U_a}{U_c + U_a} \cdot \frac{(\nu_c + \nu_a)}{\nu_c |z_c| \, \mathrm{F}} RT \frac{\partial c_2}{\partial y}, \tag{3.2.16}$$

in ideal solutions, based on equations (3.2.10)–(3.2.15). On the other hand, according to the Fick equation of diffusion (which actually defines the diffusion coefficient) we have:

$$J_2 = -D \frac{\partial c_2}{\partial y} \tag{3.2.17}$$

Literature on page 248.

(since the diffusion of the solvent has been disregarded, the subscript in D_{12} is omitted).

Comparing these latter two equations, the diffusion coefficient of a binary electrolyte in ideal solution becomes:

$$D = \frac{U_c U_a}{U_c + U_a} \cdot \frac{\nu_c + \nu_a}{\nu_c |z_c|} \frac{RT}{F} =$$

$$= \frac{U_c U_a}{U_c + U_a} \cdot \frac{|z_c| + |z_a|}{|z_c||z_a|} \frac{RT}{F} \tag{3.2.18}$$

(since $\nu_c |z_c| = \nu_a |z_a|$). This equation (the so-called *Nernst equation*) is identical with that originally deduced by Nernst in his first theory on diffusion of electrolytes.

If the solution does not behave ideally, it should be considered, as shown by Hartley [57], that chemical potential is defined not by the concentration but by the mean activity ($a_{ca} = \gamma_{ca} c_{ca}$) of the electrolyte, and the mean activity coefficient γ_{ca} varies with concentration. Ionic mobilities also change with concentration. Thus, the diffusion coefficient (if U_c^0 and U_a^0 are the limiting values of the mobilities extrapolated to infinite dilution) is:

$$D = \frac{U_c^0 U_a^0}{U_c^0 + U_a^0} \cdot \frac{|z_c| + |z_a|}{|z_c||z|} \cdot \frac{RT}{F} \left(1 + \frac{\partial \ln \gamma_{ca}}{\partial \ln c_{ca}}\right). \tag{3.2.19}$$

This is the so-called *Nernst–Hartley equation* which can be given in a more simple form too, if the diffusion coefficient D^0 denotes that valid in ideal solution:

$$D = D^0 \left(1 + \frac{\partial \ln \gamma_{ca}}{\partial \ln c_{ca}}\right) = D^0 B \tag{3.2.20}$$

where the factor arising from changes of the activity coefficient is denoted by B. In ideal solutions:

$$\frac{\partial \ln \gamma_{ca}}{\partial \ln c_{ca}} \to 0.$$

The activity coefficients can be determined independently of diffusion, thus the correction factor:

$$\left(1 + \frac{\partial \ln \gamma_{ca}}{\partial \ln c_{ca}}\right) = \left(1 + c_{ca} \frac{\partial \ln \gamma_{ca}}{\partial c_{ca}}\right) = \frac{\partial \ln a_{ca}}{\partial \ln c_{ca}}$$

can be calculated from other measurements. On the other hand, measuring D and determining D^0 by extrapolation, the activity coefficient can be calculated from diffusion measurements.

The diffusion coefficient D directly determined experimentally changes with electrolyte concentration, and this change exceeds the experimental error significantly (as an example, see Table 3.4). The change in the term D/B is much less, but particularly in solutions containing polyvalent ions,

Table 3.4

Diffusion coefficients of some electrolytes given in $cm^2/s^{-1} \cdot 10^{-5}$ units at 25 °C

c_{ca}, mole $\cdot$ 1^{-1}	KCl		LiCl		CaCl$_2$		LaHCl$_3$	
	D	D/B	D	D/B	D	D/B	D	D/B
0	1·993	1·993	1·366	1·366	1·335	1·335	1·293	1·293
0·001	1·964	1·998	1·342	1·366	1·249	1·320	1·175	1·307
0·002	1·954	2·001	1·335	1·366	1·225	1·319	1·245	1·316
0·005	1·934	2·004	1·323	1·368	1·179	1·310	1·105	1·331
0·007	1·925	2·005	1·317	1·368	—	—	1·084	1·327

it also exceeds the experimental error. It can be concluded from this that the change of the diffusion coefficient with concentration is caused not only by changes in the activity coefficients, since changes in ionic mobility can also influence it. In other words, the assumption $L_{ik} = 0$ is not rigorously correct even in dilute solutions. Thus the effect of ions on one another should also be considered, which is described by the kinetic theory of electrolyte diffusion.

Investigating the dependence of the Onsager phenomenological coefficient L_{12}, related to the diffusion of binary electrolytes, on concentration, Miller [58] established that the term L_{12}/I (I is the ionic strength of the solution) as a function of $\sqrt{c}$ tends to the same slope in the case of electrolytes of various valences. This empirical correlation has also been deduced theoretically by Lorenz [59].

Essentially *the kinetic theory* of the diffusion of electrolytes takes into account the consequences of the electric interaction between ionic charges. The electric attraction and repulsion of ions are, to a certain extent, ordering effects which, acting against the disordering effect of the thermal motion, alter the statistical distribution of ions in solution. This effect has been described by the Debye-Hückel theory and its advanced form outlined by Bjerrum, Onsager, Falkenhagen, and others (cf. Sections 4.2.3 and 5.1) giving the best approximation to reality so far.

Without going into details here, we refer only to the fact that the interaction of ions by means of their coulombic forces appears in two effects in connection with transport phenomena. Both effects can be attributed to the fact that, in the vicinity of each ion, ions of opposite charge are in excess. Thus, each ion in solution is surrounded by an *ion atmosphere* of opposite charge. Of course, the ion atmospheres of the individual ions penetrate one another. Around a resting ion, the ion atmosphere is spherically symmetrical, therefore the resultant force on the central ion is zero in the statistical average. If, however, the ion is moving under the effect of a gradient of chemical or electric potential, the ion atmosphere will be deformed, since there is not time enough for the complete development of

Literature on page 248

the equilibrium charge distribution in front of and behind the moving ion because of the finite velocity of its movement. Thus, the opposite charge density in front of the moving ion is lower than that behind it, which results in a retardation force; this appears in enhanced frictional resistance and in decreased mobility. This is the *relaxation effect** (see Section 4.2.3.2).

The other phenomenon arising from electric interaction of ions is the *electrophoretic effect;* the fact that the water molecules in the immediate vicinity of ions move with them and other water molecules are also carried along by them is responsible for this phenomenon. Namely, the ions move in a flowing medium that carries them more or less away. Due to the presence of the ion atmosphere, positive ions restrict the movement of negative ions to a greater extent than that of positive ions, while negative ions have a greater effect on the positive ones in their ion atmosphere. If positive and negative ions move in opposite directions, the electrophoretic effect also appears in the increase of frictional resistance and in the decrease in mobility. On the other hand, the co-current movement of ions of opposite charge can give rise to an increased mobility by this effect.

The *relaxation effect* is appreciable only in the case when positive and negative ions move in opposite directions under the effect of an external potential gradient. In the course of diffusion, however, ions of opposite charge move in identical directions and the relaxation effect does not markedly influence the velocity of diffusion. This can be attributed to the fact that in diffusion the mobilities of ions vary with electrolyte concentration to a far smaller degree than in the case of electric conduction.

The *electrophoretic effect* in diffusion also alters the mobilities of ions, but in a different way than in electric conduction. In diffusion, the interaction of the ionic charges slows down the ions of higher mobility, while it accelerates those of lower mobility. Consequently, the electrophoretic effect increases the mobility of ions of lower mobility and has a reversed effect on the faster ones in the given electrolyte. If the mobilities of different ions are identical, the electrophoretic effect disappears.

The effect of electrophoretic interaction on diffusion was discussed by Onsager and Fuoss [60] on the basis of the Debye–Hückel theory of strong electrolytes (see Section 5.1), and it has been further developed by Stokes [61]. We refer to the original papers for the details of the theory, but the essential results can be summarized as follows.

If the electrophoretic effect does not change the mobility too much (as is usual), this effect can be regarded as altering the mobilities of the cation and anion by factors $(1 + \delta_c)$ and $(1 + \delta_a)$, respectively in the equation of the diffusion coefficient, equation (3.2.19) (δ_c, $\delta_a \ll 1$). Taking into account the electrophoretic effect, the diffusion coefficient is:

$$ D = \frac{U_c^0(1 + \delta_c)U_a^0(1 + \delta_a)}{U_c^0(1 + \delta_c) + U_a^0(1 + \delta_a)} \cdot \frac{|z_c| + |z_a|}{|z_c||z_a|} \cdot \frac{RT}{F}\left(1 + \frac{\partial \ln \gamma_{ca}}{\partial \ln c_{ca}}\right). \qquad (3.2.21) $$

* The name indicates that this effect can be attributed to the finite relaxation time needed for the rearrangement of the disturbed equilibrium distribution.

The relationships for diffusion can be more easily compared with those for electric conduction, when in this latter equation the relative mobilities, i.e. the *equivalent conductivities* (λ_c, λ_a) *of ions*, are considered instead of the absolute mobilities. According to the definition:

$$\lambda_c = F U_c$$

and

$$\lambda_a = F U_a.$$

(3.2.22)

Inserting these values into equation (3.2.21), we obtain the relationship between the diffusion coefficient of the electrolyte and the equivalent conductivities of ions:

$$D = \frac{\lambda_c^0(1 + \delta_c)\lambda_a^0(1 + \delta_a)}{\lambda_c^0(1 + \delta_c) + \lambda_a^0(1 + \delta_c)} \cdot \frac{|z_c| + |z_a|}{|z_c||z_a|} \cdot \frac{RT}{F^2}\left(1 + \frac{\partial \ln \gamma_{ca}}{\partial \ln c_{ca}}\right).$$

(3.2.23)

The δ quantities related to the electrophoretic effect can be calculated from the theory of strong electrolytes. Because of the complicated nature of the conditions, an approximate method should be applied in the calculation, expressing the effect by means of expansion into a series retaining the first n terms, depending on the actual requirements. If the velocities of the cation and anion are altered by values Δv_c and Δv_a, respectively, produced by a given force under given conditions, and the common velocity, of the motion of the two kinds of ions in diffusion is $\bar{v}$; further, if t_c and t_a are the transference numbers of the cation and anion, respectively, we have:

$$\delta_c = \frac{\Delta v_c}{\bar{v}} = \frac{F^2}{N}\Sigma A_n \frac{\dfrac{z_c^{2n}}{t_c^0 \Lambda^0} + \dfrac{z_c^n z_a^n}{t_a^0 \Lambda^0}}{a^n(z_c - z_a)},$$

(3.2.24)

and

$$\delta_a = \frac{\Delta v_a}{\bar{v}} = \frac{F^2}{N}\Sigma A_n \frac{\dfrac{z_c^n z_a^n}{t_c^0 \Lambda^0} + \dfrac{z_a^{2n}}{t_a^0 \Lambda^0}}{a^n(z_c - z_a)}$$

(3.2.25)

according to the complicated approximate calculations, not discussed in detail here [60, 61], where Λ^0 is the equivalent conductivity of the electrolyte extrapolated to infinite dilution. In these expressions, a is the 'radius' of the ion, A_n is a quantity depending on the dielectric constant of the solution, on the characteristic distance $\varkappa$ (see Section 5.1.3), on the value of a, and on temperature. (It should be noted that $z_c - z_a = |z_c| + |z_a|$). For the diffusion coefficient, this theory yields the following equation:

$$D = (D^0 + \Sigma \Delta_n)\left(1 + \frac{\partial \ln \gamma_{ca}}{\partial \ln c_{ca}}\right) =$$

$$= (D^0 + \Sigma \Delta_n)\left(1 + c_{ca}\frac{\partial \ln \gamma_{ca}}{\partial c_{ca}}\right),$$

(3.2.26)

Literature on page 248

where D^0 is the Nernst-type limiting value of the diffusion coefficient in infinitely dilute solutions given by equation (3.2.18), or expressed by ionic conductivities:

$$D^0 = \frac{|z_c| + |z_a|}{z_c z_a} \cdot \frac{RT}{F^2} \frac{\lambda_c^0 \lambda_a^0}{\lambda_c^0 + \lambda_a^0}. \tag{3.2.27}$$

For members of the series Δ_n, representing the electrophoretic effect, the following equation applies:

$$\Delta_n = kTA_n \frac{(z_c^n t_a^0 + z_a^n t_c^0)^2}{a^n |z_c z_a|}. \tag{3.2.28}$$

For the restrictions in such a calculation, see Stokes's paper [6]. Due to these restrictions, only the first two terms ($n = 2$) in the expansion into series can be taken into account in symmetrical binary electrolyte solutions ($z_c = z_a = z$), while for asymmetrical types, one is limited to the first term ($n = 1$). Thus, the diffusion coefficient of symmetrical types of electrolytes in dilute solutions is given by equation (3.2.26) as follows:

$$D = (D^0 + \Delta_1 + \Delta_2)\left(1 + \frac{\partial \ln \gamma_{ca}}{\partial \ln c_{ca}}\right). \tag{3.2.29}$$

This is the so-called *Onsager–Fuoss equation*.

For aqueous solutions of electrolytes consisting of monovalent cations and anions, at 25 °C:

$$\Delta_1 = -8.07 \times 10^{-6} \frac{(t_a^0 - t_c^0)^2}{1 + \varkappa a} \sqrt{c} \tag{3.2.30}$$

and

$$\Delta_2 = +8.77 \times 10^{-21} \frac{\Phi_2(\varkappa a)}{a^2}, \tag{3.2.31}$$

where $\Phi_2(\varkappa a)$ is a complex function of the characteristic distance and ionic radii (its value can be obtained from tables [62]), furthermore, c and a are expressed in mole litre^{-1} and cm units, respectively.

In solutions of asymmetrical electrolytes, the diffusion coefficient is given by Stokes [61] as:

$$D = (D^0 + \Delta_1)\left(1 + \frac{\partial \ln \gamma_{ca}}{\partial \ln c_{ca}}\right) \tag{3.2.32}$$

and

$$\Delta_1 = -8.07 \times 10^{-6} \frac{(z_c t_a^0 + z_a t_c^0)^2}{|z_c z_a|} \frac{\sqrt{I}}{1 + 0.3291 \times 10^8 \, a\sqrt{I}}, \tag{3.2.33}$$

where I is the ionic strength of the solution.*

---

* According to Onsager and Fuoss [60], equation (3.2.30) also applies to asymmetrical electrolytes.

Comparison with experiments. The checking of the theoretical relationships concerning the diffusion coefficient of electrolytes is rather difficult on the basis of experimental data, since the diffusion coefficient depends on both the concentration gradient of free enthalpy and–to a smaller extent–on the electrophoretic effect. Theoretically, the effect of these two factors can be easily separated, since that of the free enthalpy can be calculated from changes of the mean activity coefficients of electrolytes in equations (3.2.26), (3.2.29) and (3.2.32) regarding the diffusion coefficient, while the activity coefficient can be obtained from equilibrium measurements independent of diffusion. However, since the electrophoretic effect is small in comparison with that of free enthalpy, the concentration dependence of the activity coefficient should be accurately known in order to separate the electrophoretic effect on the diffusion coefficient reliably. Otherwise, the uncertainty arising from the experimental error of the activity coefficient measurement can conceal the influence of the electrophoretic effect. The comparison is also made difficult by the fact that the above relationships regarding the diffusion coefficient apply to dilute solutions ($c_{ca} < 0.01$ mole/litre) only. In more concentrated solutions, the changes in viscosity and volume caused by diffusion should also be taken into account, as well as the hydration of ions.

In dilute solutions of **1–1** electrolytes, the electrophoretic effect is small (Fig. 3.6). The calculated values of the diffusion coefficient modified by this effect show good agreement with the observed values (Table 3.5; the ionic size a taken is the value giving the best fit in the activity coefficient calculations).

The theoretical calculation is more uncertain for electrolytes of asymmetrical valences. The conditions are also made complicated by the appreciable extent of ion pair formation even in dilute solutions containing ions of higher valence. The mobility of ion pairs is, in general, somewhat greater than that of free ions, since ion pair formation is accompanied by a decrease in hydration, so that the size of the diffusing ion pair is smaller than the sum

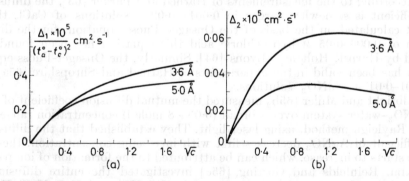

Fig. 3.6 Dependence of electrophoretic corrections $\varDelta_1$ and $\varDelta_2$ on concentration and ionic size in aqueous solutions of 1–1 electrolytes at 25 °C

Literature on page 249

Table 3.5

The value ($(D^\circ + \Delta_1 + \Delta_2) \times 10^5$ cm^2 s^{-1} in aqueous solutions
at 25 °C for 1–1 electrolytes according to Robinson and Stokes

c_{ca}, mole · l^{-1}	LiCl ($a = 4\cdot32$ Å)		NaCl ($\alpha = 3\cdot97$ Å)		KCl ($\alpha = 3\cdot36$ Å)	
	calculated	found	calculated	found	calculated	found
0	1·366	—	1·610	—	1·993	—
0·001	1·366	1·366	1·611	1·611	1·995	1·998
0·002	1·366	1·366	1·612	1·613	1·996	2·001
0·005	1·366	1·368	1·614	1·617	1·999	2·004
0·01	1·368	1·369	1·616	1·618	2·003	2·009

Table 3.6

Diffusion coefficients D of 2–1 and 1–2 electrolytes in aqueous solutions
according to Robinson and Stokes

c_{ca}, mole · l^{-1}	CaCl$_2$		Li$_2$SO$_4$		Na$_2$SO$_4$	
	calculated	found	calculated	found	calculated	found
0	1·330	—	1·041	—	1·230	—
0·001	1·257	1·249	0·989	0·990	1·162	1·175
0·002	1·223	1·225	0·978	0·974	1·141	1·160
0·005	1·185	1·179	0·955	0·950	1·097	1·123

of the sizes of two separated ions. There is an uncertainty in the calculation
with respect to the value of a, as well. In spite of all this, the deviation
of the calculated values from the measured ones is not too high in several
cases (see examples in Table 3.6).

According to the measurements of Harned and Parker [63], the diffusion
coefficient is somewhat larger in 0·001 – 0·02 M solutions of CaCl$_2$ than
that calculated on the basis of the Onsager–Fuoss equation; for the diffusion of 0·006–0·05 M hydrochloric acid this equation has been found to
hold by Harpst, Holt and Lyons [64]. Similarly, the Onsager-Fuoss equation has been valid in the experiments of Harned and Shropshire [65a] in
0·001–0·01 M NaNO$_3$ solutions.

Albright and Miller [65b] measured the mutual diffusion coefficient of the
AgNO$_3$–water system over a broad (0·05 – 8 mole/l) concentration range by
the Rayleigh method, using laser light. They established that the diffusion
coefficient of AgNO$_3$ decreases first with increasing concentration, then it
also starts to increase, which can be attributed to the formation of ion pairs.

Kim, Reinfelds and Gosting [65c] investigated the entire diffusional
behaviour of two ternary systems (H$_2$O–KCl–HCl and H$_2$O–NaCl–HCl).
They evaluated their results by the Miller method [65d] and outlined a
general picture of the diffusion of two electrolytes containing a common
ion. The fact that the ternary transport properties can be calculated from

those of the corresponding binary systems indicates that the mixing of two electrolytes does not alter significantly the properties predominating in the transport processes, even if proton transport is also involved. Proton transport is, according to several observations, sensitive to the changes in the liquid structure of water (cf. Sections 4.2.2.2 and 4.7.3); however, in dilute solutions the alterations of the structure of water due to the Na^+ and K^+ ions do not modify dramatically the mobility of the proton ('hydrogen ion') and the respective transport mechanism.

* * *

It occurs sometimes, mainly in electrochemical practice, that diffusion of electrolytes is also influenced by an external electric field. For such conditions, the theory of *polarized diffusion* has been elaborated by Laity [66].

Relationship between diffusion coefficients of ions and the structural change caused by them in solutions. In ordinary measurements on diffusion of dissolved electrolytes, the diffusion coefficients of the individual ions cannot be determined directly. For the limiting case of infinite dilution, however, the independent diffusion of the ith type of ions can be obtained from the Nernst equation of the diffusion coefficient which is similar to equation (3.2.18):

$$D_i = \frac{U_i^0 RT}{z_i F} = \frac{\lambda_i^0 RT}{z_i F^2} , \qquad (3.2.34)$$

where U_i^0 is the absolute mobility of the ith ions and $\lambda_i^0 = U_i^0 F$ is the limiting value of the equivalent conductivity in infinite dilution. Since this latter can be obtained from conductivity measurements by means of extrapolation (cf. Section 4.2.3), the diffusion coefficient of the ith ion can be calculated and is, in approximation, identical with the trace diffusion coefficient (cf. Section 3.4.5), disregarding the electrostatic interaction between ions. The calculated diffusion coefficients of some ions, obtained in this way, are summarized in Table 3.7.

On the basis of the simple hydrodynamic theory of diffusion, it could be expected that the larger the atomic (or molecular) weight of the ion, the smaller its mobility. However, in Table 3.7, it can be seen that the diffusion coefficient of the Li^+ ion with the lowest atomic weight is the smallest and, for example, in the column of alkali metal ions the diffusion coefficient increases with increasing atomic weight. This phenomenon is well known in connection with equivalent conductivity (cf. Section 5.2.4.1), and, in general, it is explained by the fact that ions of identical charge become the more hydrated, the smaller their radius. On the basis of the detailed analysis of the conditions, however, Vdovenko, Gurikov and Legin [67a] showed that this order of the diffusion coefficients cannot be simply interpreted by the increased size of the hydrated ion in diffusion. We simply refer to the fact that the activation energy of diffusion of ions, ΔH_i, (calculated

Literature on page 249

Table 3.7

Diffusion coefficients of some ions calculated from their conductivities
and the activation energies ΔH_i of their diffusion at 25 °C

Ion	Ionic mass	λ_i^0, cm · ohm^{-1} · mole^{-1}	D_i, cm^2 · s^{-1}	ΔH_i kcal · mole^{-1}
Li$^+$	6·9	33·6	1·03	4·57
Na$^+$	23·0	50·1	1·33	4·36
K$^+$	39·1	73·5	1·95	3·96
Cs$^+$	132·9	77·2	2·05	3·84
Ag$^+$	107·9	61·9	1·65	—
NH$_4^+$	18·0	73·5	1·96	—
F$^-$	19·0	55·4	1·47	—
Cl$^-$	35·4	76·3	2·03	4·08
I$^-$	126·9	76·8	2·04	3·98
NO$_3^-$	62·0	71·5	1·90	—
ClO$_4^-$	99·5	67·3	1·79	—
(H$_2$O)	18·0	—	1·93	4·16

from the temperature coefficients) does not differ too much from the
activation energy of self-diffusion of pure water (Table 3.7), which could
not be understood if the difference in the diffusion coefficients arose simply
from differences in the sizes of species. In the interpretation of this phenomenon, the combined effect of different factors should be considered.
Of these, the changes in the water structure caused by the field and size
of ions play an important role (Section 1.4.2). Furthermore, a special
condition emphasized by Frank *et al.* [68], should also be considered,
namely that ions are usually surrounded by a double hydrate sphere (Section 5.2). In the inner shell A around the ion, the water molecules are
bound strongly by ion-dipole forces and their mobility is lower than that in
pure water. In the adjacent layer B the molecules are more disordered,
their translational and rotational movement being freer than in water regions
far from ions where the ions have not disturbed the structure. Thus, in
layer B, the water structure is distorted to a certain extent.

According to Vdovenko, Gurikov and Legin [67a], the Samoilov theory
of thermal motion and diffusion can be extended to ions of double hydrate
spheres too. If E_A and E_B denote the energy barrier between two adjacent
equilibrium positions of water molecules in the inner and outer layers,
i.e. the activation energy of their displacement, the average lifetime of
molecules in these two layers will be:

$$\tau_A = \tau_A^0 \exp\left[\frac{E_A}{RT}\right],$$

and   (3.2.35)

$$\tau_B = \tau_B^0 \exp\left[\frac{E_B}{RT}\right],$$

where τ_A^0 and τ_B^0 are constants defined by the frequency of vibration about the equilibrium position. In approximation, ions are displaced in the solution by two means:

(i) they move without their hydrate spheres, or, equivalently, the water molecules move away from the vicinity of the ion to a farther point; this requires an activation energy E_A;

(ii) the ions are displaced together with layers A bound tightly to them.

The joint jump of these ion hydrate complexes to an adjacent equilibrium position means overcoming a high energy barrier; therefore, it is more probable that thermal motion and diffusion of hydrated ions takes place in several elementary steps involving translational and rotational movements. The authors mentioned above suppose that in each of these steps a maximum of one water molecule moves together with the ion. In the diffusion of a hydrated ion (ion + layer A), two stages should in fact be distinguished: the displacement of some molecules from their equilibrium positions located nearest to the inner hydrate complex in layer B, and the movement of the hydrated complex occupying the hole so formed. The activation energy of the second step is small, thus it can be assumed that the activation energy in diffusion of the the hydrated complex agrees with the value of E_B which controls the exchange of molecules within layer B. Both types of displacement of the ions contribute to diffusion in the ratio of the corresponding statistical weights (C_A, C_B):

$$D_i = C_A \exp\left[-\frac{E_A}{RT}\right] + C_B \exp\left[-\frac{E_B}{RT}\right], \qquad (3.2.36)$$

if these two processes can be regarded as independent.

On the other hand, the self-diffusion coefficient D_w^* of water is:

$$D_w^* = C_0 \exp\left[-\frac{E_w}{RT}\right], \qquad (3.2.37)$$

where E_w is the height of the barrier between two adjacent equliibrium positions of molecules in pure water, and C_0 is a constant.

From the temperature dependence of the diffusion coefficient of ions and the self-diffusion coefficient of water E_A and E_B can be calculated on introducing some approximating assumptions. Analysing the results, it can be concluded that hydrate spheres A and B exist around each ion. For water molecules bound strongly in layer A, E_A is larger than E_w, while for water molecules bound loosely in the second hydrate sphere B, E_B is smaller than E_w. Even around ions, as strongly hydrated as Li^+, a second layer of distorted structure can also be found. On the other hand, around large 'negatively hydrated' ions, e.g. Cs^+, I^-, (Sections 1.4.2 and 5.2.1) the mobility of the closest water molecules is smaller than in pure

Literature on page 249

water.* According to this, water molecules of both lower and higher mobilities than that in pure water can be found around each ion. If the decrease in mobility is higher in layer A than the increase in mobility in layer B (i.e. with respect to the whole hydrate sphere, the mobility of the water molecules decreases), the phenomenon of positive hydration occurs. If, however, the increase in the mobility in layer B predominates, negative hydration takes place. According to the calculations regarding the values of E_A and E_B, Li^+ and Na^+ ions are displaced with positive hydration, while K^+, Cs^+, Cl^- and I^- ions exhibit negative hydration. For Ag^+ ions of a radius of $1 \cdot 13$ Å, E_A is approximately equal to E_B; these, as well as NO_3^- ions, do not on average change the mobility of water molecules in their vicinity, ClO_4^- ions, however, move with positive hydration, according to these measurements. E_B has about the same value for monovalent ions of radii larger than that of K^+ ion which indicates that, around large ions, the structural change in layer B is mainly due to the electrostatic field that changes identically with the distance around each monovalent ion. Around small ions the water molecules in layer A are nearer to one another and to the ion than in the case of the large ones. Due to this latter condition, the mutual repulsion between water molecules in layer A is greater, which facilitates the departure of these molecules from layer A. Investigating the order of various effects in connection with hydration, some conclusions can be drawn regarding the actual hydrate sphere with a predominant role in the given phenomenon.

Conclusions regarding the structure of electrolyte solutions can also be drawn from studies on self-diffusion, as emphasized recently by Chemla [67a]. In his work, the structural factors influencing the self-diffusion were investigated by the isotope-labelling technique. He established quantitative relationships between the transport coefficients of the components of the solution and the association of ions.

3.2.2 EFFECTS OF ASSOCIATION OR INCOMPLETE DISSOCIATION ON DIFFUSION IN DILUTE SOLUTIONS

In electrolyte solutions, the mutual interaction of ions modifies their statistical distribution by various modes without destroying their relative independence. In many cases, however, ions of opposite charge form some closed units, in addition to this. This can take place partly on the effect of electrostatic attraction (ion association), so that the products are mainly ion pairs which more or less retain their hydrate spheres, and no covalent bond is formed between them. This phenomenon often appears in not too dilute solutions of strong electrolytes. On the other hand, they can form neutral molecules by means of covalent bonds between ions of opposite

---

* According to Samoilov, positive hydration is transformed into the negative type at an ionic radius of about $1 \cdot 1$ Å; this size is intermediate between that of Na^+ and K^+ ions.

charge, which is the reverse process to the dissociation of molecules; this predominates in solutions of weak electrolytes.

The formation of ion pairs or molecules in electrolyte solutions alters the diffusion conditions in at least two respects. Partly, the activity of ions is decreased, and the gradient of free enthalpy is changed; partly, their frictional resistance is decreased by the formation of one larger species from smaller ones. If the change in free enthalpy is taken into account by the change in the activity coefficient with concentration, determined experimentally in the given solution, as was done in the calculations in the previous sections, association has no effect on this factor. This is because the ion pairs and the non-dissociated molecules are in equilibrium with the ions, and the chemical potential is identical in the equilibrium states. Thus, in both states, the gradient of free enthalpy is the same. However, the decrease in the frictional resistance acts towards the increase of the diffusion coefficient (unlike electric conductivity which is decreased by association).

The formation of ion pairs or molecules alters the electrophoretic effect as well, because the ion concentration decreases from c_{ca} to αc_{ca} in solutions of binary electrolytes, if α is the degree of dissociation of the electrolyte.

If one assumes that D_{ca}^0 is the diffusion coefficient of ion pairs and non-dissociated molecules in ideal solutions, the diffusion coefficient of a partly dissociated or associated symmetrical electrolyte will be, according to equation (3.2.29):

$$D = \{\alpha(D^0 + \varDelta_1 + \varDelta_2) + 2(1 - \alpha)D_{ca}^0\}\left(1 + \frac{\partial \ln \gamma_{ca}}{\partial \ln c_{ca}}\right). \qquad (3.2.38)$$

This equation shows quite a good agreement with experiment, which is proved by the investigations of Harned and Hudson [69], for example, on zinc sulphate solutions.

According to the experiments, the ion pair formation is small in dilute solutions of $1-1$ electrolytes, so the estimation of D_{ca}^0 is difficult. In more concentrated solutions, however, it becomes significant, as indicated by the studies of Wishaw and Stokes [70] on concentrated ammonium nitrate solutions.

The conditions are similar in solutions of weak electrolytes which dissociate to a small extent only. Of course, the electrophoretic effect is negligible in them, while the diffusion of non-associated molecules plays an important role. With respect to this latter phenomenon, dissolved molecules can affect the solvent appreciably. This is shown, for example, by diffusion experiments on citric and acetic acid in aqueous solutions investigated by Müller and Stokes [71] and Vitagliano and Lyons [72], respectively; these yielded a value of 0.657×10^{-5} and 1.201×10^{-5} cm^2 s^{-1} for D_{ca}^0 of non-dissociated citric and acetic acid, respectively, while the Nernst limiting value of the diffusion coefficient of the monocitrate ion (RT λ^0/F^2) is 0.81×10^{-5}, and that of the acetate ion is 1.088×10^{-5}. The diffusion co-

Literature on page 249

efficient of the citric acid molecule is thus much lower than that of the monocitrate ion, while for the acetic molecule it is higher than for the acetate ion. Considering also that the mobility of the acetate ion is small for its size and, further, that the activity coefficients of acetate salts are relatively large, a strong interaction of the acetate ion with water molecules or with the structure of water can be assumed. The relatively high mobility of the monocitrate ion, however, indicates that this ion has a strong structure-breaking effect on water.

Measuring the diffusion coefficient in phosphoric acid solutions in the 0.1–6.0 mole/1 concentration range, Edwards and Huffman [73] have established that the measured values are in good agreement with the calculated ones when taking into account partial dissociation and the relationship $D = kT/6\pi\eta r$. In the calculations on friction using the value of $kT/4\pi\eta r$, however, a significant deviation can be observed. They concluded from this (Section 3.1.2.1) that at the 'surface' of the diffusing phosphoric acid molecules no slipping occurs; thus the diffusing particles are surrounded by a layer of water molecules moving together with them, i.e. they are hydrated.

From the diffusion coefficients of CdI_2 solutions in the 0.01–1.0 M concentration range, Garland, Tong and Stockmayer [74] concluded that the apparent mobility of the diffusing particles changes appreciably with concentration. On the basis of the data obtained–in agreement with other observations–they calculated the mobility and stability of CdI^+ and CdI_2 in approximation.

A more detailed theory on the diffusion of strong electrolytes in solutions, in which the interaction leading to association is significant, has been elaborated by Kiprianov [75].

Wendt [76a] investigated the diffusion conditions in ternary systems: in solutions containing only strong electrolytes, as well as in solutions containing both strong and weak electrolytes. From the approximative equations deduced, the relationships between the four phenomenological diffusion coefficients and the limiting values of ionic conductivities and the concentration can be obtained. From the known or estimated values of the phenomenological coefficient and the chemical potential gradient, the four diffusion coefficients of the system can be calculated. For the systems $NaCl-KCl-H_2O$, $LiCl-KCl-H_2O$ and $LiCl-NaCl-H_2O$ containing 1–1 electrolytes, the calculated values are in good agreement with the experimental ones; however, for the system $H_2SO_4-Na_2SO_4-H_2O$ containing 2–1 electrolytes, significant deviations appear.

On the basis of the Onsager theory regarding the conduction (L_{ij}) and frictional (R_{ij}) coefficients Pikal [76b] demonstrated that the term describing the formation of ion pairs in the limiting law is a natural consequence of the electrostatic interaction, without requiring the introduction of an adjustible parameter to account for the mobility of ion pairs. The equation derived from this theory containing terms $c^{1/2}$, $c \lg c$, c and $c^{3/2}$, is in good agreement with the experience.

3.2.3 DIFFUSION OF BINARY ELECTROLYTES IN CONCENTRATED SOLUTIONS

In principle, similar considerations apply to the diffusion of electrolytes in concentrated solutions as for that of non-electrolytes (Section 3.1.2.4). However, due to often significant hydration of ions, the motion of water in the hydrate sphere and the change in viscosity are also to be taken into account, in addition to the counter-diffusion of the solvent. One can start with equation (3.1.144) of Hartley and Crank to calculate the diffusion coefficient, following the deduction of Robinson and Stokes [77]. If component 1 is free water (not bound in hydrate spheres), while component 2 is the hydrated electrolyte whose 1 mole is bound to h moles of water, the mutual diffusion coefficient will be (relating the flux to a plane separating two parts in which the volume remains constant):

$$D = \frac{\partial \ln x_e \gamma_e}{\partial \ln x_e} \frac{\eta_w^0}{\eta} (x_w D_{ew}^0 + x_e D_{ww}^0), \qquad (3.2.39)$$

where the subscripts w and e refer to free water and the hydrated electrolyte, respectively. In this expression, x_e is the ratio of the number of diffusing electrolyte particles and the total number of diffusing entities. In fact, diffusion of ions (and not of electrolyte molecules) takes place in electrolyte solutions; however, ions of opposite charge cannot separate macroscopically from each other because of the interaction of the electric charges of ions, so they diffuse together with identical velocity, therefore we can use the concentration, partial molar volume, etc., of the electrolyte as a whole in the calculations on diffusion. The term $\partial \ln x_e \gamma_e = \partial \ln a_e$ is, however, an exception,* because the dissolved electrolyte becomes hydrated and dissociated, which results in an increase in the number of entities in comparison with the 'non-dissociated' electrolyte. If one 'molecule' gives ν ions in the course of the total dissociation, $a_e' = a_{ca}^\nu$ is the conventional mean activity of the non-hydrated salt. Due to hydration in aqueous solutions, we have:

$$\mathrm{d} \ln a_e = \mathrm{d} \ln a_e' + h\, \mathrm{d} \ln a_w, \qquad (3.2.40)$$

where a_w is the activity of water. According to the Gibbs–Duhem equation, however, we obtain (m is the molarity of the solution):

$$\mathrm{d} \ln a_w = -\frac{m}{55 \cdot 51} \mathrm{d} \ln a_e' \qquad (3.2.41)$$

and

$$\mathrm{d} \ln a_e = (1 - 0 \cdot 018\, hm)\, \mathrm{d} \ln a_e' =$$
$$= (1 - 0 \cdot 018\, hm)\, \nu\, \mathrm{d} \ln a_{ca}. \qquad (3.2.42)$$

---

* Due to the logarithmic form of this expression, $\mathrm{d} \ln a_e = 1/a_e\, (\mathrm{d} a_e)$ is independent of the units of activity and concentration used.

Literature on page 249

To interpret equation (3.2.39) correctly, the limiting values of D_{ew}^0 and D_{ww}^0 should be elucidated. In the limiting case of dilute solutions ($x_e \to 0$), this equation is reduced to Nernst's one in equation (3.2.18). According to this:

$$D_{ew}^0 = \frac{D^0}{\nu}, \qquad (3.2.43)$$

where D^0 is the Nernst-type limiting value of the diffusion coefficient, and:

$$D^0 = \frac{U_c U_a}{U_c + U_a} \cdot \frac{|z_c| + |z_a|}{|z_c||z_a|} \cdot \frac{RT}{F} =$$

$$= \frac{\lambda_c \lambda_a}{\lambda_c + \lambda_a} \cdot \frac{|z_c| + |z_a|}{|z_c||z_a|} \cdot \frac{RT}{F^2}. \qquad (3.2.44)$$

Taking into account the electrophoretic effect, i.e. considering a moderately dilute solution in which this effect cannot be neglected, but the diffusion of the solvent in the opposite direction is still not significant, we have:

$$D_{ew} = \frac{D^0 + \Delta_1 + \Delta_2}{\nu}. \qquad (3.2.45)$$

However, D_{ww}^0 is the diffusion coefficient of water in an infinitely dilute solution where the water structure is not altered appreciably by the dissolved electrolyte, i.e. this is the self-diffusion of pure water and it can be written as:

$$D_{ww}^0 = D_w^*, \qquad (3.2.46)$$

where D_w^* is the self-diffusion coefficient of pure water at the given temperature and pressure. Further, the mole fractions can be expressed as follows:

$$x_w = \frac{55 \cdot 51 - hm}{55 \cdot 51 - hm + m}, \quad x_e = \frac{m}{55 \cdot 51 - hm + m}; \qquad (3.2.47)$$

thus

$$\frac{\partial \ln x_e}{\partial \ln m} = \frac{1}{1 + 0 \cdot 018 \, (1 - h)m}. \qquad (3.2.48)$$

With respect to these, equation (3.2.39) can be transformed into:

$$D = D_{ew}^0 \cdot \frac{\partial \ln x_e}{\partial \ln m} \cdot \frac{\partial \ln m}{\partial \ln x_e} \left(x_w + x_e \frac{D_{ww}^0}{D_{ew}^0} \right) \frac{\eta_w^0}{\eta} =$$

$$= D^0 \frac{\partial \ln a_{ca}}{\partial \ln m} \, (1 - 0 \cdot 018)m \left\{ 1 + 0 \cdot 018 \, m \left(\nu \frac{D_w^*}{D^0} - h \right) \right\} \frac{\eta_w^0}{\eta}. \qquad (3.2.49)$$

Table 3.8

Comparison of calculated and measured diffusion coefficients of KCl at 25 °C

m	$D_{measured}$	$\dfrac{\eta}{\eta_0}$	f(D)	$f(D)\dfrac{\eta}{\eta_0}$
0	1·610 ($= D^0$)	1·000	(1·00)	(1·00)
0·01	1·547	1·001	1·001	1·002
0·05	1·506	1·004	0·997	1·001
0·10	1·484	1·009	0·989	0·998
0·30	1·477	1·027	0·982	1·009
0·50	1·474	1·046	0·965	1·009
1·00	1·482	1·094	0·927	1·014
2·00	1·511	1·205	0·838	1·010
3·00	1·538	1·341	0·752	1·008
4·00	1·567	1·509	0·678	1·023

In this expression,* the electrophoretic effect is neglected. This latter can be taken into account by substituting $(D^0 + \Delta_1 + \Delta_2)$ for D^0. In equation (3.2.49) concentration and flux are related to the hydrated solute. However, in the evaluation of the experimental results, hydration is neglected in calculating the concentration in the usual way. The volume concentration of the electrolyte (e.g. in $mole \times cm^{-3}$ units), however, is not influenced by hydration. In the usual calculations, diffusion coefficients and flux are based on the number of moles per cm^3 of the anhydrous solute.

Table 3.8 shows data of KCl solutions at 25 °C which can serve as an example for comparing the theoretical relationship in equation (3.2.49) with the measured data. The data applied in the calculation are: the self-diffusion coefficient of water D_w^* is equal to $2\cdot4 \times 10^{-5}$ cm^2 s^{-1}, the activity coefficients have been taken from tables containing data of equilibrium measurements; in the calculation of the electrophoretic effect, $a = 3\cdot97$ Å. The deviation of the theoretical values from the measured ones is given by the following ratio:

$$f(D) = \frac{D_{measured}}{(D^0 + \Delta_1 + \Delta_2)\left(1 + m\dfrac{\partial \ln \gamma_{ca}}{\partial m}\right)}. \qquad (3.2.51)$$

It is seen in the Table that the viscosity of the solution changes markedly with concentration and this influences the calculation of the diffusion coefficient. However, the calculation method used to obtain the term correcting for viscosity is questionable. In the expression given above, η is the macro-

---

* In equation (3.2.49) – and in all the others similar to that – the term containing activity can also be written as follows:

$$\frac{\partial \ln a_{ca}}{\partial \ln m} = 1 + m\frac{\partial \ln \gamma_{ca}}{\partial m}. \qquad (3.2.50)$$

Literature on page 249

scopical viscosity of the solution. Its use in the calculations can hardly be justified however, since the water structure in the immediate vicinity of ions is evidently altered by their electric fields. Thus, in the frictional resistance of the motion of ions, not (or not only) the 'average' viscosity is predominating, but the microviscosity of the water layers adjacent to the ion.

The value of constant a is also uncertain or somewhat arbitrary in the calculation of the electrophoretic effect. .

On the basis of diffusion and activity coefficient measurements, the value of h can be calculated from equation (3.2.49). It is doubtful, however, how this quantity actually corresponds to the hydration number, since the viscosity factor is uncertain and this influences the results significantly. For example, using experimental data measured in NaCl solutions of concentration up to 1 M, $h = 1.1$ is obtained on calculating with a value of of η/η_0 corresponding to macroscopical viscosity, while, if the change in viscosity is neglected, we obtain $h = 3.5$. Neither of them can be regarded as the real value of the hydration number. Anyway, it is questionable whether factor h corresponds only to the consequences of hydration. In addition to hydration proper, other short-range interactions between water and ions can result in or contribute to the appearance of the term $(1 - \text{const} \times m)$ in the expression of the diffusion coefficient. In solutions of acids, it is particularly unjustified to interpret h as the hydration number, since the 'hydrogen ion' migrates only partly by hydrodynamic movement of H_3O^+ units; in a considerable part of its migration, the proton jumps from the H_3O^+ ion to the neighbouring water molecule.

With respect to all these uncertainties and some other difficulties, the theory on diffusion of electrolytes in concentrated solutions reviewed above is only the first approximation of the description of this complicated phenomenon, although it probably has considered the most important interactions involved. However, for a more reliable treatment of the real conditions, further significant development is necessary.

Recently, an activation theory has been elaborated by Bhatia, Gubbins and Walker[78a] which describes satisfactorily the temperature and concentration dependence of the mutual diffusion coefficient in concentrated aqueous potassium hydroxide solutions. On the basis of the theory, the diffusion coefficients can also be compared with other transport coefficients.

3.2.4 EFFECT OF A MAGNETIC FIELD ON THE DIFFUSION OF ELECTROLYTES

The diffusion of dissolved electrolytes can also be influenced by an external magnetic field. Lielmezs and Musbally [78b] have shown, on the basis of the theory of the thermodynamics of irreversible processes, that in a solution of 1−1 electrolytes having a concentration of c_s, the diffusion coefficient D measurable experimentally under the effect of a magnetic induction B is:

$$D = D^\circ \left[1 + \frac{(v_1 \cdot B) - (v_2 \cdot B)}{\mu_{ss} \, \text{grad} \, c_s} \right], \qquad (3.2.52)$$

where D° is the experimentally measurable diffusion coefficient in the absence of the magnetic field, v_1 and v_2 are the average velocities of the displacement of the two kinds of ions, μ_s is the chemical potential of the electrolyte and $\mu_{ss} = \partial\mu_s/\partial c_s$.

Dumargue, Humeau and Penot [78c] also started from the macroscopic theory of irreversible processes, and derived general equations for the diffusion occurring in the presence of magnetic induction. In connection with the isothermal convection diffusion occurring in solutions of binary electrolytes, they elaborated the principles of a new experimental technique designed for the measurement of the local velocities in electrolyte solutions, based on a theoretical investigation of the laminary flow.

3.3 DIFFUSION OF MIXTURES OF SOLUTES

Under natural and industrial conditions, as well as in the laboratory, systems containing more than two components or solutions of more than one solute present in non-uniform concentrations occur frequently. In such solutions, the conditions are much more complicated than in a two-component system. Even in dilute solutions, the diffusion of a given solute cannot be described by the Fick law using one single diffusion coefficient; more data are required, since the mass flux of each component is influenced by the other components as well. Although multicomponent systems often occur and they are very important, their diffusion conditions have been investigated scientifically mainly since the 1950's.

When treating the diffusion conditions in multicomponent systems–similarly to other studies–the determination of the number of components requires special attention. The number of independent kinds of species should be regarded as the *number of components* in this respect, too. According to this, a solution of one electrolyte is a binary system, although there are two kinds of ions and the solvent present, i.e. three kinds of species altogether. The three kinds of species, however, are not independent of one another, because there is an unequivocal correlation between the concentration of the positive and negative ions according to the law of electroneutrality. In the system $NaCl-KCl-H_2O$, there are Na^+, K^+, and Cl^- ions and water molecules, i.e. four kinds of species, but the system is still a ternary one, owing to the correlation:

$$\sum_i c_i z_i = 0$$

holding for the concentrations of the three kinds of ions in accordance with the law of electroneutrality (c_i is the concentration of ions, in g eq $\times$ l^{-1} units).

If two dissolved electrolytes have no common ion, the system is a quaternary one because the law of electroneutrality yields only one relationship for any number of kinds of ions. In the presence of more than two kinds of ions, the law of electroneutrality does not define unequivocally the ratio

Literature on page 249

of the concentrations of the individual kinds of ions, it only sets certain limits for them.

In comparison with the number of kinds of particles in the solution, the number of components is also decreased when the different species undergo a chemical reaction or equilibrate with each other. For example, in the system $NaCl-H_2O$, there are five kinds of species: Na^+, Cl^-, OH^- and H^+ ions and H_2O. But hydrogen and hydroxyl ions are in a dissociation equilibrium with each other according to the equation:

$$H_3O^+ + OH^- \rightleftharpoons 2H_2O.$$

This equilibrium, together with the law of electroneutrality, reduces the number of components to three.

According to these, only the *number* of components is given unequivocally, however, the laws of nature do not determine unambiguously which kinds of particles should be considered independent. Usually, it is convenient to choose components so as to obtain a combination containing solvent molecules and one or more neutral electrolytes.

3.3.1 FUNDAMENTALS OF THE GENERAL THEORY

The general theory of diffusion in multicomponent systems has been outlined by Onsager [79], Prigogine [80], Hooyman, de Groot and Mazur [81] and de Groot [82].

Treating a multicomponent system, the fluxes of each component (usually different from one another) must be considered, since the fluxes of the individual components are influenced not only by the gradients of their own chemical potentials, but by those of the other components, too. Namely, cross-effects are operative, in accordance with the thermodynamics of irreversible processes (cf. Section 3.1.2.3). If diffusion is one-dimensional, the mass fluxes are:

$$J_i = -\sum_1^k D_{ij}\frac{\partial c_i}{\partial y}, \quad i = 1, 2, \ldots, k. \tag{3.3.1}$$

These equations define k^2 diffusion coefficients which correspond to the Onsager phenomenological coefficients L_{ij} (cf. Section 3.1.2.3). However, the diffusion coefficients defined in this way are not independent of one another. Partly, the number of independent mass fluxes is only $(k-1)$ in a k-component system, while the kth one is already determined by the others, which reduces the number of coefficients to $(k-1)^2$. On the other hand, the Onsager reciprocity relation gives the relation $\frac{1}{2}(k-1) \times \times (k-2)$ for the phenomenological coefficients. According to all these facts, the number of independent phenomenological coefficients is $\frac{1}{2}k(k-1)$.

The general theory of diffusion in multicomponent systems is very complicated and it is not easy to obtain theoretical relationships directly applicable to experimental data. Therefore, we limit our treatment to some relationship on ternary systems.

There are two independent mass fluxes in ternary systems ($k = 3$). If the one-dimensional flux is expressed in mole cm^{-2} s^{-1} units, and it is related to a reference plane fixed with respect to the number of moles (i.e. the net number of moles crossing the plane is zero), the phenomenological equations of diffusion in the system, at constant temperature and pressure, are:

$$J_1 = - L_{11} \left(\frac{\partial \mu_1}{\partial y} - \frac{\partial \mu_3}{\partial y} \right) - L_{12} \left(\frac{\partial \mu_2}{\partial y} - \frac{\partial \mu_3}{\partial y} \right), \tag{3.3.2}$$

and

$$J_2 = - L_{21} \left(\frac{\partial \mu_1}{\partial y} - \frac{\partial \mu_3}{\partial y} \right) - L_{22} \left(\frac{\partial \mu_2}{\partial y} - \frac{\partial \mu_3}{\partial y} \right). \tag{3.3.3}$$

The gradients of chemical potential can be expressed by at least two of the three concentration gradients. In the case of solutions it is advisable to eliminate the concentration gradient of the solvent.

Introducing the following term according to Lamm [83]:

$$B_{ij} = 1 + \left(\frac{\partial \ln \gamma_i}{\partial x_i} \right)_{T,P,x_j} \tag{3.3.4}$$

the following equations hold for the changes of activities:

$$x_1 \, \partial \ln a_1 = - B_{12} \, \partial x_3 - B_{13} \, \partial x_2,$$
$$x_2 \, \partial \ln a_2 = - B_{21} \, \partial x_3 - B_{23} \, \partial x_1, \tag{3.3.5}$$
$$x_3 \, \partial \ln a_3 = - B_{31} \, \partial x_3 - B_{32} \, \partial x_3,$$

further

$$B_{23} + B_{32} = B_{31} + B_{13} = B_{21} + B_{12}. \tag{3.3.6}$$

Taking into account these equations and regarding the solvent as the third component, the mass fluxes of the two solutes are:

$$J_1 = - RT \left[\left\{ L_{11} \left(\frac{B_{32}}{x_3} + \frac{B_{12}}{x_1} \right) + L_{12} \left(\frac{B_{31}}{x_3} + \frac{B_{21} - B_{23}}{x_2} \right) \right\} \frac{\partial x_1}{\partial y} + \right.$$
$$\left. + \left\{ L_{11} \left(\frac{B_{12} - B_{13}}{x_1} + \frac{B_{31}}{x_3} \right) + L_{12} \left(\frac{B_{31}}{x_3} + \frac{B_{21}}{x_2} \right) \right\} \frac{\partial x_2}{\partial y} \right], \tag{3.3.7}$$

and

$$J_2 = - RT \left[\left\{ L_{21} \left(\frac{B_{32}}{x_3} + \frac{B_{12}}{x_1} \right) + L_{22} \left(\frac{B_{31}}{x_3} + \frac{B_{21} - B_{23}}{x_2} \right) \right\} \frac{\partial x_1}{\partial y} + \right.$$
$$\left. + \left\{ L_{21} \left(\frac{B_{12} - B_{13}}{x_1} + \frac{B_{31}}{x_3} \right) + L_{12} \left(\frac{B_{31}}{x_3} + \frac{B_{21}}{x_2} \right) \right\} \frac{\partial x_2}{\partial y} \right]. \tag{3.3.8}$$

Literature on page 249

However, in these relationships, only four of the six factors B_{ij} are independent of one another.

The fluxes of two components in a ternary system, related to the plane on both sides of which the number of moles are unchanged, are, similarly to equation (3.3.1):

$$J_1 = -D_{11}c_1\frac{\partial x_1}{\partial y} - D_{12}c_1\frac{\partial x_2}{\partial y}, \qquad (3.3.9)$$

and

$$J_2 = -D_{21}c_2\frac{\partial x_1}{\partial y} - D_{22}c_2\frac{\partial x_2}{\partial y}. \qquad (3.3.10)$$

If these latter two equations are compared with equations (3.3.7) and (3.3.8) obtained from the thermodynamic deduction, it appears that the four diffusion coefficients are different, in general. Thus, in a three-component system, diffusion can be described by four diffusion coefficients. Since, however, the fluxes of only three components are considered, i.e. the phenomenon can be described by three independent coefficients, there must be a relationship between the four diffusion coefficients. This relationship is related to the thermodynamic coefficients B_{ij}. Further experimental investigations are required to determine the four different diffusion coefficients, as well as the three independent thermodynamic coefficients. In such experiments, the validity of the Onsager reciprocity relation ($L_{12} = L_{21}$) can also be checked. The investigations of Miller [84], Dunlop [85], and others have shown that this relation is valid within the experimental errors.

Siver [86] has outlined a method to investigate the interaction of diffusion flows in multicomponent systems by means of isotopically labelled compounds that can be applied in both parallel and counter-current flows. His experiments on several ternary systems verified the Onsager correlation. Diffusion in multicomponent systems has been studied by Pamfilov, Lopushanskaia and Tsvetkova [87] on the basis of the general equations of mass transport. They deduced general phenomenological equations of diffusion flows in which the self coefficients and those of the cross-effects are expressed explicitly by means of state parameters and thermodynamic functions. The correlation of these coefficients with measurable quantities made it possible to characterize the mutual influence of the diffusion flows. They have demonstrated the dependence of the phenomenological coefficients on temperature, pressure and concentration. Schönert [88] also investigated in detail the concentration dependence of the transport coefficient of multicomponent systems in solutions in which the concentration of one of the components is negligibly low.

The simultaneous diffusion of a non-electrolyte (penta-erythritol) and NaCl has been investigated by Woolf [89a], and the values D_{11}, D_{22}, D_{12} and D_{21} have been calculated for this ternary system. In conclusion, he emphasized that knowledge of the tracer diffusion coefficient (cf. Section 3.4) is also required for complete documentation of diffusion in a ternary (in this case, it is the tracer diffusion coefficient of penta-erythritol in aqueous NaCl solutions).

In their investigations concerning the effect of salts on the diffusion of non-electrolytes dissolved in water, Tham and Gubbins [89b] have found that salts, in general, lower the diffusion coefficients of non-electrolytes, and this effect is more marked with increasing charge and radius of the ion. In most cases the experimental results are in agreement with the conclusion drawn on the basis of the kinetic theory of multicomponent systems consisting of rigid spheres. The deviations have been attributed to attractive forces that have a considerable influence on the dynamics of collisions.

On the basis of diffusion coefficient measurements in solutions of various substances of different molecular weight (methanol, acetamide, sugar, tannine) in glycerol–water mixtures, Marinin [90] has established that the ratio $D\eta/T$ decreases with increasing temperature, while it increases with increase in the viscosity of the solvent. In solvent mixtures for which $\eta > 0\cdot05$–2 poise, this latter change is linear:

$$\frac{D\eta}{T} = A + B\eta, \qquad (3.3.11)$$

where A and B are constants. With increasing size of the diffusing molecule, this dependence is reduced to the Stokes-Einstein equation. In a given solvent mixture, the temperature coefficient of the diffusion coefficient does not depend on the size of the diffusing molecules, and it is approximately equal to the temperature coefficient of the viscosity of the solvent.

3.3.2 DIFFUSION OF ELECTROLYTE MIXTURES

The theory of diffusion of two or more electrolytes in common solutions starts with the principles described in Section 3.2.3, but the influence of cross-effects is more complicated. Also under such conditions, the law of electroneutrality is valid, i.e. the positive and negative charges cannot be separated from each other in macroscopical volumes. However, in the present case, it does not follow from this, even for ions of identical valence, that they diffuse with the same velocity due to the interactions, since, if there are more than two kinds of ions in the solution with different individual mobilities, electroneutrality is not necessarily realized by identical velocities of the pairs of ions, but, incidentally, every type of ions may migrate with different velocities. However, the predominating interaction between diffusing ions of different mobilities should be such, under these conditions too, as to produce velocities suitable to maintain electroneutrality. The other, thermodynamic, interactions are of smaller importance.

Due to the different migration velocities of the individual kinds of ions, the relaxation effect, arising from the non-instantaneous development of the ion atmosphere, influences diffusion in solutions of electrolyte mixtures, in addition to the electrophoretic effect. However, even under such condi-

Literature on page 249

tions, diffusion is much less altered by the relaxation effect than electric conduction, so it can be neglected in several cases.

The simple Fick law is not valid for the diffusion of electrolyte mixtures either, since the cross-effects should also be taken into account. If the treatment is limited to ternary systems, i.e. to solutions of two electrolytes, the two mass fluxes describing diffusion will be:

$$J_1 = - D_{11} \frac{\partial c_1}{\partial y} - D_{12} \frac{\partial c_2}{\partial y}, \tag{3.3.12}$$

and

$$J_2 = - D_{21} \frac{\partial c_1}{\partial y} - D_{22} \frac{\partial c_2}{\partial y}, \tag{3.3.13}$$

according to Dunlop et al. [91], for one-dimensional diffusion (in a volume-fixed reference system), where c_1, and c_2 are the concentrations of the two electrolytes. In these equations, D_{11} and D_{22} are related to the effect of the self concentration gradients of the individual components on the diffusion of this component. This is the predominating effect, and D_{11} and D_{22} are the *main diffusion coefficients*. D_{12} and D_{21} stand for the cross-effects, i.e. they take into account the effect of the concentration gradient of the other component, and influence diffusion to a smaller but not negligible extent. Dunlop [92] has called attention to the fact that the dimensions of the main diffusion coefficients are always area/time (similarly to other cases of diffusion), but the dimensions of D_{12} and D_{21} are the normal ones only, if c_1 and c_2 are expressed in the same concentration units; their numerical values, however, depend, even in this case, on whether the concentration is measured in the number of grams or moles in unit volume.

Diffusion coefficients change with concentration. Since it is evident that the flux component 1 becomes zero, if $c_1 \to 0$, irrespective of how large the concentration gradient of component 2 is, consequently $D_{12} \to 0$ if $c_1 \to 0$. Similarly, $D_{21} \to 0$ if $c_2 \to 0$.

Equations (3.3.12) and (3.3.13) have been studied in detail by O'Donnel and Gosting [93a] in the system $NaCl-KCl-H_2O$. These measurements were carried out in dilute solutions and, in addition to the primary effects on the main diffusion coefficients, only the restrictions for the values of the four diffusion coefficients obtained from the requirement of electroneutrality and from the agreement of the ionic mobilities in infinite dilution have been considered in the interpretation of the results. The secondary effects arising from changes in the activity coefficients and from electrophoretic and relaxation effects have been disregarded.

When two kinds of monovalent cations (denoted by 1 and 2) and only one kind of anion (denoted by 3) are present in the aqueous solution, and the following abbreviations are introduced:

$$\tilde{D}_i = \frac{RT \lambda_i^0}{F^2} \times 10^{-7} \qquad (i = 1, 2, 3) \tag{3.3.14}$$

and
$$\chi = (\tilde{D}_1 + \tilde{D}_3)c_1 + (\tilde{D}_2 + \tilde{D}_3)c_2, \qquad (3.3.15)$$
then the theory of Gosting [93a] gives the following relationships:

$$D_{11} = + \tilde{D}_1\left\{1 - \frac{c_1(\tilde{D}_1 - \tilde{D}_3)}{\chi}\right\}$$

$$D_{12} = - \tilde{D}_1\frac{c_1(\tilde{D}_2 - \tilde{D}_3)}{\chi}$$

$$D_{21} = - \tilde{D}_2\frac{c_2(\tilde{D}_1 - \tilde{D}_3)}{\chi} \qquad (3.3.16)$$

$$D_{22} = + \tilde{D}_2\left\{1 - \frac{c_2(\tilde{D}_2 - \tilde{D}_3)}{\chi}\right\}.$$

According to the data available for ionic conductivities:

$$\text{for Na}^+ \text{ ion:} \quad \tilde{D}_1 = 1\cdot334 \times 10^{-5}$$

$$\text{for K}^+ \text{ ion:} \quad \tilde{D}_2 = 1\cdot957 \times 10^{-5}$$

$$\text{for Cl}^- \text{ ion:} \quad \tilde{D}_3 = 2\cdot032 \times 10^{-5}.$$

According to the theoretical relationships, all the four diffusion coefficients are constant if c_1/c_2 is also constant. D_{12} and D_{22} are otherwise not too sensitive to concentration, since $\lambda^0_{K^+}$ and $\lambda^0_{Cl^-}$ hardly differ from each other, i.e. $\tilde{D}_2 \approx \tilde{D}_3$. However, D_{11} and D_{21} depend appreciably on the ratio of the concentrations of the two electrolytes, since D_1 markedly differs from D_3. It is seen from the relationships above that $D_{12} \to 0$, if $c_1 \to 0$, independently of the value of c_2 and $D_{21} \to 0$, if $c_2 \to 0$, independently of the value of c_1. The experimental results are in agreement with these conclusions. Further, in agreement with the theory, if $c_1 \to 0$, D_{22} tends to the binary diffusion coefficient of a KCl solution of the same concentration, in the case of a given value of c_2; while D_{11} tends to the diffusion coefficient of Na$^+$ in such a KCl solution which contains only traces of Na$^+$ ions (i.e. the tracer diffusion coefficient of Na$^+$). While $D_{12} \to 0$, D_{21} tends to:

$$D_{21} \to - \frac{\tilde{D}_2(\tilde{D}_1 - \tilde{D}_3)}{\tilde{D}_2 + \tilde{D}_3}. \qquad (3.3.17)$$

Thus, if trace amounts of Na$^+$ ions diffuse in KCl solution, the simple form of Fick's first law describes the flux of Na$^+$ ions reliably, since c_1 is negligibly small, D_{12} is practically equal to zero. For the flux of K$^+$ ions, however, the simple Fick law is not valid under such conditions either,

Literature on page 249

because $D_{12} \neq 0$. The concentration gradient of Na^+ ions–whatever value it may have–influences the flux of K^+ ions to a certain extent. For self-diffusion, which can be considered the special case of tracer diffusion in certain respects (see Section 3.4.5), the conditions are similar. For example, the diffusion of radioactive Na^+ ions added in traces to a NaCl solution is described reliably by the simple Fick law, but the cross-effect cannot be neglected in the expression of the flux of inactive Na^+ ions of relatively high concentration: the flux of inactive Na^+ ions is altered by the concentration gradient of radioactive Na^+ ions, present only as a trace.

Measuring the four diffusion coefficients in aqueous $MgCl_2 - NaCl$ solutions and utilizing some activity coefficients obtained in other studies, Wendt and Shamim [93b] have calculated the phenomenological transport coefficients. Their investigations confirmed the validity of the Onsager relationship.

Although the theories based on primary interactions describe the main characteristics of the change in the diffusion coefficients in agreement with experiments, some smaller deviations between the calculated and experimentally observed data indicate that the effects quoted above as secondary ones cannot always be neglected.

* * *

The ratio $D\eta/T$ shows a similar dependence in solutions of electrolyte mixtures to that in simple solutions. For example, Arvia, Marchiano and Podestá [94] measured the diffusion of iron(II) and iron(III) cyanide ions in potassium hydroxide solutions, and observed that the ratio $D\eta/T$ constant in good approximation between 24 and 50 °C, and it is independent of concentration, as well.

For the diffusion coefficients of solutions containing three kinds of ions, general equations have been obtained by Schönert [95] with respect to the electrophoretic and relaxation effects.

* * *

The diffusion rate of *hydrogen ions* in electrolyte solutions plays some role in the theory of electrolysis determining, for example, the limiting current of the electrolytical hydrogen evolution. Glietenberg, Kutschker and Stackelberg [96] measured the diffusion coefficient of hydrogen ions in LiCl, NaCl, KCl and $NaClO_4$ solutions at 0 and 25 °C. The value of D_{H^+} decreases with increasing electrolyte concentration. This increase is greatest in LiCl solutions and the least in KCl solutions, which has been explained as the combined result of the simultaneous effects of ionic fields and electrostatic interactions between ions on the water structure. The product $D_{H^+}\eta$ significantly depends on the nature of the cation; however, the effect of anions does not seem to be appreciable: the exchange of Cl^- ions to ClO_4^- ions, which are much larger in size, does not alter the product $D_{H^+}\eta$ markedly.

Correlations between the diffusion coefficients and the conduction and frictional coefficients. In multicomponent systems, the diffusion mass flow is related to the experimental diffusion coefficients according to the equation of diffusion generalized by Onsager and Fuoss [97]:

$$J_i = \sum_{j=1}^{q} D_{ij} \frac{\partial c_j}{\partial y} \qquad (i = 1, 2, \ldots, q), \qquad (3.3.18)$$

where q is the number of components, and the subscript ij, as earlier, indicates the effect of the flux of the jth kind on the ith kind of ions. A system consisting of q components has q^2 diffusion coefficients. In the case of electrolyte diffusion, q is the number of independent components obtained after taking into account the law of electroneutrality. If the flux is referred to fixed volume:

$$\sum_{i=1}^{q} \overline{V}_i J_i = 0,$$

$$\sum_{i=1}^{q} \overline{V}_i \left(\frac{\partial c_i}{\partial y} \right)_t = 0,$$

(3.3.19)

(if $\overline{V}_i$ is the partial molar volume of the ith component).

Instead of the conventional diffusion coefficients directly measurable experimentally, under such conditions diffusion can also be described by the phenomenological proportionality of fluxes and thermodynamic forces. This can be based on the transport phenomenological coefficients introduced in Section 3.1.2.3. According to the fundamental relationships of the thermodynamics of irreversible processes, the flux of the ith component is:

$$J_i = \sum_{j=1}^{q} L_{ij} \frac{\partial \tilde{\mu}_j}{\partial y} \qquad (i = 1, 2, \ldots, q), \qquad (3.3.20)$$

where $\tilde{\mu}_j$ is the electrochemical potential of the jth component. The rather complicated relationships between D_{ij} and L_{ij} implied in these two types of descriptions have been deduced by Woolf, Miller and Gosting [98]. In some respects the relationships become simpler when the coefficients L_{ij} depending on the reference system are replaced in the calculations by the *frictional coefficients* R_{ij} introduced by Onsager and developed further by Laity [99]; their interpretation can be obtained from the following equation:

$$\frac{\partial \tilde{\mu}_i}{\partial y} = \sum_{j=1}^{q} R_{ij} c_j (\overline{v}_i - \overline{v}_j), \qquad (3.3.21)$$

where $\overline{v}_i$ and $\overline{v}_j$ are the average velocities of ions i and j, respectively, and:

$$J_i = c_i \overline{v}_i. \qquad (3.3.22)$$

The reciprocity relation also applies to the frictional coefficients:

$$R_{ij} = R_{ji}. \qquad (3.3.23)$$

Literature on page 249

Frictional coefficients R_{ij} are independent of the reference system of fluxes, because they correspond to the *difference* in velocity of the components.

The coefficients L_{ij} and R_{ij} are also valid for transport phenomena other than diffusion, because every transport process is accompanied by the net displacement of the species with respect to one another, and, in each of them, the interaction and friction between the species is of decisive importance. Some correlations regarding various transport processes of electrolytes have been deduced by Laity [100].

3.3.3 DIFFUSION OF ELECTROLYTES IN SOLUTIONS OF NON-ELECTROLYTES

Several studies have dealt with diffusion of electrolytes in solvent mixtures (mainly in mixtures of alcohols and water) and in solutions of non-electrolytes (e.g. sugar solutions). The majority of these considered the solvent mixture or the aqueous solution of non-electrolytes as a single component with respect to diffusion, and they investigated the diffusion of the electrolyte as that of the second component, mainly in relation to the viscosity of the solution. For *pseudo-binary* systems in this sense, the theoretical laws of diffusion of electrolytes hold with about the same approximation as for real binary aqueous solutions.

In this way, for example, Harned and Shropshire [101] have found a good agreement between the theoretically and experimentally determined diffusion coefficients in 0·25 and 0·75 m sugar solutions for the diffusion of KCl in the 0·004–0·025 mole/l concentration range. Investigating the diffusion of KCl in 10, 20 and 38·6 weight per cent aqueous methanol solutions, Kulkarni and Lyons [102] also observed that the activity coefficients calculated from the diffusion coefficients agree satisfactorily with those determined by other methods when regarding the system as a pseudo-binary one. In this respect, the solvent mixture can be considered to be a uniform component. Such investigations, however—owing to their nature—take into account only the concentration changes of the diffusing electrolyte and implicitly assume that the relative concentration of the non-electrolytes remains constant. Considering only the relationship between viscosity and the diffusion coefficient, this point of view is justified. In detailed studies on the phenomena, however, the effect of the electrolyte diffusion on the distribution of the dissolved non-electrolyte in the solution should also be investigated.

Studying the effect of diffusing electrolytes on the dissolved non-electrolyte, Erdey-Grúz *et al.* [103] and particularly Hunyár [104] have established that the diffusing electrolyte alters the concentration of the non-electrolyte. In a solution with initially uniform concentration of the non-electrolyte, the diffusion of the electrolyte results in a concentration gradient of the non-electrolyte. In their experiments they investigated the effect of diffusion of LiCl, KCl, $N(CH_3)_4Cl$, HCl, KBr, $Li(COOCH_3)$, $BaCl_2$ and $MgCl_2$ in solutions of propenol, saccharose, acetic acid, propionic acid, butyric acid, crotonic acid, succinic acid, malic acid and citric acid which were initially in a uniform concentration throughout the system. (These weak acids can be considered approximately non-electrolytes with respect to diffusion owing to the low degree of dissociation.) In the experiments performed by the diaphragm method–also checked by the Oeholm method with no diaphragm present–they have observed that the initially uniform concentration of the non-electrolyte increases with time in the direction of the diffusion of the electrolyte. This is particularly noteworthy, since diffusing ions carry along some water due to their hydration, which would decrease the concentration of the non-electrolyte in the direction of the diffusion flux of the electrolyte. If the latter still increases, it can be concluded that ions somehow transport non-electrolyte molecules in the course of the diffusion, and their amount is relatively greater than the amount of water molecules in the hydrate spheres of the ions. An effect opposite to that of organic non-electrolytes appears in the case of the diffusion of KCl in a solution of arsenous acid: the concentration of As_2O_3 decreases in that part of the solution into which the electrolyte diffuses. This is an evident consequence of the fact that ions carry along more water than arsenous acid. The decrease in the concentration of arsenous acid in the direction of the diffusion, however, is much smaller than could be expected by taking into account the hydration numbers assumed on the basis of other investigations. Thus, it seems that ions also transport arsenous acid molecules, but their amount is less than that of the water molecules involved in this transport.

It became evident from these experiments that the amount of non-electrolyte transported by the electrolyte in its diffusion, calculated with respect to one mole of electrolyte crossing the diaphragm, decreases with the diffusion time (see Table 3.9 as an example). It can be concluded

Table 3.9

The amount of propenol transported in diffusion of electrolytes at 25 °C. Initial electrolyte concentration: $1 \cdot 0$ mole l^{-1}

Electrolyte	Initial concentration of propenol mole $\cdot l^{-1}$	Diffusion time day	Number of moles of propenol transported per 1 mole of electrolyte
KCl	2	1	0·033
KCl	2	2	0·031
KCl	2	4	0·023
KCl	0·2	2	0·003
KCl	2·0	2	0·031
KCl	5·8	2	0·098
KCl	10·0	2	0·13
KCl	25·0	2	0·31
LiCl	2·0	2	0·033
$N(CH_3)_4Cl$	2·0	2	0·015
HCl	2·0	2	0·005
KBr	2·0	2	0·025
$Li(CH_3COO)$	2·0	2	0·041

Literature on page 250

from this that the concentration gradient of the non-electrolyte resulting from diffusion of the electrolyte gives rise to a counter-diffusion; this becomes more significant with increasing difference in the concentration of the non-electrolyte at the two sides of the diaphragm, i.e. with increasing duration of diffusion. From this back-diffusion, it follows that non-electrolyte molecules are bound to the ions so weakly that they contribute to the chemical potential of the 'free' non-electrolyte, since back-diffusion of the non-electrolyte is due to the difference in its chemical potential.

The difference in the amount of non-electrolyte carried along in the diffusion of the electrolyte and that transported back in the counter-diffusion can be calculated, as given by Erdey-Grúz and Hunyár [105]. Let the thickness of the diffusion diaphragm be h, its cross-section q, and consider a ternary solution divided into two parts of identical volume v at the two sides of the diaphragm. Solute 1 (non-electrolyte) has initially identical concentrations on both sides; in one of the solutions substance 2 is dissolved in an initial concentration c_2. Each mole of the latter will carry along a net number n of moles of component 1 in the course of its diffusion (i.e. this is the amount of substance 1 greater than that of water transported as a result of hydration). Owing to the concentration gradient produced in this process, counter-diffusion of the transported solute will start. The amount of substance apparently transported across the diaphragm in time t (i.e. the difference in the amount of matter carried along by solute 2 and that transported back in counter-diffusion) is:

$$m_1 = \frac{n D_2 c_2 v}{2(D_2 - D_1)} \left\{ \exp\left[-\frac{2 D_1 q t}{hv} \right] - \exp\left[-\frac{2 D_2 q t}{hv} \right] \right\}, \quad (3.3.24)$$

where D_1 and D_2 are the diffusion coefficients of the two solutes. The transported amount of substance 2 diffusing primarily across the diaphragm in the same time is:

$$m_2 = \frac{c_2 v}{2} \left\{ 1 - \exp\left[-\frac{2 D_2 q t}{hv} \right] \right\}. \quad (3.3.25)$$

The ratio m_1/m_2 decreases with time. Extrapolating these ratios to $t = 0$, the value of n can be calculated, i.e. the amount of substance 1 transported by solute 2. These quantities are given in Tables 3.9 and 3.10 The experiments on KCl diffusing in propenol solutions confirmed the above relationships to a good approximation.

Tables 3.9 and 3.10 provide information on the quantities of several non-electrolytes carried along in the diffusion of different electrolytes in the experiments of Erdey-Grúz et al. The amount of the non-electrolyte per mole of diffusing electrolyte is proportional to the initial concentration of the former. The diffusing electrolyte transports more non-electrolyte than water even in the most dilute non-electrolyte solutions. It can be established that the value of n, i.e. the net number of moles of non-electrolyte transported by 1 mole of the diffusing electrolyte, decreases in the order KCl < LiCl < BaCl$_2$ < MgCl$_2$ (except propenol). It also increases

Table 3.10

The amount of non-electrolyte transported in the course of diffusing
electrolytes

Initial electrolyte concentration = 1·0 M, that of the non-electrolyte = 0·2 M,
the diffusion time = 2 days

Non-electrolyte	Diffusing electrolyte	Number of moles of non-electrolyte transported by 1 mole of electrolyte		
	KCl	LiCl	BaCl₂	MgCl₂
Propenol	0·024	0·025	0·068	0·061
Acetic acid	0·018	0·037	—	0·083
Propionic acid	0·021	0·041	0·075	0·090
Butyric acid	0·025	0·049	0·085	0·099
Crotonic acid	0·025	0·046	—	0·095
Succinic acid	0·010	0·035	—	0·080
Malonic acid	0·010	0·042	—	0·12
Citric acid	0·0070	0·042	—	0·10
Saccharose	0·008	0·016	0·025	0·035
Arsenic trioxide	−1·2	—	—	—

with increasing molecular weight of non-electrolytes of similar structure.
The amount of non-electrolytes transported by HCl is small in comparison
with other 1–1 electrolytes which, perhaps, is connected with the fact that
hydrogen ions migrate not only by the hydrodynamic movement of H_3O^+,
but by the prototropic mechanism as well (cf. Section 4.2.2.2).

The cause of transport of non-electrolytes by diffusing ions has not been
elucidated satisfactorily. It can hardly be assumed that non-electrolyte
molecules are solvated directly to ions to such a degree, since, on one hand,
non-electrolytes decrease the solubility of electrolytes (e.g. in a solution
containing 33 weight per cent of propanol, the solubility of KCl is half
of that in water); on the other hand, it is unlikely that all the ions investigated would become more strongly solvated than hydrated in the presence
of even small amounts of various non-electrolytes. It seems more likely
that the water molecules in the hydrate sphere form hydrogen bonds with
the non-electrolyte molecules; this is promoted by the destruction of
the water structure in the presence of ions. Furthermore, association of
molecules of non-electrolytes with each other, present in much lower concentration than water, may also contribute to this phenomenon, since
these molecules, being relatively independent, can be bound to water molecules within the hydrate spheres more easily than other water molecules
built into the lattice-like structure. The experiments of Freundlich and
Krüger [106] and Hale and De Vries [107] also indicate the transport of
non-electrolyte molecules by ions.

Literature on page 250

3.4 SELF-DIFFUSION AND TRACER DIFFUSION

According to our knowledge, in liquids the relatively independent particles (atoms, molecules, ions) vibrate about their temporary equilibrium positions until they gain, due to fluctuations, an energy sufficient to overcome the energy barrier separating two adjacent equilibrium positions and jump to this adjacent position. Apart from other factors, the frequency of the jumps depends on the height of the energy barrier separating two neighbouring equilibrium positions of the species, which is determined by the liquid structure. With no external force acting, the direction of these jumps is disordered, i.e. they have equal probability in all directions. In this way, the molecules leave their equilibrium positions at a given time and they are strolling with a velocity depending on their environment. This phenomenon is the *self-diffusion*. In connection with this, mass flux can be discussed in a similar sense as in connection with normal diffusion, and the *self-diffusion coefficient* can be defined similarly to the ordinary diffusion coefficient.

Self-diffusion is closely related to viscous flow, since the elementary step of viscous flow, taking place on the effect of a shearing force, proceeds also by means of jumps of the molecules of the liquid from their temporary equilibrium positions to neighbouring ones. However, the direction of these jumps is not entirely disordered owing to the effect of the external shearing force; on the statistical average, those in the direction of the shearing force are favoured as compared with those in other directions.

With respect to all these relationships, studies on self-diffusion can provide some information to help understand the liquid structure and the phenomena involved in transport processes.

Strictly speaking, self-diffusion cannot be investigated experimentally, since it is impossible to distinguish a given atom or molecule from the others while keeping its properties completely unchanged in order to follow its strolling caused by the fluctuations. Self-diffusion, however, can be approached by replacing a very small portion of the atoms of the liquid by an isotope which can be observed separately owing to its specific properties. Such a specific property can be radioactivity, or atomic weight (in mass spectrometric measurements), or its spectrum which differs from that of the other isotopes, etc.

The replacement of some liquid molecules by their isotopes does not leave the liquid structure unchanged. Even if the effect of radioactivity is disregarded (which is usually justified), the change in molecular weight caused by the replacement with the isotope somewhat modifies the energetic and geometrical conditions in the liquid. This effect is the smaller the less the relative difference in the atomic weights of the isotope used for indication and that replaced: in the most cases it is negligible. However, because of the great relative difference in the atomic weights of the hydrogen isotopes, the diffusion of the HDO and HTO molecules is not entirely identical with the self-diffusion of water. The extent of the deviation can be estimated by comparing the diffusion of the molecules HDO, HTO and $H_2^{18}O$ labelled with different isotopes. The comparison of the diffusion of HDO and $H_2{}^{18}O$ can serve as a basis to judge whether or not the exchange of atoms between molecules has any importance beside the diffusion of whole molecules.

The diffusion of molecules labelled with isotopes is, in fact, the limiting case of *tracer diffusion* in a uniform liquid. Tracer diffusion means the diffusion of a single component present in very low concentration in a solution of identical composition all over the system. In other words, the component present in traces in the solution has a concentration gradient, while the concentration gradient of all the other components of high concentration in comparison with the former is zero. Tracer diffusion takes place when, for example, Na^+ ions of negligibly low concentration diffuse in a KCl solution of uniform concentration. Tracer diffusion can be utilized, for example, for studies on some aspects of the structure of electrolyte solutions. In fact, self-diffusion is that special case of tracer diffusion when the component present in traces differs only in the atomic or molecular weight or, eventually, in radioactivity from the component present in higher concentration, while its chemical properties are almost identical.

A mixture of isotopes or isotopically labelled molecules is always ideal in the thermodynamic sense. Thus, the gradient of the chemical potential is the gradient of $RT \ln c$ for each isotope. The driving force of the interdiffusion of isotopes is, in fact, the term arising from the entropy of mixing in the free enthalpy of mixing. The conditions are essentially the same in the case of tracer diffusion of ions as well.

3.4.1 FUNDAMENTALS OF THE GENERAL THEORY OF SELF-DIFFUSION

The self-diffusion coefficient D^* of liquids can be defined on the basis of the Fick law. In general, if c^* is the concentration of the isotopically labelled species (molecules or atoms), and J^* is its flux, in the case of one-dimensional diffusion:

$$J^* = -D^* \frac{\partial c^*}{\partial y}, \qquad (3.4.1)$$

and

$$\frac{\partial c^*}{\partial t} = D^* \frac{\partial^2 c^*}{\partial y^2}. \qquad (3.4.2)$$

D^* defined in this way approximates the better the intrinsic self-diffusion coefficient, the smaller the difference in the properties of the isotope-labelled species and those of the others forming the bulk of the liquid.

After introducing appropriate simplifying assumptions, the hydrodynamic theory of diffusion gives an approximative calculation of the self-diffusion coefficient. According to Collins and Raffel [108], in liquids consisting of

Literature on page 250

elastic spherical molecules in cubic packing where the energy distribution is of a Boltzmann-type, the self-diffusion coefficient is:

$$D^* = \frac{kT}{2{\cdot}77\,\pi\,d_c\,\eta_c},\tag{3.4.3}$$

where d_c is the collision diameter of molecules, while η_c is the collision part-viscosity defined by them, which is about $0{\cdot}25{-}0{\cdot}5$ times the viscosity for several organic liquids.

For the self-diffusion coefficient, the kinetic theory of Eyring [109] leads to the following relationship:

$$D^* = \left(\frac{N}{V^0}\right)^{1/3}\frac{kT}{\eta} = \left(\frac{3}{4\pi}\right)^{1/3}\frac{kT}{\eta r} = \frac{kT}{0{\cdot}5\,\pi\eta r},\tag{3.4.4}$$

where r is the radius of the diffusing species, V^0 is the molar volume of the liquid, and η is viscosity. This equation is based on the assumption (probably not accurately correct) that the packing factor $\chi = 1$. Equation (3.4.4) is similar to equation (3.1.38) deduced from the hydrodynamic theory, only the numerical factor is different. According to Li and Chang [110], the main reason for the difference is that the Eyring theory simplifies the conditions too much and disregards the fact that the molecules move between neighbours that are not immobile but are also in motion. Taking this into account:

$$D^* = \frac{a-b}{2a}\left(\frac{N}{V^0}\right)^{1/3}\frac{kT}{\eta},\tag{3.4.5}$$

where the number of all neighbours is a, and the number of neighbours in one plane is b. For simple cubic packing $a = 6$, $b = 4$, and the self-diffusion coefficient is (if $\chi = 1$):

$$D^* = \left(\frac{3}{4\pi}\right)^{1/3}\frac{kT}{6\eta r} = \frac{kT}{3{\cdot}2\,\pi\eta r}.\tag{3.4.6}$$

This equation already does not differ too much from equation (3.1.38) of the hydrodynamic theory. Ottar [111] modified this theory further and according to him:

$$D^* = \frac{1}{4\sqrt{2}}\left(\frac{N}{V^0}\right)^{1/3}\frac{kT}{\eta} \approx \frac{1}{5{\cdot}7}\left(\frac{N}{V^0}\right)^{1/3}\frac{kT}{\eta}.\tag{3.4.7}$$

In the theory of absolute reaction rates, viscosity can be calculated from the heat of evaporation. Thus, on the basis of the facts discussed in Section 2.2, the absolute value of the diffusion coefficient can also be calculated, in principle, for liquids of simple structure:

$$D^* = \frac{\lambda^2}{v_f^{1/3}}\left(\frac{kT}{2\pi m}\right)^{1/2}\exp\left[-\frac{\varepsilon_0}{kT}\right] = \left(\frac{V^0}{N}\right)^{2/3}\frac{1}{v_f^{1/3}}\left(\frac{kT}{2\pi m}\right)^{1/2}\exp\left[-\frac{\Delta H_{evap}}{nRT}\right],\tag{3.4.8}$$

where n is a number between 3 and 4. The practical application of this relationship, however, is limited by the uncertainties in the calculation of free volume v_f and in the value n.

In a further development of the general theory of self-diffusion, McLaughlin's [112] main criticism was that, although in the Eyring theory the displacement of liquid molecules requires the presence of holes in their neighbourhood, no factor directly accounting for the presence of holes appears in the expression of self-diffusion. According to the advanced theory, the self-diffusion coefficient also depends on the distance a separating the displaced molecule from the nearest hole, further, it depends on the force function of the intermolecular interaction, too. If ε denotes the constant appearing in the Lennard–Jones intermolecular potential* that measures the energy required to form a hole and V_0 is a constant also related to the force function, while ε_0 is the energy barrier between adjacent equilibrium positions of the displaced molecule, the self-diffusion coefficient is, according to McLaughlin:

$$D^* = \frac{kT}{2\pi m} \frac{a^2}{v_f^{1/3}} \exp\left[-\frac{\varepsilon \left(\dfrac{V_0}{V}\right)^4}{kT} \right] \exp\left[-\frac{\varepsilon_0}{kT} \right], \qquad (3.4.9)$$

in liquids consisting of spherically symmetrical molecules. This relationship differs from that of the original Eyring theory equation (3.4.8) in the factors accounting for the holes.

Recently, Rahman [113a] has dealt in detail with the relationship between the structure and self-diffusion of monoatomic liquids by a method that became applicable using a digital computer, which we only refer to here.

Recently a review has been published by Hertz [113b] on several qualitative and semiquantitative theories regarding the relationship between the structural properties of liquids and their molecules and the self-diffusion coefficients observed. At the same time, some relationships have been pointed out between self-diffusion and the structure of liquids. Recently, O'Reilly [113c] has dealt with the theory of self-diffusion with respect to the rotational relaxation of liquid molecules.

3.4.2 SELF-DIFFUSION OF WATER

The details of the theory on self-diffusion have not been elucidated entirely for monoatomic liquids either. There are even more questionable assumptions regarding water and other liquids of complex internal structure due to significant intermolecular interactions.

---

* The Lennard–Jones intermolecular potential is:

$$\Phi(r) = 4\varepsilon \left\{ \left(\frac{\sigma}{r}\right)^{12} - \left(\frac{\sigma}{r}\right)^6 \right\}.$$

Literature on page 250

Samoilov's theory [114] casts light on important features of the self-diffusion of water. It emphasizes that one of the characteristics of liquids is the average residence time τ of the molecules in a given equilibrium position ('cell') before jumping to the neighbouring one. Its reciprocal value j is the number of activated jumps of an individual molecule into neighbouring equilibrium positions per unit time. According to Wirtz [115], this is:

$$j = j_0 \exp\left[-\frac{\varepsilon}{kT}\right] = j_0 \exp\left[-\frac{E}{RT}\right], \qquad (3.4.10)$$

where ε is the activation energy of a single jump, $E = N\varepsilon$, $j_0 = 2 K\nu$ (ν and K are the frequency and the coefficient of the vibration about the equilibrium position of the molecule, respectively). The activation energy consists of two parts: ε_1 being necessary for the formation of a hole, and ε_f required for the jump into the hole formed:

$$\varepsilon = \varepsilon_1 + \varepsilon_f.$$

In a single jump, the molecules cover a distance l between two adjacent equilibrium positions. This corresponds to the abscissa of the first maximum in the radial distribution curve, and it can be regarded as the 'diameter' of the molecule.

However, this picture is still deficient, as stated by Samoilov, even if the rotation of the molecules is omitted, since in the description of the motion of a molecule, it is not sufficient to consider the vibration within one 'cell' (about one equilibrium position) and the jump into the neighbouring cell only. It should also be taken into account that the cell itself can also be displaced while the molecule vibrates in it for time τ, i.e. the equilibrium position is strolling as well, owing to the movement of adjacent molecules. In water and some other liquids, the order of the molecules is so extended that the cell is probably also displaced mainly by activated jumps. If the individual molecules jump j_1 times to a distance of l_1 per second, and the whole cell jumps j_2 times to a distance of l_2 per second, the self-diffusion coefficient will be:

$$D^* = \frac{1}{6} j_1 \bar{l}_1^2 + \frac{1}{6} j_2 \bar{l}_2^2,$$

and, if E_1 and E_2 are the activation energies of the jumps of the individual molecules and the whole cell, respectively, we have:

$$D^* = A_1 \exp\left[-\frac{E_1}{RT}\right] + A_2 \exp\left[-\frac{E_2}{RT}\right], \qquad (3.4.11)$$

where A_1 and A_2 are constants. The ratio of the frequencies of these two types of jumps resulting in self-diffusion depends on the values of E_1, E_2, A_1 and A_2.

The structure of water contains several cavities; it is rather porous (see Section 1.3.3), and the lattice-like arrangement of the molecules results in structural cavities where a water molecule can fit in. This fact appreciably facilitates the self-diffusion as compared with other liquids built up from more tightly packed molecules. Since there is a structural cavity near each molecule of a volume sufficient for a water molecule, the jumps in the direction of these cavities need not be preceded by hole formation. It can be shown that there are some directions even for the jumps of the 'cell' as a whole where the preformation of a hole is unnecessary. The structural cavities–within the short-range ordered regions–are arranged in definite directions, which leads to differences in the probabilities of the jumps of a molecule in the various directions (of course, this does not cause any anisotropy in self-diffusion, because the ordered regions as units are arranged in a disorder).

Due to the presence of structural cavities, the energy of hole formation is less important in the self-diffusion of water. Since the molecules are usually displaced into existing structural cavities in water, i.e. the space to be occupied by the molecules in self-diffusion is already given, self-diffusion does not require any increase in the activation volume. This is in accordance with the conclusion of Cuddeback, Koeller and Drickamer [116] who stated that the activation volume calculated from the pressure dependence of the self-diffusion coefficient is negative. It can be shown, further that the activation energy required to overcome the potential barrier is the same for the displacement of molecules as for that of cells, i.e. in equation (3.4.11), $E_1 = E_2 = E$. According to this, the self-diffusion coefficient is:

$$D^* = A_1 \exp\left[-\frac{E}{RT}\right] + A_2 \exp\left[-\frac{E}{RT}\right] =$$

$$= A \exp\left[-\frac{E}{RT}\right] \tag{3.4.12}$$

$(A = A_1 + A_2)$. It can be assumed, in addition, that the individual molecules and the cells cover the same average distance l in one jump.

Extending the Samoilov theory, Gurikov discussed the nature of the cavities making self-diffusion possible and the state of the molecules in the cavities in detail [117a]. According to him, it should be supposed that water molecules in the cavities are in interaction with those in the crystal-like lattice. Although some spectroscopic and neutron diffraction data indicate that some of the water molecules rotate freely, this explanation of the observations is not convincing. As suggested by Gurikov, it seems to be reasonable to suppose that water molecules within the cavities form hydrogen bonds with those in the lattice surrounding the cavity.

The self-diffusion based on the translational mobility of molecules in the structural cavities of water depends on the ratio of the molecules in the cavities, i.e., on the degree of occupation of the cavities. Constructing a model for the motion of molecules and using a computer, Neronov [117b] has calculated the self-diffusion coefficient of water from the velocity of

Literature on page 250

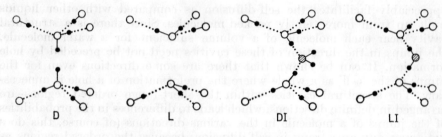

D   L   DI   LI

○ oxygen atoms of the lattice
⊘ oxygen atoms in the cavities
• protons

Fig. 3.7 Relative positions of H$_2$O molecules at orientational defects of water

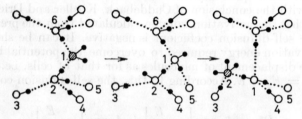

Fig. 3.8 Possible mechanism of the transfer of *DI*
defects when the corresponding *D* defect is transferred
from position O$_2$–O$_6$ to position O$_1$–O$_3$

the molecules in the cavities, which is in the same order of magnitude as
the values observed.

In a more detailed discussion of the mechanism of self-diffusion in water,
Haas [118] started with the theory applied by him to the self-diffusion in
ice. According to this, two types of cavities (so-called *D* and *L* defects)
should be postulated in order to interpret the dielectric relaxation in ice.
The appearance of the *D* and *L* defects is connected with the rotation
of molecules, and it is not accompanied by displacement of oxygen atoms
(Figs 3.7 and 3.8). Since the dielectric relaxation time is approximately
equal to the average duration of an elementary step of diffusion, Haas
introduced the concept of *DI* and *LI* defects: these defects are produced
by molecules located in the cavities of the tetrahedral lattice and bound to
two molecules in the lattice around the cavity by hydrogen bonds, while
the latter molecules are connected with the neighbouring *D* and *L* defects.
The transfer of the *DI* and *LI* defects is accompanied by the displacement
of *D* and *L* defects which results in dielectric relaxation; this is accompanied
by the exchange of molecules in the lattice and in the cavities. This exchange
can be assumed to be the elementary step of self-diffusion. This theory is
in agreement with the approximate equality of the activation energies of
self-diffusion, viscosity, and dielectric relaxation in water [119].

Rather different results have been obtained by several authors [120] for the self-diffusion coefficient of water, which can be explained by the difficulties in the experimental methods and the unsatisfactory accuracy. In the presumably most reliable measurements the self-diffusion coefficient of water was found to be $2 \cdot 57 \times 10^{-5}$ cm^2 s^{-1} at 25 °C by Wang [121], while Devell's careful measurements [122a] gave 2.25×10^{-5} cm^2 s^{-1} under the same conditions.

Some assumptions have been published regarding the appreciable altering effect that isotopes (D, T, or ^{18}O) used as indicators exerted on the self-diffusion coefficient of water. However, it has been proved by Devell [122a] that no such effect occurs within the limits of the experimental error; the measurements applying deuterium and tritium as tracer elements have given the same value within the limits of experimental error. The diffusion velocity of HTO has also been investigated by him in $H_2O - D_2O$ mixtures of various concentrations, and it has been established that, in a mixture whose mole fraction is 0·02 for deuterium (this concentration is far higher than that of the mixture required for the measurement of the self-diffusion coefficient by means of deuterium), the self-diffusion coefficient is the same as in pure H_2O. In a mixture of mole fraction 0·5, $D^* = 2 \cdot 11 \times 10^{-5}$ cm^2 s^{-1}, and even in a mixture of mole fraction 0·9 it decreases only to $D^* = 1 \cdot 90 \times 10^{-5}$ cm^2 s^{-1}. Thus, self-diffusion is less influenced by isotopical labelling than has been supposed. In contradiction with these, according to Mills [122b], the self-diffusion coefficient of water depends all the same slightly on the mass of the molecule used for labelling. In ordinary water at 218·3 °K:

$D^* = 2 \cdot 31 \times 10^{-5}$ cm^2 s^{-1},   when the mass number of the labelled molecule is 18,

$D^* = 2 \cdot 27 \times 10^{-5}$ cm^2 s^{-1} when the mass number of the labelled molecule is 19,

$D^* = 2 \cdot 235 \times 10^{-5}$ cm^2 s^{-1},   when the mass number of the labelled molecule is 20.

Mills [122c] investigated the diffusion of HDO in H_2O and D_2O over broad concentration and temperature ranges, using tritium for indication, and calculated the self-diffusion coefficient of pure H_2O and D_2O from these data. The measurements have made possible the examination of the mass-dependence of self-diffusion under strictly comparable conditions. The dependence of the rate of self-diffusion on the mass of the diffusing entity, seems to follow a more complex relationship than the simple square-root function, probably because of the different interaction potentials of the different isotope species. This problem requires, however, still further studies.

The self-diffusion of water varies with temperature: according to the Arrhenius equation, the $\ln D^*$ vs. $1/T$ function is linear. The activation energy calculated by Wang using this relationship for the self-diffusion measured by means of $H_2{}^{18}O$ is 4·5 kcal/mole. Andreev [123] obtained

Literature on page 250

a somewhat different result investigating the diffusion of a H_2O–HDO system, and found the activation energy to be 4·16 kcal/mole over the 15–45 °C range which is lower than the activation energy for the diffusion of Li^+ and Na^+ ions (4·57 and 4·36 kcal/mole), but higher than that of K^+, Rb^+, Cs^+, Cl^-, Br^-, and I^- ions (3·96, 3·87, 3·84, 4·08, 4·00, and 3·98 kcal/mole, respectively).

On the basis of self-diffusion experiments using $H_2{}^{18}O$ indication, Wang [124a] has established that the ratio $D^*\eta/T$ is constant within the limits of experimental error in the range of 5–25 °C. This indicates that the volume of the diffusing species is constant, and even in the case of the tetrahedral hydrogen-bonded structure of water, diffusion can be regarded as the movement of individual H_2O molecules. The activation energy of self-diffusion being sufficient to break two hydrogen bonds per activated molecules is also in accordance with this, since each hydrogen bond exists between two water molecules.

In polar liquids of voluminous lattice-like structure similar to water, the mechanism of self-diffusion and dielectric relaxation is essentially the same. It follows from this that the relationship of the dielectric relaxation time τ with the self-diffusion coefficient and the average distance λ between two successive equilibrium positions of the diffusing molecule is:

$$D^* = \frac{\lambda^2}{\tau}. \tag{3.4.13}$$

Since the density of water changes only by 0·3 per cent between 5 and 25 °C, λ is almost constant in this temperature range, as shown by Wang [121]. Calculating with a relaxation time value determined independently, the product $D^*\tau$ is really constant within the limits of experimental error, indicating that the deduction given above is in agreement with the experiments. On the basis of equation (3.4.13), diffusing water molecules cover a distance $\lambda = 3·7$ Å in one elementary step. We mention here that the $O-O$ distance between adjacent H_2O molecules in ice I is 2·76 Å, while the distance between the second neighbours is 4·51 Å.

The self-diffusion coefficient of supercooled water has also been studied. Pruppacher [124b] carried out measurements between $+30$ and -25 °C by means of a new tritium labelling technique, and his results were in good agreement with the values measured with ^{18}O tracer, at temperatures higher than 0 °C. At the temperature of the freezing point, no singularity was observed in the supercooled liquid, but the activation energy of self-diffusion steeply increased with decreasing temperature (4.49, 5.70 and 8.20 kcal/mole at $+30$ °C, 0 °C and -25 °C, respectively). These results are interpreted on the basis of the Eyring theory, leading to the conclusion that the mechanisms of self-diffusion, viscous flow and dielectric relaxation are essentially identical. According to the measurements of Gillen, Douglass and Hoch [124c], the activation energy of the self-diffusion of water is 11 kcal/mole at -31 °C. The temperature dependence of the self-diffusion coefficient was found to be similar to the temperature dependence of the ratio of broken hydrogen bonds, as determined from the Raman and infrared spectral data. This is considered to be the empirical basis of the concept attributing a significant role to the hydrogen bonds and their rupture in the mechanism of diffusion in water.

3.4.3 SELF-DIFFUSION IN MIXTURES

When taking into account self-diffusion, the diffusion conditions become rather complicated in mixtures. On one hand, the components of the mixture diffuse on the effect of their own concentration gradients and–owing to cross-effects–of the concentration gradients of the other components as well; furthermore, molecules, of each component also change their positions due to self-diffusion. However, self-diffusion is usually only slightly influenced by the cross-effects. Diffusion due to the concentration gradients of the individual components themselves and interdiffusion are predominating.

The phenomena of self-diffusion in mixtures can be approached in different ways [125]. Following the arguments of Carman and Stein [126], one can start with the description of self-diffusion of mixtures by taking into account equations (3.1.28) and (3.1.50) to obtain the diffusion coefficient of isotope 1* labelling the component 1 in the mixture:

$$D_1^* = kTU_1^* \frac{\partial \ln a^*}{\partial \ln c_1^*} = kTU_1^* \frac{\partial \ln x_1^* \gamma_1^*}{\partial \ln c_1^*} \qquad (3.4.14)$$

(the values marked by an asterisk refer to the isotope).

If there is no macroscopic concentration gradient in the mixture, the activity coefficient γ_1 is constant all over the system, and the concentration c_1^* is:

$$c_1^* = \frac{x_1^*}{(x_1 + x_1^*) V_1 + x_2 V_2}, \qquad (3.4.15)$$

where the denominator is practically independent of x_1^* ($\ll x_1$), thus:

$$\frac{\partial \ln x_1^* \gamma_1^*}{\partial \ln c_1^*} = 1.$$

It follows from this:

$$D_1^* = kTU_1^*. \qquad (3.4.16)$$

Thus, under such conditions, the simpler Nernst–Einstein equation is valid. Since the resistance accompanying the motion of molecules in liquids mainly results from the viscosity of the medium:

$$D_1^* = \frac{kT}{\alpha_1 \eta}, \qquad (3.4.17)$$

Literature on page 250

15

where α_1 is a quantity with the dimensions of length, determined mainly by the size of the diffusing species. If Stokes' Law is valid, $\alpha_1 = 6\pi r_1$ (r_1 is the radius of the spherically shaped diffusing species). On the basis of this relationship, it can be assumed that the following correlation holds independently of temperature and composition in ideal mixtures:

$$\frac{D_1^* \eta}{T} = \text{const,}$$

and similarly:

$$\frac{D_2^* \eta}{T} = \text{const,} \tag{3.4.18}$$

and:

$$\frac{D_1^*}{D_2^*} = \text{const.} \tag{3.4.19}$$

According to Hartley and Crank [127], the following equation can be deduced for the interdiffusion and self-diffusion coefficients:

$$D_{12} = \frac{\partial \ln a_1}{\partial \ln x_1} (x_2 D_1^* + x_1 D_2^*). \tag{3.4.20}$$

Recently, this expression has been compared by McCall and Douglas [128] with observations obtained by studying deuterium-indicated self-diffusion in benzene–cyclohexane, acetone–chloroform, acetone–benzene and acetone–water mixtures by means of NMR spectroscopy. Their results are in fair agreement with the Hartley–Crank theory; however, the accurate quantitative comparison shows some deviations. In order to eliminate these, the authors mentioned have outlined a theory to relate the mutual and the self-diffusion coefficients with the velocity of molecules.

Since dilute solutions are always ideal, it follows that:

$$\frac{\partial \ln a_1}{\partial \ln x_1} = 1. \tag{3.4.21}$$

Under such conditions, the limiting value of the interdiffusion coefficient becomes equal to the self-diffusion coefficient of the component present in low concentration at both ends of the concentration range. Thus:

$$\text{when } x_1 \to 0, \quad \text{then } D_{12}^0 = D_1^* \tag{3.4.22}$$

$$\text{when } x_1 \to 1, \quad \text{then } D_{12}^1 = D_2^*. \tag{3.4.23}$$

Further, if it can be supposed that α_1 and α_2 are independent of concentration in the expression of friction, we have:

$$D_1^* \eta = D_{01}^* \eta_1 = D_{12}^0 \eta_2 \tag{3.4.24}$$

and:

$$D_2^* \eta = D_{02}^* \eta_2 = D_2^1 \eta_1, \tag{3.4.25}$$

where D_{01}^* and D_{02}^* are the self-diffusion coefficients, while η_1 and η_2 are the viscosities of pure components 1 and 2. On the basis of these correlations, the above expression for D_{12} can be transformed so as to contain only the limiting values:

$$D_{12}\,\eta = \frac{\partial \ln a_1}{\partial \ln x_1}\,(x_2\eta_2 D_{12}^0 + x_1\eta_1 D_{12}^1). \qquad (3.4.26)$$

On the other hand, if D_{12}^0 and D_{02}^* are known:

$$D_{12}\,\frac{\eta}{\eta_2} = \frac{\partial \ln a_1}{\partial \ln x_1}\,(x_2 D_{12}^0 + x_1 D_{02}^*). \qquad (3.4.27)$$

The limiting law above was proved experimentally by Carman and Stein [126] for mixtures of ethyl iodide and n-butyl iodide, and it has been established that the product $D^*\eta/T$ is approximately constant for both components. According to the experiments of Mills and Kennedy [129], in aqueous electrolyte solutions containing large ions the term $D^*\eta/T$ is approximately independent of temperature, but in those of small ions this relationship is not valid.

The correlation between the self-diffusion and interdiffusion coefficients of binary systems discussed above and deduced by other methods as well have been assumed to hold in first approximation in the whole concentration range. However, self-diffusion in mixtures is, in fact, a much more complicated phenomenon which is indicated by the specific features of the empirical relationships too. The self-diffusion of liquids is appreciably altered by adding a new component, even in a very low concentration, i.e. in dilute solutions.

The effect of solutes on self-diffusion of liquids can be attributed mainly to three interactions as suggested by Wang [130]: the spatial hindering effect of dissolved molecules on the motion of the solvent species, the solvation, and the disturbance of the liquid structure by dissolved molecules. If the solute molecules are large in comparison with the solvent particles, their diffusion and thermal strolling, resulting in self-diffusion, takes place with a relatively low velocity compared with that of the solvent molecules. Consequently, the slow solute molecules hinder the motion of the solvent entities because they have to go round the large molecules; this slows down self-diffusion by increasing the pathway of diffusion. Hence, the self-diffusion coefficient calculated from data taken in solutions is smaller than the value measured in the absence of solutes. This *obstruction (hindering) effect* depends on the volume fraction (Φ) of the solute in the solution, and on the shape and size of the molecules. If D_1^* denotes the apparent self-diffusion coefficient, D_0^* is the self-diffusion coefficient of the pure solvent, and the solute molecules have a rotational ellipsoidal shape, one has:

$$D_1^* = (1 - \alpha\Phi)D_0^*, \qquad (3.4.28)$$

Literature on page 250

where α is a constant depending on the size of the molecule. If the solute alters the self-diffusion of the liquid only by the obstruction effect, the ratio D_1^*/D_0^* changes linearly with the volume fraction of the solute. Such a correlation has been found by, for example, Biancheria and Kegeles [131a] for the variation of the self-diffusion coefficient of glycol amide in the presence of polymers.

The influence of the obstruction effect due to solute molecules of different sizes and shapes on the self-diffusion was investigated by Bezrukov, Vorontsov-Veljaminov and Ushakova [131b], on the basis of the theory of random walks in a three-dimensional lattice. According to this, in solutions with a low (~ 0.01) volume fraction, D_1^* depends on the volume fraction as predicted by the analytical theory of Wang [130]. With increasing concentration, however, the relative self-diffusion coefficient D_1^*/D_0^* approaches the value 2/3. The dependence of D_1^* on the concentration has also been calculated in the case of solvation. On the basis of the self-diffusion coefficient measured in solutions of ovalbumin, the average residence time of water molecules in the hydrate spheres of the protein molecules was estimated; this was found to be about twenty times higher than the average residence time of a molecule beside its neighbour in pure water.

In several cases, the self-diffusion of the solvent is altered by hydration (or, in general, by solvation) of the solute. The *hydration effect* (solvation effect), arising from this, can be attributed to the fact that solvent molecules in the hydrate sphere do not take part in self-diffusion, which lowers the self-diffusion coefficient. In addition, the size of solute molecules increases on hydration, which also increases the obstruction effect and decreases the self-diffusion coefficient. The combined result of these two effects on the self-diffusion coefficient determined experimentally is, according to Wang [130]:

$$\frac{D_{exp}^*}{D_0^*} - \Delta_1 = 1 + \Delta_2 - \{\alpha(v_2\varrho_1 + H) + H\}w_2, \qquad (3.4.29)$$

where Δ_1 and Δ_2 are small correction factors that can be calculated theoretically, v_2 is the specific volume of the solute, ϱ_1 is the density of the solvent, w_2 is the weight fraction of the solute in the solution, while H is the mass of the solvent bound in the solvate sphere related to unit mass of solvent-free solute. This equation seems to be valid for the self-diffusion of water in ovalbumin solutions, as established by Wang, Anfinsen and Polestra [132].

In liquids of significant internal structure ('associated' liquids), self-diffusion can also be altered markedly by *small molecules* distorting the structure in their vicinity. Owing to this *structure-breaking effect*, the self-diffusion coefficient of the solvent is lowered by the solute to a smaller degree than could be expected on the basis of the increase in macroscopic viscosity. This structure-breaking effect influences self-diffusion in a direction opposite to that of obstruction and solvation effects. If the solute distorts the structure of the solvent, $D\eta/T$ is not constant but increases with increasing concentration of the solution; this effect becomes more pronounced with decreasing temperature (increase in temperature itself is a structure-breaking effect). This effect becomes appreciable mainly in electrolyte solutions.

In a modified kinetic theory of self-diffusion, Ottar [133a] suggested that not only single liquid molecules can diffuse but their groups consisting of n molecules can as well. He found the diffusion coefficient of these groups to be:

$$D_n = \frac{p\lambda^2}{6\sqrt{n}} \, ,$$

where p is the number of jumps of groups into neighbouring positions per second, while λ is the distance between two adjacent equilibrium positions of solvent molecules. It has been supposed that in aqueous solutions, the solute molecules form n_s hydrogen bonds with water molecules depending on their structure, and the activation energy of diffusion is required to break them. At any one moment, only a fraction of the possible hydrogen bonds (n_s) exists and this fraction changes continuously. Ottar assumed, further, that the diffusion coefficient determined experimentally corresponds to the motion of solute molecules surrounded by n_s water molecules. So, according to the Eyring theory, the diffusion coefficient will be:

$$D_{n_s} = \frac{\lambda^2}{6\sqrt{n_s}} \cdot \frac{kT}{h} \exp\left[-\frac{E_k}{RT}\right],$$

where E_k is the activation energy needed to break these bonds. The relationship of viscosity with the self-diffusion coefficient of the liquid is thus:

$$\frac{1}{\eta} = \frac{4\sqrt{2}\, N^{2/3} V^{1/3}}{RT} D^*.$$

This equation is in good agreement with the observations for several liquids (e.g. water, ethyl bromide, benzene).

Starting with the theory of viscous flow, Albright [133b] has deduced a correlation between the tracer diffusion coefficients of the components and the viscosity of the liquid in multicomponent liquid mixtures. In this way, the viscosity of the mixture is, in good approximation:

$$\eta = RT \sum_i (\xi_i c_i V_i^{-2/3}/D_i^*),$$

where c_i, v_i and D_i^* are the concentration, molar volume, and tracer diffusion coefficient of the ith component, respectively, while the factors ξ_i are empirical constants.

Measuring the self-diffusion coefficient of dioxan in dioxan–water mixtures, Erdey-Grúz and Lévay [133c] have established that the self-diffusion coefficient exhibits a minimum at 20 mole per cent dioxan content, while the activation energy of self-diffusion has a maximum at 15 mole per cent concentration for dioxan. The product $D^*\eta$ is approximately constant

Literature on page 250

starting from pure dioxan up to about 40 mole per cent dioxan content, then it decreases with further increase in water concentration. They have concluded from this that the liquid structure is not altered appreciably by water within the range of 0 to 60 mole per cent for water; the structure characteristic of water starts to be built up by the development of hydrogen bonds between water molecules in mixtures with water contents higher than the former range. This is in agreement with the investigation of Erdey-Grúz and Nagy-Czakó [133d] regarding the transference number of hydrogen ions.

The conditions are slightly different in mixtures of water and methanol. According to the investigations of Erdey-Grúz, Fodor-Csányi, Lévay and Szilágyi-Győri [133e], the self-diffusion coefficient of methanol has a minimum at 15–20 mole per cent methanol content, while the activation energy of self-diffusion shows a maximum in a mixture containing 15 mole per cent of methanol. On the other hand, the product $D^*\eta$ increases in the range of 100 to about 45 mole per cent methanol content, it decreases after passing a maximum and, finally, it goes through a minimum at about 8 mole per cent at low temperatures. This behaviour, being in agreement with several other phenomena, indicates that there are complicated–and partly structure-making–interactions between water and methanol.

In respect of the self-diffusion of water, the conditions are somewhat different in mixtures of water and methanol. According to the investigations of Erdey-Grúz, Inzelt and Fodor-Csányi [133g], the self-diffusion coefficient of water decreases first with increasing concentration of methanol, then in the mixture containing about 20% methanol it passes through a flat minimum and starts to increase again monotonically; that is, it changes similarly to the self-diffusion coefficient of methanol in the same mixtures. The variation in the product $D^*\eta$ calculated for water is, however, significantly different from that obtained for methanol: the former has no minimum, but increases from the beginning of the addition of methanol, and a flat maximum is observed at about 45% methanol content. These variations in the self-diffusion of water can perhaps be explained by assuming a stabilization of the liquid structure of water on the effect of some methanol, which is indicated by several other phenomena, too. However, at higher methanol concentrations the structure-destroying effect becomes more emphasized, and this will be predominating in mixtures of medium methanol concentration. The stabilization of water molecules observed again in mixtures with high methanol content can probably be attributed to the fact that water molecules are incorporated into the chains formed by the methanol molecules.

The self-diffusion coefficient of iodine was studied in water–methanol and water–dioxan mixtures by Erdey-Grúz and Lévay [133h]. The self-diffusion coefficient of I_2 depends on the water content in different ways in these two solvent mixtures, which can be explained by differences in the effects of methanol and dioxan altering the liquid structure of water; further the interaction of I_2 is also different with solvent molecules of different kinds.

3.4.4 SELF-DIFFUSION OF WATER IN ELECTROLYTE SOLUTIONS

Dissolved electrolytes alter the self-diffusion coefficient of water, and, in principle, all the three effects discussed in the previous section influence this change. The ratio of the three types of effects, however, depends on the conditions, mainly on the size, charge, and hydration of ions. For example, in Wang's measurements [134] in dilute NaCl solutions it has been observed that the self-diffusion coefficient of water decreases initially with increasing concentration; this can be explained by the fact that the obstruction and hydration effects of the strongly hydrated Na^+ ion overcompensate the structure-breaking effect. In dilute KI solutions, however, the self-diffusion coefficient of water increases with increasing concentration, and this is partly due to the strong structure-breaking effect in the immediate environment of the large I^- ion that exceeds the obstruction and hydration effects.

Podolsky's theory [135] tried to explain the variations in the self-diffusion coefficient of water caused by the effect of dissolved electrolytes. This theory is based on the Eyring theory of diffusion (cf. Section 3.1.2.2). According to equation (3.1.59), the distance λ covered by the diffusing molecule in one elementary step (activated jump) plays an important role in the expression of the diffusion coefficient. In fact, transport phenomena can only be characterized by one single λ and by one activation energy in pure liquids. In solutions, the individual activated jumps of molecules may have different activation energies depending on their temporary environment, even if λ is the same all over the system, which, however, is not necessarily so. According to Podolsky, the activation energy can be obtained as a sum of two terms: one of them is related to the pure solvent, while the other refers to the displacement in an environment distorted by the solute. The average of these two energies should be inserted into the corresponding equations. Starting from this assumption, the following results can be obtained for the relation between the self-diffusion coefficient D_1^* of the solution and that of the pure solvent, D_{01}^*:

$$D_1^* = D_{01}^* \left(1 + \frac{B_k + B_j}{1 + r} c\right), \qquad (3.4.30)$$

where B_k and B_j are empirical constants related to the ions present in the solution, while r depends on the difference in the effects exerted by the ions on solvent molecules and those of the solvent molecules on one another. The B values are related to the number of water molecules bound to ions which can be utilized to gain some information on the hydration of ions, considering the variation of the diffusion coefficient with concentration.

The results of the Podolsky theory are in good agreement with the self-diffusion coefficient of water determined by Wang [134] in 1 M KI, KCl, and NaCl solutions (using $H_2^{18}O$ for indication). Studies of Tamás, Lengyel and Giber [136] yielded similar results for the self-diffusion of water

Literature on page 251

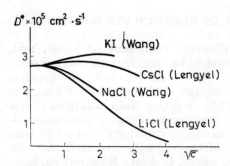

Fig. 3.9 Variation of the self-diffusion coefficient of water in solutions of some electrolytes

in LiCl and CsCl solutions over a broad concentration range applying the same indicator (Fig. 3.9). The diffusion coefficient increases at first on the effect of KI and CsCl in dilute solutions but afterwards it decreases. In NaCl and LiCl solutions, it decreases initially indicating that the obstruction and hydration effects are predominating in solutions of these strongly hydrated electrolytes. Measuring the self-diffusion coefficient of water in concentrated KCl, NaI, CsI, CsF, and $MgCl_2$ solutions, Tamás and Ujszászy [137] have found the Podolsky theory to give a good quantitative basis for the description of the relationship between the self-diffusion coefficient and viscosity; they have also calculated some hydration numbers from their experimental results.

The effect of dissolved ions on the self-diffusion of water has been discussed in detail by Devell [122a] on the basis of Wang's suggestions [130]. The effect of electrolytes also involves two decelerating factors, and one accelerating one, similarly to the general case of mixtures (Section 3.4.3). One of the decelerating factors is the obstruction effect; the solvent molecules must go round the ions, which move more slowly under the conditions of self-diffusion than water molecules do. Thus, the water molecules must move along a longer pathway to cover a certain distance in the presence of ions than in pure water. This decreases the self-diffusion coefficient D^* of water in solutions as compared with that (D_0^*) of pure water:

$$D^* = D_0^* (1 - \alpha\Phi), \tag{3.4.31}$$

where Φ is the volume fraction of the retarding species (hydrated ions), while α is a factor depending on the shape of the retarding particle. For spherical species, $\alpha = 1 \cdot 5$. In solutions containing hydrated ions of strong electrolytes, the mobility of ions is not much less than that of water molecules, thus it can be assumed that the retardation effect is small.

The other effect hindering self-diffusion is the hydration of ions. If the exchange between free water molecules and those bound in hydrate spheres is relatively slow with respect to the velocity of the diffusional displacement, the molecules in the hydrate spheres hardly participate in self-diffusion. Further, if dissolved ions are supposed to be almost immobile with respect to self-diffusion, the change in the self-diffusion coefficient due to hydration will be:

$$D^* = D_0^* \left\{ 1 - f \left(1 - \frac{v_h}{v_f} \right) \right\}, \tag{3.4.32}$$

where v_h and v_f are the frequencies of the exchanges involving hydrated and free water molecules respectively in the course of diffusion, and f is the fraction of water molecules bound in hydrate spheres. If the interaction between bound and free water molecules is slow, one has:

$$D^* = D_0^*(1 - f). \qquad (3.4.33)$$

Besides these two effects which retard self-diffusion, an accelerating effect occurs if dissolved ions destroy the water structure around the hydrate spheres, decreasing the local viscosity in the immediate vicinity of ions. This phenomenon is closely related to that called negative hydration by Samoilov.

According to the above discussion, the fact that the self-diffusion coefficient of water is decreased by Mg^{2+} salts, and to a greater extent than that observed in the case of Na^+ salts, can be explained by the stronger hydration of Mg^{2+} ions as compared with Na^+ ions; hence their obstruction and direct hydration effects are also larger. On the basis of its effect on self-diffusion of water–in agreement with other conclusions obtained from other phenomena–the obstruction effect of the Li^+ ion is greater, while its structure-breaking effects on water are smaller than those of the Na^+ ion. On the other hand, the Br^- ion has a greater structure-breaking effect than the Cl^- ion. Thus, the self-diffusion coefficient of water in LiBr solutions is approximately the same as in NaCl solutions of identical concentration. From measurements in $MgSO_4$ solutions, it can be concluded that the obstruction and hydration effects of SO_4^{2-} ions are predominating over the structure-breaking effect. In concentrated solutions, KCl and KI decrease the diffusion coefficient because the majority of water molecules are already bound in hydrate spheres; this is why the increase in the ratio of dissolved molecules cannot exert a further structure-breaking effect on water.

The Wang–Devell theory regarding the alteration of the self-diffusion of water by dissolved electrolytes has been proved by the experiments of Jones, Rowlands and Monk [138] in $MgCl_2$, $CaCl_2$, $SrCl_2$ and $BaCl_2$ solutions in the $0{\cdot}5$–$1{\cdot}0$ M concentration range. In these solutions, the self-diffusion coefficient decreases linearly as a function of the molar concentration of the salt:

$$D^* = D_0^*(1 - hc). \qquad (3.4.34)$$

If it is assumed that the hydration effect of self-diffusion prevails in these solutions of strongly hydrated ions, equation (3.4.34) leads to the relationship $hc = f$, where h is the average hydration number of the ions. If, however, the decrease in the self-diffusion coefficient is attributed to the obstruction effect, equation (3.4.31) gives $hc = \alpha\Phi$, and in this respect h is the measure of the size of the hydrated ion. In the case of small and polyvalent ions the hydration effect is preferred, while for large ions of small charge the obstruction effect is preferred, but the data available are not sufficient for an unequivocal decision.

Literature on page 251

Investigating the self-diffusion of water in KNO_3 and $LiSO_4$ solutions, Andreev [139] established that the self-diffusion coefficient is hardly increased by K^+ and NO_3^- ions, however, SO_4^{2-} ions retard self-diffusion significantly (by 19 per cent at 15 °C). On studying self-diffusion in NH_4ClO_4, Ca $(ClO_4)_2$, and $Ba(ClO_4)_2$ solutions, Yemelianov and Agishev [140] established that the self-diffusion of water is enhanced by perchlorate and ammonium ions, i.e. they are negatively hydrated. The negative hydration of NH_4^+ ions is more extensive than that of K^+ ions, but it is smaller than that of Cs^+ ions.

It can be established, in general, that the number of water molecules with mobilities altered by positive or negative hydration of cations is approximately independent of the nature of the anion in dilute aqueous solutions.

Studying the relationship between self-diffusion and the structural properties of solutions, McCall and Douglas [141] emphasized that the non-uniform, widely complex nature of the hydration phenomenon should be taken into account in this respect too. Particularly, one should consider that, in addition to the geometry of the arrangement of water molecules around ions, the average lifetime of the hydrate sphere is also important, i.e. the rate of exchange of water molecules in it. The average lifetime of the hydrate sphere of some ions is several hours, while other ions may have hydrate spheres of such a short lifetime that relaxation has no influence on certain processes. The model which supposes only a definite number of water molecules to be influenced by ions, while all the others remain unaltered, is also oversimplified. In connection with the effect of dissolved electrolytes on self-diffusion, it should be considered as well that, although the ion hydrate concept can usefully be applied for the description of several phenomena, it in fact, covers very complex effects, representing them in a very simplified way.

McCall and Douglas investigated the effect of several electrolytes on the self-diffusion of water using the NMR method. The electrolytes studied formed hydrates of short lifetime, and it can be assumed that during the time of the measurement (about 0·1 s), the hydrate molecules were exchanged several times with the free ones. Thus, water molecules of various states participate in self-diffusion, and the self-diffusion coefficient measured is, in fact, only an average value. All the ions investigated by them diffuse more slowly than water (the slowest are Na^+ and Ca^{2+} ions), and most of them decrease the self-diffusion coefficient of water; this decrease is nearly linear with increasing concentration in the 0·1–1 M concentration range. Self-diffusion of water is decreased by LiBr, HNO_3, H_2SO_4, HI, $KOOCCH_3$, LiCl, KF, $CaCl_2$, K_2CO_3, NaBr, $ZnCl_2$, KCl, LiOH, NaOH, KOH, NaCl, in this order, while it is increased by KI, KBr, NaI, CsI, KNO_3. The relative change in the self-diffusion coefficient $(D^* - D_0^*)/D_0^*$ can be approximately expressed as the sum of two terms characteristic of the two types of ions; these are related to the terms expressing the effect of ions on viscosity.

The effect of hydrogen ions on the self-diffusion of water has been studied by Yemelianov, Nikiforov and Kucheryavenko [142] by spin-resonance methods in HCl, HBr, and HI solutions. According to their measurements, the H_3O^+ ion causes a smaller decrease in the mobility of water molecules than Li^+ and Na^+ ions do. On this basis, H_3O^+ seems to be hydrated positively, but to a smaller degree than Li^+ and Na^+ ions. However, in view of the activity coefficients, Li^+ and H_3O^+ ions are hydrated approximately to the same extent. This contradiction can be resolved by considering that the exchange of protons between H_3O^+ and H_2O molecules (which results in the prototropic conduction mechanism of the hydrogen ion, see Section 4.2.2.2), takes place continuously without the action of electric field, but, of course, has no favoured direction. Thus, it can be supposed that during its lifetime, the hydration around the H_3O^+ ion is approximately the same as around the Li^+ ion, but, due to the jump of the proton (H^+) from H_3O^+ to an adjacent water molecule, the hydrated complex can be displaced without any appreciable displacement of water molecules. Within the time of the measurement (0·05 s), one proton jumps to adjacent water molecules several times, which decreases the effect of positive hydration and increases the mobility of water molecules in acid solutions as compared with salt solutions under comparable conditions. Cl^-, Br^- and I^- ions increase the mobility of water molecules according to the measurements of Yemelianov et al., i.e. they are hydrated negatively.

Self-diffusion of other liquids is influenced by electrolytes in a similar way to that of water. For example, Yemelianov and Gaysin [143] studied the effect of NH_4I, KI, NaI, LiI and BaI_2 on the self-diffusion of methanol. From the variation of the self-diffusion coefficient and viscosity, they have concluded that the ions are solvated, and solvation is the more extended, the smaller the ions and the larger their charge.

3.4.5 TRACER DIFFUSION OF IONS

In practice (e.g. polarography and several other electrode processes), it may occur that ions present in low concentration diffuse in a solution of another electrolyte which is present in high but gradientless concentration. This phenomenon called *tracer diffusion* occurs when, for example, Na^+ ions present in negligibly low concentration diffuse in a KCl solution of uniform and relatively high concentration. In general, tracer diffusion can be directly investigated by radioactive tracer elements. If the diffusing ion is almost chemically identical with one of the ions present in uniform concentration (e.g. K^* isotope in a KCl solution), tracer diffusion becomes identical with self-diffusion.

Under the conditions of tracer diffusion, the total concentration and the ionic strength of the electrolyte are almost the same all over the solution. Thus, its activity coefficient are also the same everywhere, and the gradient of the chemical potential of the diffusing tracer ion can be expressed with satisfactory accuracy by the concentration gradient, even in more concentrated solutions. In this way, the thermodynamic term is omitted in the expression of the diffusion coefficient.

Literature on page 251

The tracer diffusion coefficient can also be determined separately for the individual kinds of ions. Unlike ordinary diffusion of electrolytes, the restriction that ions of opposite charge should move with the same velocity is not valid for this case. Because of the high concentration of the 'supporting electrolyte' present in uniform concentration, the diffusing tracer ion does not produce any appreciable diffusion potential, consequently the velocity of its diffusion is not modified significantly by the cross-effect arising from the electric field.

With respect to other effects of the electric forces between ions, tracer diffusion of ions also differs from the normal diffusion of electrolytes. The concentration dependent electrophoretic effect becomes negligibly small, since the concentration of the diffusing tracer ion is very low. On the other hand, the relaxation effect is important, because the tracer ion diffuses in a relatively immobile medium which results in the deformation of the ion atmosphere and a decrease of mobility. (In normal diffusion of binary electrolytes, each ion diffuses with the same velocity, so the ion atmosphere remains spherically symmetrical).

On the basis of the relationships discussed in Section 3.3.1 for electrolyte diffusion, an individual tracer diffusion coefficient can be defined for the fluxes of the individual kinds of ions.

The individual diffusion mass flux of the ith ion at constant temperature is:

$$J_i = -L_{ii} \frac{\partial \mu_i}{\partial y} = -D_i^* \frac{\partial c_i}{\partial y} = -D_i^* \frac{c_i M_i}{1000} \frac{\partial m_i}{\partial y}, \qquad (3.4.35)$$

where M_i is the molecular weight, c_i is the concentration in mole cm^{-3} units, and m_i is molality. According to the Nernst equation, the tracer diffusion coefficient in the limiting case of infinitely dilute solution is:

$$D_i^* = \frac{RTU_i^*}{F|z_i|}. \qquad (3.4.36)$$

In solutions of monovalent ions, with respect to equation (3.2.22), this becomes:

$$D_i^{*0} = 2 \cdot 662 \times 10^{-7} \lambda_i^0 \text{ cm}^2 \text{ s}^{-1} \qquad (3.4.37)$$

at 25 °C, where λ_i^0 is the limiting value of the equivalent conductivity of the ith ion.

If, as usual, the supporting electrolyte is present in a relatively high concentration, the tracer diffusion coefficient differs from the value given in equation (3.4.36).

On the basis of the Onsager theory [144] outlined for the diffusion of several electrolytes simultaneously present in a solution, Gosting and Harned [145a] have given the concentration dependence of tracer diffusion and self-diffusion coefficients, taking into account the fact that the transference number of the tracer ion is negligibly small. According to the calculations based on the theory of strong electrolytes (cf. Section 5.1), one has:

$$D_i^* = U_i kT \left[1 - \frac{\varkappa z_i^2 e^2}{3\varepsilon kT} \left\{ 1 - \sqrt{d(U_i)} \right\} \right], \qquad (3.4.38)$$

where $\varkappa$ is the so-called characteristic distance, defined in Section 5.1.2, which is proportional to the square root of ionic strength $I\left(=\dfrac{1}{2}\sum_j c_j z_j^2\right)$, e is the elementary electric charge, ε is the dielectric constant of the solution, while $d(U_i)$ is a quantity depending on the mobilities and charges (valences) of all ions present. If a solution of tracer ion i containing two other ions (denoted by 2 and 3) is considered and λ_i^0, λ_2^0, and λ_3^0 are the limiting values of the equivalent conductivities of the ions:

$$d(U_i) = \frac{|z_i|}{|z_2| + |z_3|} \left(\frac{|z_2|\,\lambda_2^0}{|z_i|\,\lambda_2^0 + |z_2|\,\lambda_i^0} + \frac{|z_3|\,\lambda_3^0}{|z_i|\,\lambda_3^0 + |z_3|\,\lambda_i^0} \right). \quad (3.4.39)$$

If each ion is monovalent, and ion i is an isotope of the type 2 ion, than $\lambda_i^0 = \lambda_2^0$, and one has:

$$d(U_i) = \frac{\lambda_2^0 + 3\lambda_3^0}{4(\lambda_2^0 + \lambda_3^0)} = \frac{1 + 2t_3^0}{4}, \quad (3.4.40)$$

where t_3^0 is the limiting value of the transference number of type 3 ions (with a charge opposite to that of the tracer ion) in infinite dilution. When the ions are, incidentally, of the same mobility, or their equivalent conductivities are identical (e.g. this is approximately true in KCl solutions):

$$d(U_i) = 0.5$$

and

$$1 - \sqrt{d(U_i)} = 0.293.$$

Measuring the self-diffusion coefficient of the ammonium ion by means of labelled $^{15}NH_4^+$ in NH_4Cl and $(NH_4)_2SO_4$ solutions, Tanaka and Hashitani [145b] have established that in dilute solutions the results are in good agreement with those calculated on the basis of the Onsager–Gosting–Harned equation, and the activation energy of self-diffusion is 16 kJ/mole.

In order to emphasize the relationship between the value of the tracer diffusion coefficient D_i^* corresponding to a solution of ionic strength I and the value of D_i^{*0} corresponding to infinite dilution, equation (3.4.38) can be formulated as follows:

$$D_i^* = D_i^{*0} \left[1 - \frac{\varkappa z_i^2 e^2}{3\varepsilon kT} \{ 1 - \sqrt{d(U_i)} \} \right]. \quad (3.4.41)$$

Inserting the value of $\varkappa$ obtained from the theory of strong electrolytes and the universal constants, it becomes:

$$D_i^* = D_i^{*0} \left[1 - \frac{2.801 \times 10^6}{(\varepsilon T)^{3/2}} \{ 1 - \sqrt{d(U_i)} \} z_i^2 \sqrt{I} \right]. \quad (3.4.42)$$

Literature on page 251

For aqueous solutions of 1–1 electrolytes at 25 °C, we can write:

$$D_i^* = D_i^{*0}[1 - 0.7816\{1 - \sqrt{d(U_i)}\}\sqrt{c}]. \tag{3.4.43}$$

These latter two equations are the *Onsager limiting laws* for tracer diffusion and self-diffusion modified by Gosting and Harned, which can be expected to be valid only in solutions of low ionic strength due to the simplifying assumptions applied in their deduction (not discussed here in detail). The limiting value of the tracer diffusion expressed according to the Nernst equation in terms of the equivalent conductivity of the tracer ion is:

$$D_i^{*0} = \frac{RT\,\lambda_i^0}{|z_i|\,F^2}, \tag{3.4.44}$$

and, at 25 °C:

$$D_i^{*0} = 2.661 \times 10^{-7}\frac{\lambda_i^0}{|z_i|}. \tag{3.4.45}$$

The validity of the Onsager limiting law of self-diffusion given above can be expected only in very dilute solutions. Measurements of high precision carried out by Mills and Godbole [146a] have shown that the experimental values are in good agreement with the calculated ones for the self-diffusion of Na^+ ions in NaCl solutions in the 0.0002–0.09 mole/l^{-1} concentration range (Fig. 3.10). The agreement can be extended to more concentrated solutions, when correcting the equation for an ionic size of 4 Å. However, the experimental method applied is, in principle, questionable so the results are not reliable.

The value of the tracer diffusion coefficient obtained from equation (3.4.42) cannot in general be unequivocally compared with the majority of the observations. This is because no measurements which are correct in principle have been carried out with sufficient accuracy in solutions of such low ionic strength that the Onsager law could be expected to be valid. The results of measurements in more concentrated solutions deviate significantly from those expected on the basis of the Onsager law, and even their extrapolation to low concentrations does not tend to agree. Fig. 3.11 shows the tracer diffusion of some ions as typical examples obtained in the measurements of Mills [147], Stokes, Woolf and Mills [148] and Woolf [149] in KCl solutions of concentration c. The dotted lines correspond to the calculated limiting values of the Onsager law. According to the experimental data obtained in more concentrated solutions of the supporting electrolyte, the ratio D_i^*/D_i^{*0} is always higher than the theoretically calculated one, and nothing indicates that in dilute solutions the experimentally determined D_i^*/D_i^{*0} vs. $\sqrt{c}$ curves would approach the theoretically calculated ones even in their slopes. The cause of the great deviations has not been satisfactorily elucidated so far. The theory of strong electrolytes, on which the Onsager law is based, regards ions as point-like charges and this certainly plays a role in this deviation. When taking into account finite ionic diameters, the theoretically calculated values increase, but it cannot be expected that the large deviation from the measured values could be eliminated in this way.

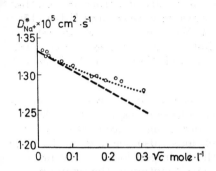

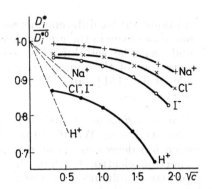

Fig. 3.10 The self-diffusion coefficient of Na⁺ ion in NaCl solutions of low concentration

The dotted line corresponds to the Onsager law, while the pointed curve represents the correlation corrected by an ionic size of 4 Å

Fig. 3.11 D_i^*/D_i^{*0} values for the tracer diffusion of some ions in KCl solution vs. concentration of the solution

Measuring the self-diffusion coefficient in aqueous solutions of NaF and NaCl by the radioactive isotope method, Turq, Lantelme and Chemla [146b] found the Onsager limiting law to be valid in good approximation in gelatinous solutions.

The difference between experimentally determined and theoretically calculated values of tracer diffusion coefficients requires further accurate investigations. The viscosity of the supporting electrolyte solution is probably responsible for the differences, too, but this effect cannot be predominating. We refer to the fact that although the viscosity of NaCl solutions hardly varies with concentration, the tracer diffusion coefficients still change significantly as a function of concentration, as shown in Fig. 3.10. It can be supposed that these changes are connected mainly with the electric interaction of ions.

The importance of electric interactions has been indicated by the results of Samoilov, Goncharov, Jaskichev, Markova and Alekseieva [146c] concerning the self-diffusion of ³⁶Cl⁻ and ¹³¹I⁻ ions in solutions of alkali metal and ammonium chlorides. The activation energy of self-diffusion depends on the concentration of the macro component of the solution. The activation energies of the self-diffusion of the anion and the cation approach each other with increasing concentration. The concentration dependence of the activation energy observed in dilute solutions has been attributed to the stabilizing effect of ions on the liquid structure of water due to the formation of contact ion pairs between the labelled anion and the cation of the macro-component.

It should be noted in connection with the comparison of observed and calculated data that the experimental technique of tracer diffusion measurements is very delicate, and the data measured under the same apparent

Literature on page 251

conditions but by different methods often differ markedly from one another –sometimes even in their general trends. For example, Wang and Miller [150] found that the self-diffusion coefficient of Na^+ ions plotted against the square root of the ionic strength of the NaCl solution passes a maximum, while Mills [151], and Nielsen, Adamson and Cobble [152] have not observed this.

Tracer diffusion of hydrogen ions is of particular interest. According to the studies of Woolf [149], the tracer diffusion coefficient of the hydrogen ion decreases more rapidly with increasing concentration of the supporting electrolyte than that of other ions (also see Fig. 3.11). This is probably influenced by the fact that the hydrogen ion (i.e. the proton) diffuses not only by means of the hydrodynamic motion of the oxonium ion (H_3O^+), but by the prototropic mechanism as well (see Section 4.2.2.2), i.e. a proton can jump from the H_3O^+ ion to the neighbouring water molecule without the displacement of this H_2O molecule. It is also to be noted in connection with this that the decreasing effect of LiCl on the tracer diffusion of the hydrogen ion is particularly high, much higher than that of the K^+ ion. This can probably be attributed to the fact that the K^+ ion, with a size almost identical with that of water molecules, hardly deforms the water structure. On the other hand, the small Li^+ ion of high surface charge density exerts a structure-breaking effect on water even in such a low concentration in which the effect of K^+ ions is negligible. The orientation of water molecules built into the hydrate spheres of Li^+ ions differs from those bound together only by hydrogen bonds, and this hinders the proton tranfer between water molecules. Hence, some effects with no marked influence on other ions will appear in the tracer diffusion of hydrogen ions.

Tracer diffusion depends on the viscosity of the liquid, too. This is revealed by the fact that the product $D^*\eta$ changes with the concentration or ionic strength of the supporting electrolyte less than D^* itself. According to Mills and Kennedy [153], the product $D^*\eta$ varies linearly with $\sqrt{c}$ for K^+ and Rb^+ ions. As established by these authors, the tracer diffusion of these and other 'non-hydrated' ions is influenced mainly by two factors. On one hand, the solvent molecules should be removed from the pathway of the migration of diffusing ions; the resistance arising from this is proportional to the macroscopic viscosity. On the other hand, the ion must gain an energy sufficient to move out of the ion atmosphere which, however, depends on the potential of the ion atmosphere at the location of the diffusing ion. In ion atmosphere there is an excess of ions with charges opposite to that of the central ion. The larger the hydrated counter-ion, the farther the average charge of the ion atmosphere is from the central ion, and–under otherwise identical conditions–the smaller the binding potential to the central ion, the easier is the departure of the central ion from the ion atmosphere. This is in agreement with the observation that, after correcting for viscosity, the tracer diffusion of the iodide ion is faster in the presence of LiI than in KI, NaI or RbI solutions. Investigating the self-diffusion of Na^+ and Rb^+ ions, Mills [154] has concluded that the effect of viscosity is predominating in the concentration dependence, but a correction by the ratio η/η_0 overcompensates this effect in 0·1–4 M NaCl and LiCl solutions. In addition, he has pointed out that the tracer diffusion coefficients of cations are more distinct in dilute solutions than those of anions. This fact is believed to be brought into agreement with the Onsager law by taking into account ionic sizes.

The self-diffusion coefficient of ions varies with concentration similarly to tracer diffusion. For example, measuring the self-diffusion coefficient of Na+ ions, Mills [155] obtained the results shown in Fig. 3.12 for 0·2–4 M NaCl solutions.

The tracer diffusion coefficient of iodide ions in alkali halide solutions has been measured by Stokes, Woolf and Mills [156] and by Lengyel, Tamás and Vértes [157], while that of the Cl− ion has been obtained by Mills [158] (Figs 3.13 and 3.14). The tracer diffusions of these two ions depend similarly on the nature of the supporting electrolyte. This is well brought out when the ratio $D*/D*^0$ is plotted against concentration (Fig. 3.15). It is seen that the tracer diffusion coefficients of these two anions depend on the nature of the supporting electrolyte much more than on that of the diffusing ions. The order KCl−NaCl−LiCl is probably determined by the effect of ionic size on the relaxation effect.

The concentration dependence of the tracer diffusion coefficient of ions varies similarly in mixtures of water with organic liquids. For example, Marcinkowsky, Phillips and Kraus [159] measured the self-diffusion coefficient of Na+ ions in the presence of NaCl and NaClO₄ dissolved in mixtures of methanol, ethanol or propanol with water. In every mixture it is lower than in water, the lowest value being found in propanolic mixtures under comparable conditions. The product $D^*_{Na}\,\eta/\eta_0$ decreases with increasing alcohol concentration in all the three aqueous alcohol mixtures, and it can

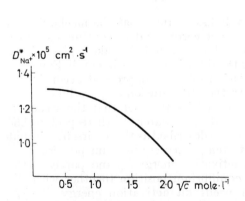

Fig. 3.12 Self-diffusion coefficient of Na+ ions in NaCl solutions of medium concentration

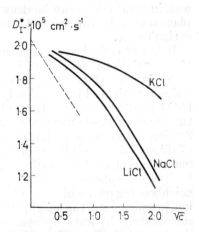

Fig. 3.13 Tracer diffusion coefficient of I− ions in alkali chloride solutions

Literature on page 251

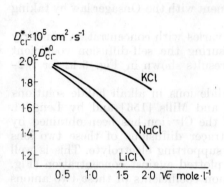

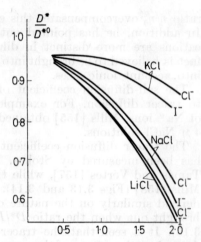

Fig. 3.14 Tracer diffusion coefficient of
Cl⁻ ions in alkali chloride solutions

Fig. 3.15 Variation of ratio D^*/D^{*0}
for the tracer diffusion of I⁻ and
Cl⁻ ions in alkali halide solutions

be described by a single curve as a function of concentration. In ethylene-
glycol–water mixtures, $D^*_{Na} \eta/\eta_0$ is independent of the concentration of
the solvent mixture, while in glycerol–water mixtures it increases with
increasing glycerol concentration.

Since ions move relatively independently in tracer and self-diffusion,
their velocity is not controlled by that of ions of opposite charge; thus
certain conclusions can be drawn for their individual properties from these
phenomena. In this respect, the activation energy of the diffusion of ions
is significant.

It has already been recognized by Polissar [160] that the mechanism of
the translational motion of ions does not correspond to continuous movement in the classical hydrodynamical sense, but it takes place in a
series of activated jumps. Analysing the diffusion of ions in detail, Samoilov
[161] pointed out that the self-diffusion of ions can proceed according to
two types of mechanisms similarly to the self-diffusion of pure liquids (see
Section 3.4.2). This similarity is supported by the fact that the structure
of dilute solutions hardly differs from that of water. With respect to the
main characteristics of diffusion, it can be described as the ion itself, without
its hydrate sphere, jumping from its temporary equilibrium position to a
neighbouring one, which requires an activation energy E_1, and partly as the
ion jumping into the neighbouring equilibrium position together with the
surrounding hydrate sphere, which needs an activation energy E_2 (see
Section 3.4.2). Thus, the self-diffusion coefficient of the ith kind of ions is:

$$D^*_i = A_1 \exp\left[-\frac{E_1}{RT}\right] + A_2 \exp\left[-\frac{E_2}{RT}\right] \qquad (3.4.46)$$

(A_1 and A_2 are constant coefficients in this respect). According to Samoilov, the activation energy of the jump of individual water molecules and that of 'cells' formed by several water molecules are identical with each other in pure water, i.e. the self-diffusion coefficient of water is:

$$D^*_{\text{water}} = A \exp\left[-\frac{E}{RT}\right].$$ (3.4.47)

It can be supposed that the activation energy required for the jump of the whole hydrated ion into the neighbouring equilibrium position in dilute solutions is approximately equal to the activation energy of the jump of a 'cell' in water, i.e. $E_1 \approx E_2 \approx E$, since, in dilute solutions, water molecules surrounding the hydrated ion are arranged approximately in the same way as in pure water. Namely, water molecules in the hydrate sphere shield most of the electric charge of the ion influencing the next water molecules. The jump of a 'bare' ion from its temporary equilibrium position to the neighbouring one, however, requires an activation energy different from that of pure water. The difference in energy $E_1 - E = \Delta E$ arises from the alteration of the energy barrier between two adjacent equilibrium positions by the ion, and, at the same time, this can be a measure of the hydration of ions (for immediately neighbouring water molecules). Thus the self-diffusion coefficient of ions is:

$$D^*_i = A_1 \exp\left[-\frac{E + \Delta E}{RT}\right] + A_2 \exp\left[-\frac{E}{RT}\right].$$ (3.4.48)

The ΔE values calculated from measurements on self-diffusion coefficients at different temperatures are collected in Table 3.11.

In solutions containing ions of various valences of the same element labelled isotopically, isotope exchange takes place in addition to self-diffusion. The theory of simultaneous diffusion and isotope exchange has

Table 3.11

Difference in the activation energy ΔE for some ions, regarded as the measure of hydration

Ion	ΔE, kcal $\cdot$ mole^{-1}	Ion	ΔE, kcal $\cdot$ mole^{-1}
Li$^+$	$+0.73$	Cl$^-$	-0.27
Na$^+$	$+0.25$	Br$^-$	-0.29
K$^+$	-0.25	I$^-$	-0.32
Cs$^+$	-0.33		
Mg^{2+}	$+2.61$		
Ca^{2+}	$+0.45$		

Literature on page 251

been elaborated by Gossman and Sedlaček [162a], and they have proved
these results by experiments on the self-diffusion and isotope exchange
of Tl^+ and Tl^{3+} ions.

Pikal [162b] investigated the isotope effect of tracer diffusion and found
the difference in the rate of diffusion of $^{22}Na^+$ and $^{24}Na^+$ ions, diffusing
simultaneously, to be in the order of magnitude of some tenth per cent.
This is not in agreement with the assumption that the tracer diffusion
coefficient of a given species is inversely proportional to the square-root
of its mass.

The self-diffusion coefficient of both ions of KCl in water–methanol mixtures has been investigated by Erdey-Grúz, Fodor-Csányi, Lévay and Szilágyi-Győri [133e]. It has been established that the self-diffusion coefficients
of the two kinds of ions hardly differ from each other. Starting from pure
aqueous solutions, the self-diffusion coefficients of both ions decrease on
the addition of methanol, pass a smooth minimum at about 35–40 mole
per cent methanol content, then increase again in solutions of increasing
methanol content. Starting from pure aqueous solutions, the product $D^*\eta$
decreases monotonically but not uniformly with increase in the methanol
content, and, in pure methanol, it is only about one third of that in water.

Weiss and Nothnagel [133f] have studied the self-diffusion of the solvent
and of spherically symmetrical ions in H_2O and D_2O solutions of several
electrolytes. According to their measurements, the self-diffusion coefficient
depends on the temperature as predicted by an Arrhenius-type equation
at each concentration investigated. In some cases, however, the activation
energy and the frequency factor are different at low and high temperatures.
These observations are partly supported by theoretical results.

Transition of common diffusion into tracer diffusion. The diffusion conditions
can also be qualitatively altered in a certain way in a given system if
the concentration ratio of the components is essentially changed. The self-
diffusion of Br^- ions in $MgBr_2$ solutions can serve as an example for this,
and has been investigated in detail by Mills [163]. The diffusion coefficients
measured by means of radioactive bromine indication and calculated according to the Fick law are shown in Fig. 3.16. In solutions of concentrations
higher than about 10^{-1} mole/l, the diffusion coefficients measured by
means of tracers ions agree with the values calculated on the basis of the

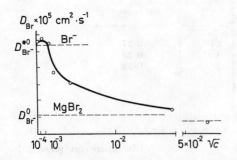

Fig. 3.16 Diffusion of $MgBr_2$ labelled
with radioactive bromine in aqueous
solutions at 25 °C

The upper dotted line corresponds to the correlation of
the Onsager limiting law derived for the self-diffusion
coefficient of Br^- ions; the lower dotted line belongs
to diffusion of binary electrolyte calculated from the
Onsager–Fuoss limiting law; o: measured values.

Onsager–Fuoss equation for binary electrolytes, within the limits of experimental error. In solutions of lower concentrations, however, the measured values are higher than those calculated in this way, and at sufficiently low concentrations, reach the value of the tracer diffusion of the Br^- ion calculated on the basis of the Onsager equation. Thus, under such conditions, Br^- ions diffuse independently of Mg^{2+} ions. This can be explained by the fact that when the concentration of $MgBr_2$ is decreased so that it becomes commensurable to the concentration of OH^- and OH_3^+ ions, the electroneutrality around the Br^- ion can be ensured not only by the slower Mg^{2+} ions, but by hydrogen ions of higher mobility, too. In sufficiently dilute solutions, the OH_3^+ and OH^- ions can be considered as the supporting electrolyte, in which Br^- ions diffuse freely as tracer ions. This also indicates that tracer diffusion is, in fact, a limiting case of diffusion in multicomponent systems.

3.5 ELECTRON DIFFUSION IN SOLUTIONS

Transport of electric charge in aqueous solutions can take place not only by diffusion or migration of ions, but—as has been pointed out by Levich [164] on the basis of an idea by Frumkin—by electron transfer as well if the solution contains an oxidation–reduction system. Under such conditions, in addition to the displacement of complete ions, electrons can be transferred from the reduced species to the oxidated ones; this is equivalent to the displacement of electric charge without the displacement of the ions themselves. In the case of uniform concentrations, this transition has no favoured direction, thus no macroscopic charge transport takes place. If, however, the chemical or electric potential has a gradient in the solution, and, therefore, the concentration of the oxidized or reduced species is not uniform in the solution but varies as a function of the position coordinate, electron transfer becomes orientated depending on the concentration gradient. In this way, a macroscopic charge transport appears which contributes to that caused by the displacement of the ions themselves.

If only a chemical potential gradient occurs in the solution, diffusion takes place. In the presence of an oxidation–reduction system, the total flux J of diffusion consists of two parts: that arising from the displacement of the ions (J_i) and that caused by electron transfer (J_e).

The theory of electron diffusion in aqueous solutions of oxidation-reduction systems has been elaborated by Dahms [165]. For an oxidation–reduction system corresponding to the general formula:

$$\underset{c_1}{Ox_1} + \underset{c_2}{Red_2} \underset{k_2}{\overset{k_1}{\rightleftharpoons}} \underset{c_3}{Ox_2} + \underset{c_4}{Red_1}$$

the rate of electron transfer in solutions depends on both concentrations $c_1, \ldots, c_4$ and rate constants k_1 and k_2 of the two opposite reactions involved

Literature on page 251

in the equilibrium. If diffusion is caused by a linear concentration gradient of direction y, and electron transfer can take place up to a distance λ from the given species, the diffusion flux due to electron transfer is, according to the theory:

$$J_e = F\lambda^2 \left\{ (k_2 c_3 + k_1 c_4) \frac{dc_2}{dy} + (k_1 c_1 + k_2 c_2) \frac{dc_4}{dy} \right\}, \qquad (3.5.1)$$

taking into account that:

$$\frac{dc_2}{dy} = -\frac{dc_1}{dy}$$

and: $(3.5.2)$

$$\frac{dc_4}{dy} = -\frac{dc_3}{dy},$$

under normal conditions.

In solutions, the ordinary diffusion of ions occurs simultaneously with electron transfer, its flux being determined by the diffusion coefficients and (in first approximation) by the concentration gradient. If the diffusion coefficients of the solutes present are approximately the same (D), the flux of the diffusion of ions is:

$$J_i = 2FD \left(\frac{dc_2}{dy} + \frac{dc_4}{dy} \right). \qquad (3.5.3)$$

Under such conditions, the total flux of diffusion will be:

$$J = J_i + J_e.$$

The participation of electron diffusion in the total diffusion flux is greater the higher the concentrations and the rate constants. Since it can be assumed that λ is of the order of magnitude of 10^{-8}–10^{-7} cm, the flux caused by electron transfer is appreciable only if the rate constants are considerably larger than 10^8 cm³ mole⁻¹ s⁻¹. For example, according to Dahms' calculations, in the process:

$$\text{Fe(o-phen)}_3^{3+} + e^- \rightleftharpoons \text{Fe(o-phen)}_3^{2+}$$

the electron flux becomes:

$$J_e = 5 \times 10^{-6} J_i$$

($k_1 = k_2 = 10^8$ cm³ mole⁻¹ s⁻¹, $\lambda = 5 \times 10^{-8}$ cm, $D = 5 \times 10^{-6}$ cm² s⁻¹, $c = 10^{-4}$ mole cm⁻³). That is, the contribution of electron diffusion to the total flux is negligibly small under such conditions. However, in the reaction of tris-4,7-dimethyl-1,10-phenanthroline iron(II) and hexachloroiridate(IV), $k_1 = 10^{12}$ and $k_2 = 4 \times 10^{-12}$, according to Halpern, Legard and Lumry [166], thus:

$$J_e = 0.13 J_i$$

when the following values are inserted $\lambda = 10^{-7}$ cm, $c_3 = c_4 = 10^{-4}$, and $D = 10^{-5}$, i.e. electron diffusion contributes appreciably to the total flux.

On the basis of measurements on ferrocene-ferricinium solutions, Ruff [167] has confirmed experimentally the existence of electron diffusion.

REFERENCES

to Chapter 3

[1]   As a review for diffusion, see for example H. J. V. Tyrrel, Diffusion and Heat Flow in Liquids (London, 1961); W. Jost, Diffusion (Darmstadt, 1957); W. Jost, Diffusion in Solids, Liquids, Gases (New York, 1952); R. A. Robinson and R. H. Stokes, Electrolyte Solutions, pp 45 and 284 (London, 1959); D. A. Frank-Kamenetskii, Diffusion and Heat Transfer in Chemical Kinetics (New York, 1969).

[2]   H. P. Neumann, Ber. Bunsenges., 1968, 72, 1205.

[3]   G. S. Hartley and J. Crank, Trans. Faraday Soc., 1949, 45, 801.

[4]   K. F. Alexander, Z. Phys. Chem., 1954, 203, 181.

[5]   For details, see W. Jost, Diffusion (Darmstadt, 1957); H. J. V. Tyrrel, Diffusion and Heat Flow in Liquids (London, 1961).

[6]   Yu. L. Rozenstok, Zh. Fiz. Khim., 1965, 39, 1135.

[7]   W. Sutherland, Phil. Mag. 1905, 9, 781.

[8]   A. Einstein, Ann. Phys., 1906, 19, 289; 1911, 34, 591.

[9]   T. Loflin and E. McLaughlin, J. Phys. Chem., 1969, 73, 186.

[10] R. J. Bearman, J. Chem. Phys. 1960, 32, 1308.

[11] C. H. Byers and C. J. King, J. Phys. Chem., 1966, 70, 2499.

[12] L. G. Longsworth, J. Phys. Chem., 1963, 67, 689.

[13] F. Perrin, J. Phys. Radium. 1936, 7, 1.

[14] R. O. Herzog, R. Illig and H. Kudar, Z. Phys. Chem., 1933, 167, 329.

[15] J. H. Wang, J. Phys. Chem., 1954, 58, 686.

[16] F. C. Collins and H. Raffel, J. Chem. Phys., 1954, 22, 1728; 1955, 23, 1454.

[17] P. Walden, Z. Elektrochem., 1906, 12, 77.

[18] A. Polson, J. Phys. Chem., 1950, 54, 649.

[19] L. G. Longsworth, J. Amer. Chem. Soc., 1953, 75, 5705.

[20] G. S. Hartley, Phil. Mag., 1931, 12 (7), 473.

[21] L. Onsager and R. Fuoss, J. Phys. Chem., 1932, 36, 2689.

[22] A. R. Gordon, J. Chem. Phys., 1937, 5, 522.

[23] See e.g. L. J. Gosting and D. F. Akeley, J. Amer. Chem. Soc., 1952, 74, 2058; F. J. Gutter and G. Kegeles, J. Amer. Chem. Soc., 1953, 75, 3893; J. H. Wang, J. Phys. Chem., 1954, 58, 686.

[24] A. R. Gordon, J. Amer. Chem. Soc., 1950, 72, 4840.

[25] S. Broersma, J. Chem. Phys., 1958, 28, 1158.

[26] G. S. Hartley, and J. Crank, Trans. Faraday Soc., 1949, 45, 801.

[27] C. Carman and L. Z. Stein, Trans. Faraday Soc., 1956, 52, 619.

[28] See e.g. H. S. Harned, Disc. Faraday Soc., 1957, 24, 7; C. L. Sandquist and P. A. Lyons, J. Amer. Chem. Soc., 1954, 24, 7; P. A. Lyons, J. Amer. Chem. Soc., 1954, 76, 4641.

[29] For details see: S. Glasstone, K. J. Laidler and H. Eyring, The Theory of Rate Processes, p. 516 (New York, 1941); J. F. Kincaid, H. Eyring and A. E. Stearn, Chem. Rev., 1941, 28, 301; R. M. Mozo, Statistical Mechanical Theories of Transport Processes (London, 1967).

[30a] Cf. J. H. Arnold, J. Amer. Chem. Soc., 1930, 52, 3937.

[30b] S. Nir, and W. D. Stein, J. Chem. Phys., 1971, 55, 1958.

[31] H. S. Taylor, J. Chem. Phys., 1938, 6, 331.

[32] See e.g. S. Glasstone, K. J. Laidler and H. Eyring, The Theory of Rate Processes. p. 153 (New York, 1941).

[33]   A. E. STEARN, E. M. IRISH and H. EYRING, *J. Phys. Chem.*, 1940, **44**, 981.
[34]   H. EYRING and J. HIRSCHFELDER, *J. Phys. Chem.*, 1937, **41**, 249; J. O. HIRSCH-
       FELDER, *J. Chem. Ed.*, 1939, **16**, 540.
[35]   L. ONSAGER and R. M. FUOSS, *J. Phys. Chem.*, 1932, **36**, 2687; further see A. R.
       GORDON, *J. Chem. Phys.*, 1937, **5**, 522; 1939, **7**, 89, 963; P. VAN RYSSELBERGHE,
       J. Amer. Chem. Soc., 1938, **60**, 2326.
[36]   R. E. POWELL, W. E. ROSEVEARE and H. EYRING, *Ind. Eng. Chem.*, 1941,
       33, 430.
[37]   J. C. M. LI and PIN CHANG, *J. Chem. Phys.*, 1955, **23**, 518.
[38a] G. M. PANCHENKOV, *Zh. Fiz. Khim.*, 1950, **24**, 1390; *Dokl. AN SSSR*, 1958,
       118, 755; *J. Chim. Phys.*, 1957, **5**, 931.
[38b] I. RUFF and V. FRIEDRICH, *J. Phys. Chem.*, 1971, **75**, 3297.
[38c] I. RUFF, V. FRIEDRICH, K. DEMETER and K. CSILLAG, *J. Phys. Chem.*, 1971,
       75, 3303.
[38d] I. RUFF, V. J. FRIEDRICH and K. CSILLAG, *J. Phys. Chem.*, 1972, **76**, 162.
[38e] I. RUFF and V. J. FRIEDRICH, *J. Phys. Chem.*, 1972, **76**, 2957.
[38f] S. LENGYEL, *J. Phys. Chem.*, In press.
[38g] P. BLANCKENHAGEN, *Ber. Bunsen Ges.*, 1972, **76**, 891.
[38h] D. FRENKEL, G. H. WEGDAM and J. VAN DER ELSKEN, *J. Chem. Phys.*, 1972,
       57, 2691.
[38i] M. K. AHN, S. J. KNAK JENSEN and D. KIVELSON, *J. Chem. Phys.*, 1972,
       57, 2940.
[39]   I. GYARMATI, Non-equilibrium Thermodynamics (Berlin 1970); P. VAN RYSSEL-
       BERGHE, Thermodynamics of Irreversible Processes (Paris, 1963); I. PRIGO-
       GINE, Introduction to the Thermodynamics of Irreversible Processes (Oxford,
       1968); S. R. DE GROOT, The Thermodynamics of Irreversible Processes (Amster-
       dam, 1963); R. HAASE, *Ergebn. exakt. Naturwiss.*, 1952, **26**, 56; R. HAASE,
       Thermodynamik der irreversiblen Prozesse (Darmstadt, 1963).
[40]   H. D. ELLERTON and P. J. DUNLOP, *J. Phys. Chem.*, 1967, **71**, 1291.
[41]   D. G. MILLER, *J. Phys. Chem.*, 1965, **69**, 3374.
[42]   A. S. CUKROWSKY and B. BARANOWSKI, *Z. Phys. Chem.*, 1969, **240**, 167, 253,
       285.
[43]   A. V. PAMFILOV and A. I. LOPUSHANSKAIA and L. B. TSVETKOVA, *Zh.
       Fiz. Khim.*, 1968, **42**, 2810.
[44]   H. SCHÖNERT, *J. Phys. Chem.* 1969, **73**, 62.
[45a] H. SCHÖNERT, *Z. Phys. Chem. N. F.* 1968, **62**, 50.
[45b] T. K. KETT and D. K. ANDERSON, *J. Phys. Chem.*, 1969, **73**, 1262, 1268.
[46]   R. P. WENDT, *J. Phys. Chem.*, 1965, **69**, 1227.
[47]   G. S. HARTLEY and J. CRANK, *Trans. Faraday Soc.*, 1949, **45**, 801.
[48]   B. F. WISHAW and R. H. STOKES, *J. Amer. Chem. Soc.*, 1954, **76**, 2065.
[49]   H. S. HARNED, *Disc. Faraday Soc.*, 1957, **24**, 7.
[50]   G. J. JANZ, G. R. LAKSHIMINARAYANAN, M. P. KLOTZIN and G. E. MAYER,
       J. Phys. Chem., 1966, **70**, 536.
[51]   F. SAUER and V. FREISE, *Z. Elektrochem.*, 1962, **66**, 353.
[52]   O. LAMM, *J. Phys. Chem.*, 1947, **51**, 1063; *Acta Chem. Scand.*, 1952, **6**, 1331;
       1954, **8**, 1120; 1957, **11**, 362; *Trans. Faraday Soc.*, 1960, **56**, 767.
[53]   F. A. L. DULLIEN, *Trans. Faraday Soc.*, 1963, **59**, 856.
[54]   D. J. DUNLOP, *J. Phys. Chem.*, 1964, **68**, 26.
[55]   J. G. ALBRIGHT, *J. Phys. Chem.*, 1963, **67**, 2628.
[56a] R. R. IRANI and A. W. ADAMSON, *J. Phys., Chem.*, 1960, **64**, 199.
[56b] K. NAKANISHI and T. OZASA, *J. Phys. Chem.*, 1971, **74**, 2956.
[56c] J. H. HILDEBRAND, *Science*, 1971, **174**, 490; J. H. HILDEBRAND, J. M. PRAUS-
       NITZ and R. L. SCOTT, Regular and Related Solutions, Chapter 3 (New York,
       1970); J. H. HILDEBRAND and R. H. LAMOREAUX, *Proc. Nat. Acad. Sci. U.S.*,
       1972, **69**, 3428.
[56d] J. H. DYMOND, *Trans. Faraday Soc.*, 1972, **68**, 1789.
[57]   G. S. HARTLEY, *Phil. Mag.*, 1931, **12 (8)** 473.
[58]   D. G. MILLER, *J. Phys. Chem.*, 1960, **64**, 1598.
[59]   P. B. LORENZ, *J. Phys., Chem.*, 1961, **65**, 704.
[60]   L. ONSAGER, and R. FUOSS, *J. Phys. Chem.*, 1932, **36**, 2689.

[61]   R. H. Stokes, J. Amer. Chem. Soc., 1953, **75**, 2533, 4563; R. A. Robinson
       and R. H. Stokes, Electrolyte Solutions, p. 291 (London, 1959); cf. A. W.
       Adamson, J. Phys. Chem., 1954, **58**, 514.
[62]   See e.g. R. A. Robinson and R. H. Stokes, Electrolyte Solutions, p. 145
       (London, 1959).
[63]   H. S. Harned and H. W. Parker, J. Amer. Chem. Soc., 1955, **77**, 265; cf.
       P. A. Lyons and J. F. Riley, J. Amer. Chem. Soc., 1954, **76**, 5216.
[64]   J. A. Harpst, E. Holt and P. A. Lyons, J. Phys. Chem., 1965, **69**, 2333.
[65a] H. S. Harned and J. A. Shropshire, J. Amer. Chem. Soc., 1968, **80**, 2618.
[65b] J. G. Albright and D. G. Miller ,J. Phys. Chem., 1972, **76**, 1853.
[65c] H. Kim, G. Reinfelds and L. J. Gosting, J. Phys. Chem., 1973, **77**, 934.
[65d] D. G. Miller, J. Phys. Chem., 1966, **70**, 2639; 1967, **71**, 616.
[66]   R. W. Laity, J. Phys. Chem., 1963, **67**, 671.
[67a] V. M. Vdovenko, Yu. V. Gurikov and E. K. Legin, Radio Khim., 1966, **8**, 323.
[67b] M. Chemla, Revue Roumaine de Chimie, 1972, **17**, 63.
[68]   H. S. Frank and M. W. Evans, J. Chem. Phys., 1945, **13**, 507; H. S. Frank
       and W. Y. Wen, Disc. Faraday Soc., 1957, **24**, 133.
[69]   H. S. Harned and R. M. Hudson, J. Amer. Chem. Soc., 1951, **73**, 3781, 5880.
[70]   B. F. Wishaw and R. H. Stokes, J. Amer. Chem. Soc., 1954, **76**, 2065.
[71]   G. T. A. Müller and R. H. Stokes, Trans. Faraday Soc., 1957, **53**, 642.
[72]   V. Vitagliano and P. A. Lyons, J. Amer. Chem. Soc., 1956, **78**, 4538.
[73]   O. W. Edwards and E. O. Huffman, J. Phys. Chem., 1959, **63**, 1830.
[74]   C. W. Garland, S. Tong and W. H. Stockmayer, J. Phys. Chem., 1965, **69**,
       1718.
[75]   V. A. Kiprianov, Zh. Fiz. Khim., 1961, **35**, 2389.
[76a] R. P. Wendt, J. Phys. Chem., 1965, **69**, 1227.
[76b] M. J. Pikal, J. Phys. Chem., 1972. **76**, 3124.
[77]   R. A. Robinson and R. H. Stokes, Electrolyte Solutions, p. 320 (London,
       1959).
[78a] R. N. Bhatia, K. E. Gubbins and R. D. Walker, Trans. Faraday Soc., 1968,
       64, 2091.
[78b] J. Lielmezs and G. M. Musbally, Electrochim. Acta, 1972, **17**, 1609.
[78c] P. Dumargue, P. Humeau and F. Penot, Electrochim. Acta, 1973, **18**, 447.
[79]   L. Onsager, Ann. N.Y. Acad. Sci., 1945, **46**, 241.
[80]   I. Prigogine, Bull. Acad. Belg. Cl. Sci., 1948, **34**, 930.
[81]   G. J. Hooyman, S. R. de Groot and P. Mazur, Physica, 1955, **21**, 360; G. J.
       Hooyman, Physica, 1956, **22**, 751.
[82]   S. R. de Groot, J. chim. phys., 1957, **54**, 851.
[83]   O. Lamm, J. Phys. Chem., 1957, **61**, 948; cf. H. J. V. Tyrrel, Diffusion and
       Heat Flow in Liquids, p. 40 (London, 1961).
[84]   D. G. Miller, J. Phys., Chem., 1959, **63**, 570; Chem. Rev., 1960, **60**, 15.
[85]   P. J. Dunlop, J. Phys. Chem., 1959, **63**, 612.
[86]   P. Ya. Siver, Zh. Fiz. Khim., 1962, **36**, 1947.
[87]   A. V. Pamfilov, A. I. Lopushanskaia and L. B. Tsvetkova, Zh. Fiz. Khim.,
       1968, **42**, 2810.
[88]   H. Schönert, J. Phys. Chem., 1969, **73**, 62.
[89a] L. A. Woolf, J. Phys. Chem., 1963, **67**, 273.
[89b] M. K. Tham and K. E. Gubbins, Faraday Transact. I. **1972**, 1339.
[90]   V. A. Marinin, Zh. Fiz. Khim., 1955, **29**, 1564; 1956, **30**, 129.
[91]   R. L. Baldwin, P. J. Dunlop and L. J. Gosting, J. Amer. Chem. Soc., 1955,
       77, 5235; P. J. Dunlop and L. J. Gosting, J. Amer. Chem. Soc., 1955, **77**, 5238.
[92]   P. J. Dunlop, J. Phys. Chem., 1957, **61**, 994.
[93a] I. J. O'Donnell and L. J. Gosting, The Structure of Electrolytic Solutions
       (ed. by W. J. Hamer), p. 160 (New York, London, 1959).
[93b] R. P. Wendt and M. Shamim, J. Phys. Chem., 1971, **74**, 2770.
[94]   A. J. Arvia, S. L. Marchiano and J. J. Podestá, Electrochim. Acta, 1967,
       12, 259.
[95]   H. Schönert, Z. Phys. Chem. N. F., 1966, **51**, 196.
[96]   D. Glietenberg, A. Kutschker and M. Stackelberg, Ber. Bunsenges.,
       1968, **72**, 562.

[97]   L. ONSAGER and R. M. FUOSS, *J. Phys. Chem.*, 1932, **26**, 2689.
[98]   L. A. WOOLF, D. G. MILLER and L. J. GOSTING, *J. Amer. Chem. Soc.*, 1962, **84**, 317.
[99]   R. W. LAITY, *J. Phys. Chem.*, 1959, **63**, 80.
[100] R. W. LAITY, *J. Chem. Phys.*, 1959, **30**, 682.
[101] H. S. HARNED, J. A. SHROPSHIRE, *J. Amer. Chem. Soc.*, 1960, **80**, 5652; 1960, **82**, 799.
[102] M. V. KULKARNI and P. A. LYONS, *J. Phys. Chem.*, 1965, **69**, 2336.
[103] T. ERDEY-GRÚZ (in collaboration with A. HUNYÁR, É. POGÁNY and A. VÁLI), *Acta Chim. Acad. Sci. Hung.*, 1948, **1**, (3), 7; POGÁNY É, *Magyar Kémiai Folyóirat*, 1942, **48**, 85.
[104] Á. HUNYÁR, Thesis (Budapest, 1937); *J. Amer. Chem. Soc.*, 1949, **71**, 3552.
[105] T. ERDEY-GRÚZ and A. HUNYÁR, *Acta Chim. Acad. Sci. Hung.*, 1948, **1**, (3), 27.
[106] H. FREUNDLICH and D. KRÜGER, *J. Phys. Chem.*, 1939, **43**, 981.
[107] C. H. HALE and T. DE VRIES, *J. Amer. Chem. Soc.*, 1948, **70**, 2473.
[108] F. C. COLLINS and H. RAFFEL, *J. Chem. Phys.*, 1955, **23**, 1454; 1954, **22**, 1728.
[109] S. GLASSTONE, K. J. LAIDLER and H. EYRING, Theory of Rate Processes, p. 516 (New York, 1941).
[110] J. C. M. LI, and P. CHANG, *J. Chem. Phys.*, 1955, **23**, 518.
[111] B. OTTAR, *Acta Chem. Scand.*, 1955, **9**, 344; *J. chim. phys.*, 1957, **54**, 856.
[112] E. MCLAUGHLIN, *Trans. Faraday Soc.*, 1959, **55**, 29.
[113a] A. RAHMAN, *J. Chem. Phys.*, 1966, **45**, 2585.
[113b] H. G. HERTZ, *Ber. Bunsenges.*, 1971, **75**, 183.
[113c] D. E. O'REILLY, *Ber. Bunsenges.*, 1971, **75**, 208.
[114] O. YA. SAMOILOV, *Dokl. AN SSSR*, 1955, **101**, 125; B. K. PROKORENKO, O. YA. SAMOILOV and I. Z. FISHER, *Dokl. AN SSSR*, 1959, **125**, No. 2, 356.
[115] K. WIRTZ, *Z. Naturforsch.*, 1948, **3a**, 672.
[116] R. B. CUDDEBACK, R. C. KOELLER and H. G. DRICKAMER, *J. Chem. Phys.*, 1953, **21**, 589.
[117a] YU. V. GURIKOV, *Zh. Strukt. Khim.*, 1964, **5**, 188.
[117b] YU. I. NERONOV, *Zh. Strukt. Khim.* 1971, **12**, 8.
[118] C. HAAS, *Phys. Letters*, 1962, **3**, 126.
[119] cf. J. H. SIMPSON and H. Y. CARR, *Phys. Rev.*, 1958, **111**, 1201; J. A. SAXTON, *Proc. Roy. Soc.* 1952, **A 213**, 473; J. H. WANG, C. V. ROBINSON and I. S. EDELMAN, *J. Amer. Chem. Soc.*, 1953, **75**, 466.
[120] Cf. e.g. the critical review of H. J. V. TYRRELL, Diffusion and Heat Flow in Liquids (London, 1961).
[121] J. H. WANG, *J. Phys. Chem.*, 1954, *J. Phys. Chem.*, 1954, **58**, 686.
[122a] L. DEVELL, *Acta Chem. Scand.*, 1962, **16**, 2177; a review on the results of earlier measurements, as well.
[122b] R. MILLS, *Ber. Bunsenges.*, 1971, **75**, 195.
[122c] R. MILLS, *J. Phys. Chem.*, 1973, **77**, 685.
[123] S. N. ANDREEV, *Zh. Strukt. Khim.*, 1964, **5**, 371.
[124a] J. H. WANG, *J. Phys. Chem.*, 1965, **69**, 4412.
[124b] H. R. PRUPPACHER, *J. Chem. Phys.*, 1972, **56**, 101.
[124c] K. T. GILLEN, D. C. DOUGLASS and N. J. R. HOCH, *J. Chem. Phys.*, 1972, **57**, 5117.
[125] See e.g. R. W. LAITY, *J. Phys. Chem.*, 1959, **63**, 80; 1959, **30**, 682; O. LAMM, *J. Phys. Chem.*, 1947, **51**, 1063; 1957, **61**, 948.
[126] P. C. CARMAN and L. H. STEIN, *Trans. Faraday Soc.*, 1956, **52**, 619.
[127] G. S. HARTLEY and J. CRANK, *Trans. Faraday Soc.*, 1949, **45**, 801.
[128] D. W. MCCALL and D. C. DOUGLAS, *J. Phys. Chem.*, 1967, **71**, 987.
[129] R. MILLS and J. W. KENNEDY, *J. Amer. Chem. Soc.*, 1953, **75**, 5696.
[130] J. H. WANG, *J. Amer. Chem. Soc.*, 1954, **76**, 4755; *J. Phys., Chem.*, 1954, **58**, 686.
[131a] A. BIANCHERIA and G. KEGELES, *J. Amer. Chem. Soc.*, 1957, **79**, 5908.
[131b] O. F. BEZRUKOV, P. N. VORONTSOV-VELJAMINOV and E. M. USHAKOVA, *Zh. Strukt. Khim.*, 1972, **13**, 388.
[132] J. H. WANG, C. B. ANFINSEN and F. M. POLESTRA, *J. Amer. Chem. Soc.*, 1954, **76**, 4763; for the calculation of *H*, see: J. H. WANG, *J. Amer. Chem. Soc.*, 1955, **77**, 258.

[133a] B. OTTAR, *Acta Chem. Scand.*, 1955, **9**, 344.
[133b] J. G. ALBRIGHT, *J. Phys. Chem.*, 1969, **73**, 1280.
[133c] T. ERDEY-GRÚZ and B. LÉVAY, *Acta Chim. Acad. Sci. Hung.*, 1971, **69**, 215.
[133d] T. ERDEY-GRÚZ and I. NAGY-CZAKÓ, *Acta Chim. Acad. Sci. Hung.*, 1971, **67**, 283.
[133e] T. ERDEY-GRÚZ, P. FODOR-CSÁNYI, B. LÉVAY and E. SZILÁGYI-GYŐRI, *Acta Chim. Acad. Sci. Hung.*, 1971, **69**, 423.
[133f] A. WEISS and K. H. NOTHNAGEL, *Ber. Bunsenges.*, 1971, **75**, 216.
[133g] T. ERDEY-GRÚZ, GY. INZELT and P. FODOR-CSÁNYI, *Acta Chim. Acad. Sci. Hung.*, 1973, **77**, 173.
[133h] T. ERDEY-GRÚZ and B. LÉVAY, *Acta Chim. Acad. Sci. Hung.*, In press.
[134] J. H. WANG, *J. Phys. Chem.*, 1954, **58**, 686.
[135] R. J. PODOLSKY, *J. Amer. Chem. Soc.*, 1958, **80**, 4442.
[136] J. TAMÁS, S. LENGYEL and J. GIBER, *Acta Chim. Acad. Sci. Hung.*, 1963, **38**, 225.
[137] J. TAMÁS and K. UJSZÁSZY, *Acta Chim. Acad. Sci. Hung.*, 1966, **49**, 377.
[138] R. J. JONES, L. G. ROWLANDS and C. B. MONK, *Trans. Faraday Soc.*, 1965, **61**, 1384.
[139] G. A. ANDREEV, *Zh. Fiz. Khim.*, 1965, **39**, 2586.
[140] M. I. YEMELIANOV and A. Sh. AGISHEV, *Zh. Strukt. Khim.*, 1965, **6**, 909.
[141] D. W. MCCALL and C. D. DOUGLAS, *J. Phys. Chem.*, 1965, **69**, 2001.
[142] M. I. YEMELIANOV, E. A. NIKIFOROV and K. S. KUCHERYAVENKO, *Zh. Strukt. Khim.*, 1968, **9**, 954.
[143] M. I. YEMELIANOV and N. K. GAYSIN, *Zh. Strukt. Khim.*, 1967, **8**, 1010.
[144] L. ONSAGER, *Phys. Rev.*, 1931, **37**, 405; 1931, **38**, 2265; *Ann. N. Y. Acad. Sci.*, 1945, **46**, 241.
[145a] L. C. GOSTING and H. S. HARNED, *J. Amer. Chem. Soc.*, 1951, **73**, 159.
[145b] K. TANAKA and T. HASHITANI, *Trans. Faraday Soc.*, 1971, **67**, 2314.
[146a] R. MILLS and E. W. GODBOLE, *J. Amer. Chem. Soc.*, 1960, **82**, 2395.
[146b] P. TURQ, F. LANTELME and M. CHEMLA, *Electrochim. Acta*, 1969, **14**, 1081.
[146c] O. YA. SAMOILOV, V. V. GONCHAROV, V. I. JASHKICHEV, V. G. MARKOVA and L. S. ALEKSEIEVA, *Zh. Strukt. Khim.*, 1972, **13**, 384.
[147] R. MILLS, *J. Phys. Chem.*, 1957, **61**, 1258, 1631.
[148] R. H. STOKES, L. A. WOOLF and R. MILLS, *J. Phys. Chem.*, 1957, **61**, 1634.
[149] L. A. WOOLF, *J. Phys. Chem.*, 1960, **64**, 481.
[150] J. H. WANG and S. MILLER, *J. Amer. Chem. Soc.*, 1952, **74**, 1611.
[151] R. MILLS, *J. Amer. Chem. Soc.*, 1955, **77**, 6116.
[152] J. M. NIELSEN, A. W. ADAMSON and J. W. COBBLE, *J. Amer. Chem. Soc.*, 1952, **74**, 446.
[153] R. MILLS and J. W. KENNEDY, *J. Amer. Chem. Soc.*, 1953, **75**, 5696.
[154] R. MILLS, *J. Phys. Chem.*, 1959, **63**, 1873.
[155] R. MILLS, *J. Phys. Chem.*, 1957, **61**, 1258.
[156] R. H. STOKES, L. A. WOOLF and R. MILLS, *J. Phys. Chem.*, 1957, **61**, 1634.
[157] S. LENGYEL, J. TAMÁS and A. VÉRTES, *Acta Chim. Acad. Sci. Hung.*, 1963, **37**, 279.
[158] R. MILLS, *J. Phys. Chem.*, 1957, **61**, 1631.
[159] A. E. MARCINKOWSKY, H. O. PHILLIPS and K. A. KRAUS, *J. Phys. Chem.*, 1965, **69**, 3968; 1968, **72**, 1201.
[160] M. J. POLISSAR, *J. Chem. Phys.*, 1938, **6**, 833.
[161] O. YA. SAMOILOV, *Zh. Fiz. Khim.*, 1955, **29**, 1582.
[162a] A. GOSSMAN and J. SEDLAČEK, *Z. Phys. Chem.*, 1969, **240**, 26.
[162b] M. J. PIKAL, *J. Phys. Chem.*, 1972, **76**[7] 3038.
[163] R. MILLS, *J. Phys. Chem.*, 1962, **66**, 2716.
[164] V. G. LEVICH, Advances of Electrochemistry (ed. by P. DELAHAY), 1966, **4**, 314.
[165] H. DAHMS, *J. Phys. Chem.*, 1968, **72**, 362.
[166] J. HALPERN, R. J. LEGARD and R. LUMRY 1963, *J. Amer. Chem. Soc.*, 1963, **35**, 680.
[167] I. RUFF, *Electrochim. Acta*, 1970, **15**, 1059.

4. ELECTROLYTIC CONDUCTION

4.1 MACROSCOPIC RELATIONSHIPS OF ELECTROLYTIC CONDUCTION

4.1.1. GENERAL

The components in mixtures or solutions can be transferred from one point in space to another by the effects of the viscous flow of the material system as a whole, of diffusion brought about by differences in chemical potential, and of migration caused by differences in electric potential at constant temperature. The flux produced by these three types of transport processes is:

$$J_i = c_i \overline{V} - D_i \Delta\mu_i - z_i U_i' c_i F \Delta\psi, \qquad (4.1.1)$$

considering the ith component of concentration c_i in dilute solution and at constant temperature, where $\overline{V}$ is the velocity of the liquid as a whole, ψ is the electrostatic potential (its negative gradient is the electric field strength), and U_i' is the mobility of the ith type of the solute species.

In equation (4.1.1) for total flux, the first term arises from the viscous flow discussed in Chapter 2, the second term represents diffusion as reviewed in Chapter 3, while the third term is due to ionic migration caused by electric force (i.e. electrolytic conduction) which is the subject of this chapter.

According to the general definition, the *mobility* U_i' of the species is the average translational velocity [cm s^{-1}] gained by the ith kind of species under the effect of unit [1 Newton mole^{-1}] force of any origin. The dimensions of U_i' are [mole cm^2 joule^{-1} s^{-1}]. Since the charge of one mole of ions is $z_i F$, the product of this and the electric field strength, $-\Delta\psi = -\left(\dfrac{\partial\psi}{\partial x} + \dfrac{\partial\psi}{\partial y} + \dfrac{\partial\psi}{\partial z}\right)$, is equal to the force acting on the ions. The product of this term and the mobility gives the average velocity of the ith kind of ions with respect to the solvent (or to the whole solution, considered immobile). The product of this velocity and concentration c_i [g-ion cm^{-3}] yields the contribution to ionic migration caused by the electric force to flux J_i [g-ion cm^{-2} s^{-1}] of the ith kind of ions. In electrochemistry, the ionic mobility (often called absolute mobility) is usually related to the electric force. In this sense, the mobility U_i is the average velocity of the ith type of ions gained on the effect of an electric force of 1 V cm^{-1}. Thus the dimensions of U_i are [cm s^{-1}/V cm^{-1}], i.e. [cm^2 s^{-1} V^{-1}]. The relationship between the mobilities measured in these two types of units is accordingly:

$$U_i = z_i F U_i'. \qquad (4.1.2)$$

Depending on the conditions, not all the three transport processes necessarily contribute to the mass transfer in every case. If the liquid as a whole

Literature on page 326

is macroscopically at rest, viscous flow does not influence the flux; if the chemical potential (the activity) of the ith substance is constant over all the material system, diffusion plays no role; while if the electric charge of the particles of the ith substance is zero, electric field does not directly contribute to the flux. In practice, however, disregarding the flow of liquids in closed systems–the isolated appearance of only one type of transport process occurs rarely. For example, diffusion alters the concentration of the solution, and also its density, which results in convection, i.e. viscous flow. Electric current passing through an electrolyte causes changes in the concentration in the vicinity of the electrodes giving rise to diffusion; in most cases, convection will also contribute to mass transfer. In addition to this, in the course of electrolyte diffusion–owing to the different mobilities of different ions–an electric potential difference is produced in the solution which influences the flux even when there is no external electric field in the actual system. Nevertheless, for the purpose of theoretical discussion, the separation of the three types of transport processes seems to be justified, since the essential relationships are even emphasized in this way, and the difficulties arising in their unified treatment can be avoided.

Transport processes brought about by the electric potential gradient–like other transport processes–can be discussed from various aspects. The phenomena of electric conduction can be described by means of the correlations existing between macroscopically measurable characteristics, disregarding the molecular mechanism of the process. For practical purposes, this is usually sufficient, but, of course, it gives no explanation of the molecular mechanism of the phenomena. On the other hand, the conduction processes can be treated by molecular statistics, which gives a picture of the molecular mechanism of electric conduction. This information on the mechanism of conduction provided by molecular statistics is of differing importance depending on the starting assumptions and the methods of the theory applied. The Debye–Hückel theory considers the solvent as a continuum in which ions migrate according to the laws of hydrodynamics, and it mainly investigates the influence of the electrostatic interaction of the ionic charges on mobilities. In order to reveal the conduction mechanism in a more detailed manner, the molecular structure of the solvent should be considered, too, and the elementary steps of the displacement of ions should be elucidated and analysed by a more sophisticated statistical mechanical model. The theory based on the relationships between macroscopical properties has given satisfactorily reliable results for the majority of practical purposes. The theories on the mechanism of electrolytic conduction, however, have led to conclusions showing quantitative agreement with the experiments only in the most simple limiting cases so far; they still contribute appreciably to the interpretation of the main characteristics of the fundamental properties of electric conduction.

The theoretical importance of studies on electrolytic conduction in research on the mechanism of transport processes is stressed by the fact that it is this transport process which can be measured with the highest accuracy.

4.1.2 ELECTROLYTIC CONDUCTIVITY, TRANSFERENCE NUMBER, AND IONIC MOBILITY

In the general expression of transport processes (equation 4.1.1), the third term represents mass transport due to electric potential gradient, i.e. electrolytic conduction. If macroscopically immobile solutions in which there is no chemical potential or concentration gradient are investigated at constant temperature, the flux of the ith kind of solute will be:

$$J_i = z_i U_i' c_i \mathrm{F} \varDelta \psi. \tag{4.1.3}$$

The electric current density resulting from mass flux is:

$$i = \mathrm{F} \sum_i z_i J_i. \tag{4.1.4}$$

This relationship shows that electrolytic current is the flux of electrically charged particles, i.e. of ions.

The concentration of ions can vary in a solution, since ions can be formed or destroyed, e.g., in electrode processes or other chemical reactions. If the rate of ion formation in a solution is v_k mole cm^{-3} s^{-1}, the so-called continuity equation is valid; it expresses the mass balance at a certain point in the solution:

$$\frac{\partial c_i}{\partial t} = - \varDelta J_i + v_k. \tag{4.1.5}$$

In common electrochemical systems, formation and disappearance of ions take place only at the electrodes. Under such conditions, $v_k = 0$ in the bulk of the solution, provided that the phenomena are investigated at such a distance from the electrodes that the changes due to electrode processes have no influence.

A further limitation of transport processes in electrolyte solutions is the electroneutrality condition:

$$\sum_i z_i c_i = 0. \tag{4.1.6}$$

According to experience, this condition always holds in macroscopical regions of homogeneous solutions within the limits of observability. Deviations occur only at the interfaces between phases and in close vicinity to inhomogeneities (up to a distance of about 10–100 Å from them) as a result of electrochemical double layers formed at these places.

In practice, the electric potential gradient is usually linear in good approximation (let its direction be direction y). Furthermore, if there is only one binary electrolyte present in the solution with cations and anions with charges z_c and z_a ($= -z_c$) respectively, the flux of cations and anions is:

$$J_c = - z_c U_c' c_c \mathrm{F} \frac{\partial \psi}{\partial y} = - c_c U_c \frac{\partial \psi}{\partial y}, \tag{4.1.7}$$

Literature on page 326

and

$$J_a = + z_a U_a' c_a \mathrm{F} \frac{\partial \psi}{\partial y} = + c_a U_a \frac{\partial \psi}{\partial y}. \tag{4.1.8}$$

On the other hand, from equation (4.1.4), the current density is:

$$i = \mathrm{F}(z_c J_c + z_a J_a) =$$

$$= \mathrm{F}(-z_c c_c U_c + z_a c_a U_a) \frac{\partial \psi}{\partial y} =$$

$$= \mathrm{F}z(c_c U_c + c_a U_a) \frac{\partial \psi}{\partial y}, \tag{4.1.9}$$

where $z = |z_c| = |z_a|$.

When the specific conductivity of the solution is denoted by $\varkappa$, one has:

$$i = \varkappa \frac{\partial \psi}{\partial y}; \tag{4.1.10}$$

according to Ohm's Law, and comparing this equation with equation (4.1.9), we obtain:

$$\varkappa = \mathrm{F}z(c_c U_c + c_a U_a). \tag{4.1.11}$$

In electrochemistry, in order to emphasize the essential feature of these correlations, the *equivalent conductivity* is usually applied instead of specific conductivity; it is defined as:

$$\Lambda = \frac{\varkappa}{c_{eq}} \, \Omega^{-1} \, \mathrm{cm}^2, \tag{4.1.12}$$

when c_{eq} corresponds to the concentration given in g-eq cm^{-3} units. If concentration is expressed in g-mole cm^{-3} units (c_m), we obtain the *molar conductivity* defined as follows:

$$\Lambda_m = \frac{\varkappa}{c_m} \, \Omega^{-1} \, \mathrm{cm}^2. \tag{4.1.13}$$

For binary electrolytes, it becomes:

$$\Lambda_m = z\Lambda. \tag{4.1.14}$$

If fraction α of the dissolved electrolyte is present as ions, i.e. α is the degree of dissociation we have:

$$zc_c = zc_a = \alpha c_{eq}. \tag{4.1.15}$$

According to this and from equation (4.1.11), we obtain:

$$\Lambda = \alpha \mathrm{F}(U_c + U_a). \tag{4.1.16}$$

In electrochemistry, usually instead of the mobilities U their products with the Faraday number are used:

$$\lambda_c = \mathbf{F}U_c \, \Omega^{-1} \, \text{cm}^2,$$

$$\lambda_a = \mathbf{F}U_a \, \Omega^{-1} \, \text{cm}^2, \tag{4.1.17}$$

where λ_c and λ_a are the *equivalent conductivities (relative mobilities, or shortly, ionic conductivities) of the ions* in a solution of given composition and concentration. The introduction of these quantities is advantageous, since, in the case of total dissociation, the sum of the equivalent conductivities of the ions in the actual solution yields the equivalent conductivity of the solution.

In the above meaning, α is only the portion of the dissolved electrolyte present as ions migrating relatively independently, irrespective of the nature of the free uncharged species, being molecules bound together by covalent chemical bonds or ion pairs kept together (associated) by means of electrostatic attraction forces. Weak electrolytes are mostly present as non-dissociated molecules involved in covalent bonds in their solutions, so α is the degree of dissociation corresponding to their chemical dissociation equilibrium. In dilute solutions, strong electrolytes are dissociated completely or almost completely and there are no covalent molecules or, at least, their ratio is negligibly small. It often happens, however, that a small fraction of ions form ion pairs (occasionally triplets) bound together by electrostatic attraction forces, which also decreases conductivity. The degrees of both dissociation and association vary with concentration, but they follow different relationships. Consequently, the dependence of conductivity on state parameters (e.g. the dielectric constant of the solvent) is different for non-dissociating molecules and for ion pair formation. The degree of dissociation increases with increasing dilution, while the degree of association decreases with it, and in infinite dilution (when extrapolated to zero concentration) the degree of dissociation is unity, while that of association is zero. According to this, the limiting value of equivalent conductivity of electrolytes extrapolated to zero concentration is:

$$\Lambda^0 = \lambda_c^0 + \lambda_a^0. \tag{4.1.18}$$

In this limiting value, the conductivity of each ion is independent of that of other ions present in the solution (this is *the Kohlrausch law of independent migration of ions*).

The limiting value of the equivalent conductivity (mobility) of ions can be calculated from the equivalent conductivity of the corresponding binary electrolyte and the limiting value of the transference number, according to equation (4.1.35). The limiting values can be extrapolated from the results of measurements on the correlation between conductivity and concentration (cf. Section 4.2. and 4.2.3). Recently, Newman, Schober and Lawyer [1b] showed that ionic mobilities can be calculated directly from changes of

Literature on page 326

conductivity under the effects of electric and magnetic fields (ionic Hall effect). However, since the Hall potential is small, great experimental difficulties have to be overcome in these measurements.

The equivalent conductivity of electrolytes or ions depends too on the conditions, on the nature and structure of the solvent, on the concentration of the electrolyte and other solutes incidentally present, on the composition of the solution, on temperature, pressure, etc. This dependence is a consequence of interactions between ions, as well as between ions and the solvent molecules. In the studies on the interactions, the investigation of the concentration dependence of conductivity is particularly important at constant temperature and pressure. Owing to the interaction of ions, the mobility of a given kind of ions depends on the nature and concentration of the other kinds of ion present in a solution of finite dilution even under conditions otherwise identical. Thus, in solutions of finite dilution, the law of independent migration of ions is not valid. A significant part of the theories of strong electrolytes deals with the concentration dependence of conductivity.

The dependence of conductivity on concentration. The equivalent conductivity of strong electrolyte solutions increases to a limiting value with a decrease in concentration (or with an increase in dilution $V = 1/c$), according to the experiments at constant temperature and pressure. This limiting value is approached up to 1–2 per cent, by that obtained in very dilute but still well-measurable solutions. For the most dilute solutions, the empirical *square root law of Kohlrausch* is valid in rather good approximation. According to this, the equivalent conductivity of an electrolyte solution increases linearly with the decrease of concentration:

$$\Lambda = \Lambda^0 - A\sqrt{c} . \tag{4.1.19}$$

In this concentration range A is independent of concentration and the chemical nature of the electrolyte, but depends on its valence type. The higher the valence of the ions in the electrolyte, the larger is the value of A.

According to the square root law, the limiting value of conductivity Λ can be determined with an accuracy of about 0·1 per cent in the most favourable cases by extrapolating from conductivity data measured in very dilute solutions (about 10^{-4} mole 1^{-1}).* Extrapolation is the more reliable, the broader the concentration range covered in the measurements.

* In dilute solutions the technical limit in measuring the accurate value of conductivity caused by the electrolyte investigated is determined by the fact that the total conductivity of the solution is measured directly, and also includes contributions of the conductivity of the solvent itself and of the contaminants which can hardly be avoided. The conductivities arising from these factors should be accounted for in corrections. The percentage contributions of the conductivities of the solvent and the contaminants to the specific conductivity directly measured increases with decreasing concentration of the electrolyte investigated, so there is a limit in concentration below which Λ cannot really be determined with sufficient precision because of the uncertainty in the necessary correction.

Therefore, it is an important aim of studies on the conductivity of electrolytes to establish a correlation which allows extrapolation of increased accuracy as compared with the simple square root law. According to experiments, the validity of the square root law can be extended to a broader concentration range by an additional linear term:

$$\Lambda = \Lambda^0 - A\sqrt{c} - Bc. \tag{4.1.20}$$

Exact knowledge of the limiting value Λ^0 of conductivity is important, since the limiting values of the individual ionic conductivities can be calculated according to equation (4.1.35) if the transference number of the actual electrolyte is known; important fundamental data of electrochemical transport processes can thus be obtained.

The following empirical relationship applies to the conductivity of dilute solutions of weak binary electrolytes, in good approximation:

$$\lg \Lambda = A - \frac{1}{2} \lg c, \tag{4.1.21}$$

where A is constant.

The concentration dependence of the conductivity of more concentrated solutions of strong electrolytes was found to be more complicated and more individual. The validity of the square root law can be extended to about $0 \cdot 1$ mole l^{-1} concentration by applying a third concentration term:

$$\Lambda = \Lambda^0 - A\sqrt{c} - Bc - Dc^{3/2}. \tag{4.1.22}$$

The constants A, B and D of this equation should be determined empirically from data measured in the given electrolyte.

Experience indicates that, in several cases, conductivity of strong electrolytes varies with concentration according to the *cube-root law* in more concentrated solutions:

$$\Lambda = \Lambda^0 - A\sqrt[3]{c}. \tag{4.1.23}$$

The limiting value of the equivalent conductivity of weak electrolytes cannot be obtained from extrapolation of directly measured data, because Λ is far from the limiting value Λ^0, even in the most dilute but still well-measurable solutions. However, Λ^0 can be calculated on the basis of equation (4.1.18) from the limiting values of the respective ionic conductivities. Since most of the weak electrolytes are an acid or a base, while salts are strong electrolytes (even the salts of weak acids and bases) with few exceptions, the λ^0 values of the ions of weak electrolytes can be calculated from data measured in the corresponding strong electrolytes, e.g. that of the ions of acetic acid from those of hydrochloric acid and sodium acetate, or that of ammonium hydroxide from those of ammonium chloride and potassium hydroxide.

Literature on page 326

17*

Conductivity of solutions containing more than two kinds of ions. The relationships applying to solutions containing several kinds of ions (electrolytes dissociating to more than two kinds of ions or a mixture of electrolytes) are similar to those holding for binary electrolytes. It follows from the proper generalization of the above relationships that the specific conductivity of the solution in the presence of several types of ions is:

$$\varkappa = \Sigma c_i |z_i| \lambda_i, \tag{4.1.24}$$

taking into account equations (4.1.11) and (4.1.17), where c_i is the concentration of the ith type of ions expressed in g-ion cm^{-3} units, λ_i is its equivalent conductivity (relative mobility) in the actual solution. The molar conductivity of the whole solution is:

$$\Lambda_m = \frac{1}{c_m} \Sigma c_i |z_i| \lambda_i \tag{4.1.25}$$

(c_m is the concentration of the electrolyte solution in [mole cm^{-3}] units). In analogy with equation (4.1.13), the equivalent conductivity is:

$$\Lambda = \frac{1}{n_e c_m} \Sigma c_i |z_i| \lambda_i, \tag{4.1.26}$$

when the number of positive and negative types of ions is v_c and v_a in the solution, and $n_e = \Sigma v_c z_c = \Sigma v_a z_a$.

The requirement of the validity of the above equations is the validity of Ohm's Law in electrolytes, i.e. that both the number (concentration) and the mobility of ions should be independent of the electric field strength. Under normal conditions, in general, this requirement is satisfied, but appreciable deviations can appear in the case of high field strength or high frequencies (cf. Sections 4.2.3.6. and 4.2.3.7).

Transference number. In electrolyte solutions, electricity is transported partly by cations and partly by anions migrating in opposite directions. The current or the current density (when related to a cross-section of 1 cm²) is the sum of the absolute values of the amounts of electricity transported per second by oppositely charged ions in opposite directions, since, with respect to the electric current, the displacement of positive electric charges in one direction is equivalent to the displacement of the same amount of negative charge with the same velocity in the opposite direction. Thus, the effect of electric current outside the conductor does not depend on whether it is transported by ions in electrolytes (bipolar conduction) or by electrons in metals (unipolar conduction). Within the conductor, however, the mechanism of conduction is essentially different in the two types of substances.

In solutions a quantity of electrochemical importance is the ratio of the participation of the various types of ions of the given electrolyte in the transport of electricity. This ratio is the *transference number* (t). If in a solution the current density is i and the portion transported by the ith kind of ions is i_i, the transference number of this kind of ions is, by definition:

$$t_i = \frac{i_i}{i}. \tag{4.1.27}$$

Under linear potential conditions (if in the solution potential gradient exists only in direction y, and in all other directions perpendicular to this it is zero), the total current density is:

$$i = \mathbf{F} \, \Sigma \, c_i |z_i| \, U_i \frac{\partial \psi}{\partial y}, \tag{4.1.28}$$

taking into account equations (4.1.2), (4.1.3) and (4.1.4), in the case of dilute solutions. The fraction of the total current density transported by the ith kind of ions is:

$$i_i = \mathbf{F} \, c_i |z_i| \, U_i \frac{\partial \psi}{\partial y}. \tag{4.1.29}$$

Consequently, the transference number of the ith type of ions is:

$$t_i = \frac{c_i |z_i| \, U_i}{\Sigma \, c_i |z_i| \, U_i} = \frac{c_i |z_i| \, \lambda_i}{\Sigma \, c_i |z_i| \, \lambda_i}, \tag{4.1.30}$$

from equation (4.1.17); on the basis of equation (4.1.26) we have:

$$t_i = \frac{c_i |z_i| \, \lambda_i}{n_e c_m \Lambda} = \frac{c_i |z_i| \, \lambda_i}{c_m \Lambda \, \Sigma \, \nu_c z_c}. \tag{4.1.31}$$

Further, as it follows from the definition of the transference number, the sum of the transference numbers of all the ions present in the solution is equal to unity:

$$\Sigma \, t_i = 1. \tag{4.1.32}$$

If only one binary electrolyte is present in the solution, the transference numbers of its ions will be:

$$t_c = \frac{\lambda_c}{\lambda_c + \lambda_a},$$

$$t_a = \frac{\lambda_a}{\lambda_c + \lambda_a} = 1 - t_c. \tag{4.1.33}$$

Extrapolating to the limiting value of the infinitely dilute solution, it becomes:

$$t_c^0 = \frac{\lambda_c^0}{\lambda_c^0 + \lambda_a^0} = \frac{\lambda_c^0}{\Lambda^0}, \tag{4.1.34}$$

Literature on page 326

from which the limiting value of ionic equivalent conductivity can be derived:

$$\lambda_c^0 = t_c^0 \Lambda^0,$$

$$\lambda_a^0 = t_a^0$$

$$\Lambda^0 = (1 - t_c^0)\Lambda^0. \tag{4.1.35}$$

The extrapolation of the limiting value of transference number is easier than that of equivalent conductivity, because it varies less with concentration (cf. Section 4.2.3.5). However, knowledge of it is important, because the limiting values of ionic mobilities (conductivities) can be calculated from it; these data are fundamental quantities in electrochemical transport processes.

The transference number, giving the proportion of electricity transported by the individual types of ions, is characteristic of the given electrolyte under given conditions, but not of the individual types of ions separately. In solutions of binary electrolytes the transference number of the cation is also dependent on the anion, and vice-versa. It is seen in equation (4.1.30) that the transference number of any type of ions can almost be decreased at will by increasing the concentration of electrolytes consisting of foreign ions to a sufficient degree in the solution. In this way, the ions present in relatively low concentration can be excluded almost completely from the conduction of electricity; this is of great importance, e.g., in polarography.

Transference numbers depend on the actual conditions (solvent, concentration, temperature, electric potential gradient, etc.) in a similar way to ionic mobilities. However, since transference numbers are determined by the ratio of mobilities, they are altered by the conditions less than the mobilities themselves (cf. Section 4.2.3.5).

Transference numbers can be directly determined experimentally by the Hittorf method from the change in the concentration of the electrolyte in the course of electrolysis, or by the method of a moving boundary, observing optically the displacement of the boundary between electrolyte solutions of identical conductivities under the effect of potential gradient [1a]. In several cases, the transference number can be calculated indirectly as well from the electromotoric force of corresponding galvanic cells [2].

In Table 4.1 the transference numbers of some ions in various electrolyte solutions are summarized. (The concentration dependence of the transference number can be found in Table 4.5.)

For the exact theoretical interpretation of the transference numbers measured directly in the experiments, it should be stated exactly what the movement of the ions is referred to. This involves difficulties similar to those occurring in the precise determination of a reference system in diffusion. In Section 3.1.2, these problems have already been dealt with in detail, so here it seems sufficient to emphasize the special aspects occurring in connection with the transference number [3]. In principle, the most practical way would be to refer the displacement of ions to the solvent, however, this can be realized only in extrapolating to the limiting case of infinitely dilute solutions. In the moving boundary method, the data are referred to a coordinate system fixed with respect to the wall of the vessel containing

Table 4.1

The transference number of some ions in various electrolytes
extrapolated to infinite dilution at 25 °C

Ion	Electrolyte	t, °C
H^+	HCl	0·821
Li^+	LiCl	0·336
Na^+	NaCl	0·396
Na^+	Na_2SO_4	0·386
K^+	KCl	0·491
NH_4^+	NH_4Cl	0·491
Ag^+	$AgNO_3$	0·464
Ca^{2+}	$CaCl_2$	0·438
La^{3+}	$LaCl_3$	0·427
Ce^{3+}	$CeCl_3$	0·460

the solution. These data, however, also imply the displacement of the solution as a whole caused by the change in the volume at the electrodes. On the other hand, the Hittorf method relates the change in concentration due to the different mobilities of ions to a given amount of the solvent.

Considering the so-called Hittorf-type transference numbers measured directly, it should be taken into account that ions are generally hydrated (cf. Section 5.2) and they migrate together with a certain number of water molecules (this number cannot be determined unequivocally). Consequently, the total amount of water present in the solution cannot be regarded as a reference system that is unaffected by the electric current, because the water molecules in the hydrate sphere which move together with the ions also move on the effect of the electric potential gradient. Only the water outside the hydrate spheres can be considered as an immobile medium when electric current acts, so only the 'intrinsic' transference numbers referred to this will reflect the real conditions. Then, if the hydration numbers of the cation and anion are denoted by h_c and h_a (i.e. one mole of ions transports this amount of water), and t_c' and t_a' are the intrinsic transference numbers of the cation and anion, respectively, in binary electrolyte solutions W moles of water will be transported by F coulombs of electricity in the direction of positive current:

$$W = t_c' h_c - t_a' h_a. \qquad (4.1.36)$$

Buchböck [4] and Washburn [5] calculated the difference in the amount of water transported by cations and anions by measuring the transference number in an electrolyte solution also containing non-electrolyte (e.g. sugar). Non-electrolytes do not migrate under the effect of the potential gradient, and the ions were assumed not to carry along non-electrolyte molecules either. The concentration of this non-electrolyte was altered by the water transported by the ions, and this amount of water was

Literature on page 326

calculated from the change in concentration. If these assumptions hold, the intrinsic transference number can be calculated from Hittorf-type transference numbers (t_c, t_a) and from the measurement of the amount W of transported water:

$$t'_c = t_c + \frac{c}{55 \cdot 5} W;$$

$$t'_a = t_a + \frac{c}{55 \cdot 5} W. \qquad (4.1.37)$$

where c is the concentration of electrolyte. Thus the intrinsic transference numbers could be obtained from the Hittorf-type ones, if the amount of water moving together with ions in the hydrate spheres could be determined. Recently, however, Erdey-Grúz et al., and Longsworth, as well as Hale and De Vries (see Section 5.2.4.2) pointed out that ions carry along not only water molecules but also non-electrolyte molecules, so the amount of water

Table 4.2

The mobility of some ions in aqueous solution at 25 °C extrapolated to infinite dilution (Ω^{-1} cm^2 g-eq^{-1})

Ion	λ_i^0	Ion	λ_i^0
H$^+$	349·8	F$^-$	55·4
Li$^+$	36·68	Cl$^-$	76·35
Na$^+$	50·10	Br$^-$	78·14
K$^+$	73·50	I$^-$	78·84
Rb$^+$	77·81	OH$^-$	198·3
		ClO$_3^-$	64·6
Ag$^+$	61·90	ClO$_4^-$	67·36
NH$_4^+$	73·55	BrO$_3^-$	55·74
N(CH$_3$)$_4^+$	44·92	CN$^-$	78
$\frac{1}{2}$ Mg^{2+}	53·05	NO$_3^-$	71·46
$\frac{1}{2}$ Ca^{2+}	59·50	CH$_3$COO$^-$	40·90
$\frac{1}{2}$ Ba^{2+}	63·63	C$_6$H$_5$COO$^-$	35·8
$\frac{1}{2}$ Zn^{2+}	56·6	H$_2$PO$_4^-$	36
$\frac{1}{2}$ Cd^{2+}	54	$\frac{1}{2}$ SO$_4^{2-}$	80·02
		$\frac{1}{2}$ S$_2$O$_6^{2-}$	93
$\frac{1}{3}$ Al^{3+}	63	$\frac{1}{2}$ WO$_4^{2-}$	69·4
		$\frac{1}{2}$ CrO$_4^{2-}$	85
$\frac{1}{3}$ La^{3+}	69·7	$\frac{1}{2}$ HPO$_4^{2-}$	57

Table 4.3

The concentration dependence of the mobilities of some ions g-eq·l⁻¹

The concentration dependence of the mobilities of some ions g-eq$\cdot$l^{-1}

Ion	λ_i					
	0·0001	0·0005	0·001	0·01	0·05	0·1
H^+	315	312	311	307	301	294
Li^+	33·2	32·8	32·5	30·8	28·8	27·5
Na^+	43·2	42·8	42·4	40·5	37·9	36·4
K^+	64·1	63·7	63·3	60·7	57·2	55·1
Cs^+	67·4	66·9	66·6	63·7	60	58
Ca^{2+}	50·4	49·0	48·0	41·9	35·2	32·0
Ba^{2+}	54·0	53·5	51·5	44	—	—
OH^-	172	171	171	167	161	157
F^-	46·2	45·8	45·5	43·2	40	38
Cl^-	64·9	64·4	64·0	61·5	57·9	55·8
Br^-	67·0	66·5	66·1	63·7	60·6	59·1
I^-	65·6	65·3	64·9	62·7	60·1	—
NO_3^-	61·3	60·8	60·4	57·6	53·3	50·8
ClO_3^-	54·5	54·0	53·6	50·9	46·5	44·0
SO_4^{2-}	66·5	66·0	63·8	55·5	45	40

transported due to hydration cannot be calculated unequivocally from the change in the concentration of the non-electrolyte.

Extrapolating to infinite dilution, the limiting value of the Hittorf-type transference number is identical with the intrinsic one. In more concentrated solutions, however, the deviation becomes significant. In connection with the transference number extrapolated to infinitely dilute solutions, it should also be taken into account, however, that these numbers do not refer to bare ions, but to hydrated ones that move as kinetic entities and whose sizes can be appreciably larger than the ionic radii measurable in ionic crystals.

Transference numbers are also influenced by the incomplete dissociation or ionic association of the electrolyte in solution. The non-dissociated molecules and ion pairs are dipole-like species, and they participate in the solvation of ions together with the other dipole molecules in the solution (e.g. ion pairs can form ion triplets with other ions). The molecules and ion pairs in the solvate sphere migrate under the effect of the potential gradient, however, without contributing to the current (i.e. without transporting electricity). Under such conditions, the amount of electricity transported by the electrolyte cannot be calculated unequivocally from the transported amount of electrolyte, and only a 'net transference number' can be calculated from the measurements. This problem was discussed in detail by Spiro [6] and Haase [7].

Mobility of ions. The equivalent conductivity gives the *sum* of the mobilities of ions in the solution, when extrapolating to infinite dilution of strong electrolyte solutions. Unlike some other quantities related to the sum of components which yield the individual values only on applying some more

Literature on page 326

or less arbitrary assumptions, these are eliminated in calculating the individual ionic mobilities. This is due to the fact that the *ratio* of ionic mobilities can also be measured directly, by means of the transference numbers.

With the knowledge of the sum and the ratio of mobilities, the ionic mobilities themselves (the equivalent conductivity of ions) can be calculated from equation (4.1.35).

The mobilities calculated in this way for some ions are given in Table 4.2. Table 4.3 shows some data on the concentration dependence of mobility. For conclusions on the state of dissolved ions and their effect on the solvent in their vicinity obtained from the magnitude of ionic mobilities, see Section 5.2.

4.2 FUNDAMENTALS OF THE THEORY OF ELECTROLYTIC CONDUCTION

4.2.1 HYDRODYNAMIC THEORIES OF ELECTROLYTIC CONDUCTION, I

The simplest theory of the transport of electricity in electrolyte solutions considers the ions to be rigid spheres carrying electric charge and moving in the solvent (supposed to be a continuous medium), under the influence of a force arising from the potential gradient.

According to Newton's law, ions are accelerated by a force f and this effect is inversely proportional to their mass. However, since the ions are in a viscous medium, their movement gives rise to a frictional force which increases with their velocity. The force acting on the ions is the product of their charge and the potential gradient. If—as is usual—this force is constant, then, owing to the acceleration caused by it, the velocity of ions increases rapidly to make the frictional force, acting against the movement, equal to the motive force. When this state is attained, the resultant of the forces influencing the motion of ions becomes zero and ions will move further with uniform velocity, as long as the motive or the frictional force remains unchanged. The order of magnitude of the relaxation time required to reach this steady state is, according to the laws of hydrodynamics, $m/6\pi\eta r$ in a medium of viscosity η and for a sphere having mass m and radius r. When the spheres are assumed to have radii of the order of magnitude $r \approx 10^{-8}$ cm and mass $m \approx 10^{-22}$ g (see below) moving in a medium of viscosity $\eta \approx 10^{-2}$ poise, the relaxation time is in the order of magnitude of about 10^{-14}–10^{-13} s. This is such a low value that it can be neglected in almost every case.

The frictional force f' acting on the ions is proportional to the average velocity $\bar{v}_i$ in the direction of the force:

$$f' = f_{s,i}\bar{v}_i, \tag{4.2.1}$$

where $f_{s,i}$ is the frictional resistance of the ions. Since, in steady state, we have:

$$f_{s,i}\,\bar{v}_i = |z_i|\,e\,\frac{\partial \psi}{\partial y}, \tag{4.2.2}$$

supposing a linear potential gradient, in steady state we obtain:

$$\bar{v}_i = \frac{|z_i|\,e}{f_{s,i}}\,\frac{\partial \psi}{\partial y}, \tag{4.2.3}$$

i.e. the migration velocity of ions–after a negligibly short relaxation time–is proportional to the gradient in electric potential, assuming that the frictional resistance is independent of the latter. This assumption holds, in general, with satisfactory approximation except when the potential gradient is very high or it varies with a very high frequency.

Since the mobility of ions, expressed in the customary units of electrochemistry is equal to the velocity [cm s^{-1}] gained under the influence of a potential gradient of 1 V cm^{-1}, we have:

$$U_i = \frac{10^7\,|z_i|\,Ne}{Nf_{s,i}} = \frac{10^7\,|z_i|\,F}{Nf_{s,i}}\ \text{cm}^2\,\text{s}^{-1}\,\text{V}^{-1}, \tag{4.2.4}$$

when referring to 1 mole of ions on the basis of equation (4.2.3) (measuring the charges in coulombs, the tension in volts, and the frictional resistance in c.g.s. units). The molar conductivity of the ith type of ions is, on the other hand:

$$\lambda_i = FU_i = \frac{10^7\,|z_i|\,F^2}{Nf_{s,i}}, \tag{4.2.5}$$

$\bar{v}_i$ is the average velocity of displacement in the direction of the electric lines of force which contributes to the velocity of the random thermal motion. In electrochemistry, however, thermal motion can be neglected in relationships regarding macroscopical processes, since it has no favoured direction and $\bar{v}_i$ can be taken simply as the velocity of the ions.

Equation (4.2.3) is the grounds for the validity of Ohm's Law for electrolytes as well. It can be attributed to this that there is no striking or simple correlation between mass and the mobility or conductivity of ions. (For example, the mass of the I$^-$ ion is almost four times as much as that of the Cl$^-$ ion, while their mobilities hardly differ from each other.) According to the laws of movements with great friction, the velocity is determined not directly by the mass of the moving body, but by its frictional resistance–in addition to the motive force.

However, the traditional opinion that the mobility of the dissolved ions is independent of their mass is not rigorously correct, as indicated by recent investigations. Fischer and Hessler [8], and Trosin [9] have pointed out that the mobility of ^{41}K$^+$ ion is by about 2·3 per cent higher than that of ^{39}K$^+$ at 25 °C and in dilute solution, and Bakulin [10] has proved the difference in the mobilities of ^{7}Li$^+$ and ^{6}Li$^+$ ions, establishing that the difference increases with concentration, but is independent of temperature.

Literature on page 326

Up to rather high concentrations (about 10 g-eq/1000 g), it is also independent of the nature of the anion. In very concentrated $LiNO_3$ solutions, however, the relative difference in the mobilities of the two isotopes is smaller than in LiCl solutions. To explain this, he assumed that an appreciable number of groups similar to those existing in the crystal lattice of $LiNO_3 \cdot 3H_2O$ are present in more concentrated solutions of $LiNO_3$, while the structure of the chloride solutions is similar to the hydrate $LiCl \cdot 2H_2O$, i.e. the structures of the concentrated solutions of these two salts differ from each other. The change of ionic mobilities with mass has also been confirmed by the investigations of Horne and Birkett [11].

The dependence of the frictional constant of ionic migration on the dielectric relaxation of the liquid has been studied by Boyd [12] and Zwanzig [13] for dilute solutions, while for more concentrated solutions this analysis has been carried out by Fernández-Prini [14]. They concluded that part of the concentration dependence of ionic mobility which represents dielectric relaxation contains no term having an exponent lower than $c^{3/2}$.

Ionic mobilities or conductivities are determined by the frictional resistance $f_{s,i}$, according to the hydrodynamic theory. The influence of ionic sizes on conductivity is also due to this. According to Stokes' Law, the frictional resistance of a sphere with radius r moving in a continuous medium of viscosity η on the surface of which the liquid is not sliding, is:

$$f_{s,i} = 6\pi\eta r,$$

and, according to equation (4.2.2), we have:

$$\lambda_i = \frac{10^7 |z_i| F^2}{N \, 6\pi\eta r}. \tag{4.2.6}$$

Rigorously, this equation is not valid for solutions of common ions, because the size of the liquid molecules is not negligibly small in comparison with their sizes, thus the liquid can be considered a continuum only in approximation. For ions migrating under the effect of an electric potential gradient, the problem of frictional resistance is, in most respects, identical with that occurring in the case of solutes diffusing under the influence of a gradient in chemical potential (see Section 3.2.1).

Although Stokes' Law is not rigorously valid for electric conduction, the real conditions are still reliably reflected in the main characteristics to apply it to the approximative interpretation of experimental facts and in the unified description of the phenomena.

On the basis of Stokes' Law, the radius of a species migrating under the effect of electric potential gradient, which is usually called Stokes radius (r_{St}), can be calculated from ionic conductivities. Applying equation (4.2.6) that takes into account the frictional resistance according to Stokes' Law, the radius of the ion migrating in the solution is:

$$r_{St} = \frac{10^7 |z_i| F^2}{N \, 6\pi\lambda_i\eta} = \frac{0{\cdot}820 |z_i|}{\lambda_i\eta}, \tag{4.2.7}$$

when the unit of r_{St} is Å ,that of λ_1 is $cm^2 \, \Omega^{-1} \, eq^{-1}$, while that of η is poise. In Table 4.4 the Stokes radii of some ions are given together with the values of r_{cr} calculated on the basis of crystallographic data of the corresponding ionic crystals. It can be seen in the Table that the Stokes radii are much larger than the crystallographic ones and the order of the radii calculated by these two means is also opposite in the same column of the periodic system (i.e. for similar ions).

Unlike other ions, the Stokes radii of hydroxide and hydrogen ions are much smaller than the crystallographic ones. In aqueous solutions, the 'hydrogen ion' is actually present as an oxonium ion (H_3O^+), and its crystallographic radius can be determined, for example, by X-ray diffraction investigation of perchloric acid monohydrate crystals which contain H_3O^+ and ClO_4^- ions in the lattice. The strikingly small value of the Stokes radius –which is a consequence of the unusually high equivalent conductivities in comparison with other ions–indicates that hydrogen and hydroxide ions contribute to the transport of electricity not only by simple hydrodynamic motion but by another mechanism too (see Section 4.2.2.2).

Although Stokes' Law is not rigorously valid for the movement of ions in solution, it still does not differ too much from reality, according to several experiments. This is why the above statement on the relationship between ionic conductivity and radii can be considered real to a certain extent and, applying it as a working hypothesis, further conclusions can be drawn from it.

The difference between the ionic radii in crystals and solutions is usually interpreted in that way that ions are hydrated in solution, i.e. due to their electrostatic field, they bind some dipole molecules of water to themselves so strongly that these molecules will migrate together with the ions, forming a kinetic entity from the point of view of translational movement (cf. Section 5.2.4). According to this view, the deviation of the Stokes radii from the crystallographic ones can be attributed to the fact that the electric field strength at the surface of the ion is higher, the smaller the radius of the ion–in the case of identical ionic charges–and the effect on the dipole molecules of water increases with increasing field strength. Thus, the small Li^+ ion binds so many water molecules as to make the volume of the hydrated ion formed larger than the large but 'bare' Cs^+ ion, at the surface of which water molecules are hardly bound with respect to the small field strength. In the strong electric field of small ions, deformation polarization also contributes to orientation polarization (the ordering of the permanent dipole axes) of water molecules; due to this, the field of water molecules around the ion (in addition to that of the ion) will also affect the neighbouring water molecules, and a hydrate sphere of several layers can be formed.

Accepting this assumption, the difference between the volumes calculated from the crystallographic radius of the 'bare' ion and the Stokes radius of the hydrated ion is equal to the volume v_h of the hydrate sphere (see Section 5.2.4.1). By an additional assumption regarding the packing of water mole-

Literature on page 326

Table 4.4

Pauling's crystallographic radii and Stokes type radii of some ions at 25 °C

Ion	r_{cr}, Å	r_{St}, Å
Li+	0·60	2·37
Na+	0·97	1·83
Mg²+	0·65	4·46
Ca²+	0·99	3·07
La³+	1·15	3·95

cules in the hydrate sphere and on the basis of the volume of water molecules, the number of water molecules bound to an ion, i.e. the hydration
number, can be calculated [15a]. Supposing that water molecules are rigid
spheres arranged according to closest packing in the hydrate sphere, the
lower limit of hydration number calculated in this way can be obtained.
If, however, electrostriction is taken into account, and the water molecules
are assumed to be so compressed by the electric field of the ions that the
mutual separation of atoms decreases to the bond length, the upper limit of
the hydration number calculated can be so obtained. The hydration numbers
based on ionic mobilities, however, differ significantly from those calculated
by some other methods. This comparison is made in Table 5.2.

The calculation of the radius or volume of dissolved ions from conductivity data on the basis of Stokes' Law is unconvincing, since it is
assumed in the calculation that the macroscopic viscosity of the solvent
or solution is the decisive factor in the frictional resistance of ionic migration.
In reality, however, the translational movement of ions depends primarily
on the viscosity of the liquid layers in their vicinity, and it may hardly
be doubted that this is not equal to the bulk viscosity of water far from ions
and unaffected by their field. Viscosity depends on the liquid structure and
this structure is altered by the field of the ions. Others consider that the
phenomena summarized as resulting from hydration do not merely arise
from the joint movement of water molecules and ions forming a kinetic
entity; they can be attributed mainly to changes in the translational
mobility of water molecules surrounding the ions under the effect of their
ionic fields (cf. Section 5.2.3). Thus, the alteration in microviscosity in their
immediate vicinity is reflected in the ionic mobility.

On the basis of our present knowledge, it cannot be judged unequivocally
how reliable is the description of reality by the various models of ionic mobility. Nevertheless, it can be stated that in the calculations the use of the
average viscosity measured macroscopically is, in fact, unjustified. On the
other hand, several facts indicate that migrating ions transport water
molecules even if the joint movement of the whole hydrate sphere of the
ion is less probable. No satisfactorily sound basis is available, however, to
judge the ratio of the influences of the ionic size and the change in microviscosity on conductivity. Owing to this uncertainty, extension of the
validity of Stokes' Law by applying some corrections to account for
the discrete molecular structure of the solvent, instead of considering it a

continuum, does not contribute too much to the interpretation of ionic migration.

The radii of alkali and halide ions have been calculated by Reff [15b] on the basis of Stokes' Law, taking into account the change in the liquid structure of water under the effect of the electric field of ions.

Some information can be obtained on the limits of validity of Stokes' Law from the investigations on the conductivity of tetra-alkyl ammonium ions carried out by Kraus [16]. On the basis of reasonable assumptions regarding the molecular structure, the radius r_m of these ions can be calculated approximately, and compared with those obtained from Stokes' Law. If it is assumed that, in the course of their migration, these large monovalent ions do not transport water molecules, the difference between the radii calculated in these two ways can indicate the degree of approximation provided by Stokes' Law. In Fig. 4.1, it can be seen that the radii calculated from Stokes' Law are lower than the true ones for small ions, and it seems that this rule can be considered valid for ions having radii larger than about 5–6 Å. Some further studies are required, however, to elucidate what part of this difference between the ionic radii calculated from Stokes' Law and by other methods can be attributed to the non-continuous nature of the liquid, and how much arises from other factors, e.g. from the difference in microviscosity and the macroscopical viscosity. Kay and Evans [17a] have studied the influence of the structure-changing effect of tetra-alkyl ammonium ions on conductivity as a function of the temperature and pressure (cf. Section 1.4.2).

On the basis of the conductivity of tetra-alkyl ammonium ions in H_2O and D_2O, their radii calculated from Stokes' Law are identical–as obtained by Kurmgaltz [17b]–from which it can be concluded that hydrophobic hydration takes place in these solutions involving the formation of lattice-like structures. The Stokes radii calculated from the conductivity of tetra-alkyl ammonium and other ions larger than the $(C_3H_7)_4N^+$ ion measured in various organic solvents are independent of the nature of the solvent. This indicates that the Stokes radii of these ions are equal to their real radii under such conditions and no appreciable solvation occurs.

Boyd [18] and Zwanzig [19] explained the deviation of Stokes' Law from the experimental data in solutions of ions of small radii on the basis of dielectric relaxation effects, while Passeron [20] attributed the energy dissipation in friction arising in the course of ionic migration to a mechanism other than the generally accepted one. According to him, the moving ion actually

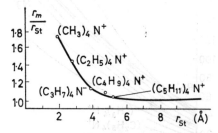

Fig. 4.1 The dependence of the ratio of the radii r_m/r_{St} (calculated on the basis of the molecular structure and Stokes' Law) on the ion radius

Literature on page 327

slips between the water molecules, and the water dipoles located sufficiently close to the ion are orientated towards it, i.e. they are turned in the electric field of the ion, and, due to this, they gain some potential energy. After the ion has passed these water molecules, their dipoles return again to their normal position due to thermal collisions in which there is no favoured orientation and, simultaneously, the energy gained previously is dissipated as thermal energy. The correction of Stokes' Law calculated theoretically on the basis of this model agrees satisfactorily with the correction r_m/r_{St}, determined experimentally for tetra-alkyl ammonium ions mentioned above. The problem of the constancy of the product $\Lambda^0\eta^0$ and its dependence on viscosity has been discussed from some other aspects by Brummer and Hills [21] and Gorenbein [22] (cf. also Section 5.2.4).

The temperature dependence of ionic mobility is large: at 100 °C, the equivalent conductivity of ions is about five to six times higher than that at 0 °C (Fig. 4.2). The temperature coefficient of conductivity is between 1·8 and 2·4 per cent for most of the ions, only that of the hydrogen ion is singificantly lower (1·4 per cent). This change arises mainly from changes in the viscosity of water, which is shown by the experimental fact that the product of conductivity and viscosity $\lambda^0\eta^0$ depends on the temperature much less than either λ^0 or η^0 do.

According to the empirical *Pisarzhevskii–Walden rule*:

$$\lambda^0\eta^0 = \text{const.} \qquad (4.2.8)$$

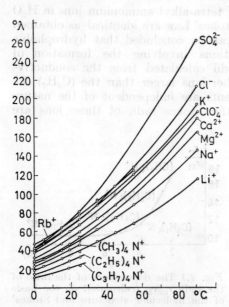

Fig. 4.2 The variation with temperature of the limiting value of the equivalent conductivity of some ions

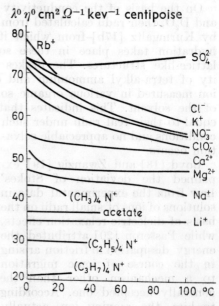

Fig. 4.3 The variation with temperature of the product $\lambda^0\eta^0$ for some ions

In reality, however, only a few ions obey this rule over a broad temperature range (Fig. 4.3). The 'Walden product' $\lambda^0\eta^0$ decreases, in general, with increasing temperature, but this change does not exceed 25 per cent between 0 and 100 °C. This can also be pictured as the decrease of η^0 in the product $\lambda^0\eta^0$ overcompensating the increase in λ^0 due to increasing temperature.

The Pisarzhevskii–Walden rule is a consequence of Stokes' Law implied in equation (4.2.6), provided that the radius of the migrating ion is unaltered by temperature, and the change in conductivity is caused by changes in macroscopic viscosity. In general, none of these criteria can be supposed to be correct. The thermal energy increasing with temperature certainly affects the size of the hydrate sphere and the microviscosity caused by the effect of the ionic field on the water structure does not necessarily show the same temperature dependence as macroviscosity. Because of this, it is not surprising that $\lambda^0\eta^0$ is generally not constant, as shown by experiments Nevertheless, investigation of the variation in the product $\lambda^0\eta^0$ with temperature—and with other parameters—is useful in the interpretation of this behaviour since the effect of macroscopic viscosity, large in every case, is excluded from this to a certain extent so the effects of other factors can be treated more easily and their investigation becomes less complicated.

4.2.2 KINETIC THEORIES OF ELECTROLYTIC CONDUCTION, I

The hydrodynamic theories of the transport of electricity by dissolved ions reviewed above give relationships that are important practically, and, within certain limits, lead to conclusions which are in agreement with experiments, but they do not reveal the molecular mechanism of the process. In this latter respect—as in the case of diffusion—the kinetic statistical theories should be applied. The most satisfactory explanation of the mechanism of electrolytic conduction can be expected from theories considering ions to be discrete particles, and also accounting for the fact that the liquid is not a continuum either, but consists of molecules. Such a theory can elucidate how ions break through the solvent, i.e. what elementary processes are involved in ionic migration. The solution of this problem has been given by the Eyring theory which, advanced by some recent studies, has yielded the deepest understanding of the mechanism of the transport of electricity so far, although real conditions are only approximately reflected because of the simplifying assumptions introduced to make calculation possible. Nevertheless, some important special problems in electrolytic conduction are not accounted for sufficiently well and this is why the kinetic statistical theory of electrolytic conduction suggested by Debye, Hückel and Onsager is still important, although it considers the liquid as a continuum and only the consequences of the electrostatic interaction of ionic charges are dealt with in it. This latter theory (cf. Section 4.2.3) gives valuable information particularly on transport phenomena and thermodynamic properties in the most dilute electrolyte solutions (cf. Section 5.1); in moderately dilute and

Literature on page 327

more concentrated solutions its results are less reliable. The complicated nature of the structure of liquids and particularly of water, as well as our limited knowledge on it, also contributes significantly to the fact that our knowledge on ionic migration, involving several problems because of the complex interactions, is not satisfactory in many aspects, although very broad and thorough research work has been carried out in this field.

4.2.2.1 ABSOLUTE REACTION RATE THEORY OF THE TRANSPORT OF ELECTRICITY

In electrolyte solutions, ions, migrating under the effect and in the direction of the electric potential gradient, transport electricity. The conductivity of the solution is determined by the charge of the ions and the velocity of their translational movement. The fundamental process is, thus, a molecular rate process which can be treated on the basis of the absolute reaction rate theory according to Eyring [23], being in several respects similar to the diffusion taking place on the effect of the gradient in chemical potential (cf. Section 3.1.2.2).

The Eyring theory is suitable for pointing out the most essential characteristics and starts with the fact that in the elementary step of conduction, the ions are transferred from their equilibrium positions to adjacent ones under the effect of the electric field. This requires an activation energy because of the energy barrier existing between two equilibrium positions located at a distance λ from each other (Fig. 4.4). If the concentration is not uniform in the solution, a concentration difference $\lambda \dfrac{dc}{dy}$ occurs between two adjacent equilibrium positions. Producing an electric field in the solution (switching on a potential difference between two electrodes), the shape of the free enthalpy curve is altered (dotted line in Fig. 4.4): the potential gradient ($\Delta\psi$) facilitates the translational movement of ions in one direction ('forwards'), while it hinders the motion in the other direction ('backwards'). If work w is required for the elementary process of ionic migration, i.e. to transfer the ion from its equilibrium position to an adjacent one, which means that w is the difference in free enthalpy between positions at a distance λ from each other, one portion of this work (αw) arises from the increase in the free enthalpy level of the initial state, while the other portion, $(1-\alpha)w$, is due to the lowering of the free enthalpy level of the final state in the elementary process. The factor α is determined by the electronic structures in the initial and final, as well as in the activated states. The number of ions crossing the energy barrier in unit time, i.e. the average velocity of the forward movement increased by the electric field is:

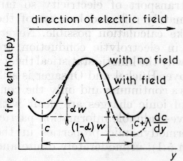

Fig. 4.4 The free enthalpy curve of ionic migration

$$\bar{v}_f = N\,\lambda\,ck\,\exp\left[\frac{\alpha w}{kT}\right], \qquad (4.2.9)$$

where k is the specific rate of the process (rate constant) which, in the ideal case, is:

$$k = \frac{kT}{h}\exp\left[-\frac{\Delta G^{\ddagger}}{RT}\right] \qquad (4.2.10)$$

according to equation (3.1.77), when no electric field acts. The average velocity of movements in the opposite direction (backwards) decreased by the field is:

$$\bar{v}_b = N\,\lambda\left(c + \lambda\frac{dc}{dy}\right)k\,\exp\left[-\frac{(1-\alpha)w}{kT}\right]. \qquad (4.2.11)$$

The resultant average velocity (from the left to the right-hand side in Fig. 4.4) is:

$$\bar{v} = \bar{v}_f - \bar{v}_b =$$
$$= N\,\lambda\,ck\,\exp\left[\frac{\alpha w}{kT}\right] - N\,\lambda\left(c + \lambda\frac{dc}{dy}\right)k\,\exp\left[-\frac{(1-\alpha)w}{kT}\right]. \quad (4.2.12)$$

Expanding the exponential terms into series and accounting for the fact that, under normal conditions, $kT \gg w$, all the terms containing w/kT at a power higher than one can be neglected in first approximation. Under such conditions, the resultant rate is:

$$\bar{v} = N\,\lambda\,ck\left(1 + \frac{\alpha w}{kT}\right) - N\,\lambda\left(c + \lambda\frac{dc}{dy}\right)k\left\{1 - \frac{(1-\alpha)w}{kT}\right\} =$$
$$= \frac{N\,\lambda\,ckw}{kT} - N\,\lambda^2 k\frac{dc}{dy}\left\{1 - \frac{(1-\alpha)w}{kT}\right\}. \qquad (4.2.13)$$

Since $kT \gg w$, consequently $\dfrac{(1-\alpha)w}{kT} \ll 1$, thus it is negligible as com-

pared with unity, so the average velocity of ions becomes:

$$\bar{v} = \frac{N\,\lambda\,ckw}{kT} - N\,\lambda^2 k\frac{dc_i}{dy}. \qquad (4.2.14)$$

If e denotes the charge of the electron, z_i is the valence of the ith type of ions, $\Delta\psi$ is the electric potential gradient, then the work required to displace an ion a distance λ is:

$$w = \lambda z_i e\,\Delta\psi. \qquad (4.2.15)$$

Literature on page 327

18*

Taking this into account in calculating the average resultant velocity of the translational movement of the ith type of ions present in concentration c_i, we have:

$$\bar{v}_i = \frac{\mathrm{N}\,\lambda^2 c_i k z_i \mathrm{e}\,\varDelta\psi}{kT} - \mathrm{N}\,\lambda^2 k\,\frac{\mathrm{d}c_i}{\mathrm{d}y}. \tag{4.2.16}$$

According to equation (3.1.58), $\lambda^2 k$ is equal to the diffusion coefficient D_i, thus:

$$\bar{v}_i = \frac{\mathrm{N}\,D_i c_i z_i \mathrm{e}\,\varDelta\psi}{kT} - \mathrm{N}\,D_i\,\frac{\mathrm{d}c_i}{\mathrm{d}y}. \tag{4.2.17}$$

When the velocity of ionic migration due to the electric field is significant, the second term of this equation is negligibly small in comparison with the first one, i.e. the migration velocity of the ith type of ions can be written as:

$$\bar{v}_i = \frac{\mathrm{N}\,D_i c_i z_i \mathrm{e}\,\varDelta\psi}{kT}. \tag{4.2.18}$$

On the other hand, expressing the migration velocity with ionic mobility U_i cm s^{-1}/V cm^{-1}, we obtain:

$$\bar{v}_i = \mathrm{N}\,c_i U_i\,\varDelta\psi. \tag{4.2.19}$$

These latter two equations give for the mobility of the ith type of ions:

$$U_i = \frac{D_i z_i \mathrm{e}}{kT}, \tag{4.2.20}$$

while the equivalent conductivity ($\lambda_i = \mathrm{F} U_i$) of the ith type of ions is:

$$\lambda_i = \frac{D_i z_i \mathrm{F}^2}{RT} \tag{4.2.21}$$

since $\mathrm{Ne} = \mathrm{F}$ and $\mathrm{Nk} = \mathrm{R}$ (R is expressed in Volt Coul. mole^{-1} degree^{-1} units).

According to experiments, the activation energies of migration of the majority of ions–calculated on the basis of the temperature dependence of mobilities–are approximately equal; in aqueous solutions they are about 4·0–4·2 kcal/mole at 25 °C. (Only hydrogen and hydroxide ions show some significant deviation from this value, their activation energies being much smaller: that of the hydrogen ion is 2·8 kcal/mole.) The activation energy of migration of the majority of ions does not differ too much from that of viscous flow. Eyring concluded from this that ionic migration velocity is, in fact, determined by the rate of jumps of the solvent molecules into the adjacent equilibrium positions. This is the elementary process of viscous flow as well. However, since solvent molecules move in the strong electric field of ions in the course of ionic migration, their activation energies are evidently not entirely the same as that of viscous flow of the pure solvent, and the rate constant k of the motion of ions is not completely identical with that of viscous flow. This is in agreement with experimental evidence, and it can be understood that the deviation is higher, the smaller the size of the ion and the higher its charge (i.e. the higher the electric field strength at the boundary of the ion).

In spite of the deviations expected in the above sense, it can be assumed in first approximation that the rate constant k for ionic migration is the same as that for viscous flow. If, furthermore, ions are not much larger than solvent molecules, it can be supposed that λ is the same in both processes (in this respect, λ represents the distance between two adjacent equilibrium positions of the particles). Thus, in equation (4.2.21) regarding ionic conduction, the diffusion coefficient can be inserted from equation (3.1.59), and we obtain:

$$D_i = \frac{\lambda_1}{\lambda_2 \lambda_3} \cdot \frac{kT}{\eta}, \tag{4.2.22}$$

i.e.

$$\lambda_i = \frac{\lambda_1}{\lambda_2 \lambda_3} \cdot \frac{kT}{\eta} \cdot \frac{z_i F^2}{RT}, \tag{4.2.23}$$

from which:

$$\lambda_i \eta = \frac{\lambda_1}{\lambda_2 \lambda_3} \cdot \frac{k z_i F^2}{R}. \tag{4.2.24}$$

Here, k is measured in erg degree^{-1}, R in Volt Coul. mole^{-1} degree^{-1}, so $k/R = 1 \cdot 66 \times 10^{-17}$. Further, since $F = 96 \cdot 500$ Coul., the above equation becomes:

$$\lambda_i \eta = 1 \cdot 55 \times 10^{-7} \frac{\lambda_1}{\lambda_2 \lambda_3} z_i. \tag{4.2.25}$$

According to the investigations on the self-diffusion of water: $\dfrac{\lambda_2 \lambda_3}{\lambda_1} \approx$

$\approx 1 \cdot 4 \times 10^{-7}$; consequently, the product $\lambda_i \eta$ does not differ too much from unity, which agrees with experimental results in its order of magnitude.

Of course, in reality, the rate constant of the translational movement of ions is probably somewhat lower than that of viscous flow, because the electric field of ions increases the free enthalpy of activation. In this way, $\lambda_i \eta$ is somewhat lower than the value obtained from equation (4.2.25), which gives an even better agreement between theoretically calculated value and the experimental one.

The ratios $\lambda_2 \lambda_3 / \lambda_1$ do not differ much from one another in different solvents, thus the product $\lambda_i \eta$ is almost independent of the solvent. This corresponds to the empirical *Pisarzhevskii–Walden rule* that is approximately valid for several solvents.

A large number of studies have dealt with the mechanism of ionic migration and electrolytic conduction; some of them have perhaps provided a

Literature on page 327

better understanding of one or other side of this problem. In our present view, however, the fundamentals of the Eyring theory are correct, though many questions regarding the details must be answered in the future. Recently, Horne et al. [24] have dealt in detail with the mechanism of transport of electricity in solutions, confirming the opinion that, in solutions of 1–1 electrolytes, the rate-determining step of ionic migration and electric conduction is vacancy formation in the solution in the vicinity of the moving ion. However, it is not known reliably how many water molecules are carried along by an ion when jumping into a vacancy. Starting with the Frank–Wen theory of water structure (cf. Section 1.3.3.8), it is assumed that in the water clusters including the ions, the cation replaces a water molecule. A vacancy, however, can be formed in the clusters, as well as between monomeric water molecules.

Some conclusions have been drawn by Swain and Evans [25] regarding the mechanism of ionic migration from the difference in conductivities of electrolytes in common and heavy water. The conductivity differences are attributed to the difference in the vibrational zero-point energies of hydrated H_2O and D_2O molecules. Supposing that alkali and halide ions hydrate four water molecules in the ground state, the two limiting cases of the exchange mechanism of the hydrate sphere of the migrating ion have been investigated:

(i) an ion moving 'from the left- to the right-hand side' gives a water molecule to the solvent at its left-hand side, then it is displaced slightly in the right-hand direction and takes a new water molecule into its hydrate sphere (elimination–addition mechanism which assumes three water molecules bound to the ions in the transition state);

(ii) first, the ion moves to the right-hand side carrying along but expanding its solvate sphere, then an additional water molecule is built in and a water molecule is lost at the left-hand side (addition–elimination mechanism which involves five water molecules bound to the ion in the transitional state).

Since, according to their investigations, the average vibrational frequency decreases in the transitional state, which indicates that water molecules in the hydrate sphere are bound in the transition state more loosely than in the ground state, they concluded that the addition–elimination mechanism is involved in ionic migration. In this case, the five water molecules of the transitional hydrate sphere are bound more loosely than in the ground state, which is assumed to involve four water molecules. If ions migrated according to the elimination–addition mechanism, the free water molecules of the hydrate sphere would be bound more tightly (their vibrational frequency would be higher) in the transition state than in the ground state.

4.2.2.2 THEORY OF PROTOTROPIC ('ANOMALOUS') CONDUCTION

The limiting value of the equivalent conductivity of ions is about 30–80 cm^2 Ω^{-1} at 25 °C in aqueous solutions, except that of hydrogen and hydroxide ions which have much higher conductivities ($\lambda_{H^+}^0 = 350$, $\lambda_{OH^-}^0 = 199$). These high conductivities–as emphasized particularly by Darmois

[26]–cannot be explained by hydrodynamic migration of hydrogen and hydroxide ions, since, calculating on the basis of Stokes' Law which is approximately valid for hydrodynamic migration, the radii of these ions ($r_{H+} \approx 0.2$ Å, $r_{OH-} \approx 0.4$ Å) are unreasonably small in comparison with the 3–5 Å values obtained for the radii of other hydrated ions in solutions, calculated in this way. The unreality of the radius of hydrogen ion calculated from Stokes' Law is evident, since no H^+ ions, i.e. bare protons, exist permanently in solution. However, in aqueous solution, the oxonium ion (H_3O^+) exists and in other solvents containing hydroxyl groups a similar derivative (e.g. in methanol, methyloxonium ion, $CH_3OH_2^+$, in ethanol, ethyloxonium ion, $CH_3CH_2OH_2^+$) is produced because of the formation of a donor–acceptor bond between protons and the lone-pair electrons of solvent molecules.

The oxonium ion can also be detected in the solid state. On the basis of crystallographic investigations on perchloric acid hydrate crystals, Volmer [27] established that there are H_3O^+ and ClO_4^- ions in the crystal lattice of this substance and this result has also been confirmed by Wicke, Eigen and Ackermann [28a]. The exact dimensions of the oxonium ion were determined recently by Lundgren and Williams [28b] by measuring the scattering of neutrons on a single crystal of p-toluenesulphonic acid monohydrate, in which the separately located cations are discrete H_3O^+ ions. The position of protons (hydrogen atoms) could thus also be accurately determined. According to these studies, the H_3O^+ ion has a pyramidal shape with an O–H distance of 1·01 Å, the angle in the $O\!\!\begin{smallmatrix}H\\H\end{smallmatrix}$ group being 110·4°, and the oxygen atom is located at a distance of 0·32 Å above the plane formed by the three protons. The size of the oxonium ion does not differ too much from that of water molecules and it can be assumed to be an intermediate between sodium and potassium ions. In the electrochemistry of aqueous solutions, when referring to the hydrogen ion, in fact the oxonium ion is always considered. Oxonium ions fit into the short-range lattice-like structure of liquid water almost in the same way as H_2O molecules do. H_3O^+ ions form bonds with the adjacent water molecules like other ions, and their electric field alters the water structure analogously. In other words, the oxonium ion is also hydrated. This is in agreement with several equilibrium properties of acid solutions; for example, the activity coefficients in solutions of LiCl, LiBr, LiI and $LiClO_4$ do not differ too much from those of the corresponding acids under comparable conditions, which indicates that, in the thermodynamic sense, the sizes of the hydrated Li^+ ion and hydrated oxonium ion are not much different.

The hydrodynamic mobility of the oxonium ion can be calculated approximately from the self-diffusion coefficient of water at 25 °C, $D^* \approx 2.25 \times 10^{-5}$ cm² s⁻¹ from the equation:

$$D^* = \frac{RT\,\lambda^0}{F^2}. \tag{4.2.26}$$

Literature on page 327

In this way, the equivalent conductivity of the oxonium ion, migrating only hydrodynamically, would be $\lambda^0 \approx 85$ cm^2 Ω^{-1} (instead of 350), i.e. it would not differ much from the equivalent conductivity of the K$^+$ ion ($\lambda^0 = 73\cdot5$).

Similarly, the equivalent conductivity of the hydroxide ion cannot be explained by hydrodynamic migration. The size of this ion is almost the same as that of the fluoride ion, so a pure hydrodynamic migration would result in an equivalent conductivity of about $\lambda^0 \approx 55$ (instead of 199).

On the basis of experiments on this problem [29a], it must be stated unequivocally that hydrogen and hydroxide ions–being the most important two ions because of the great practical importance of water–conduct electricity by a mechanism other than that of other ions. However, it can be supposed that, in addition to this particular mechanism, hydrodynamic migration of H$_3$O$^+$ and OH$^-$ ions also contributes to the transport of electricity.

The peculiar nature of the conduction mechanism of the two ions of water itself is beyond doubt; however, its actual characteristics are still argued about, and the corresponding studies have still not given any results unequivocally acceptable in all respects.

The theories developed to interpret the anomalous conductivity of hydrogen and hydroxide ions start with the fact that protons move not only by migrating together with the whole ion under the influence of the electric field, but also according to a *proton exchange (prototropic) mechanism* involving the transfer of protons from water molecules or ions to neighbouring water molecules or ions, respectively. This process can be described very schematically as follows:

$$\overset{+}{H-O}-H + O-H \rightarrow H-O + H-\overset{+}{O}-H, \qquad (4.2.27)$$

and

$$\overset{-}{O} + H-O \rightarrow O-H + \overset{-}{O}. \qquad (4.2.28)$$

The water molecule formed in the protonic jumps has an unfavourable orientation for the next protonic jump, therefore they must rotate before accepting a new proton. The mechanism of prototropic conduction is, thus, similar to a certain extent to that assumed by Grotthus for electrolytic conduction at the beginning of the last century. According to the theory of prototropic conduction, the protonic jumps between ions and water molecules occur continuously, without the existence of an electric potential gradient as well; however, the electric field makes these processes ordered so that they contribute to the transport of electricity.

In the conduction of electricity by the hydroxide ion, the conclusion of Zatsepina [29a] made on the basis of a survey of many data, should probably also be considered; according to this, the hydroxide ion exists as a H$_3$O$_2^-$ ion after attachment to one molecule of water by means of a strong hydrogen bond; in this complex, the distance between the two oxygen atoms is $2\cdot7$ Å.

The equations given above for conduction taking place by means of protonic jumps represent only the overall process, but they show no details of the mechanism. According to the theory of Hückel [30] the orientation of H_3O^+ ions in the direction of the electric field results in frequent protonic jumps in the direction of the lines of force. With respect to the relatively high inertia of the oxonium ion, it follows from the theory that this effect does not appear under high-frequency a.c. current, and the conductances of hydrogen and hydroxide ions decrease to the normal values. However, this is not observed in the experiments.

As suggested by Bernal and Fowler [31] and Huggins [32], prototropic conduction is connected with the short-range lattice-like structure of water that enables undisturbed proton exchange to take place along several molecules, i.e. over a larger distance. Electric field only orders the protonic jumps without requiring rotation of groups of higher inertia. According to this model, the anomalous conductivity should decrease with increasing pressure, since pressure causes depolymerization of water. In contrast with this, the conductivity of hydrogen and hydroxide ions was found to increase with increasing pressure. The theory of Bernal and Fowler supposes that this high mobility is due to the small thickness of the energy barrier between neighbouring equilibrium positions, and the proton moves in the direction of the electric field by means of the quantum mechanical tunnel effect. Since the tunnel effect depends strongly on the mass of the particles, the conductivity of D_3O^+ ions ought to be much smaller than that of H_3O^+ ions according to the theory mentioned above. However, the difference in the mobilities of the two kinds of ions is much smaller ($\lambda_{H+}/\lambda_{D+} = 1\cdot4$) than that expected on the basis of the theory (at least $\lambda_{H+}/\lambda_{D+} = 6$). Hence, it can be assumed that the energy barrier is rather broad and the tunnel effect does not prevail in the conductivity of hydrogen ions. A recent detailed analysis by Conway and Salomon [33] has also proved that the transfer of the proton from the H_3O^+ ion to a neighbouring H_2O molecule cannot be the rate-determining step in prototropic conduction, either in calculating on the basis of the tunnel effect or in applying the theory of classical proton transfer. In the classical theory, a conductivity ratio higher than the value determined experimentally would correspond to proton transfer acting as the rate-determining process. However, at high pressures and temperatures, it seems that protonic jumps can become the rate-determining factors [34].

On the basis of the absolute reaction rate theory, Stearn and Eyring [35] have dealt with the problem of prototropic conduction. In order to give a thorough theoretical treatment of the proton transfer from oxonium ions to neighbouring water molecules, the height and thickness of the energy barrier between neighbouring equilibrium positions ought to be calculated on the basis of quantum mechanics. This, however, cannot be done satisfactorily in complicated systems like aqueous solutions, all the less so since even the exact structure of water is under discussion. Therefore, Stearn and Eyring have calculated theoretically the dependence of conduc-

Literature on page 327

tivity on the height of the energy barrier which has been determined from measured conductivity values. They assumed in the calculations that in the Grotthus-type mechanism, the transfer of the proton from the oxonium ion to an adjacent water molecule is the rate-determining step, and that rotation of the water molecules to reach a favourable position is fast. The theory gives a reasonable description of the prototropic mechanism in many respects; however, in comparison with the experiments, it cannot be considered satisfactory.

The problem of prototropic conduction has been approximated from another side starting with the association theory of water (suggested by Eucken, cf. Section 1.3.3.7) by Gierer and Wirtz [36]. They assumed that proton transfer occurs not between separated molecules and ions, but within a complex consisting of water molecules bound together by hydrogen bonds ('water polymer') in such a way that the covalent bond of one of the protons of the oxonium ion built into the complex is transformed into a hydrogen bond, and this proton transforms a water molecule originally bound by a hydrogen bond into an oxonium ion:

$$\left[\cdots \text{H}-\overset{+}{\underset{|}{\overset{|}{\text{O}}}}-\text{H} \cdots \overset{\text{O}-\text{H}}{\underset{|}{\text{H}}} \cdots \right] \rightarrow \left[\cdots \text{H}-\overset{\text{O}}{\underset{|}{\text{H}}} \cdots \text{H}-\overset{+}{\underset{|}{\overset{|}{\text{O}}}}-\text{H} \cdots \right]. \quad (4.2.29)$$

Hydrogen bonds are always being formed and broken between adjacent water molecules (producing associates) independently of the external electric field. According to this theory, the H_3O^+ ions participate in the decomposition and formation of associates to the same extent as H_2O molecules do. An external electric field orientates this process to a certain degree in the direction of the lines of force: more H_3O^+ ions join the complexes on the side in the direction of the field, and at the opposite side more H_3O^+ ions leave them than in the other directions. In this mechanism, the formation of hydrogen bonds or the transfer of protons along these bonds is the rate-determining step. The theory of Gierer and Wirtz, however, cannot distinguish between these two possibilities.

In order to explain the difficulties occurring in the interpretation of the infra-red absorption spectra of aqueous solutions, Wicke, Eigen and Ackermann [37] started with $H_9O_4^+$ complexes in their theory, and they assumed that the proton jumps from the H_3O^+ ion to a water molecule in the secondary hydrate sphere around it. The rate-determining process is the rearrangement of the electron shell of the new H_3O^+ ion. According to this theory the proton, which moves rather freely, vibrates several times between the primary H_3O^+ and the secondary hydrate sphere before the centre of the complex as a whole is shifted to a distance equal to the length of a hydrogen bond. The conduction mechanism of hydroxide ions is similar, according to Ackermann [38], the proton transfer taking place in the $H_7O_4^-$ complex.

A detailed theoretical investigation of the mechanism of prototropic conduction has been outlined by Conway, Bockris and Linton [39], particularly with respect to the rearrangement of water molecules after the protonic jump. Calculating the transport velocity for several possible mechanisms, they concluded that the rate-determining step is the rotation of solvent molecules, acting as electric dipoles in the first solvate sphere, which is required to make possible continuous proton transfer. In comparison with the rate of this rotation of the solvent molecules, the proton transfer itself is a fast process along the hydrogen bonds between oxonium ions and the neighbouring solvent molecules. The same is true of the rate-determining process in conduction taking place by means of hydroxide ions. The force causing this rotation of water molecules is partly due to the effect of ions on the dipole molecules of the solvent in the conduction by means of H_3O^+ ions, and partly to the repulsion between hydroxyl groups facing each other in the course of the preceding proton transfer. According to the calculations regarding this phenomenon, part of the force causing rotation arises from this latter effect in aqueous solutions. On the other hand, in proton transfer in the course of conduction by means of OH^- ions, two OH bonds will not face each other, thus the force causing rotation of water molecules arises only from the effect of OH^- ions on the dipole molecules of the solvent. Since, as supposed in the theory, the mobilities are proportional to the square root of the forces acting in water, it can be expected that part of

conductivity of hydroxide ions due to proton transfer is $\sqrt{\dfrac{1}{4}} = \dfrac{1}{2}$ times

that of the hydrogen ion, which corresponds approximately to the results of experiments. The evaluation of the experimental data in this respect, however, can only be made in approximation because the prototropic part of conductivity cannot be measured separately. Probably no appreciable deviation from reality is caused when the difference in the equivalent conductivities of hydrogen and potassium ions is considered, in first approximation, to be the prototropic portion of the conductivity of the hydrogen ion, and that of hydroxide ion is considered to be the difference in the conductivities of hydroxide and fluoride ions (these differences are sometimes called *extra-conductivities*). The Conway–Bockris–Linton theory elucidates several problems, but it still cannot be regarded as proved. Neither the studies of Perrault [40a], Broadwater and Evans [40b] nor the measurements of Lin [40c] on the entropy of hydrogen ion transport give a final solution to the problem.

According to the theories on conductivity by hydrogen ions, H_3O^+ ions are formed only temporarily in connection with proton transfer. With respect to this, it seems to be justified to ask whether the hydrodynamic migration of oxonium ions also contributes to the conductivity of 'hydrogen ions'. This can be decided on the basis of the ratio of the time required for proton transfer and the average lifetime of H_3O^+ ions. Proton transfer needs a time of about 10^{-14} s, according to the corresponding theories, while the average lifetime of H_3O^+ is about 10^{-11}–10^{-12} s. Thus, the proton is bound to a given water molecule for most of its lifetime (about 99 per cent), i.e. it is present as an H_3O^+ ion, and the transfer from one water molecule to another takes only a relatively short time. Further, since the H_3O^+ ion, resting with respect to the electric field in the moment of its forma-

Literature on page 327

tion, reaches the final velocity corresponding to the field strength within a time in the order of magnitude of 10^{-14} s under the normal conditions of conductivity measurements, as given by the hydrodynamic calculations, it should be concluded that the H_3O^+ ions take part in the conduction of electricity without appreciable relaxation under normal conditions. Thus, the electric conductivity measured represents approximately the sum of the velocity of the hydrodynamic migration of oxonium ions and of the frequency of proton transfers. Similar conclusions hold for the hydroxide ion as well. Since, with respect to several properties, the hydrodynamic mobility of H_3O^+ ions does not differ too much from that of K^+ ions [41a], or that of OH^- ions from that of F^- and Cl^- ions, the calculation of conductivity due to the prototropic mechanism is approximately correct when subtracting the equivalent conductivities of K^+, Cl^-, or F^- ions, from the experimentally determined equivalent conductivities of the hydrogen and hydroxide ions respectively, measured under the same conditions.

Recently, on the basis of their theory elaborated for transfer diffusion (cf. Section 3.1.2.2), Ruff and Friedrich [41b] arrived at the conclusion that the very rapid second-order exchange reaction

$$H_3O^+ + H_2O \rightleftharpoons H_2O + H_3O^+$$

completely provides for the conduction of electricity by 'hydrogen ions', and the hydrodynamical migration of H_3O^+ ions has no considerable role. However, Lengyel [41c] reviewing an erroneous assumption in the theory, has shown that the rate of the transfer mechanism involving the above exchange reaction is lower than the value calculated by Ruff and Friedrich, therefore the contribution of the hydrodynamical migration of H_3O^+ ions to the conduction of electricity, besides the prototropic mechanism, can be assumed in this theory, too.

In pure aqueous solutions, the equivalent conductivity, i.e. the hydrodynamic mobility of H_3O^+ and OH^- ions, cannot be measured directly. It becomes, however, measurable in approximation, when prototropic conduction is eliminated by adding some foreign solutes which do not conduct electricity or, at least, efficiently decrease its possibility. Investigating experimentally the transference number of electrolytes dissolved in water–methanol mixtures, Erdey-Grúz and Majthényi [42] have established that, in a 0·02 mole l^{-1} hydrochloric acid solution containing 90 mole per cent methanol, the transference number of hydrogen ion is $t_H^{25°} = 0·560$ and $t_H^{5°} = 0·538$ at 25 and 5 °C, respectively, while the values measured in pure aqueous solutions are $t_H^{25°} = 0·827$, and $t_H^{5°} = 0·840$. Thus, under such conditions, the equivalent conductivity of the hydrogen ion is only slightly higher than that of the Cl^- ion, i.e. the conductivity of hydrogen ions arises primarily from their hydrodynamic migration. This can evidently be attributed to the fact that methanol destroys the lattice-like structure of water and water polymers bound together by hydrogen bonds, and thus the possibility of proton transfer from one water molecule to another is eliminated or, at least, it is greatly decreased. By a further increase in the methanol concentration, the transference number of hydrogen ions again starts to increase in hydrochloric acid solutions, since proton transfer starts between methanol molecules:

$$CH_3OH_2^+ + CH_3OH \rightarrow CH_3OH + CH_3OH_2^+ . \qquad (4.2.30)$$

In the solution corresponding to the minimum in the transference number, this process is not significant, since water has a higher proton affinity than methanol [37]; consequently the equilibrium

$$H_3O^+ + CH_3OH \rightleftharpoons H_2O + CH_3OH_2^+ \qquad (4.2.31)$$

is strongly shifted to the left-hand side even in the 90 mole per cent alcoholic solution that makes the hydrodynamic migration of the H_3O^+ ions almost undisturbed. However, by a further decrease in the concentration of water, the equilibrium shifts to the right-hand side so that prototropic conduction by means of methanol molecules becomes possible to an increased extent.

According to Erdey-Grúz and Majthényi [42], the transference number of hydroxide ions also decreases on the addition of methanol to aqueous solutions of KOH; the values of $t_{OH}^{25°} = 0.733$ and $t_{OH}^{5°} = 0.737$ corresponding to pure aqueous solutions will become $t_{OH}^{25°} = 0.497$ and $t_{OH}^{5°} = 0.475$ in 99 mole per cent methanolic solutions. Under such conditions, the equivalent conductivity of hydroxide ions is, thus, lower than that of potassium ions. The transference number of hydroxide ions decreases continuously with increasing methanol content, so the prototropic conduction mechanism of hydroxide ions by means of methanol molecules does not take place.

The studies of Erdey-Grúz, Kugler and Majthényi [43] have shown that the effect of other monovalent alcohols on the transference number and conductivity of hydrogen and hydroxide ions is similar in aqueous solutions although the effect of polyvalent alcohols exhibits other features.

Dioxan also restricts prototropic conduction in aqueous solutions, according to Erdey-Grúz and Nagy-Czakó [44a], to such an extent that the transference number of hydrogen ions rapidly falls with increasing dioxan content. In aqueous hydrochloric acid solutions containing more than 50 mole per cent of dioxan; in the presence of 60 mole per cent of dioxan, the conductivity of hydrogen ions is lower than that of chloride ions ($t_H^{25°} = 0.42$). Thus, dioxan destroys the water structure more efficiently than alcohols, which is not surprising in view of its molecular structure. Prototropic conduction through dioxan molecules does not seem to occur.

Other phenomena also indicate that H_2O molecules separated from the normal liquid structure of water do not participate in prototropic conduction. For example, Gusev and Palei [44b] as well as Gusev [44c] investigated the effect of non-electrolytes on the conductivity of aqueous solutions of electrolytes. They have inferred that water molecules, being in the hydrate spheres of ions (that is, separated from the original liquid structure of water to a certain extent), do not participate in the prototropic conduction mechanism.

Literature on page 327

From the effect of non-electrolytes dissolved in water on prototropic conduction, it can be concluded that the presence of water is not sufficient for the transport of electricity according to this mechanism; it is also required that some complexes (associates, polymers, lattice-like regions) formed from a certain number of water molecules should be present through which proton transfer takes place undisturbed. The decrease in the size of such complexes hinders the conduction by the prototropic mechanism, and monomeric water molecules seem to be unsuitable for prototropic conduction. This is also supported by the fact that the ratio $(\varLambda^{\circ}_{HCl} - \varLambda^{\circ}_{KCl})/\varLambda^{\circ}_{KCl}$, which provides some approximate information on the ratio of prototropic conduction, is 2·26, 1·84, and 1·07 at 0, 25, and 100 °C, respectively. In this way, the ratio of prototropic conduction in the transport of electricity is strongly decreased by increasing temperature, which is probably due to the loosened connection between water molecules at elevated temperatures, i.e. the size of the regions bound together by hydrogen bonds decreases and the liquid structure of water is destroyed. With increasing pressure, however, the ratio of prototropic conduction increases. According to Hamann [45], the ratio given above is 2·15 at 25 °C and 3000 atm. Thus, pressure has a structure-making effect on water within certain limits.

For the role of monomeric water molecules in the prototropic conduction mechanism, Horne and Courant [46a] have drawn dissimilar conclusions from their conductivity measurements in aqueous hydrochloric acid solutions between −1 and 10 °C. They have established that the activation energy of protonic conduction is approximately constant, unlike that of other ions which exhibit a maximum at the temperature of the density maximum. According to them, vacancies determining the hydrodynamic migration can be formed in both the Frank–Wen-type molecular clusters and the monomeric water between them, while the rotation of water molecules required for prototropic conduction can occur only in the monomeric state. In general, this latter mechanism is predominating, but, below 2 °C, the ratio of hydrodynamic conduction increases. Investigating the effect of high pressures on conductivity, they have concluded that pressure breaks hydrogen bonds, which facilitates the rotation of the H_2O and H_3O^+ species. It is also in connection with this that the conductivity of HCl solutions increases with pressure more rapidly than that of KCl solutions. They assumed that, under a pressure of 1500 bar, the quantum mechanical tunnel effect will be the rate-determining step of conduction.

In connection with the anomalous conduction mechanism of hydrogen ions, Naberuzhin and Shuiskii [46b] have recently investigated what theoretical conclusions can be obtained on the basis of two theoretically possible assumptions. According to one the proton takes a series of jumps along the hydrogen bonds, while, in the other assumption, it is assumed that protons run across the whole quasi-crystalline micro-region in one single elementary action similarly to the Grotthus model. They have concluded that neither of these models is in contradiction with experiments so far, i.e. information at present does not allow a decision as to which of the two alternatives reflects reality better.

The ratio of conduction by the prototropic mechanism to that by the hydrodynamic one depends nevertheless in a complicated way on temperature, pressure, and on concentration as well. According to the investigation of Lown and Thirsk [46d] in the 25–200 °C and 1–3000 atm. range, the conduction due to proton transfer is reduced progressively with increasing concentration in the 1–6·6 mole l^{-1} range in LiOH, NaOH and KOH solutions, and the hydrodynamic migration becomes more and more predominating. This has been explained by the fact that an increasing portion of the water molecules is bound in the hydrate spheres of the ions with increasing concentration, and these molecules thus become unsuitable for the conduction mechanism via proton transfer of hydroxide ions. The prototropic conduction mechanism of KOH solutions breaks down at a concentration lower than that of LiOH solutions, which can be ascribed to the more extended ionic association in the latter case. In KOH solutions, the Walden product becomes more and more independent of temperature and pressure with increasing concentration which also indicates the suppression of the prototropic mechanism. The conductivity of phosphoric acid solutions increases with increasing pressure, if the concentration is low, while the conductivity decreases with increasing pressure in medium concentrations; however, it becomes independent of pressure in solutions in the 11·7–15·7 mole l^{-1} concentration range, up to 3000 atm. pressure. This can be attributed to the fact that water molecules bound in the hydrate spheres do not participate in the prototropic conduction mechanism. Lown and Thirsk also pointed out that there is a certain similarity between the prototropic and hydrodynamic conduction mechanisms, namely, both of them are accompanied by the rupture of hydrogen bonds. Maybe this explains the (not trivial) fact that the Debye–Hückel–Onsager theory holds in good approximation for dilute solutions of both KCl and HCl.

On the basis of recent quantum mechanical investigations on the transfer of protons between water molecules by means of the tunnel effect, Weidemann and Zundel [46c] established that the anomalous conductivity of acid solutions can be attributed to 'structural migration' of $H_5O_2^+$ and $H_9O_4^+$ groups in water. Essentially, structural migration is the transfer of the proton, causing an excess in charge, from an H_2O molecule to a neighbouring one within these groups. They have shown that the polarizability of the hydrogen bond involved in the tunnel effect in the $H_5O_2^+$ group is about 100 times higher than that of the free H_3O^+, therefore the proton transfers within the $H_5O_2^+$ groups–which occur frequently, with no favoured direction in the absence of an external field–are orientated by the external field much more than the transfers to free H_2O molecules. In this way–unlike earlier models–the rate-determining step of prototropic conduction is protonic tunnelling within the group $H_5O_2^+$ and no free H_3O^+ ions participate in it.

According to the general picture regarding the prototropic and hydrodynamic mechanisms, the structure-breaking effects on water decrease the participation of proton transfer in conduction and increase the hydro-

Literature on page 327

dynamic mobility, whereas the structure-making effects have an opposite influence. The investigations of Roberts and Northey [46e] on solutions containing Cd^{2+} ions, however, are not in agreement with this model. Furthermore, they have observed that in the presence of alkali halides the effect of anions on protonic mobility is larger than that of cations.

Hydrodynamic theory of anomalous conduction. The common basis of the theories on the anomalously high equivalent conductivity of hydrogen and hydroxide ions described above is the assumption that this phenomenon can be somehow attributed to the relayed transport of protons by water molecules. These theories explain several experimental facts, but there are several observations that are not sufficiently interpreted by them. One of them is, for example, that the vibration frequencies of H_3O^+ and OH^- are not too different from each other, according to Bader [47]; this is not in accordance with the view that the rate-determining step in anomalous conduction is the rotation of the species. It is also to be noted that H_3O^+ has a pyramidal shape, as stated by Richards and Smith [48] by reason of NMR measurements, and its electronic structure (the configuration of the maximum in charge density) differs appreciably from that of water, while the structure of OH^- is more similar to that. Thus, it could be expected that the activation energy of the formation of H_3O^+ is higher than that of the OH^- ion, which, however, is in contradiction with the ratio of mobilities observed empirically for these two types of ions. The fact established by Eigen and De Maeyer [49] that the mobility of hydroxide ions in ice is lower than that of hydrogen ions by about two orders of magnitude is also to be interpreted. Starting with these conditions, Zatsepina [50a] has developed a new theory of anomalous conduction assuming that the high conductivity of hydrogen and hydroxide ions is brought about, not by the relayed prototropic conduction mechanism, but by some structural reasons which are responsible for the high hydrodynamic mobility of these ions. Of the ions formed in autoprotolysis (dissociation) of water according to the following equation:

$$2H_2O \rightleftharpoons H_3O^+ + OH^-$$

the electronic structure of H_3O^+ ions differs essentially from that of H_2O molecules in ice because of the pyramidal shape of the former ions; therefore, the hydroxonium ion is hydrophobic, non-hydrating, and does not form hydrogen bonds with water. In the hexagonal structure of ice, each H_2O molecule is surrounded by six structural cavities with sizes larger than those of water molecules. The formation of an H_3O^+ ion different in its structure from that of water is accompanied by the rupture of all the hydrogen bonds binding it to the lattice. In this way, the H_3O^+ ion gets into a structural cavity, and—owing to its hydrophobic nature—it can reach the neighbouring cavity easily. Thus, it can migrate with a high mobility from one cavity to another until it recombines to water on meeting an OH^- ion. Since the electronic structure of the OH^- ion is similar to that of the H_2O molecule, not all of its hydrogen bonds are broken and the ion remains essentially in the crystal lattice. Its displacement requires a high activation energy, since it involves breaking the hydrogen bonds fixing it to the lattice. According to Zatsepina, the hydrogen ion migrates similarly in liquid water and in ice and it is supposed to be hydrophobic and non-hydrating even under such conditions. Though it is in contradiction with the opinion of several research workers, the interpretations regarding the corresponding experimental data are so varied and contradictory that the above assumption cannot be excluded. Due to the small size of the non-hydrated H_3O^+ ion, it can migrate with a rather high mobility through the structural cavities in liquid water, since its hydrophobic nature prevents it reaching such quasiequilibrium positions in which it could reside for a relatively long time.

According to Zatsepina [50], the hydroxide ion forms a $H_3O_2^-$ ion with one water molecule ($H-O-H-O-H^-$) in which the $O-O$ distance is much smaller than in a usual hydrogen bond. The electronic configuration of the $H_3O_2^-$ ion is also different from that of the H_2O molecule, and this ion is also hydrophobic and forms almost no hydrogen bonds. The $H_3O_2^-$ ion migrates 'freely' according to the hydrodynamic mechanism, and its decreased mobility, as compared with that of H_3O^+ ion, can be attributed only to its larger mass.

Zatsepina's theory elucidates several relationships, but several assumptions are implied in it which have not been confirmed so far. Although it can be accepted only with reservations, it nevertheless indicates that the interpretation of the anomalous conductivity on the basis of a prototropic mechanism is not the only possibility, all the same this theory cannot be considered to be unequivocally refuted.

Conduction by protons can also take place in the solid state. After surveying the available data, however, Bruinink [50b] came to the conclusion that there is no unquestionable evidence for protonic conduction in the majority of the so-called protonic conductors. Solids in which the protons migrate rapidly would be, however, very advantageous for use in fuel cells.

4.2.3 KINETIC THEORIES OF ELECTROLYTIC CONDUCTION, II.
THEORY OF ELECTROSTATIC INTERACTION OF IONS

The molecular theory of ionic migration provides some information on the mechanism of transport of electricity in electrolyte solutions, although its conclusions on the actual properties agree only in the order of magnitude with the experimental data. The accurate calculation of the decisive quantities in the conduction of electricity has not been possible on the basis of this theory.

With respect to the transport of electricity by electrolytes, the degree of dissociation and the limiting value of ionic mobility (the equivalent conductivity of ions) in infinitely dilute electrolyte solutions (λ_i^0) are the most important parameters. The degree of dissociation or, from another point of view, the degree of association, represents the fraction of the dissolved electrolyte suitable for the transport of electricity (i.e. present

Literature on page 328

in ionic form); on the other hand, mobility with a specific velocity characteristic measures the frictional force, i.e. it accounts for the amount of electricity transported per second by a given amount of ions under the effect of a given electric potential gradient.

The limiting value of ionic mobility in infinitely dilute solutions is determined by the interactions between a given type of ion and the solvent, and this limiting value is independent of the other ions present simultaneously, for a given ion and solvent. This condition is expressed by the *Kohlrausch law of independent ionic migration*. This statement, however, does not hold for solutions of finite dilution because, under such conditions, mobility is also influenced by the electrostatic interaction of the dissolved ions and not only by the interaction between ions and the solvent.

The limiting value of mobility is, after all, an empirical quantity which still cannot be determined directly in experiments, since the equivalent conductivity varies with concentration even in the most dilute solutions suitable for accurate measurements because of certain technical reasons. The value of λ_i^0 can be determined only by extrapolating the equivalent conductivity and transference number of the electrolyte in finite concentrations, to zero concentration. For reliable extrapolation, however, the concentration dependence of these parameters should be known. In this respect the most important one is the concentration dependence of equivalent conductivity (i.e. the sum of ionic mobilities), since the transference number, being the ratio of the ionic mobilities, changes with concentration less than equivalent conductivity does.

4.2.3.1 WEAK AND STRONG ELECTROLYTES

According to Section 4.1.2, the equivalent conductivity of solutions of binary electrolytes in an arbitrary concentration c is:

$$\Lambda = \alpha(\lambda_c + \lambda_a), \tag{4.2.32}$$

where both the degree of dissociation α and the equivalent conductivity of ions depend–though differently–on concentration.

In solutions, weak electrolytes are present mainly as non-dissociated molecules containing covalent bonds. The majority of weak electrolytes are either an acid or a base (HA or BOH, where A denotes the anion of the acid and B the cation of the base); salts of weak electrolyte character occur very rarely. In the solution of a weak acid, only a small portion of the molecules dissociate under the effect of the proton affinity of water:

$$\text{HA} + \text{H}_2\text{O} \rightleftharpoons \text{H}_3\text{O}^+ + \text{A}^-.$$

Such a dissociation is a chemical process, and the equilibrium constant of the law of mass action is valid for this equilibrium (considering dilute solutions in which the activity and concentration of water can be regarded as constant):

$$K = \frac{[\text{H}_3\text{O}^+][\text{A}^-]}{[\text{HA}]} = \frac{\alpha^2 c}{1 - \alpha}. \tag{4.2.33}$$

If the dissociation constant is small enough, ($K < 10^{-4}$ mole 1^{-1}), the concentration of ions in dilute solution ($c < 10^{-1}$ mole 1^{-1}) will be so low that their mobilities will be independent of concentration and equal to the limiting value corresponding to infinite dilution, in first approximation. Under such conditions:

$$\Lambda = \alpha(\lambda_c^0 + \lambda_a^0). \tag{4.2.34}$$

Since:

$$\Lambda^0 = \lambda_c^0 + \lambda_a^0, \tag{4.2.35}$$

the degree of dissociation of the solution is, in first approximation:

$$\alpha = \frac{\Lambda}{\Lambda^0}. \tag{4.2.36}$$

Although the Λ^0 value of weak electrolytes cannot be determined by direct extrapolation either, it can be calculated from the ionic mobilities determined in measurements on the corresponding strong electrolytes. Thus, on the basis of the law of mass action, we obtain:

$$K = \frac{\Lambda^2 c}{\Lambda^0(\Lambda^0 - \Lambda)}. \tag{4.2.37}$$

This relationship (the *Ostwald dilution law*) describes the conductivity of solutions of weak electrolytes as a function of concentration. Since dissociation of covalent molecules is a chemical process, the dissociation constant depends on the chemical composition of the electrolyte.

The ratio of the concentration of ions moving relatively freely and the total concentration of the electrolyte can be altered not only by the formation or dissociation of covalent molecules, but also by ion pair formation, due to electrostatic attraction between oppositely charged ions (*association*, cf. Section 5.1.5). Ion pair formation should be accounted for mainly in solutions of strong electrolytes, but it takes place only to a small extent in dilute solutions. Hence, this process is often considered not from the point of view of neutral species (speaking of the degree of dissociation), but from the point of view of an ionic state giving the degree of association, δ. In many cases, it is difficult to distinguish between dissociation and association in such a sense [1]. The law of mass action is valid also for association, because the thermodynamic relationships this law is based on are independent of the nature of the forces keeping the particles together.

The conductivity of weak electrolytes is important with respect to the equilibrium conditions rather than to the transport processes, thus mainly the solutions of strong electrolytes will be dealt with here.

In aqueous solutions of about 0·01 M concentration and at normal temperature, the equivalent conductivities of strong 1–1 electrolytes are by about five to seven per cent lower than the limiting values corresponding to infinite dilution. It can be concluded from this that, under such condi-

Literature on page 328

tions, strong electrolytes are present predominantly in the form of rather independent ions. One of the fundamental problems of the theories of strong electrolyte solutions is the ratio of the changes of conductivity in dilute solutions arising from the change in the degree of dissociation or association and that due to changes in ionic mobilities.

In first approximation, this question can be answered unequivocally. Namely, if the change in conductivity were resulting primarily from the change in the degree of dissociation, the equivalent conductivity in dilute solutions should show a linear concentration dependence according to the Ostwald dilution law, since, if Λ does not differ too much from Λ^0, Λ will be only slightly altered by further dilution, and Λ^2 is approximately constant, i.e.:

$$\Lambda \approx \Lambda^0 - \text{const } c . \tag{4.2.38}$$

However, this is in contradiction with the Kohlrausch empirical law according to which the equivalent conductivity of dilute solutions of strong electrolytes depends linearly on the square root of concentration (cf. equation (4.1.19)). This contradiction is also revealed by the fact that the K values calculated on the basis of the Ostwald dilution law are not constant, even in approximation, and even in the most dilute solutions of strong electrolytes. This leads to the conclusion that strong electrolytes are almost completely dissociated in the most dilute solutions and the change of equivalent conductivity with dilution is caused mainly by the change in ionic mobilities. The question whether ionic association in a small degree still takes place in dilute solutions can be answered only in very careful and accurate investigations (cf. Section 4.2.3.4). Anyhow, this process can become significant only in more concentrated solutions.

In addition to conductivity, some other phenomena also prove that strong electrolytes are almost completely dissociated in dilute solutions, and no appreciable influence on the properties of these solutions is exerted by non-dissociated molecules or ion pairs incidentally present. The most direct evidence for this are the optical properties of dilute solutions of electrolytes. It has been shown by the studies of Halban [51], Kortüm [52] and others that, in solutions of typical strong electrolytes, no lines or bands can be found in the absorption or Raman spectra that could be attributed to non-dissociated molecules. On the other hand, in dilute solutions containing coloured ions (Cu^{2+}, Co^{2+}, Ni^{2+}, MnO_4^-, CrO_4^{2-}), light absorption is proportional to the total concentration of the electrolyte, which is only possible in the case when the electrolyte is almost completely dissociated. According to Bjerrum and Nielsen [53], the molar extinction coefficient of strong electrolytes is independent of concentration in the $10^{-4}-10^{-2}$ mole l^{-1} range, while the equivalent conductivity changes by several per cent. In solutions of the same concentration of weak electrolytes, however, the molar extinction coefficient varies appreciably, which confirms that the concentration of non-dissociated molecules changes in comparison with that of the ions. However, for the detection of molecules present in low concentration, optical methods are not sensitive enough. In solutions of ions of higher valence, the formation of covalent molecules and ions (e.g. $PbCl^+$, $CaCl^+$)

can be appreciable, as shown by the spectrophotometric measurements of, for example, Fromherz [54], and Monk [55] and their co-workers.

The majority of substances behaving as strong electrolytes in solution (mainly salts) form ionic crystals or melts in the pure state. The dissolution of these so-called *true electrolytes* is, thus, not more than the separation of ions originally present and held together by coulombic forces by the action of solvent molecules. These solutions are usually quoted as *ideal electrolytes*; the electrostatic interaction of ions predominates in them, thus ion pairs can be formed in a low degree, but no covalent molecules are, in general, formed. In solvents of low dielectric constant, however, ionic association can be appreciable even in dilute solutions.

Compounds belonging to another group of substances behaving as strong electrolytes in solution are mainly strong acids, consisting of covalent molecules in the pure state, which are ionized only in the course of the chemical interaction with the solvent i.e. the decomposition of covalent molecules and the formation of new ones; they are known as *potential electrolytes*. For example, ionization is brought about by the reaction of covalent hydrochloric acid molecules with the covalent molecules of water:

$$HCl + H_2O \rightleftharpoons H_3O^+ + Cl^-.$$

These two types of strong electrolytes hardly differ from each other with respect to the concentration dependence of conductivity in their most dilute aqueous solutions. The square root law is equally valid in good approximation for the most dilute solutions of both hydrogen and potassium chloride. In both groups, the prevailing effect (in addition to hydration) is the electrostatic interaction of ionic charges, which decreases ionic mobilities with increasing concentration. However, in somewhat more concentrated solutions, some noticeable deviations can appear. For example, in the concentration range 0·001 to 0·5 mole l^{-1}, the conductivity of HCl solutions changes linearly with $\sqrt[3]{c}$ which is a consequence of the prototropic conduction mechanism, as suggested by Horne [56a]. The high degree of ionization in aqueous solutions of strong acids has been proved by the Raman and NMR measurements of Akitt, Covington, Freeman and Lilley [56b] which indicated that perchloric acid is completely dissociated even in a solution of 6 mole l^{-1} concentration at 25 °C, and the degree of dissociation is not lower than 90 per cent even in 12 mole l^{-1} concentration.

Free uncharged complexes (or those of smaller charge) can be formed to a lesser extent from ions in diluted solutions of strong electrolytes as well. In solutions of true electrolytes, these are ion pairs kept together mostly by electrostatic attraction, while in solutions of potential electrolytes covalent molecules can also be formed instead of or in addition to these. Both processes are accompanied by a decrease in the number of charge carriers and the law of mass action is valid for both processes, thus they influence the concentration dependence of conductivity in a similar way. Hence, on the basis of measuring the concentration dependence of conductivity, the formation of

Literature on page 328

a real molecule or that of an ion pair (association) cannot be distinguished unequivocally. This is also difficult to do by other methods because most of the properties suitable for this (in addition to conductivity, e.g. the osmotic and potentiometric parameters) also depend on the general electrostatic interaction of the ions, which makes the conditions very complicated. Optical parameters, however, sometimes make it possible to distinguish between these two types of ionic associations, because optical properties are significantly altered by covalent bonds which can usually be measured well.

Summarizing all the ideas in connection with these phenomena, it can be established that the concentration dependence of the equivalent conductivity of the most dilute aqueous solutions of strong electrolytes can be attributed mainly to the alteration of ionic mobilities resulting from the electrostatic interaction of their charges. The theory on the conductivity of dilute solutions of strong electrolytes and on its extrapolation to infinite dilution is based on the theory of electrostatic interaction of ions. The acceptability of this idea is not altered by the statistical theory of electrolytes outlined recently by Martinov [57], either.

4.2.3.2 FUNDAMENTALS OF THE DEBYE—HÜCKEL—ONSAGER THEORY OF STRONG ELECTROLYTES; THE VARIATION OF CONDUCTIVITY WITH CONCENTRATION; THE LIMITING LAW

The molecular state of solutions of strong electrolytes differs from that of non-electrolyte solutions, owing to the effects of the electrostatic field of the ions. The electric field modifies the water structure locally, and it can attach water molecules to the ions, i.e. it can produce hydration (cf. Section 5.2.). In sufficiently dilute solutions, however, where the excess of water molecules is very high in comparison with the ions, the hydration conditions do not change markedly with electrolyte concentration, and their effect is reflected in the limiting value of the equivalent conductivity (Λ^0 and λ^0). Under such conditions, only the average effect of the mutual repulsion and attraction of ions varies appreciably with concentration, since the effective range of the interactions responsible for this is much wider than for those causing hydration.

In electrolyte solutions, coulombic interactions, keeping ions of identical charge apart from one another and attracting those of opposite charge, operate against the thermal motion that acts towards uniform distribution of ions. The still up-to-date theory regarding the distribution due to these various effects has been elaborated by Debye and Hückel [58]. According to their theory (for details, see Section 5.1), the characteristic differences between the properties of dilute solutions of strong electrolytes and those of non-electrolytes can be attributed to the fact that in the vicinity of each ion, ions of opposite charge are in excess on the statistical average: every ion is surrounded by an *ionic atmosphere* of opposite average charge. Each ion included in this ionic atmosphere is, of course, the centre of another ionic atmosphere and thus, they will penetrate into one another. Macroscopically, the charge distribution is uniform, each region of the solution which is large enough with respect to the molecular sizes and fixed (e. g. to the vessel containing the solution) is electrically neutral on the statistical average. The charge distribution corresponding to the ionic atmosphere can be revealed only in a coordinate system fixed with respect to the individual ions and moving together with them.

In the absence of an external electric field, i.e. when the ion is immobile (more exactly, it exhibits only thermal motion), its ionic atmosphere is spherically symmmetrical, so the resultant effect on the central ion is zero. If, however, ions are translated under the effect of an external electric force, the conditions are altered. These alterations can be classified into two groups. One of them can be attributed to the fact that the central ion and ions of opposite charge in the ion cloud migrate in opposite directions under the effect of the electric potential gradient, and they also transport some of the water molecules present in their environment. Since each ion is at the same time a central ion and a member of an ionic atmosphere, this effect is responsible for the fact that ions move in a medium that is not immobile, but is one flowing in a direction opposite to that of their translational movement. Consequently, when measuring their migration velocity (and simultaneously, their conductivity) in an external coordinate system fixed, e.g. to the vessel, the value obtained is lower than that expected in the absence of this phenomenon *(electrophoretic effect)*.

The other group of consequences of the translational movement of ions under the effect of an external field can be attributed to the deformation of the ionic atmosphere, due its finite relaxation time, and its resultant effect on the central ion which is a force acting in a direction opposite to that of the movement of the ion itself *(relaxation effect)*. Namely, the ion migrates towards places to which its own ionic atmosphere has not been extended prior its displacement; thus, a new ionic atmosphere of the ion under consideration should be formed here, while behind it, a part of the original ionic atmosphere should diminish because the ions involved will diffuse away. As the relaxation time is finite, both processes are delayed and the concentration of oppositely charged ions is lower before the migrating ion and higher behind it as compared with the equilibrium state (i.e. the spherically symmetrical charge distribution); under the given conditions, the ionic atmosphere is deformed.

Both electrophoretic and relaxation effects decrease the mobility of ions, i.e. they appear formally as an increased frictional resistance.

In order to calculate the *electrophoretic effect*, Onsager and Fuoss [59] applied the distribution function of ions deduced in the Debye–Hückel theory, and they started by considering (cf. Section 5.1.2) that, if the solution as a whole is immobile, the forces f_c and f_a acting on cations and anions, respectively, and the force f_w acting on water molecules should be equal., i.e. if the numbers of cations, anions and water molecules are n_a, n_c and n_w respectively, in unit volume of the bulk of the solution:

$$n_w f_w = -n_c f_c - n_a f_a.$$ (4.2.39)

Literature on page 328

The effect of a spherical shell of thickness dr at distance r from the central ion on the movement of the ion can be calculated by taking into account that, due to the ionic atmosphere, the concentration of ions here differs from that in the bulk of the solution far from this ion. If ψ_j denotes the potential caused by other ions within the spherical shell, the concentration of cations and anions at this point is:

$$n_c' = n_c \exp\left[-\frac{z_c e\psi_j}{kT}\right];$$

$$n_a' = n_a \exp\left[-\frac{z_a e\psi_j}{kT}\right], \tag{4.2.40}$$

according to the Debye–Hückel theory, applying the Boltzmann distribution function.

If the change in the local concentration of water molecules is neglected (which can be done in very dilute solutions), the resultant force acting on the spherical shell mentioned above will be:

$$(n_c'f_c + n_a'f_a + n_wf_w)\, 4\pi r^2 dr. \tag{4.2.41}$$

Eliminating the term n_wf_w, on the basis of equation (4.2.39), the resultant force will be:

$$\{(n_c' - n_c)f_c + (n_a' - n_a)f_a\}\, 4\pi r^2 dr. \tag{4.2.42}$$

The theory assumes that the spherical shell moves with a velocity corresponding to the ratio of this force and the Stokes-type frictional force $6\pi\eta r$. In this way, each spherical shell–to be found at various distances r–contributes to the change in the velocity of the central ion in electrophoresis. The change in the velocity Δv_c of the central ion (e.g. cation) caused by this will be:

$$\Delta v_c = \frac{2}{3\eta} \int\limits_{r=a}^{\infty} \{(n_c' - n_c)f_c + (n_a' - n_a)f_a\} r \, dr, \tag{4.2.43}$$

where a is the minimum distance between two ions (the 'average diameter' of ions).

Substituting the values obtained above from the Bolzmann distribution function for n_c' and n_a', in first approximation:

$$\Delta v_c = -\frac{1}{6\pi\eta} \cdot \frac{\varkappa a}{1 + \varkappa a} \cdot \frac{z_cf_c - z_af_a}{a(z_c - z_a)}, \tag{4.2.44}$$

where $\varkappa$ is the reciprocal value of the characteristic distance introduced in the Debye–Hückel theory (the 'radius of the ionic atmosphere', cf. Section 5.1.3).

With respect to conductivity,* the forces f_c and f_a are the products of the electric field strength at the location of the ion and of the ionic charge.

---

* With respect to diffusion, f_c and f_a arise from the gradient in chemical potential and the diffusion potential.

Rigorously, electric field strength consists of two factors: the strength X of the external electric field and an 'inner' contribution ΔX produced by the ionic atmosphere deformed in the course of the relaxation effect. Thus, the driving forces are:

$$f_c = (X + \Delta X)z_c e$$

and

$$f_a = (X + \Delta X)z_a e. \tag{4.2.45}$$

Recently, a detailed analysis of the calculation method of the electrophoretic effect has been published by Friedman [60] pointing out some principal weaknesses in the original calculation methods. His theory also accounts for the pairwise ion–ion interaction and for the long-range hydrodynamic interaction (not additive in pairs) and considers both the finite kinetic energy and the inertia of the liquid as well. Under the usual experimental conditions, this theory yields results practically identical with those of the Debye–Hückel–Onsager theory, except the Wien effect for which a very small contribution–disregarded so far–appears. However, Friedman's theory also starts with a very simplified model of the real conditions in electrolyte solutions.

The relaxation effect has been calculated by Onsager [61] according to the suggestion of Debye and Hückel. The result of the calculations, which are complicated in spite of the numerous simplifying assumptions, is that for the additional electric field strength ΔX resulting from the deformation of the ionic atmosphere under the effect of external electric field strength X the following relationship is valid:

$$\frac{\Delta X}{X} = \frac{|z_c z_a|}{3\varepsilon kT} \cdot \frac{q\varkappa}{1 + \sqrt{q}}, \tag{4.2.46}$$

where ε is the dielectric constant of the solvent and:

$$q = \frac{|z_c z_a|}{|z_c| + |z_a|} \cdot \frac{\lambda_c^0 + \lambda_a^0}{|z_a|\lambda_c^0 + |z_c|\lambda_a^0} =$$

$$= \frac{|z_c z_a|}{(|z_c| + |z_a|)(z_a t_c^0 + |z_c| t_a^0)}, \tag{4.2.47}$$

where t_c^0 and t_a^0 are the limiting values of the transference number of cation and anion, respectively. For symmetrical electrolytes: $|z_c| = |z_a|$ and $q = 1/2$.

If U_c^0 and U_a^0 are the absolute mobilities of the cation and anion, respectively, the velocities of ions related to the solvent will be

$$v_c' = z_c e U_c^0 (X + \Delta X) = X z_c e U_c^0 \left(1 + \frac{\Delta X}{X}\right), \tag{4.2.48}$$

$$v_a' = z_a e U_a^0 (X + \Delta X) = X z_a e U_a^0 \left(1 + \frac{\Delta X}{X}\right)$$

Literature on page 328

when they move under the effect of force $X + \Delta X$. Since in infinite dilution the ion cloud disappears, the velocity of ions in an external electric field of strength X will be:

$$v_c^0 = X z_c \mathrm{e} U_c^0,$$

and

$$v_a^0 = X z_a \mathrm{e} U_a^0. \tag{4.2.49}$$

Introducing the value of $\Delta X / X$ into equation (4.2.48), we obtain from equation (4.2.46)

$$v_c' = v_c^0 \left(1 + \frac{z_c z_a \mathrm{e}^2}{3 \varepsilon kT} \cdot \frac{q \varkappa}{1 + \sqrt{q}} \right), \tag{4.2.50}$$

$$v_a' = v_a^0 \left(1 + \frac{z_c z_a \mathrm{e}^2}{3 \varepsilon kT} \cdot \frac{q \varkappa}{1 + \sqrt{q}} \right).$$

This is the velocity of ions corrected with respect to the relaxation effect, according to Onsager. Owing to the simplifying assumptions applied in the deduction, these relationships are valid only approximately and only in the case when $\varkappa a \ll 1$ and $\Delta X \ll X$; however, this holds only for very dilute solutions.

The theory of the relaxation effect has been developed further by several authors. Falkenhagen, Leist and Kelbg [62] applying the Eigen-Wicke distribution function instead of the Boltzmann one and, modifying the boundary conditions, obtained the following result for 1–1 electrolytes:

$$\frac{\Delta X}{X} = - \frac{\mathrm{e}^2}{3 \varepsilon kT} \cdot \frac{q}{1 + \sqrt{q}} \cdot \frac{\varkappa}{(1 + \varkappa a) \{ 1 + \varkappa a \sqrt{q} + \varkappa^2 a^2 / 6 \}}. \tag{4.2.51}$$

Pitts [63] increased slightly the accuracy of the calculation for symmetrical electrolytes by also taking into account higher terms in the expression of the potential. For example, according to Stokes [64], the Pitts equation describes the experiments more accurately than the Fuoss–Onsager equation in the 0·004–0·02 mole l^{-1} concentration range of HCl solutions at 25 °C.

On the basis of complicated calculations, Miruchulava [65] and Fuoss and Onsager [66] have concluded that in very dilute solutions, the expression of the concentration dependence of conductivity also contains a term $c \lg c$, but in the case of higher concentrations, the application of this term is not justified (in contrast with the opinion of other researchers). On the basis of this equation of Fuoss and Onsager, Janz and Tait [67] measured the conductivity of NaI and KI in various solvents; they established that these salts are almost completely dissociated in solvents of high dielectric constant, and an average ionic diameter value of $a = 4·5$ Å should be used in the theory of the concentration dependence of conductivity. In solvents of lower dielectric constant, ion association is appreciable and the parameter a seems to have no simple physical meaning, even in approximation.

The theory of Résibois and Davis [68] gave a somewhat different interpretation of the relaxation effect; it has been further developed by Hasselle-

Schuermans [69a] taking into account the interaction between the solvent and the ions. Their calculations on the new aspects of the theory of Brownian movement confirm the results of the Onsager and Falkenhagen theory in the case of aqueous solutions in which ions are large in comparison with the solvent molecules.

On the basis of the frequency dependence of the conductivity and permittivity of concentrated $Ca(NO_3)_2$ solutions, Ambrus, Moynikan and Macedo [69b] have inferred that under the effect of a high-frequency ($10^5 - 10^6$ Hz) electric field, a low strength relaxation of another nature, neglected up to now, also appears, in addition to the relaxation related to the relatively long-range charge transport. The cause of this second relaxation effect has not been elucidated yet. It may be in connection with the re-orientation of water molecules or with localized ion jumping.

The common action of electrophoretic and relaxation effect on the equivalent conductivity and its concentration dependence can be deduced from equations (4.2.44) and (4.2.48). The velocity of cations corrected for these two effects is, in first approximation:

$$v_c = v_c' + \Delta v_c =$$
$$= (X + \Delta X)z_c e U_c^{0\prime} - (X + \Delta X)e \cdot \frac{1}{6\pi\eta} \frac{\varkappa a}{1 + \varkappa a} \frac{z_c^2 - z_c z_a}{a(z_c - z_a)}, \qquad (4.2.52)$$

where U_c^0 is the mobility of an ion (cf. equation (4.1.2)).

The velocity of the cation in an infinitely dilute solution is:

$$v_c^0 = X z_c e U_c^{0\prime}. \qquad (4.2.53)$$

The ratio of the two latter equations is:

$$\frac{v_c}{v_c^0} = \left(1 + \frac{\Delta X}{X}\right)\left\{1 - \frac{1}{z_c U_c^{0\prime}} \cdot \frac{1}{6\pi\eta} \cdot \frac{\varkappa a}{1 + \varkappa a} \cdot \frac{z_c^2 - z_c z_a}{a(z_c - z_a)}\right\}. \qquad (4.2.54)$$

Since v_c and v_c^0 are the velocities produced by same electric field strength X in the given solution, the ratio v_c/v_c^0 can be replaced by the ratio of the equivalent conductivities of the cation λ_c/λ_c^0. On the other hand, U_c^0 can be replaced by λ^0, according to the following relationship:

$$U_c^{0\prime} = \frac{N \lambda_c^0}{F^2 |z_c|}$$

(see equations (4.1.2) and (4.1.17)). In this way, the equivalent conductivities (relative mobilities) of the cation and anion in a solution of given concentration will be:

$$\lambda_c = \left(\lambda_c^0 - \frac{F^2}{6\pi\eta N} |z_c| \frac{\varkappa}{1 + \varkappa a}\right)\left(1 + \frac{\Delta X}{X}\right), \qquad (4.2.55)$$

Literature on page 328

and

$$\lambda_a = \left(\lambda_a^0 - \frac{F^2}{6\pi\eta N}|z_a|\frac{\varkappa}{1+\varkappa a}\right)\left(1 + \frac{\Delta X}{X}\right). \qquad (4.2.56)$$

The equivalent conductivity of the electrolyte is:

$$\Lambda = \lambda_c + \lambda_a =$$

$$= \left\{\Lambda^0 - \frac{F^2}{6\pi\eta N}(|z_c|+|z_a|)\frac{\varkappa}{1+\varkappa a}\right\}\left(1 + \frac{\Delta X}{X}\right), \qquad (4.2.57)$$

when the electrolyte is dissociated completely in the solution of given concentration.

The dependence of equivalent conductivity on the concentration c of the electrolyte, given in mole l^{-1} units, or the ionic strength:

$$I = \frac{1}{2}\Sigma z_i^2 c_i \qquad (4.2.58)$$

(c_i is the concentration of ions) has been deduced by Onsager applying an additional simplifying assumption that $\varkappa a \ll 1$, i.e. $(1 + \varkappa a) \approx 1$, and that ions are influenced only by an external field strength zeX instead of $ze(X + \Delta X)$, i.e. the cross-effect of relaxation and electrophoresis is negligibly small. These two restrictions and the assumption of total dissociation are justified only in the most dilute solutions, thus they provide the limiting law of conductivity of solutions of strong electrolytes:

$$\Lambda = \Lambda^0 - \frac{|z_c z_a|e^2}{3\varepsilon kT}\cdot\frac{\Lambda^0 q\varkappa}{1+\sqrt{q}} - \frac{F^2}{6\pi\eta N}(|z_c|+|z_a|)\varkappa \qquad (4.2.59)$$

(the Onsager limiting law of conductivity). In this expression:

$$\varkappa = \sqrt{\left(\frac{8\pi Ne^2}{1000\,\varepsilon kT}\right)}\sqrt{I} \qquad (4.2.60)$$

(cf. Section 5.1.3), and:

$$I = \frac{1}{2}(z_c^2 c_c + z_a^2 c_a) = \frac{c}{2}(z_c^2 \nu_c + z_a^2 \nu_a) \qquad (4.2.61)$$

(ν_c and ν_a are the number of cations and anions formed in the dissociation of one 'molecule'). Taking this into account, the Onsager limiting law becomes:

$$\Lambda = \Lambda^0 - \left\{\frac{2\cdot801\times10^6|z_c z_a|q\Lambda^0}{(\varepsilon T)^{3/2}(1+\sqrt{q})} + \frac{41\cdot25\,(|z_c|+|z_a|)}{\eta(\varepsilon T)^{1/2}}\right\}\sqrt{I}. \qquad (4.2.62)$$

In aqueous solutions $\varepsilon = 78\cdot54$, $\eta = 0\cdot008937$ at 25 °C, thus:

$$\Lambda = \Lambda^0 - \left\{0\cdot7816|z_c z_a|\frac{q}{1+\sqrt{q}}\Lambda^0 + 30\cdot32\,(|z_c|+|z_a|)\right\}\sqrt{I}. \qquad (4.2.63)$$

In solutions of strong 1–1 electrolytes $q = 1/2$, and so:

$$\Lambda = \Lambda^0 - (0\cdot2290\ \Lambda^0 + 60\cdot32)\ \sqrt{c} =$$

$$= \Lambda^0 - A_{\text{theor}}\sqrt{c}, \qquad (4.2.64)$$

when the concentration-independent factors are summarized in factor A_{theor}. This relationship is in agreement with the Kohlrausch empirical square root law given in equation (4.1.19), and makes possible the calculation of constant A from other data.

The first term in the parenthesis in equation (4.2.62) corresponds to the decrease in conductivity due to the relaxation effect, while the second term accounts for the electrophoretic effect. This limiting law does not contain individual constants depending on the electrolyte, beyond the valence of ions and Λ^0.

In aqueous solutions of 1–1 electrolytes at normal temperatures, the Onsager limiting law is in agreement with the observations, within the experimental error, up to a concentration of some thousandth mole l^{-1} [70] (Fig. 4.5). With a further increase in concentration, the empirical conductivity will be higher than that calculated from the Onsager limiting law (Fig. 4.6).

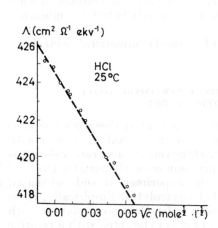

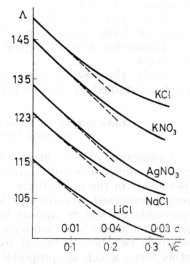

Fig. 4.5 The variation of the equivalent conductivity of dilute aqueous HCl solutions with the square root of concentration

The dotted line corresponds to the values calculated from the Onsager limiting law

Fig. 4.6 The dependence of the equivalent conductivity of dilute aqueous solutions of some 1 − 1 electrolytes on the square root of concentration

The curves represent values determined experimentally, while the dotted lines show the values calculated from the Onsager limiting law

Literature on page 328

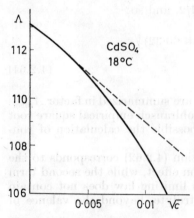

Fig. 4.7 The equivalent conductivity of dilute aqueous solutions of $CdSO_4$ as a function of the square root of concentration.

The curve represents the values observed, while the dotted line shows those calculated from the Onsager limiting law

In solutions containing ions of higher valence, the validity range of the Onsager limiting law is narrower, since in solutions of some ten thousandth mole 1^{-1} concentration, deviations already appear and the conductivity measured is generally lower than that calculated from the limiting law (Fig. 4.7).

The deviation of the Onsager limiting law from experiments indicate that the simplifying assumptions applied in the deduction are correct within the experimental error only for the most dilute solutions. The fact that in dilute solutions of strong 1–1 electrolytes exceeding the validity range of the limiting law, empirical conductivity is higher than the theoretical one, indicates that the main reason for the deviation is the omission of the term accounting for the finite size of ions, i.e. $1 + \varkappa a \neq 1$. Ion pair formation by association (which decreases conductivity) is not appreciable in the most dilute solutions, or, at least, the assumption of complete dissociation results is an error smaller than that of other simplifying assumptions. However, the negative deviations in solutions containing ions of higher valence can be attributed to significant association.

Recently, the conductivity of dilute solutions of symmetrical electrolytes have been studied by Justice [71].

4.2.3.3 DEVIATIONS FROM THE LIMITING LAW; CONDUCTIVITY OF MORE CONCENTRATED SOLUTIONS

The significance of the Onsager limiting law is that it describes the concentration dependence of equivalent conductivity in good agreement with experiments in the most dilute solutions. However, its accurate validity is strongly restricted: in concentrations higher than some thousandth mole 1^{-1} appreciable deviations appear between the experimental and calculated values even for the most simple case of 1–1 electrolytes. Empirically, these deviations can be eliminated by expanding the Onsager equation with further terms which are proportional to c, or $c^{3/2}$ or $c \lg c$, etc., with a positive or negative sign depending on the nature of the electrolyte and the concentration range. For example, the following semi-empirical correlation has been found to be valid by Hsia and Fuoss [72]:

$$\Lambda = \Lambda^0 - Sc^{1/2} + Ec\ln c + Ac + Bc^{3/2}, \qquad (4.2.65)$$

in CsBr and CsI solutions in the 0·003–0·100 mole 1^{-1} concentration range, where S, E, A and B are concentration-independent coefficients; their values can partly be calculated theoretically but their fitting to experimental data is also necessary. This and other similar equations are essentially interpolation formulae, of course, which describe the facts approximately, but by no means unequivocally; when considering that equations of somewhat different form may also be suitable to describe the phenomena with similar approximation.

The narrow validity range of the Onsager limiting law can be attributed partly to the over schematic nature of the model the Debye–Hückel theory is based on and partly to the simplifying assumptions applied in the deduction, in addition to the former ones. This is because the Debye–Hückel theory can be regarded only partly as the kinetic-statistical theory of electrolyte solutions, since the molecular structure of the solvent is not accounted for; it is treated as a continuum characterized by the macroscopic dielectric constant and viscosity. Only the distribution arising from the interaction of the electrostatic and thermal energies of ions is treated by the kinetic–statistical method, and even this treatment is not free from contradictions.* In the immediate vicinity of ions, the calculation applying the macroscopic dielectric constant can be justified evidently in the first–and rather crude–approximation only, and doubtless the micro-viscosity in the close vicinity of ions differs from the value measured macroscopically due to the electric field of ions.** In addition to its basically schematic theoretical character, some further simplifying assumptions are also implied in the theory, although their elimination would make the calculations vastly complicated. In the deduction of the Onsager limiting law, even the finite size of ions is neglected and they are treated as point-like charges. This theory also ignores the fact that the ions can form relatively stable complexes (molecules, ion pairs) as well, which also influence conductivity.

Due to the numerous simplifying assumptions, the original Debye–Hückel–Onsager theory of conductivity is valid only for the limiting case of the most dilute solutions. With the elimination of some of these simplifying assumptions or their replacement with less restricting ones, the validity of the theory can be extended to somewhat more concentrated solutions, too. The simplifying assumptions, however, cannot be avoided extensively, since this would result in such a radical alteration of the nature of the theory that its methods could no longer be applied.

No unified theory has been outlined successfully so far, though several details have already been revealed for the conductivity of more concentrated solutions, important with respect to practice and the mechanism of

---

* The Boltzmann distribution function and the Poisson law applied in the theory, in fact, contradict each other, and this contradiction can be resolved only to a certain approximation [73] (cf. Section 5.1).
** The conductivity of electrolytes is also influenced by fluctuation phenomena, therefore, the local conductivity differs from the macroscopically measurable one. The theory of this phenomenon has been elaborated by Shakhparonov [74].

Literature on page 328

transport of electricity. For example, the Onsager theory has been extended by Falkenhagen and Kremp [75] applying new statistical mechanical calculations. The equations given by the various theories for the description of conductivity have been compared with the experimental data, for example, by Fernández-Prini and Prue [76]. Recently, Kessler [77] has analysed the characteristic parameters in the theory of strong electrolytes, while the nature of the short-range ionic interactions has been studied by Kessler and Lozhkina [78].

Some extensions of the Debye–Hückel–Onsager theory of more general validity will be discussed in the following.

Consideration of the size of ions. It is assumed in the Onsager limiting law that $\varkappa a \ll 1$, i.e. the 'average ionic diameter' is negligibly small, that is, ions are point-like. This approximation can evidently hold only for the limiting case of the most dilute solutions, while, in somewhat higher concentrations, the fact $1 + \varkappa a \neq 1$ in the expression of the electrophoretic effect should also be accounted for. Thus, equation (4.2.64) is modified as follows:

$$\Lambda = \Lambda^0 - \frac{A_{\text{theor.}}}{1 + \varkappa a} \sqrt{c}, \tag{4.2.66}$$

in the case of 1–1 electrolytes. A somewhat different expression has been obtained by Stokes [79]:

$$\Lambda = \Lambda^0 - \frac{B_1 \Lambda^0 + B_2}{1 + \varkappa a} \sqrt{c}, \tag{4.2.67}$$

where B_1 and B_2 are concentration-independent coefficients. This equation agrees with experiments up to a concentration of about 0.05 mole l^{-1}.

Naturally, ionic sizes are taken into account only approximately and rather schematically in these relationships. This is so, since, on one hand, a stands for the minimum distance to which two ions can approach each other, i.e. the different size of cations and anions does not appear separately in this quantity. On the other hand, it is not clear what the minimum distance of approach means, since this is not necessarily identical with the sum of the radii of the two kinds of ions. The value of a should be determined empirically so that the relationship deduced by the Debye–Hückel theory for conductivity and the thermodynamic properties should agree as well as possible with the experiments (cf. Section 5.2.4). The a values determined empirically in this way are not identical with the sum of the ionic radii obtained from crystallographic measurements on the corresponding ionic crystals, even in solutions of binary electrolytes, but are larger than that. This deviation can be interpreted by the hydration of ions: owing to the interactions taken into account in the Debye–Hückel theory, even ions of opposite charge cannot approach each other to immediate contact, but remain separated because of their hydrate spheres. However, it has not been elucidated so far what part of the 'hydrate sphere' remains unaltered between two ions approaching each other, the less so, since the nature and interpretation of the hydrate sphere itself is still under discussion (cf. Section 5.2).

One can obtain an expression providing a better agreement with experiments, when the 'ionic diameter' a is also taken into account in the relaxation effect, according to Falkenhagen and Kelbg [80a], and Pitts [63]. In this way, the following relationship is obtained for the conductivity of 1–1 electrolytes completely dissociated in solution:

$$\Lambda = \Lambda^0 - \left\{ \frac{8 \cdot 204 \times 10^5}{(\varepsilon T)^{3/2}} \Lambda^0 + \frac{82 \cdot 5}{\eta(\varepsilon T)^{1/2}} \right\} \frac{\sqrt{c}}{1 + 50 \cdot 29\, a(\varepsilon T)^{-1/2} \sqrt{c}}$$

$$= \Lambda^0 - (A_1 \Lambda^0 + A_2) \frac{\sqrt{c}}{1 + Ba\sqrt{c}}, \qquad (4.2.68)$$

where A_1, A_2 and B are concentration-independent coefficients defined in the theory. The conductivities calculated in this way agree with the empirical ones up to a concentration of about $0 \cdot 1$ mole 1^{-1} with a standard deviation of about $0 \cdot 02$ per cent.

More significant deviations appear in solutions of higher concentrations. Duer, Robinson and Bates [80b] carried out measurements in NaF solutions, and explained the deviations from the Pitts equation by assuming a moderate association in the system.

The agreement with experiment can also be extended to higher concentrations by applying the following equation obtained for 1–1 electrolytes in the Falkenhagen theory:

$$\Lambda = \left(\Lambda^0 - \frac{A_2 \sqrt{c}}{1 + Ba\sqrt{c}} \right) \left\{ 1 - \frac{A_1 \sqrt{c}\, \Phi(\varkappa a)}{1 + Ba\sqrt{c}} \right\}, \qquad (4.2.69)$$

where

$$\Phi(\varkappa a) = \frac{e^{0 \cdot 2929\, \varkappa a} - 1}{0 \cdot 2929\, \varkappa a}. \qquad (4.2.70)$$

The viscosity of more concentrated solutions differs appreciably from that of water (η^0), too, therefore the expression given above should also be multiplied by the ratio η^0/η.

For the determination of Λ^0 by means of extrapolation, various empirical and semi-empirical equations have been suggested. Of these, Shedlovsky's one [81] shows good agreement with experiment up to about $0 \cdot 1$ mole 1^{-1} concentration in solutions of 1–1 electrolytes and can be written in the following form

$$\Lambda = \Lambda^0 - (B_1 \Lambda^0 + B_2) \sqrt{c} + bc(1 - B_1 \sqrt{c}), \qquad (4.2.71)$$

where B_1 and B_2 are constants that can be calculated theoretically, and the value of b should be determined in each case by fitting it to the experimental data. However, Fuoss and Onsager [66] pointed out that the approxi-

Literature on page 328

mately concentration-independent character of b is due to a coincidence of the particular data, as shown by the detailed theory outlined by them.

On the basis of the radial distribution function, modified by accounting for the short-range attraction forces in electrolyte solutions, the contribution of the relaxation and electrophoretic effects to conductivity has been calculated by Kremp, Ulbricht and Kelbg [82a] applying the so-called square-well potential in the statistical mechanical treatment, which yields an approximation better than the former ones. Applying the results of this theory, Ulbricht and Falkenhagen [82b] have analysed in detail the data on conductivity of solutions of tetra-alkyl ammonium ions by calculating the parameters of the square-well, too. It has been established that the explanation of the individual properties in the conductivity of tetra-alkyl ammonium ions is possible by taking into account the short-range attraction forces between ions, which also confirms the correctness of the assumption of the existence of such forces and their appearance in the non-equilibrium phenomena.

Allowance of ionic collisions. In solutions of strong electrolytes, the interaction between ions is revealed, first of all, by the decrease of ionic mobility, or equivalent conductivity with increasing concentration. However, in addition to this, the equations describing the concentration dependence of the conductivity of non-associating electrolyte solutions contain a positive–although small–term, too. Thus, there is a factor that increases mobility with increasing concentration, i.e. it acts as a factor decreasing the friction. This has been attributed to the collisions of ions by Fuoss and Onsager [83] *(kinetic or osmotic term)*. Mainly, ions in solution collide not only with solvent molecules, but with one another as well. The number of these collisions per second is affected by the deformation of the ionic atmosphere around the migrating ion caused by the external electric field. Due to the local concentration gradient of 'ions in the atmosphere', a hypothetical osmotic pressure can be considered. As a result of the asymmetry of the ionic atmosphere, for example, the concentration of anions in front of the migrating cation is somewhat lower, while, behind it, it is somewhat higher than the value corresponding to the statistical equilibrium state. Consequently, the collision frequency from behind the ion is increased by the mutual attraction of cations and anions in comparison with uncharged particles, while this increase is smaller in front of the ion which gives a small contribution to the force and velocity in the direction of the migration. The conditions are similar around migrating anions. This kinetic term of the conductivity equation varies linearly with concentration, according to the calculations.

The effect of ionic collisions on conductivity, however, has not been sufficiently clarified so far. According to Valleau [84], the effect of ionic collisions is taken into account only partly in the treatment of Fuoss and Onsager [83], since the relative mobility of the ions; has been ignored. In his theory outlined to eliminate this deficiency, Valleau [84] concluded that this effect cannot be described satisfactorily on the basis of a theory that considers the liquid to be a continuum. As it has been established in his more realistic theory, the collisional osmotic term is smaller than that calculated by Fuoss and Onsager, moreover, it can also have an opposite sign. On the other hand, however, some effects arising from ion–ion interactions, disregarded so far, can also be of importance.

4.2.3.4 DEPENDENCE OF CONDUCTIVITY ON IONIC ASSOCIATION

With increasing concentration of electrolyte solutions, an interaction decreasing the equivalent conductivity in a higher ratio than that expected on the basis of the Onsager limiting law becomes more and more effective. This can be mainly attributed to ion pair formation, though the role of some other interactions of smaller effect cannot be excluded. Ion pairs are in kinetic equilibrium with free ions; they are in permanent decomposition and re-formation in the course of the collisions with each other.

In the kinetic sense, two ions can be regarded as forming a pair when they remain together after a collision for a time significantly longer than the vibration period of the associate so formed. An ion pair consisting of two ions carrying identical absolute values of opposite charge does not participate in the transport of electricity during its life-time, thus it decreases the conductivity of the solution.

In addition to electrostatic effects, some chemical interactions can also contribute to ionic association. Taking into consideration the complicated nature of the rather unknown structure of liquids, association is, in fact, a very complex process; some of its thermodynamic and kinetic aspects have been studied by Duncan and Kepert [85] and Nancollas [86a].

The mode of association is also affected significantly by the hydration (or, in general, by the solvation) of ions. When solvation is weak, the solvent molecules are forced to leave the space between the associating ions, and *contact ion pairs* are formed. When, however, solvation is strong, the ions of the associate are separated from one another by the solvent molecules. Recently, D'Aprano and Donato [86b] studied the association conditions in aqueous solutions of alkali metal chlorates and perchlorates by the conductometric technique. In accordance with earlier observations, they found that, under comparable conditions, association is more extended, the larger the crystallographic radius of the ion and the weaker the hydration. It is emphasized that in respect of association, chlorates and perchlorates behave differently. For example, $LiClO_3$ and $NaClO_3$ associate to a greater extent than $LiClO_4$ and $NaClO_4$ do. In solutions of the corresponding K and Rb salts, however, the association of perchlorates exceeds that of the chlorates. The difference is particularly striking in the case of the Li salts. The hydration energy of the small Li ion is so high that, even in the case of ion pairs, the water molecules are not expelled from the sphere of the cation. On the other hand, it can be assumed that the perchlorate ion becomes more strongly hydrated than the chlorate ion (owing to its structure, ClO_4^- can bind more water molecules than ClO_3^-); therefore, the degree of association is lower in the solutions of $LiClO_4$ than

Literature on page 329

in those of $LiClO_3$. The conditions are different in solutions of rubidium salts: on one hand, the Rb^+ ion is presumably not hydrated, on the other hand, the perchlorate ion, owing to its regular tetrahedral shape, can approach it more closely than chlorate ions can, which facilitates association. This assumption has been confirmed by the fact that the contact distance calculated for the ion pair of $RbClO_4$ is smaller than the sum of the crystallographic radii of the two ions. According to these considerations, Li and Rb salts represent the two extreme cases, while the Na and K salts are, in this respect, intermediates. D'Aprano and Donato also pointed out that the above model of ion pair formation is in close correlation with the strengthening or destroying effect of ions on the liquid structure of water. They also refer to the fact that association alters the modifying effect of ions on the liquid structure of water, too.

In solutions of asymmetric electrolytes (consisting of ions of different valences) the associate may have an excess charge (e.g. $ZnCl^+$). Complex ions can also be formed on association of more than two ions of identical valence. Under such conditions, the associate also contributes to the transport of electricity which makes the dependence of conductivity on the state parameters more complicated, (e.g. Righellato and Davies [87], Nancollas [88], James and Monk [89a]). For some general problems of ion association we refer to Section 5.1.5.

The law of mass action is valid for ionic association, so the degree of association, i.e. the extent of the decrease in the number of free ions in comparison with the total concentration, depends on the dissociation constant. Alternatively, the association constant must be obtainable on the basis of the concentration dependence of conductivity, if a sufficiently detailed and adequate theory is available.

On the basis of the statistical model assuming ions to be rigid spheres, Kraeft and Sändig [89b] have deduced a relationship for the conductivity of solutions of symmetrical electrolytes taking into account the association according to the law of mass action; this is in good agreement with experiments up to a concentration of about 10^{-2} mole l^{-1}.

The Fuoss–Onsager equation of conductivity. In the previous sections it has been demonstrated that the apparently simple problem of transport of electricity by strong electrolytes involves, in fact, a vast complexity of interactions of several kinds. Even the Onsager limiting law holding for the most dilute solutions can only be deduced on the basis of a very much simplified model and by approximative calculations. Since the 1950's, great efforts have been made—mainly by Fuoss, Onsager and their coworkers—to outline a theory which describes the conductivity of somewhat more concentrated solutions in good approximation by accounting for possibly all the significant interactions.

The theory of the concentration dependence of equivalent conductivity has been developed recently by Fuoss and Onsager [90a] taking into account several factors in their complicated calculations. Their theory, which seems to be the most adequate at present, is based on a model greatly simplifying reality; according to it, electrolyte solutions consist of electrically charged rigid spheres in a liquid considered as a continuum both hydrodynamically and electrostatically. The introduction of the various interactions requires very complicated calculations, in which approximative treatment must also be accepted. Even in this case the final result can be treated only by means of electronic computers because of the successive approximations required.

According to the *Fuoss–Onsager equation*, the conductivity of solutions of 1–1 electrolytes having concentrations such that the relationship $\varkappa a < 0.2$ holds, is:

$$\Lambda = \Lambda^0 - Sc^{1/2}\gamma^{1/2} + E'c\gamma \ln(6 E_1' c\gamma) + Lc\gamma - K_a c\gamma f^2 \Lambda. \quad (4.2.72)$$

In this equation, γ is the portion of ions participating in the transport of electricity in the solution of given concentration, and:

$$S = \alpha \Lambda^0 + \beta_0$$

$$\alpha = 0.8204 + \frac{10^6}{(\varepsilon T)^{3/2}}$$

$$\beta_0 = \frac{82.50}{\eta(\varepsilon T)^{1/2}}$$

$$E' = E_1' \Lambda^0 + E_2'$$

$$E_1' = \frac{2.942 \times 10^{12}}{(\varepsilon T)^3}$$

$$E_2' = \frac{0.4333 \times 10^8}{\eta(\varepsilon T)^2}$$

$$L = L_1 + L_2(b)$$

$$L_1 = 3.202\, E_1' \Lambda^0 - 3.420\, E_2' + \alpha\beta_0$$

$$L_2(b) = 2E_1' \Lambda^0\, h(b) + \frac{44 E_2'}{3b} - 2E' \ln b$$

$$h(b) = \frac{2b^2 + 2b - 1}{b^3}$$

$$b = \frac{e^2}{a\varepsilon kT}$$

$$f^2 = \exp\left[\frac{-8.405 \times 10^6\, c^{1/2}\gamma^{1/2}}{(\varepsilon T)^{3/2}}\right]$$

$$K_a = \frac{1-\gamma}{c\gamma^2 f^2} = \frac{4\pi N a^3}{3000}\, e^b = \text{the association constant.}$$

Literature on page 329

There are three constants, Λ^0, K_a and L, in the Fuoss–Onsager equation which should be determined from experimental data. For this purpose, iteration treatments can be applied which, however, require the use of an electronic computer because of the cumbersome calculation work involved.

The Fuoss–Onsager equation (equation (4.2.72)) is reduced essentially to the Onsager limiting law for solutions of high dielectric constant containing the electrolyte in low concentration, while, for solutions of low dielectric constant, it yields the Ostwald dilution law.

The electrostatic theory of strong electrolytes has reached the limits of its present ability in the three-parametric equation of Fuoss and Onsager. The extremely complex nature of the real conditions is reflected in the fact that the concentration dependence of conductivity is described by a very complicated relationship despite the extended simplifying assumptions introduced, and this approximation is satisfactory only for dilute solutions (mainly in concentrations lower that about 0·01 mole l^{-1}). Although an equation can be given for the concentration dependence of the conductivity of salts of 1–1 valence type, as suggested by Pitts et al., [90b], implying only two parameters Λ^0 and a and values of physically real order of magnitude can be obtained for these two parameters from the experimental data, the theoretical fundamentals of the equation, are very questionable.

In connection with the experiments on the variation of conductivity with dielectric constant, it should also be taken into account that the composition of the solvent changes together with the dielectric constant. Thus, the interactions between dissolved ions and solvent molecules are also altered, as emphasized by Gilkerson [91]. One of the consequences of the change in the ion–solvent interaction may be the change in the size of the migrating ion which can be caused by solvation or complex formation. A possibility for separating the effect of the dielectric constant from the ion-solvent interaction has been pointed out by Hammonds and Day [92].

The first Onsager equation has been compared with experiments in several aspects. Of the papers dealing with this subject, those investigating the dependence of the ionic association constant on the dielectric constant of the solvent are of great importance. Mixtures of water and 1,4-dioxan

$$O\Big\langle{}^{CH_2-HC_2}_{CH_2-CH_2}\Big\rangle O$$

are particularly suitable for this purpose, since dioxan is miscible with water, in spite of its low dielectric constant ($\varepsilon = 2·21$ at 25 °C). In dioxan–water mixtures, the dielectric constant can be varied in the range of 78·5–2.2. However, in connection with this, it should be considered that not only the dielectric constant of the medium is altered on the addition of dioxan to water, but the liquid structure is also modified appreciably. Consequently, changes in the conductivity of aqueous electrolyte solutions due to the presence of dioxan arise not only from the change in the dielectric constant, but all the interactions influencing ionic mobility modified by the altered structure of the liquid will contribute to them. Nevertheless, the effects of the considerable changes in the dielectric constant are significant in dioxan–water mixtures.

On the basis of the mathematical analysis of data on the conductivity of aqueous electrolyte solutions containing dioxan, Fuoss and Kraus [93] concluded that the association constant of ions is:

$$K_a = A_0 \exp\left[\frac{e^2}{a\varepsilon kT}\right] \qquad (4.2.73)$$

(A_0 is a constant, a is the average ionic diameter). Thus $\lg K_a$ increases linearly with the reciprocal value of the dielectric constant. If the ion itself is also a dipole (e.g. BrO_3^- ion), a term corresponding to the ion–dipole interaction also appears in the exponent. In solutions of large ions (e.g. quaternary ammonium ions), a linear term, unknown so far, is also required to describe the concentration dependence of conductivity; this term is a consequence of the increase in viscosity caused by the large ion. The Fuoss–Kraus theory is roughly equivalent to the Bjerrum and Fuoss–Shedlovsky theories (cf. Section 5.1.5).

The association of alkali halides is negligible in water at 25 °C, according to Fuoss *et al.* [94] and it becomes significant on the addition of dioxan when reaching a dielectric constant lower than about $\varepsilon = 30$. The relationship $\lg K_a$ vs. $1/\varepsilon$ is only approximately linear, in general. Of the alkali chlorides, LiCl is the least associated one in solution as shown by Fabry and Fuoss [95], and the linear relationship is best approximated by this electrolyte (Fig 4.8). On the other hand, the association constants of rubidium halides change in the order $Cl^- > Br^- > I^-$ in solutions of given dielectric constant, and usually increase with increasing reciprocal value of the

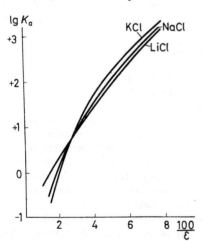

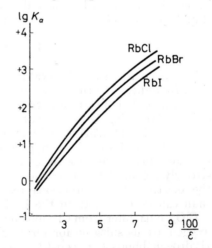

Fig. 4.8 The dependence of the association constant of some alkali metal chlorides on the dielectric constant of the solvent

Fig. 4.9 The dependence of the association constant of rubidium halides on the dielectric constant of the solvent

Literature on page 329

dielectric constant in the same way (Fig. 4.9). It has been concluded from conductivity measurements that the special structure of water is maintained up to a dioxan content of about 30 per cent by weight (8·1 mole per cent); on exceeding this value, the solvent mixture behaves as a rather structureless liquid. The ionic size a applied in the conductivity equation markedly increases with increasing dioxan content. Although this is not in contradiction with the fact that more and more water molecules are replaced by the much larger dioxan molecules in the solvate sphere of ions in the solvent mixture, it is still questionable whether a can be considered more than an empirical parameter having no simple physical meaning in such a complicated system.

Similar results have been obtained by Rizhkov and Sukhotin [96] for CsI with respect to the change of association with dielectric constant. In connection with their measurements on LiI, Atkinson and Mori [97] showed that our increasing knowledge on the conductivity of electrolytes dissolved in liquid mixtures has shown that the models applied so far, considering the solvent as a continuum, are insufficient for developing a quantitative theory of electrolyte solutions. The elaboration of a new theory which also takes into account the molecular structure of the liquid in detail is a rather urgent task.

According to the investigation of Accascina and Schiavo [98], the effect of a decrease in the dielectric constant and the presence of dioxan on $LiClO_4$ dissolved in the mixture seems to be markedly different from that exerted on LiCl. As indicated by their conductivity measurements, $LiClO_4$ is not associated to an observable extent even in a solution of dielectric constant $\varepsilon = 10$, since the concentration dependence of the equivalent conductivity can be described by the following form of the Fuoss–Onsager relationship (equation (4.2.72)) in good agreement with the experimental data:

$$\Lambda = \Lambda^0 - Sc^{1/2} + Ec \lg c + Lc, \tag{4.2.74}$$

this is valid when no observable association occurs in the electrolyte solution. In addition to this – also in contrast with the experiments on halides– the a values do not show any trend with the change in the dielectric constant, either. Excluding the case that the agreement of the experimental data with equation (4.2.74) is incidental, i.e. it arises from the circumstance that effect of association is compensated by a phenomenon unknown so far, the facts mentioned above lead to the conclusion that, in the solution of strongly solvated Li^+ ions and of large ClO_4^- ions of tetrahedral structure, the electrostatic interactions resulting in association are much weaker than can be expected on the basis of the electrostatic theory.

ClO_4^- ions differ from Cl^- ions not only in their size, but also by containing oxygen atoms suitable for forming hydrogen bonds with water molecules. Hydrogen bonds between Cl^- and I^- ions and ethanol molecules are indicated by Parker's investigations [99]. It can be assumed that they form hydrogen bonds with water molecules too, owing to the presence of the OH group. The oxygen atoms of the ClO_4^- ions provide a further possibility for this, all the more so, since the repulsion effect of the negative charge is less apparent at the boundary of these oxygen atoms. The complexes formed by means of hydrogen bonds are more stable than those formed in 'common' hydration. If we take into account, in addition, that in the solvent mixture Li^+ ions of high surface field strength can also be solvated by dioxan molecules much larger than water molecules,* the small extent of ionic association can be explained.

Among others, Badiali, Cachet and Lestrade [100] have carried out measurements on dielectric relaxation in solutions of $LiClO_4$, $NaClO_4$, $Mg(ClO_4)_2$, LiCl and LiI in alcohol–water and dioxan–water mixtures; these also indicated that the properties of halogenate ions are unlike those of halides.

Some differences in the behaviour of Li^+ and Na^+ ions with respect to association have also been shown in the studies of Campbell, Kartzmark and Oliver [101]; according to them, $LiClO_3$ is associated to a smaller extent than $NaClO_3$ in an aqueous solution containing 90 per cent by weight of dioxan, while in a 64·5 weight per cent solution of dioxan, the association of $NaClO_3$ will be smaller. It has also been established that in a 64·5 weight per cent solution of dioxan the activation energy of the conductivity and viscous flow of $NaClO_4$ is higher than in water. This is probably connected with the fact that a dioxan molecule is larger than a water molecule, thus the vacancy formation giving rise to ionic migration requires a higher energy in the presence of dioxan as compared with the case of pure water.

In connection with the conductivity of electrolytes in aqueous solutions containing dioxan, it should also be taken into account that dioxan alters the liquid structure in a complicated way. In addition to the increase in viscosity and in the activation energy of viscous flow due to the presence of dioxan, it is also indicated by the phenomenon observed by Erdey-Grúz *et al.* that, under certain conditions, the conductivity of electrolytes dissolved in dioxan–water mixtures first increases with increasing temperature, then, after passing a maximum, decreases (cf. Section 4.7.3).

The dependence of the association of strong electrolytes on the dielectric constant has been studied in mixtures of water and other non-electrolytes, mainly alcohols as well [102a]. Although these allow variation of the dielectric constant only in a shorter range, within these limits the association constant depends on the dielectric constant as expected theoretically. Nevertheless, the association states of the given electrolytes in aqueous solutions of various non-electrolytes with identical dielectric constants differ more or less from one another, which indicates that the alteration of the association constant cannot be attributed only to the dielectric constant. For example, on the basis of their measurements on the conductivity of KIO_3 dissolved in water–glycerol mixtures, Sadek, Hafez and Khalil [102b] concluded from the Fuoss–Onsager theory that the association constant of this electrolyte and parameter a increase with decreasing dielectric

* The molar volume of dioxan is 85 cm^2 while that of water is 18 cm^3.

Literature on page 329

constant, but $\lg K_a$ changes non-linearly with the reciprocal value of the dielectric constant. This is attributed by them to the fact that, in these solvent mixtures containing strong hydrogen bonds, ion pairs separated from one another by water molecules are also formed.

On the basis of conductivity measurements with quaternary (methyl-, butyl- and hexyl-) ammonium perchlorates in aqueous and non-aqueous solutions and in solvent mixtures, Accascina, Goffredi and Triolo [102c], as well as Goffredi and Triolo [102d] came to the conclusion that, under comparable conditions, the degree of association increases with the length of the side chain in aqueous solutions and in water–dioxan mixtures, but this correlation does not hold in alcohols or other solvents with hydrogen bonds. From this they have inferred that association may proceed by two different mechanisms. In alcohols and similar solvents individual ion pairs are formed by means of electrostatic interactions; this process can be regarded as an interaction of non-polarizable spherical ions in a medium considered to be a continuum, where the molecular properties of the medium need not be taken into account (i.e. specific interactions between the ions and the solvent molecules can be neglected). As the result of this type of association contact ion pairs are only formed, and association is strongly dependent on the desolvation of the anion. On the other hand, in aqueous solutions, the degree of association increases with the hydrophobic part of the cation, evidently due to the structural changes in water caused by the hydrocarbon group. Ion pair formation facilitated by the structure of water takes place to a certain extent, and this can be brought about only by the three-dimensional structural units of liquid water.

4.2.3.5 VARIATION OF THE TRANSFERENCE NUMBER OF STRONG ELECTROLYTES WITH CONCENTRATION

According to the experiments, the transference number also depends on concentration, but to a much smaller degree than conductivity or ionic mobility does, since the transference number, being the ratio of mobilities, depends on concentration only in the ratio of the concentration dependence of the mobilities (equivalent conductivities) of cations and anions in the given solution. In general, transference numbers near 0·5 (e.g. KCl) will hardly change with concentration; if the transference number of the cation is $t_c < 0·5$, it usually decreases with increasing concentration (e.g. LiCl); if, however, $t_c > 0·5$, the transference number increases with increasing concentration (e.g. HCl).

Since, in a given electrolyte, the transference number is the ratio of the equivalent conductivities of the actual ion and of the whole electrolyte, in solutions of binary electrolytes the transference number of one of the ions, for example the cation, can be calculated on the basis of the theory of electrostatic interactions (Stokes [103]). On the basis of equations (4.2.55), (4.2.57) and (4.2.67), the transference number of the cation of a binary electrolyte, provided that no appreciable association takes place, is:

$$t_c = \frac{\lambda_c}{\Lambda} = \frac{\lambda_c^0 - \dfrac{\dfrac{1}{2}|z_c|\,B_2\sqrt{I}}{1 + \varkappa a}}{\Lambda^0 - \dfrac{\dfrac{1}{2}(|z_c| + |z_a|)B_2\sqrt{I}}{1 + \varkappa a}}, \tag{4.2.75}$$

where

$$B_2 = \frac{82 \cdot 5}{\eta(\varepsilon T)^{1/2}}. \tag{4.2.76}$$

In the expression of the concentration dependence of transference number, the term corresponding to the relaxation effect is eliminated and only the difference in the electrophoretic effect of the two types of ions is responsible for the concentration dependence of transference number.

On the basis of the above equations, the transference number of the cation in electrolytes of 1–1 valence type is:

$$t_c = \frac{\lambda_c^0 - \dfrac{\dfrac{1}{2}B_2\sqrt{c}}{1 + \varkappa a}}{\Lambda^0 - \dfrac{B_2\sqrt{c}}{1 + \varkappa a}}, \tag{4.2.77}$$

while the limiting law valid for the most dilute solutions will be:

$$t_c = t_c^0 - \frac{(0\cdot5 - t_c^0)B_2\sqrt{c}}{\Lambda^0 - B_2\sqrt{c}}. \tag{4.2.78}$$

Table 4.5 gives a survey on the transference numbers of some electrolytes determined experimentally [104a] and those calculated on the basis of equation (4.2.77) using the a values given in the paper cited.

Table 4.5

The transference number of the cation of some electrolytes determined experimentally and calculated on the basis of equation (4.2.77)

c, mole · 1^{-1}	HCl, $a = 4\cdot4$ Å t_c		LiCl, $a = 5\cdot2$ Å t_c		NaCl, $a = 5\cdot2$ Å t_c		KCl, $a = 3\cdot7$ Å t_c	
	measured	calculated	measured	calculated	measured	calculated	measured	calculated
0		0·8209		0·3363		0·3962		0·4905
0·01	0·8251	0·8249	0·3289	0·3285	0·3918	0·3918	0·4902	0·4901
0·05	0·8292	0·8287	0·3211	0·3211	0·3876	0·3875	0·4899	0·4898
0·10	0·8314	0·8310	0·3168	0·3165	0·3854	0·3849	0·4898	0·4899
0·50	0·838	0·838	0·303	0·301	—	—	0·4888	0·4887
1·00	0·841	0·841	0·297	0·287	—	—	0·4882	0·4983

Literature on page 329

Even in dilute solutions of $1-1$ electrolytes, the equations derived for the concentration dependence of the transference number of electrolytes agree with the experimental values less than those given for calculating the conductivity. The accuracy of the equations suggested up to now for the description of the concentration dependence is — even in dilute solutions — poorer than that provided by the modern experimental techniques. Recently, Sidebottom and Spiro [104b] carried out computerized calculations to decide, how accurately the experimentally determined concentration dependence of the transference number can be described (1) by the equation derived for the transference number by Fuoss [104c] from the original Fuoss–Onsager equation; (2) by the supplemented formula of this equation proposed by Fuoss and Onsager [104d] taking into account the electrophoretic effect; (3) by the Pitts equation [104e] expressed for the transference number. They found that actually none of the theories can adequately describe the experiences in the case of aqueous solutions. The results obtained from numerous experimental series made in non-aqueous solutions can be brought into agreement with the equations derived from these theories only by assigning a physically unreal value to the 'ionic diameter', a (the least distance to which two ions can approach each other). Sidebottom and Spiro pointed out some possible reasons why the theories hitherto elaborated describe the concentration dependence of the transference number less accurately than that of conductivity. A satisfactory elucidation of these deviations requires still further studies.

For solutions of electrolytes of higher valence, the values calculated from the theory of electrostatic interactions differ significantly from the observed ones. The differences can be attributed partly to the over-simplified assumptions of the theory; partly, however, they arise from the fact that ionic association, and moreover, formation of complex ions of covalent bonds occur frequently in solutions of polyvalent ions. For example, according to Stokes and Levien [105], the transference number of zinc halides decreases with increasing concentration and it becomes negative in more concentrated solutions, because negatively charged anions (e.g., ZnI^{2-}) are formed in increasing quantities.* In contrast, the transference number does not show this anomaly in $Zn(ClO_4)_2$ solutions, thus this electrolyte can be considered a typical representative of the electrolytes of 2–1 valence type having simple properties. On the other hand, on the basis of the measurements on $ZnSO_4$ and $Zn(ClO_4)_2$ solutions, Dye, Faber and Karl [106] have concluded that the variation of transference numbers cannot be interpreted by the theory of electrostatic interaction, even when taking into account ion pair formation and hydrolysis, and the formation of covalent bonds should be assumed as well.

* Naturally, the transference number can be negative only formally, mainly in the case when the calculation is made as if all metal ions in the solution were present as cations without accounting for the fact that a part (or all) of them is involved in complex anions. When considering the types of ions really taking part in the transport of electricity, the transference number cannot be negative or higher than 1, according to its definition.

In concentrated solutions, only few data are available for the transference number. In concentrated aqueous solutions of RbCl and CsCl (of molality 0·25 – 4·7), Tamás, Kaposi and Schreiber [107] measured the transference number of the cation. Their results can be described well by the equation of Jones and Dole [108a] found earlier to be valid for the transference numbers in $BaCl_2$ solutions:

$$t = \frac{A}{1 + B\sqrt{c}} - 1,$$

where A and B are constants depending on the nature of the electrolyte.

Studying alkali iodides in concentrated solutions, Troshin and Malvinova [108b] have established that the transference number of the cation increases in the order $Li^+ < Na^+ < K^+$, and all the transference numbers increase with increasing concentration in the 3–7 mole l^{-1} concentration range.

In hydrochloric acid solutions, the transference number of the hydrogen ion has been investigated by Lengyel, Giber and Tamás [109], up to high concentrations. The transference number is nearly independent of concentration up to a molality of about three (at 25 °C, $t_{H^+} = 0·84$); in solutions of higher concentration it will decrease; this decrease becomes quite steep from 6·5 M concentration, and in a 14 M solution $t_{H^+} = 0·63$. The decrease in the transference number can probably be attributed to the fact that the electrolyte destroys the water structure in high concentrations, and owing to this, it hinders the conduction by the prototropic mechanism which causes a decrease in the transference number. It is to be noted that Lengyel, Giber, Beke and Vértes [110] have found a similar effect in KOH solutions: in a solution of 17 M concentration, the transference number of the hydroxide ion (0·74) is the same as in 1 M concentration. In concentrated solutions of NaOH, however, the transference number falls significantly with increasing concentration. For the interpretation of this phenomenon, further studies seem to be necessary.

4.2.3.6 VARIATION OF CONDUCTIVITY UNDER HIGH FIELD STRENGTH (THE WIEN EFFECT)

Under the usual conditions applied in practice, the maximum order of magnitude of the electric potential gradient is a few V cm^{-1}. Under such conditions, Ohm's Law is valid for electrolyte solutions, i.e. the current is proportional to the potential gradient and the proportionality factor, the resistance (or its reciprocal value: conductivity) is independent of the electric field strength. However, it has been observed by Wien [111] that a very high field strength (10^4–10^5 V cm^{-1}) increases the conductivity of electrolyte solutions *(Wien effect, the effect of field strength)*.

The factors affecting the equivalent conductivity of electrolyte solutions can be divided into three groups, and the characteristic terms applied are:

Literature on page 330

limiting values of ionic mobilities in infinite dilution (λ_c^0, λ_a^0 or Λ^0); the change in mobility due to electrostatic interaction between ions (i.e. electrophoretic and relaxation effects); and the degree of association or dissociation. High electric field strength can, in principle, affect all the three groups of factors.

The change in Λ^0 might be a consequence of the invalidity of Stokes' Law for the motion of the ions because of their high migration velocity, even in its validity range under normal conditions. According to some experiments on the movement of macroscopical bodies, the deviations from the Stokes' Law appear only when the velocity is higher than about 10^5 cm s^{-1}. The ions in salt solutions would gain this velocity on the effect of a field strength of about 10^9 V cm^{-1} which is much higher than the field strength applicable in conductivity measurements ($< 10^6$ V cm^{-1}). With respect to the Wien effect, Λ^0 can, thus, be regarded as being independent of the field strength. The highest field strength (some hundred thousand V cm^{-1} in methods applying short current impulses) that can be realized in conductivity measurements, however, will influence both the electrophoretic and relaxation effects, as well as the degree of association or dissociation.

Variation of electrostatic interaction with field strength (the first Wien effect). Under normal conditions, ions migrate very slowly. For example, at room temperature, the limiting value of the absolute mobility of K$^+$ ion is about 0.0007 cm s^{-1}. Such a slow movement also perturbs the ionic atmosphere exerting a retardation effect; however, the ion remains within the ionic atmosphere and will carry it along its way. If, however, the field strength is very high, the migration velocity of the ion will be so high that it leaves the ionic atmosphere, which does not have enough time to rearrange itself again, owing to the finite nature of relaxation time. The relaxation time is 0.5×10^{-8} s in KCl solutions of 0.01 M concentration; under the effect of a field strength of 3×10^5 V cm^{-1} the K$^+$ ion migrates with a velocity of about 210 cm s^{-1}. Within the relaxation time the K$^+$ ion covers a distance of about 12×10^{-7} cm which is about three times greater than the thickness of the ionic atmosphere ($1/\varkappa$). Thus, under such conditions, there is not enough time to rearrange the ionic atmosphere. The retardation effect of the ionic atmosphere decreases more and more and finally it ceases with increasing field strength, the electrophoretic and relaxation effects diminish, and the equivalent conductivity reaches the limiting value corresponding to infinite dilution in finite concentration.

The relative change in the conductivity of some electrolytes with field strength is given by the following expression:

$$\Delta\Lambda = \frac{\Lambda_E^0 - \Lambda^0}{\Lambda^0}, \qquad (4.2.79)$$

where Λ_E^0 is the conductivity corresponding to field strength E. This is shown in Fig. 4.10 [112]. The effect of increasing field strength starts to be observable at about 10^4 V cm^{-1}. With further increases in the field strength the initial enhancement of conductivity is rapid, then it becomes slower, reaching the value of Λ^0 at some hundred thousand V cm^{-1}. The relative increase in conductivity is higher, the higher the valence of ions involved in the electrolyte; this is, of course, connected with the fact that the electrophoretic and relaxation effects decrease the conductivity more, the higher the valence of the ions under comparable conditions.

The first approximative theory of the Wien effect has been elaborated by Falkenhagen [113] and Joos and Blumentritt [114]. The detailed and rather complicated theory for binary electrolytes was given by Wilson [115]. When the relationship regarding conductivity (equation (4.2.62)) is transformed into the following form:

$$\Lambda = \Lambda^0 - (C\,\Lambda^0 + C')\sqrt{c} \qquad (4.2.80)$$

the conductivity under a field strength E is:

$$\Lambda = \Lambda^0 - \left[C\,\frac{3\mathrm{g}(y)}{2 - \sqrt{2}}\,\Lambda^0 + \frac{C'}{\sqrt{2}}\,\mathrm{f}(y) \right]\sqrt{c}\,, \qquad (4.2.81)$$

according to the Wilson theory, where:

$$y = \frac{|z|\,\mathrm{e}E}{\varkappa \mathrm{k}T}\,,$$

while $g(y)$ and $f(y)$ are functions given in detailed tables. If $E \to 0$, then $g(y) = (2 - \sqrt{2}\,)/3$ and $f(y) = \sqrt{2}$. This relationship shows a good agreement with experiment up to a field strength of about 50 kV cm^{-1}. A theory which is also valid accurately for the highest field strength values is not available yet.

The theory extended to the Wien effect of strong electrolytes of any valence and at high field strength has been developed by Onsager and Kim [116] starting with the linearized continuity equation of Onsager and the Poisson equation.

From the theoretical point of view, the Wien effect is of great importance, because it cannot be interpreted on the basis of the classical theory of conductivity of electrolytes and hence, it can be regarded as evidence for the fact that the theory of electrostatic interactions between ions is a fundamentally correct description of the conditions in dilute electrolyte solutions.

Variation of the degree of dissociation under very high field strength (the second Wien effect). The change in the conductivity of solutions of strong electrolytes under the effect of an electric field of high strength can be well interpreted by the decreasing or disappearing retardation effect of the ionic atmosphere. In solutions of weak electrolytes, however, the increase in conductivity is much higher than in solutions of strong electrolytes of

Literature on page 330

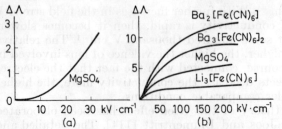
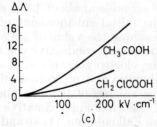

Fig. 4.10 The effect of the electric field strength on the conductivity of electrolyte solutions: (a) the initial effect, (b) the limiting effect under high field strength

Fig. 4.11 The variation of the conductivity of weak electrolytes under the effect of high electric field strength

similar valence type (e.g. Fig. 4.11 should be compared with Fig. 4.10) which already cannot be attributed to the disappearance of electrophoretic and relaxation effects. The cause of the large change in the conductivity of weak electrolytes is that collisions of ions accelerated to a high velocity with non-dissociated molecules promote their dissociation. Due to the collisions, the equilibrium:

$$\text{molecule} \rightleftharpoons \text{cation} + \text{anion}$$

is shifted towards the formation of ions, i.e. the equilibrium constant of dissociation is altered. This dissociation field effect results in a change in the dissociation constant K representing the conditions in the absence of field, to a value of K_E in a field E:

$$K_E = K \left(1 + b + \frac{b^2}{3} + \frac{b^3}{18} + \frac{b^4}{180} + \cdots \right), \qquad (4.2.82)$$

according to the theory outlined by Onsager [117], where:

$$b = 9 \cdot 64 \, \frac{E}{\varepsilon T^2},$$

for 1–1 electrolytes.

If the degree of dissociation is low, the following dependence can be deduced for the change of the conductivity of weak electrolytes under electric field:

$$\frac{\Lambda_E}{\Lambda} \sqrt{\frac{K_E}{K}} = 1 + \frac{b}{2} + \frac{b^2}{24} + \cdots \qquad (4.2.83)$$

The relative change in the equivalent conductivity is, thus, in first approximation:

$$\frac{\Lambda_E - \Lambda}{\Lambda} \approx \frac{b}{2}, \qquad (4.2.84)$$

i.e. it shows approximately a linear dependence on field strength. This relation is in good agreement with experiments for field strengths higher than 50 kV cm^{-1}. The variation under lower field strength can probably be attributed to the effect of the ionic atmosphere which has not diminished completely.

The increase of the dissociation of weak electrolytes under high electric field strength has been approximated by a theoretical method different from the earlier ones by Onsager and Liu [118]. This new method makes it possible to account for the shielding effect of ionic atmosphere on the coulombic field of ions.

Studying the change in the conductivity of electrolytes consisting of ions of higher valence under high electric field, Atkinson and Yokoi [119] have calculated the association constant of electrolytes by combining the two types of Onsager theories on the Wien effect. In solutions of $HgCl_2$ the second Wien effect has been investigated by Eigen and Eyring [120].

The degree of ionization of water is also increased by high electric field strength. On the basis of mass spectrometric investigations, Anway [121] has supposed the existence of $H_3O^+ \cdot H$ ions to be proved and calculated the distribution function of the $H_3O^+ \cdot nH_2O$ polymers. For the structure of water, he concluded that long clusters of water molecules are present in the liquid in a dipole–dipole orientation, which do not interact appreciably with the neighbouring clusters. Polymers of various n values are formed by the rupture of these long chain-like clusters.

* * *

The main features of the Wien effect can be explained on the basis of studies published so far, but some problems have still remained unsolved. One of them is the negative Wien effect: according to Spinnler and Patterson [122], the conductivity of uranyl nitrate and uranyl perchlorate solutions decreases with the increase in the external field strength in several cases, depending on the temperature and pH. They have assumed that this effect can be attributed to a change in the mechanism of conduction under high field strength. The effect of hydrogen ions has not been elucidated sufficiently in connection with the Wien effect in common solutions of acids and their salts [123].

4.2.3.7 DISPERSION OF CONDUCTIVITY (THE DEBYE—FALKENHAGEN EFFECT)

In practice, d.c. conductivity is important with respect to the transport of electricity. Nevertheless, theoretically and in the technique of measurements, in general, as well as under some other special practical conditions, the a.c. conductivity also has appreciable significance.

The resistance and conductivity of electrolyte solutions to a.c. current, of the usual frequency applied in measurement techniques and in practice is independent of frequency. Under such conditions, the duration of a single

Literature on page 330

period is long with respect to the relaxation time of the ionic atmosphere; consequently, the deformation of the ionic atmosphere corresponding to the stationary state can take place within each period of the a.c. current, thus the relaxation effect can act against a.c., as well as d.c. current. However, increasing the frequency above about 10^6 Hz, the equivalent conductivity is enhanced with further increase in the frequency, i.e. conductivity exhibits dispersion. The reason for this behaviour is, according to the theory of Debye and Falkenhagen [124], that if the duration of the half period of the a. c. current is decreased to the order of magnitude of the relaxation time there is not enough time for the development of the asymmetry of the ion cloud. Under high frequency a.c. current, the central ion vibrates in the middle of the ionic atmosphere with such a small amplitude that the ionic atmosphere remains practically spherically symmetrical and the relaxation effect vanishes. The electrophoretic effect, however, acts even under such conditions, because cations and anions are displaced in opposite directions with respects to each other in each period. Thus, due to the dispersion effect, conductivity increases in a lower ratio than it would do owing to the dissociation field effect, and the equivalent conductivity does not reach the limiting value of Λ^0 under high frequencies either.

The increase in equivalent conductivity or ionic mobility due to the increase in frequency depends on the mobility, valence, and concentration of ions (or, through these, on the relaxation time). The higher the concentration and valence of the ions, the higher the frequency of the electric field required to decrease the retardation effect of relaxation. The effect of high frequency electric field on the conductivity of electrolyte solutions has been investigated by Yermakov [125] as a function of composition.

The results of the Debye–Falkenhagen theory show a satisfactory agreement with the experiments on the effect of frequency on conductivity. It also provides evidence for the fact that the electrostatic theory of ionic interactions describes reliably and rather correctly the effects predominating under such conditions.

4.2.3.8 EFFECT OF MAGNETIC FIELD ON CONDUCTIVITY

As the migration of ions in solution is, in fact, an electric current, it is evident that an external magnetic field will also influence the transport of electricity by ions. This relationship, however, has hardly been investigated, probably because no practical importance had been attributed to electrolytic conduction under magnetic fields strong enough for this effect to be observed.

Recently, Yevseiev [126] has studied experimentally the deflection of ions migrating under the effect of an electric potential difference, in a strong magnetic field. However, according to Blumenfeld and Goldfeld [127], the combined action of electric and magnetic fields gives rise to another effect, in addition to the deflection. Namely, they have observed that the magnetic field alters the conductivity and this change persists for a while until the original state is slowly re-established after the magnetic field has stopped. To interpret this phenomenon, they assumed that water has two limiting states of different structures and magnetic properties. Their energies do not differ too much, but the energy barrier separating the two states is rather high, which hinders transformation into each other. A magnetic field primarily affects not the monomeric water molecules but the structural complexes containing several molecules which behave as dynamic entities. Under the effect of an electric potential gradient, the transformation takes place between the two states which also depends on the magnetic field. Owing to the height of the energy barrier between the two states, the perturbation of the state of water by the magnetic field cannot diminish as soon as the magnetic field is removed.

The equation describing the movement of electrically charged species (ions) in dilute solutions in a magnetic field can be deduced from a set of equations regarding the movement of harmonic oscillators. Since the mobility of the species will be direction-dependent under the magnetic field, the diffusion coefficients will be temporarily anisotropic. This anisotropy–after a time interval depending on the effective mass of the species–will disappear even in the presence of constant magnetic field.

In connection with the effect of a magnetic field on the conductivity of electrolyte solutions, only few data are yet available. Hence, further studies seem to be necessary to get a satisfactory picture about this interaction.

4.2.3.9 CONDUCTIVITY OF CONCENTRATED ELECTROLYTE SOLUTIONS

The more detailed theories of transport processes in electrolyte solutions also imply simplifying assumptions which reflect reality only in dilute solutions, even in approximation. In more concentrated solutions, the conditions are so complicated that the quantitative theory of their conductivity could not have been outlined so far.

The conductivity of concentrated solutions can be described rather well by the cube root law in equation (4.1.23), in many cases. On the other hand, it is an empirical fact, according to Wishaw and Stokes [128a], that the relationship of the Debye–Hückel–Onsager theory in equation (4.2.57) gives a more or less good approximation of the concentration dependence of the conductivity of concentrated solutions of 1–1 electrolytes, too, if one takes into account the change in viscosity η of the solution in comparison with the viscosity of pure water. Combining the constant factors in B, the equivalent conductivity of a solution of concentration c is:

$$\Lambda \frac{\eta}{\eta_0} = \left(\Lambda^0 - \frac{B \sqrt{c}}{1 + \varkappa a} \right) \left(1 + \frac{\Delta X}{X} \right). \qquad (4.2.85)$$

Although this relationship contains only one empirical parameter (a), it is in agreement with the experiments within a deviation of some few per cent up to a concentration of several mole l^{-1}, which is revealed also in

Literature on page 330

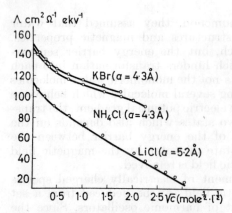

Fig. 4.12 The equivalent conductivity of concentrated solutions of some 1–1 electrolytes

The circles represent the values observed, while the curves are calculated from the relationship in equation (4.2.85)

the examples in Fig. 4.12. This agreement with experiment, however, can be considered an incidental one for concentrated solutions, rather than as a proof of the validity of the theory under such conditions. It may be an incidental inner compensation of effects acting opposite to one another.

The relationship between the conductivity and viscosity of concentrated ternary systems has been recently investigated by Lohse, Schwabe and Wolf [129].

The model of the ionic atmosphere applied in the Debye–Hückel theory of strong electrolyte solutions can evidently not reliably reflect the real conditions in solutions where water molecules are not in a large excess in comparison with the ions. The structure of dilute solutions with respect to the transport processes taking place in them can be regarded approximately as a water structure somewhat disturbed by ions. This theory, however, is not correct for concentrated solutions. The structure of saturated solutions of well-dissolving salts can be assumed to be similar to that of molten salts rather than to that of water. Thus, according to Fuoss and Onsager [130], the theory of concentrated solutions could be developed by starting with the theory of the structure of molten materials.

The structure of concentrated solutions of electrolyte-forming crystalline salt hydrates is somewhat similar to the lattice structure of solid salt crystals, as shown by the X-ray diffraction investigations of Beck [131], Mathieu and Launsbury [132], Samoilov [133] and others. The more concentrated the solution, the more it can be considered structurally as consisting of water molecules located between ions arranged more or less similarly to the crystal lattice. The structure of the most concentrated solutions can be considered to be almost a crystal lattice loosened by water molecules. The structure of concentrated electrolyte solutions is in close correlation with ionic hydration (cf. Section 5.2.), because the majority or all of the water molecules are within the hydrate sphere, moreover, in the most concentrated solutions, the number of water molecules is not sufficient for the complete development of the hydrate spheres of the ions.

Several papers have been published on the conductivity of concentrated solutions of electrolytes, recently for example Miller [134], Suryanarayana and Venkatesan [135], Chambers, Stokes and Stokes [136], Kaimakov [137], Troshin [138]; Berecz and Horányi [139] dealt with ternary solutions. Several relationships have been discovered. For example, according to Haase, Sauermann and Drücker [140], conductivity exhibits a maximum at 0·20 and 0·30 mass fraction concentrations in HCl solution and $HClO_4$ and $LiClO_4$ solutions, respectively; while Currie and Gordon [141] have established on the basis of investigations in KCl and NaCl solutions that the equivalent conductivity of the Cl^- ion in KCl solutions is much higher than in NaCl solutions $[\lambda_{Cl^-}]_{KCl} = 52\cdot6$, $[\lambda_{Cl^-}]_{NaCl} = 43\cdot55$ at 25 °C, in $2\cdot5$ mole l^{-1} concentration. On the other hand, according to Miller [134], the conductivities of NaI, $NaClO_4$ and NaSCN in more concentrated solutions tend towards one another with increasing concentration, which indicates that the conduction of electricity is successively undertaken by the Na^+ ions as the saturated solution is approached, and their mobility does not depend appreciably on the anions present. On the basis of their model on the transport by free volume, Angell [142] explained the relationships of conductivity in concentrated solutions, while Ferse and Schwabe [143] have discussed the conductivity of concentrated solutions in another approximation.

The conductivity of concentrated solutions of 1–1 electrolytes in the $0\cdot5$–$4\cdot8$ mole l^{-1} concentration range has been studied by Carman [144a] in a detailed theoretical treatment which calculates the electrophoretic effect while neglecting the relaxation effect. He has concluded that the results of diffusion and conductivity measurements in solutions of LiCl, NaCl, KCl and HCl can be well interpreted on the basis of ionic mobilities calculated from self-diffusion coefficients. The hydration numbers assumed are constant in a rather broad concentration range.

The magnetic field can result in a concentration gradient in electrolyte solutions. The possibility of the application of this phenomenon for desalination has been studied by Khalifa, Abdel-Hamid, and Abdel-Salam [128c].

4.2.3.10 CONDUCTIVITY OF SOLUTIONS OF ELECTROLYTE MIXTURES

In common solutions of two or more electrolytes, the conductivity can be obtained additively from those measured in the corresponding pure solutions of the individual electrolytes only in a very rough approximation, even if no chemical reaction in the usual sense takes place between the ions. The additivity of the equivalent conductivity in mixtures of electrolyte solutions with respect to the conductivity of the components is valid only after extrapolation to infinite dilution, i.e. the Kohlrausch law of independent ionic migration holds only for such conditions.

In finite dilution, conductivity is not additive even in the concentration range of Onsager's Limiting Law, since in the expression of this law in equation (4.2.62) only the term corresponding to the electrophoretic effect (the second one in the parenthesis) is independent of the ions of other electrolytes present in the solution. The relaxation term (the first one in the parenthesis), however, depends on the conductivities of the ions of all the electrolytes present in the solution. Thus, the mobility of a given ion is altered if some of the ions in its ionic atmosphere are exchanged for other ions, even if the ionic concentration remains unchanged. According to the theory of Onsager and Fuoss [145] outlined

Literature on page 330

for the conductivity of solutions of electrolyte mixtures, the factor $|z_a z_c| |q|$ $/(1 + \sqrt{q}\,)$ given in equation (4.2.46) for the conductivity of dilute solutions of simple electrolytes is replaced by a complicated expression for solutions of electrolyte mixtures, which depends on the mobility and valence of all ions present in the solution. In this way, in a complex solution, an electrolyte alters the conductivity of the other one (i.e. it causes a deviation from additivity) when there is a difference in the mobilities of their ions with charges of identical sign. The interaction decreases the difference in the mobilities of various ions with the same sign of charge, i.e. the slower ion is accelerated, while the faster one is slowed down. For example, in the solution of HCl + KCl, the mobility of the hydrogen ion is lower than in pure hydrochloric acid solution, while that of the potassium ion is higher than in pure potassium chloride solution.

The conditions are complicated further when more concentrated solutions are considered instead of the most dilute ones. The deviation of conductivity from additivity can be caused not only by the general electrostatic interaction of ions but by the various associations of ions as well. It is difficult to judge how much of the deviation of conductivity from additivity arises from interactions of a different nature under such conditions (with respect to this, see Davies [146] and Tate and Jones [147]).

Our knowledge of the thermodynamic properties of electrolyte mixtures is more detailed and extended than that of conductivity [148], however, it is outside the scope of the present book.

REFERENCES

to Sections 4.1 and 4.2

[1a] T. ERDEY-GRÚZ and J. PROSZT, Fizikai-kémiai Praktikum (Physico-chemical Practice), 10th edition, Vol. II, p. 175 (Budapest, 1968).
[1b] D. S. NEWMAN, C. SCHOBER and C. LAWYER, J. Electrochem. Soc., 1969, 116, 1537.
[2]   T. ERDEY-GRÚZ and G. SCHAY, Elméleti Fizikai kémia (Theoretical Physico-chemistry), 4th edition, Vol. III, p. 141 (Budapest, 1962).
[3]   For the interpretation of the transference number see e.g. the review of M. SPIRO, J. Chem. Ed., 1956, 33, 464; N. A. IZMAILOV, Elektrokhimiya rastvorov (Moscow 1966).
[4]   G. BUCHBÖCK, Z. Phys. Chem., 1906, 55, 563.
[5]   E. W. WASHBURN, J. Amer. Chem. Soc., 1909, 31, 322; Z. Phys. Chem., 1909, 66, 513.
[6]   M. SPIRO, Trans. Faraday Soc., 1959, 55, 1207.
[7]   R. HAASE, Z. Phys. Chem. N. F., 1963, 39, 27.
[8]   L. FISCHER and K. HESSLER, Ber. Bunsenges., 1964, 68, 184.
[9]   V. P. TROSIN, Zh. Fiz. Khim., 1964, 38, 2062.
[10] E. A. BAKULIN, Zh. Fiz. Khim., 1965, 39, 1065; E. A. BAKULIN and B. P. ALEKSANDROV, Zh. Strukt. Khim., 1966, 7, 179.
[11] R. A. HORNE and J. D. BIRKETT, Electrochim. Acta, 1967, 12, 1153.
[12] R. H. BOYD, J. Chem. Phys., 1961, 35, 1281; 1963, 39, 2376.
[13] R. ZWANZIG, J. Chem. Phys., 1963, 38, 1603.
[14] R. FERNÁNDEZ-PRINI, J. Chem. Phys., 1966, 45, 431; Trans. Faraday Soc., 1969, 65, 331J.
[15a] cf. e.g. H. ULICH, Trans. Faraday Soc., 1927, 23, 388; Z. Elektrochem., 1930, 36, 497; E. DARMOIS, J. Phys. Radium, 1941, 2, 2; A. EUCKEN, Z. Elektrochem., 1948, 52, 6.

[15b] I. REFF, *J. Chem. Phys.*, 1971, **54**, 4134.
[16] C. A. KRAUS, *Ann. N. Y. Acad. Sci.*, 1949, **51**, 789; H. M. DAGGETT, E. J. BAIR and C. A. KRAUS, *J. Amer. Chem. Soc.*, 1951, **73**, 799.
[17a] R. L. KAY and D. F. EVANS, *J. Phys. Chem.*, 1966, **70**, 2325.
[17b] B. S. KURMGALZ, *Zh. Fiz. Khim.*, 1971, **45**, 2559.
[18] R. H. BOYD, *J. Chem. Phys.*, 1961, **35**, 1281.
[19] R. ZWANZIG, *J. Chem. Phys.*, 1963, **38**, 1603, 1605.
[20] E. J. PASSERON, *J. Phys. Chem.*, 1964, **68**, 2728.
[21] S. B. BRUMMER and G. J. HILLS, *Trans. Faraday Soc.*, 1961, **57**, 1816, 1823; G. J. HILLS, Chemical Physics of Ionic Solutions (ed. by B. E. CONWAY and R. G. BARRADAS) p. 521 (New York, 1966).
[22] E. YA. GORENBEIN, *Zh. Fiz. Khim.*, 1961, **35**, 2156.
[23] A. E. STEARN and H. EYRING, *J. Phys. Chem.*, 1940, **44**, 955; as a review, see e.g. S. GLASSTONE, K. J. LAIDLER and H. EYRING, The Theory of Rate Processes, p. 551 (New York, 1941).
[24] R. A. HORNE, R. A. COURANT and D. S. JOHNSON, *Electrochim. Acta*, 1966, **11**, 987 R. A. HORNE and J. D. BIRKETT, *Electrochim Acta*, 1967, **12**, 1153.
[25] C. G. SWAIN and D. F. EVANS, *J. Amer. Chem. Soc.*, 1966, **88**, 383.
[26] E. DARMOIS, *J. Phys. Rad.*, 1941, **2**, 2.
[27] M. VOLMER, *Ann. Chem.*, 1924, **440**, 200.
[28a] E. WICKE, M. EIGEN and TH. ACKERMANN, *Z. Phys. Chem. N. F.*, 1954, **1**, 340.
[28b] J. O. LUNDGREN and J. W. WILLIAMS, *J. Chem. Phys.*, 1973, **58**, 788.
[29a] G. N. ZATSEPINA, *Zh. Strukt. Khim.*, 1971, **12**, 969.
[29b] For the particular nature of the conductivity of the hydrogen ion see, e.g. the review of A. GIERER and K. WIRTZ, *Ann. Phys.*, 1949, (6) **6**, 257.
[30] E. HÜCKEL, *Z. Electrochem.*, 1928, **34**, 546.
[31] J. D. BERNAL and R. H. FOWLER, *J. Chem. Phys.*, 1933, **1**, 515.
[32] M. L. HUGGINS, *J. Chem. Phys.*, 1941, **9**, 440; c.f. P. WULFF and H. HARTMANN, *Z. Elektrochem.*, 1941, **47**, 858.
[33] B. E. CONWAY and M. SALOMON, Chemical Physics of Ionic Solutions (ed. by B. E. CONWAY and R. G. BARRADAS) p. 541 (New York, 1966).
[34] Cf. R. A. HORNE, B. R. MYERS and G. R. FRYSINGER, *J. Chem. Phys.*, 1963, **39**, 2666.
[35] A. E. STEARN and H. EYRING, *J. Chem. Phys.*, 1937, **5**, 113.
[36] A. GIERER and K. WIRTZ, *Ann. Phys.*, 1949, (6) **6**, 257; c.f. R. SUHRMANN and I. WIEDERSICH, *Z. anorg. Chem.*, 1953, **272**, 167; *Z. Elektrochem.*, 1953, **57**, 93.
[37] E. WICKE, M. EIGEN and TH. ACKERMANN, *Z. Phys. Chem. N. F.*, 1954, **1**, 340; M. EIGEN, *Angew. Chem.*, 1963, **75**, 489; J. LONG and B. MUNSON, *J. Chem. Phys.*, 1970, **53**, 1356.
[38] TH. ACKERMANN, *Disc. Faraday Soc.*, 1953, **24**, 180.
[39] B. E. CONWAY, J. O'M BOCKRIS and H. LINTON, *J. Chem. Phys.*, 1956, **24**, 834.
[40a] G. PERRAULT, *Compt. rend.*, 1961, **252**, 3779.
[40b] T. L. BROADWATER and D. F. EVANS, *J. Phys. Chem.*, 1969, **73**, 3985.
[40c] J. LIN, *J. Electrochem. Soc.*, 1969, **116**, 1708.
[41a] cf. E. GRUNWALD, C. F. JUMPER and S. MEIBOOM, *J. Amer. Chem. Soc.*, 1962, **84**, 4664.
[41b] I. RUFF and V. J. FRIEDRICH, *J. Phys. Chem.*, 1972, **76**, 2954.
[41c] S. LENGYEL, *J. Phys. Chem.*, In press.
[42] T. ERDEY-GRÚZ and L. MAJTHÉNYI, *Acta Chim. Acad. Sci. Hung.*, 1958, **16**, 417; Cf. T. SHEDLOVSKY and R. L. KAY, *J. Phys. Chem.*, 1956, **60**, 151; A. R. TOURKY and S. Z. MIKHAIL, *Egypt. J. Chem.*, 1958, **1**, 1, 13.
[43] T. ERDEY-GRÚZ, E. KUGLER, and L. MAJTHÉNYI, *Electrochim. Acta*, 1968, **13**, 947.
[44a] T. ERDEY-GRÚZ and I. NAGY-CZAKÓ, *Acta Chim. Acad. Sci. Hung.*, 1971, **67**, 283.
[44b] N. I. GUSEV and P. N. PALEII, *Zh. Fiz. Khim.*, 1971, **45**, 1164.
[44c] N. I. GUSEV, *Zh. Fiz. Khim.*, 1971, **45**, 1164, 2238, 2243.
[45] S. D. HAMANN, Physico-chemical Effects of Pressure p. 123 (London, 1957).
[46a] R. A. HORNE and R. A. COURANT, *J. Phys. Chem.*, 1965, **69**, 2224; R. A. HORNE,

J. *Electrochem. Soc.*, 1965, **112**, 857; R. A. HORNE, R. A. COURANT and
D. S. JOHNSON, *Electrochim. Acta*, 1966, **11**, 987.

[46b] YU. I. NABERUZHIN and S. I. SHUISKII, *Zh. Strukt. Khim.*, 1970, **11**, 197.
[46c] E. G. WEIDEMANN and G. ZUNDEL, *Z. Naturforschung*, 1970, **25a**, 627.
[46d] D. A. LOWN and H. R. THIRSK, *Trans. Faraday Soc.*, 1971, **67**, 132.
[46e] N. K. ROBERTS and H. L. NORTHEY, *J. Chem. Soc.*, **1971**, 2572.
[47]   S. BADER, *Tetrahedron*, 1960, **10**, 182.
[48]   R. E. RICHARDS and J. A. SMITH, *Trans. Faraday Soc.*, 1951, **47**, 1261.
[49]   M. EIGEN and L. DE MAEYER, *Proc. Roy. Soc.*, 1958, **A. 247**, 505.
[50a] G. N. ZATSEPINA, *Zh. Strukt. Khim.*, 1969 **10**, 211.
[50b] J. BRUININK, *J. Appl. Electrochem.*, 1972, **2**, 239.
[51]   H. HALBAN and J. EISENBRAND, *Z. Phys. Chem.*, 1928, **132**, 401, 433; 1930,
       146, 294; H. HALBAN, G. KORTÜM and M. SEILER, *Z. Phys. Chem.*, 1935, **173**,
       449.
[52]   G. KORTÜM, *Z. Elektrochem.*, 1944, **50**, 144; G. KORTÜM and K. ANDRUSSOW,
       Z. Phys. Chem., N. F., 1960, **25**, 321.
[53]   J. BJERRUM and E. J. NIELSEN, *Acta Chem. Scand.*, 1948, **2**, 297.
[54]   H. FROMHERZ et al., *Z. Phys. Chem.*, 1931, **153**, 321; 1933, **167**, 103; 1934,
       171, 353.
[55]   C. B. MONK et al., *Trans. Faraday Soc.*, 1956, **52**, 816.
[56a] R. A. HORNE, *J. Electrochem., Soc.*, 1965, **112**, 857.
[56b] J. W. AKITT, A. K. COVINGTON, J. G. FREEMAN and T. H. LILLEY, *Trans.
       Faraday Soc.*, 1969, **65**, 2701.
[57]   G. A. MARTINOV, *Elektrokhim.*, 1965, **1**, 332, 557.
[58]   P. DEBYE and E. HÜCKEL, *Phys. Z.*, 1923, **24**, 305.
[59]   L. ONSAGER and R. M. FUOSS, *J. Phys. Chem.*, 1932, **36**, 2689; Cf. R. A. ROBIN-
       SON and R. H. STOKES, Electrolyte Solutions p. 133 and 142 (London, 1959);
       Cf. L. ONSAGER, *Angew. Chem.*, 1969, **81**, 1009.
[60]   H. L. FRIEDMAN, *J. Chem. Phys.*, 1965, **42**, 450, 459, 462; Chemical Physics
       of Ionic Solutions (ed. by B. E. CONWAY and R. G. BARRADAS) p. 487 (New
       York, 1966).
[61]   L. ONSAGER, *Phys. Z.*, 1927, **28**, 277; cf. L. ONSAGER and S. K. KIM, *J. Phys.
       Chem.*, 1957, **61**, 215. cf. L. ONSAGER, *Angew. Chem.*, 1969, **81**, 1009.
[62]   H. FALKENHAGEN, M. LEIST and G. KELBG, *Ann. Phys.*, 1952, (6) **11**, 51;
       H. FALKENHAGEN and G. KELBG, *Z. Elektrochem.*, 1954, **58**, 653; H. FALKEN-
       HAGEN, Theorie der Elektrolyte (Leipzig, 1970); For criticism of the electro-
       static theory see R. HÜMBELIN, *Chimia*, 1962, **16**, 341.
[63]   E. PITTS, *Proc. Roy. Soc.*, 1953, **A. 217**, 43; E. PITTS, B. E. TABOR, and J. DALY,
       Trans. Faraday Soc., 1970, **66**, 693.
[64]   R. H. STOKES, *J. Phys. Chem.*, 1961, **65**, 1242.
[65]   I. A. MIRUCHULAVA, *Zh. Fiz. Khim.*, 1953, **27**, 840.
[66]   R. M. FUOSS and L. ONSAGER, *J. Phys. Chem.*, 1957, **61**, 668.
[67]   G. J. JANZ and M. J. TAIT, *Canad. J. Chem.*, 1967, **45**, 1101.
[68]   P. RÉSIBOIS and H. T. DAVIS, *Physica*, 1964, **30**, 1077.
[69a] N. HASSELLE-SCHUERMANS, *J. Chem. Phys.*, 1965, **43**, 1016.
[69b] J. H. AMBRUS, C. T. MOYNIKAN and P. B. MACEDO, *J. Phys. Chem.*, 1972,
       76, 3287.
[70]   See e.g. T. SHEDLOVSKY, *J. Amer. Chem. Soc.*, 1932, **54**, 1411; 1934, **56**, 1066;
       G. KORTÜM, and H. WILSKI, *Z. Phys. Chem.*, 1953, **202**, 35.
[71]   J. C. JUSTICE, *J. Chim. Phys.*, 1968, **65**, 353.
[72]   K. L. HSIA, and R. M. FUOSS, *J. Amer. Chem. Soc.*, 1968, **90**, 3055.
[73]   See e.g. R. A. ROBINSON and R. H. STOKES, Electrolyte Solutions, p. 76
       (London, 1959).
[74]   M. I. SHAKHPARONOV, *Zh. Fiz. Khim.*, 1961, **35**, 977.
[75]   H. FALKENHAGEN and D. KREMP, *Z. Phys. Chem.*, 1965, **230**, 85.
[76]   R. FERNÁNDEZ-PRINI and J. E. PRUE, *Z. Phys. Chem.*, 1965, **228**, 373; R.
       FERNÁNDEZ-PRINI, *Trans. Faraday Soc.*, 1969, **65**, 3311.
[77]   YU. M. KESSLER, *Elektrokhim.*, 1967, **3**, 881.
[78]   YU. M. KESSLER and L. G. LOZHKINA, *Elektrochim.*, 1968, **4**, 92.
[79]   R. H. STOKES, *J. Amer. Chem. Soc.*, 1953, **75**, 4563.

[80a] H. FALKENHAGEN and G. KELBG, Z. Elektrochemie, 1954, **58**, 653; H. FALKEN-HAGEN, Elektrolyte p. 206, (Leipzig, 1953).
[80b] W. C. DUER, R. A. ROBINSON and R. G. BATES, Faraday Transact. I. 1972, **716**.
[81] T. SHEDLOVSKY, J. Amer. Chem. Soc., 1932, **54**, 1405.
[82a] D. KREMP, H. ULBRICHT and G. KELBG, Z. Phys. Chem., 1969, **240**, 65, 80.
[82b] H. ULBRICHT and H. FALKENHAGEN, Z. Phys. Chem., 1970, **243**, 305, 313.
[83] R. M. FUOSS and L. ONSAGER, J. Phys. Chem., 1957, **61**, 668; 1958, **62**, 1339; 1964, **68**, 1.
[84] J. P. VALLEAU, J. Phys. Chem., 1965, **69**, 1745.
[85] J. F. DUNCAN and D. L. KEPERT, The Structure of Ionic Solutions (ed. by W. J. HAMER) p. 380, (London, 1959).
[86a] G. H. NANCOLLAS, Chemical Physics of Ionic Solutions (ed. by B. E. CONWAY and R. G. BARRADAS) p. 197 (New York, 1966).
[86b] A. D'APRANO and J. D. DONATO, Electrochim. Acta, 1972, **17**, 1175
[87] E. C. RIGHELLATO and C. W. DAVIES, Trans. Faraday Soc., 1930, **26**, 592.
[88] G. H. NANCOLLAS, J. Chem. Soc., **1955**, 1458.
[89a] J. C. JAMES and C. B. MONK, Trans. Faraday Soc., 1950, **46**, 1041.
[89b] W. D. KRAEFT and R. SÄNDIG, Z. Phys. Chem., 1971, **247**, 343.
[90a] As a review see I. R. M. FUOSS, Chemical Physics of Ionic Solutions (ed. by B. E. CONWAY and R. G. BARRADAS) p. 463 (New York, 1966); R. M. FUOSS, L. ONSAGER and J. F. SKINNER, J. Phys. Chem., 1965, **69**, 2581; R. M. FUOSS, J. Amer. Chem. Soc., 1959, **81**, 2659; H. FALKENHAGEN and G. KELBG, Modern Aspects of Electrochemistry p. 1 (London, 1959); W. EBERLING, W. D. KRAEFT and D. KREMP, J. Phys. Chem., 1966, **70**, 3338.
[90b] E. PITTS, B. E. TABOR and J. DALY, Trans. Faraday Soc., 1970, **66**, 693.
[91] W. R. GILKERSON, J. Chem. Phys., 1956, **25**, 1199; J. B. EZELL and W. R. GILKERSON, J. Phys. Chem., 1968, **72**, 144.
[92] C. N. HAMMONDS and M. C. DAY, J. Phys. Chem., 1969, **73**, 1151.
[93] R. M. FUOSS and C. A. KRAUS, J. Amer. Chem. Soc., 1957, **79**, 3304.
[94] J. E. LIND and R. M. FUOSS, J. Phys. Chem., 1961, **65**, 999, 1414; 1962, **66**, 1727; R. W. KUNZE and R. M. FUOSS, J. Phys. Chem., 1963, **67**, 911, 914; J. C. JUSTICE and R. M. FUOSS, J. Phys. Chem., 1963, **67**, 1707.
[95] T. L. FABRY and R. M. FUOSS, J. Phys. Chem., 1964, **68**, 971, 974.
[96] E. M. RIZHKOV and A. M. SUKHOTIN, Zh. Fiz. Khim., 1962, **36**, 2205.
[97] G. ATKINSON and Y. MORI, J. Chem. Phys., 1966, **45**, 4716.
[98] F. ACCASCINA and S. SCHIAVO, Chemical Physics of Ionic Solutions (ed. by B. E. CONWAY and R. G. BARRADAS) p. 515 (New York, 1966).
[99] A. J. PARKER, J. Chem. Soc., 1966, **A2**, 220.
[100] J. P. BADIALI, H. CACHET and J. C. LESTRADE, J. chim. Phys., 1967, **64**, 1350.
[101] A. N. CAMPBELL, E. M. KARTZMARK and B. G. OLIVER, Canad. J. Chem., 1966, **44**, 935.
[102a] See e.g. C. W. DAVIES and J. C. JAMES, Proc. Roy. Soc., 1949, **195A**, 116; J. C. JAMES, J. Chem. Soc., **1950**, 1094; H. S. DUNSMORE and J. C. JAMES, J. Chem. Soc., **1951**, 2925; F. ACCASCINA, A. D'APRANO and R. M. FUOSS, J. Amer. Chem. Soc., 1959, **81**, 1058., J. L. HAWES and R. L. KAY, J. Phys. Chem., 1965, **69**, 2420; G. D. PARFITT and A. L. SMITH, Trans. Faraday Soc., 1963, **59**, 257; D. O. JOHNSTON and P. A. D. DE MAINE, J. Electrochem. Soc., 1965, **112**, 530.
[102b] H. SADEK and A. M. HAFEZ and F. Y. KHALIL, Electrochim. Acta, 1969, **14**, 1089.
[102c] F. ACCASCINA, M. GOFFREDI and R. TRIOLO, Z. phys. Chem. N. F., 1972, **81**, 148.
[102d] M. GOFFREDI and R. TRIOLO, Faraday Transact. I., 1972, 2324.
[103] R. H. STOKES, J. Amer. Chem. Soc., 1954, **76**, 1988.
[104a] L. G. LONGSWORTH, J. Amer. Chem. Soc., 1932, **54**, 274; 1935, **57**, 1185; H. S. HARNED and E. C. DREBY, J. Amer. Chem. Soc., 1939, **61**, 3113; R. W. ALLGOOD, D. J. LE ROY and A. R. GORDON, J. Chem. Phys., 1940, **8**, 418; 1942, **10**, 124; R. HAASE, G. LEHNERT and H. J. JANSEN, Z. Phys. Chem. N. F., 1964, **42**, 32; L. J. M. SMITS and E. M. DUYVIS, J. Phys. Chem., 1966, **70**, 2747.
[104b] D. P. SIDEBOTTOM and M. SPIRO, Faraday Transact. I., 1973, 1287.
[104c] R. M. FUOSS, J. Amer. Chem. Soc., 1959, **81**, 2659.

[104d] R. M. Fuoss and L. Onsager, *J. Phys. Chem.*, 1963, **67**, 628.
[104e] E. Pitts, *Proc. Roy. Soc.* A., 1953, **217**, 43; E. Pitts, B. E. Tabor and J. Daly, *Trans. Faraday Soc.*, 1969, **65**, 849.
[105] R. H. Stokes and B. J. Levien, *J. Amer. Chem. Soc.*, 1946, **68**, 333; R. H. Stokes, *Trans. Faraday Soc.*, 1948, **44**, 137.
[106] J. L. Dye, M. P. Faber and D. J. Karl, *J. Amer. Chem. Soc.*, 1960, **82**, 314.
[107] J. Tamás, O. Kaposi and P. Schreiber, *Acta Chim. Acad. Sci. Hung.*, 1966, **48**, 309.
[108a] G. Jones and M. Dole, *J. Amer. Chem. Soc.*, 1929, **51**, 1073.
[108b] V. P. Troshin and V. A. Malvinova, *Elektrokhim.*, 1971, **7**, 1047.
[109] S. Lengyel, J. Giber and J. Tamás, *Acta Chim. Acad. Sci. Hung.*, 1962, **32**, 429.
[110] S. Lengyel, J. Giber, Gy. Beke and A. Vértes *Acta Chim. Acad. Sci. Hung.*, 1963, **39**, 357.
[111] M. Wien, *Ann. Phys.*, 1927, (4) **83**, 327; 1928, **85**, 795; 1929, (5) **1**, 400.
[112] cf. H. Possner, *Ann. Phys.*, 1930, (5) **6**, 875; F. Michels, *Ann. Phys.*, 1935, (5) **22**, 735; J. Schiele, *Ann. Phys.*, 1932, (5) **13**, 811.
[113] H. Falkenhagen, *Phys. Z.*, 1929, **30**, 163; 1931, **32**, 353.
[114] G. Joos and M. Blumentritt, *Phys. Z.*, 1927, **28**, 836; M. Blumentritt, *Ann. Phys.*, 1928, (4) **85**, 812; 1929, (5), **1**, 195.
[115] W. Wilson, Dissert. Yale Univ., 1936; H. C. Eckstrom, and C. Schmelzer, *Chem. Rev.*, 1939, **24**, 367; H. Falkenhagen, Elektrolyte, p. 181 (Leipzig, 1953).
[116] L. Onsager and S. K. Kim, *J. Phys. Chem.*, 1957, **61**, 198.
[117] L. Onsager, *J. Chem. Phys.*, 1934, **2**, 599.
[118] L. Onsager and C. T. Liu, *Z. Phys. Chem.*, 1965, **228**, 428.
[119] G. Atkinson and M. Yokoi, *J. Phys. Chem.*, 1962, **66**, 1520.
[120] M. Eigen and E. M. Eyring, *Inorg. Chem.*, 1963, **2**, 636.
[121] A. R. Anway, *J. Chem. Phys.*, 1969, **50**, 2012.
[122] J. F. Spinnler and A. Patterson, *J. Phys. Chem.*, 1965, **69**, 500, 508, 513.
[123] J. F. Spinnler and A. Patterson, *J. Phys. Chem.*, 1965, **69**, 658.
[124] P. Debye and H. Falkenhagen, *Phys. Z.*, 1928, **29**, 212, 401; *Z. Elektrochemie*, 1928, **34**, 562; H. Falkenhagen, Elektrolyte p. 151 (Leipzig, 1953).
[125] V. I. Yermakov, *Zh. Fiz. Khim.*, 1960, **34**, 2258; V. I. Yermakov and P. A. Zagorets, *Zh. Fiz. Khim.*, 1963, **37**, 184.
[126] A. M. Yevseiev, *Zh. Fiz. Khim.*, 1962, **36**, 1610.
[127] L. A. Blumenfeld and M. G. Goldfeld, *Zhurn. Strukt. Khim.*, 1968, **9**, 379.
[128a] B. F. Wishaw and R. H. Stokes, *J. Amer. Chem. Soc.*, 1954, **76**, 2065.
[128b] V. B. Yevdokimov, Vestn. Moskovskogo Univ. **1970**, 174.
[128c] M. Khalifa, A. A. Abdel-Hamid and M. M. Sh. Abdel-Salam, *Z. Phys. Chem.*, 1971. **247**, 273, 333.
[129] W. Lohse, K. Schwabe and R. Wolf, *Z. Phys. Chem.*, *N. F.* 1967, **55**, 268.
[130] R. M. Fuoss and L. Onsager, *J. Phys. Chem.*, 1957, **61**, 668.
[131] J. Beck, *Phys. Z.*, 1939, **40**, 474.
[132] I. P. Mathieu and M. Launsbury, *Comptes rend.* 1949, **229**, 1315.
[133] O. Ya. Samoilov, Struktura volnykh rastvorov elektrolitov i gidratatsiya p. 108 (Moskva, 1957).
[134] M. L. Miller, *J. Phys. Chem.*, 1956, **60**, 189.
[135] C. V. Suryanarayana and V. K. Venkatesan, *Acta Chim. Acad. Sci. Hung.*, 1958, **17**, 327.
[136] J. F. Chambers, J. M. Stokes and R. H. Stokes, *J. Phys. Chem.*, 1956, **60**, 985.
[137] E. A. Kaimakov, *Zh. Fiz. Khim.*, 1964, **38**, 375.
[138] V. P. Troshin, *Elektrokhim.*, 1966, **2**, 232.
[139] E. Berecz and Gy. Horányi, *Acta Chim. Acad. Sci. Hung.*, 1961, **29**, 297.
[140] R. Haase, P. F. Sauermann and K. H. Dücker, *Z. Phys. Chem. N. F.*, 1965, **46**, 129, 140; 1965, **47**, 224.
[141] D. J. Currie and A. R. Gordon, *J. Phys. Chem.*, 1960, **64**, 1751.
[142] C. A. Angell, *J. Phys. Chem.*, 1965, **69**, 2137.

[143] A. FERSE and K. SCHWABE, *Z. Phys. Chem.*, 1965, **230**, 20.

[144a] P. C. CARMAN, *J. Phys. Chem.*, 1969, **73**, 1095.

[144b] H. S. HARNED and R. A. ROBINSON, Multicomponent Electrolyte Solutions (Oxford, 1968).

[145] L. ONSAGER and R. M. FUOSS, *J. Phys. Chem.*, 1932, **36**, 2689.

[146] C. W. DAVIES, *Endeavour*, 1945, **4**, 114.

[147] J. F. TATE and M. M. JONES, *J. Inorg. Nucl. Chem.*, 1960, **12**, 241.

[148] As a review see e.g. H. S. HARNED and B. B. OWEN, The Physical Chemistry of Electrolytic Solutions, p. 585 (New York, 1958); H. S. HARNED and R. A. ROBINSON, Multicomponent Electrolyte Solutions (Oxford, 1968); E. A. GUGGENHEIM, *Trans. Faraday Soc.*, 1966, **62**, 3446.

4.3 TEMPERATURE DEPENDENCE OF ELECTROLYTIC CONDUCTIVITY

All the factors influencing the transport of electricity by ions in solution depend on temperature, consequently equivalent conductivity also changes with temperature. The limiting value of ionic mobility varies owing to changes in the interactions between solvent molecules and ions, and between the solvent molecules themselves (i.e., changes in the liquid structure). The mutual electrostatic interaction between ions also changes, which influences the mobility in finite concentration. Finally, the equilibrium of association of ions or dissociation of molecules is also altered, resulting in an altered number of ions in a given total concentration.

Owing to the effect of thermal energy decreasing the molecular order and loosening the internal structure, the limiting value of ionic mobility increases with increasing temperature in each case. On the other hand, the hindering influence of the electrostatic interaction between ions decreases, which contributes to the increase in mobilities with temperature in finite dilution. The association of ions—essentially due to coulombic forces—also decreases with increasing temperature, which increases the number of ions and, simultaneously, the conductivity of the solution under comparable conditions. On the dissociation of covalent molecules, however, temperature has a more complex effect. Namely, electrolytic dissociation represents a chemical process between the solute and the solvent molecules in which both kinds of molecules are transformed, e.g.:

$$CH_3COOH + H_2O \rightleftharpoons CH_3COO^- + H_3O^+ . \qquad (4.3.1)$$

Depending on the actual energetic conditions, electrolytic dissociation may be endothermic or exothermic, and the degree of dissociation or the dissociation constant increases or decreases with increasing temperature. In solutions of weak electrolytes, the decrease in conductivity due to this can be so high that it overcompensates the effect of other factors increasing conductivity with temperature. In such cases, it may happen exceptionally that the conductivity of the electrolyte solution decreases with increas-

Literature on page 338

ing temperature, although this relationship is characteristic generally of metallic conduction only.

Specific conductivity varies with temperature similarly to equivalent conductivity with the additional fact that, in the ratio of these two quantities, the thermal expansion (cubic expansion coefficient) of the solution should be taken into account. Since specific conductivity is defined as:

$$\varkappa = \Lambda \frac{c}{1000}, \qquad (4.3.2)$$

when the concentration of the electrolyte in solution is expressed in g-eq l^{-1} units, it follows that:

$$\frac{\partial \varkappa}{\partial T} = \frac{1}{1000} \left\{ c \frac{\partial \Lambda}{\partial T} + \Lambda \left(\frac{\partial c}{\partial T} \right) \right\}. \qquad (4.3.3)$$

In this equation, the second term is determined by the cubic expansion coefficient according to the following relationship:

$$\frac{1}{V} \left(\frac{\partial V}{\partial T} \right) = -\frac{1}{c} \left(\frac{\partial c}{\partial T} \right),$$

i.e.

$$\left(\frac{\partial c}{\partial T} \right) = -\frac{1}{V^2} \left(\frac{\partial V}{\partial T} \right) \qquad (4.3.4)$$

since $V = 1/c$. Thermal expansion of solutions decreases the temperature coefficient of specific conductivity in comparison to that of equivalent conductivity.

Activation energy of electrolytic conduction. According to experience, the logarithm of the equivalent conductivity in dilute electrolyte solutions varies approximately linearly with the reciprocal value of the absolute temperature. From this relationship–which is similar to the Arrhenius equation of the reaction rate constant–it follows that one of the determining factors in electrolytic conduction is the activation energy E_Λ of the process. An exponential relationship is also approximately valid for electrolytic conductivity as for other transport processes and the reaction rate:

$$\Lambda = A \exp - \left[\frac{E_\Lambda}{\mathrm{R}T} \right], \qquad (4.3.5)$$

where A is constant. The activation energy of electrolytic conduction can be calculated approximately from this equation; it is about 2–8 kcal/mole in various electrolytes (e.g. 2·6 kcal/mole and 6·3 kcal/mole in $NaNO_3$ and CdI_2 solutions, respectively). However, the activation energies calculated in this way comprise several factors that will be dealt with in Section 4.4.

The activation energy of electrolytic conductivity changes with temperature; namely starting from room temperature, it decreases with increasing temperature, while with decreasing temperature it increases down to about 4 °C. This is connected with the alterations in the liquid structure caused by the temperature change, since the activation energy strongly depends on the liquid structure, as has been emphasized particularly by Horne, Myers and Frysinger [1], and Horne and Courant [2]. The decrease of the activation energy at elevated temperatures can be attributed to the fact that the water structure is progressively destroyed by the thermal motion, therefore the energy barrier between two adjacent equilibrium positions of the species is decreased and the mobility of the particles increases. Similar conclusions have also been drawn by Pamfilov and Dolgaya [3]. The fact that the activation energy of viscous flow and self-diffusion of water vary in the same way with conductivity indicates that the rate-determining mechanism of these processes is fundamentally the same: vacancy formation in the solvent with subsequent jumping of neighbouring solvent molecules into the vacancy.

The maximum of the activation energy at about 4 °C observed by Horne and Courant [2] (Fig. 4.13) in connection with the structural changes in water also results in the density maximum at this temperature. Below 4 °C, the rather loose, cavernous, tridymite-like structure characteristic of ice is predominating in water. At about 4 °C, this is increasingly transformed into a tighter quartz-like structure. In the tridymite-like water of smaller density, less energy is required for vacancy formation than in the more tight quartz-like structure. This causes the decrease of the activation energy below 4 °C. In the tridymite-like loosened structure of water, the activation energy of conductivity is commensurable with the energy required for breaking a hydrogen bond.

The observation of Horne and Johnson [4] that the activation energy of conductivity has a maximum at 4 °C in both LiCl and CsCl solutions is to be noted, though some other experiments show that dissolved LiCl is a structure-making electrolyte, while CsCl is a structure-breaking one. According to these experiments, there is no appreciable difference between structure-making and structure-breaking ions with respect to the structural changes at the temperature of the density maximum. Under such conditions, the thermal destruction of the liquid structure in dilute solutions seems not to be appreciably facilitated by otherwise structure-breaking ions, either.

From the temperature dependence of the transference number measured in

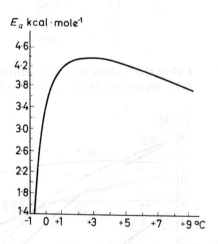

Fig. 4.13 The activation energy of the conductivity of aqueous solutions of 0·1 M KCl as a function of temperature

Literature on page 338

aqueous solutions of KCl between 0 and 115 °C, Smith and Dismukes [5] and Steel [6] concluded that the difference in the activation energy of the migration of K^+ and Cl^- ions is approximately independent of temperature (that of the Cl^- ion is higher by a value of $110-120$ cal/g-eq).

The temperature dependence of the limiting value of ionic mobility can be calculated from the values of the respective equivalent conductivity and the transference number extrapolated to infinite dilution. According to experiments, the temperature dependence of mobility can be expressed by a cubic interpolation equation:

$$\lambda^0 = \lambda^0_{25°} + a(t - 25) + b(t - 25)^2 + c(t - 25)^3, \qquad (4.3.6)$$

where λ^0 and $\lambda^0_{25°}$ are the limiting values of mobility at t and 25 °C, respectively, while a, b and c are temperature-independent parameters.

Table 4.6

The parameters of the equation describing the temperature dependence of $\lambda°$

Ion	λ^0_{25}	α	$b \cdot 10^2$	$c \cdot 10^4$
H^+	349·85	4·81595	−1·03125	−0·7670
Li^+	38·64	0·88986	+0·44075	−0·2042
Na^+	50·15	1·09160	0·47150	−0·1150
K^+	73·50	1·43262	0·40563	−0·3183
Rb^+	77·81	1·47953	0·38400	−0·4533
Cs^+	77·26	1·44790	0·38250	−0·2050
Cl^-	76·35	1·54037	0·46500	−0·1285
Br^-	78·17	1·54370	0·44700	−0·2300
I^-	76·90	1·50993	0·43750	−0·2170

Using the values given by Harned and Owen [7] in the calculations (Table 4.6), equation (4.3.6) yields values in agreement with the experiments involving an error of about 0·02 per cent for monovalent ions within a temperature interval of about 50 °C. In solutions of polyvalent ions, the extrapolation is more uncertain because of association, hydrolysis, etc., and so the validity of equation (4.3.6) is also more limited.

The limiting values of ionic mobilities increase with increasing temperature with no exceptions, but no simple rule can be found for the degree of this increase, as shown by, for example Fig. 4.14. which demonstrates the change in the difference $(\lambda^0_i - \lambda^0_{Cl^-})$ of the mobilities of Cl^- ion and some other

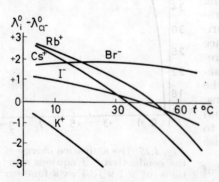

Fig. 4.14 The variation of the mobility of some ions with temperature referred to the chloride ion

ions with temperature. There is no simple relationship between the radius of the ions and the temperature coefficient of their mobilities, either. According to Owen [8], however, the value $\partial \ln \lambda^0/\partial t$ changes parallel with temperature for different ions.

The variation of the limiting value of ionic mobilities as a function of temperature is closely related to the variation of the viscosity of the solvent. This is also revealed–disregarding other factors–by the fact that the product of the equivalent conductivity and viscosity values changes with temperature much less than equivalent conductivity itself (cf. Figs 4.2 and 4.3). However, the earlier view, expressed by Kohlrausch, stating that the temperature coefficient of equivalent conductivity is approximately identical with the negative value of the temperature coefficient of the viscosity of water, does not hold good in general.

If Stokes' Law in equation (4.2.6) for frictional resistance were rigorously valid for ionic migration in electrolyte solutions, the products $\lambda^0\eta_0$ or $\Lambda^0\eta_0$ would be independent of temperature. For electrolyte solutions containing large ions, the correlation:

$$\Lambda^0\eta_0 = \text{const} \qquad (4.3.7)$$

is valid in good approximation (Table 4.7) as has already been established by Pisarzhevskii and Walden. However, solutions of smaller ions show an appreciable deviation from this rule.

Table 4.7

The product $\Lambda^0\eta_0$ in aqueous solutions of tetraethyl ammonium picrate at various temperatures

$t, °C$	Λ^0	η_0	$\Lambda^0\eta_0$
0	31	0·0179	0·556
18	53·4	0·0106	0·563
100	197	0·0028	0·56

The general validity of the Pisarzhevskii–Walden rule, however, would be expected even in the case of rigorous validity of Stokes' Law only, if the radii of the migrating ions were unaltered by temperature. Since, however, all experiments indicate that at least the small ions migrate together with their hydrate spheres and the arrangement of the water molecules in the hydrate spheres is more and more perturbed by the thermal motion when temperature is rising, and thus the size of the hydrated ion is decreased, the product $\lambda^0\eta_0$ would be expected to increase with increasing temperature because of this effect. In reality, however, this conclusion is not correct. Although the product $\lambda^0\eta_0$ is not temperature-independent for the majority of ions (see Fig. 4.3), mostly it does not increase with increasing temperature, but decreases. This indicates that the changes in

Literature on page 338

the interactions between ions and solvent molecules caused by the increase in temperature, influencing conductivity, are not limited to the reduction of the size of the hydrate sphere. This is in agreement with several other observations which indicate that hydration is not simply the binding of water molecules by the electric field of ions, but has a much more complex nature and involves the modification of the water structure in the vicinity of the ion by the volume and the electric field of the ion, as well. However, in order to discover additional details of such types of interactions, further investigations are required (cf. also Section 5.2).

Nevertheless, it can be stated that the relationship $\lambda^0 \eta_0 =$ constant holds not only for aqueous solutions of large ions presumably without hydrate spheres with rather good approximation, but for Li^+ ions, too, which are small in dehydrated state, but seem to be hydrated appreciably. Thus, hydrated Li^+ ions have a rather large size. It is also true, however, that the product $\lambda^0 \eta_0$ for Na^+, K^+ and Rb^+ ions decreases as a function of temperature with increasing slope in this order, i.e. the change is greater for the large Rb^+ ions than for the smaller Na^+ ion. This also indicates that there are many factors to be discovered with respect to the interaction between ions and solvents.

On the basis of our current knowledge, it can certainly be established that a close relationship exists between the limiting values of ionic mobilities and the viscosity of the solvent, but the form of this relationship and the determining factors have not been established accurately so far. One of the consequences of this is the insufficiency of the theory regarding the fundamental factors in electrolytic conduction, Λ^0 and λ^0, which, of course, is closely connected with the very complex nature of the structure of solutions. In order to eliminate this deficiency, detailed studies are required which take into account the effects of various factors on the relationship between equivalent conductivity and viscosity to be able to develop a satisfactory theory for λ^0.

Studying the temperature and pressure dependence of the Walden product ($\Lambda^0 \eta_0$) in aqueous solutions of tetra-alkyl ammonium halides, Kay and Evans [9] concluded that ionic mobility is influenced appreciably by the connection of the hydrocarbon chain with the water structure. They supposed that the influence of the water structure on conductivity is higher in tetra-alkyl ammonium iodide solutions than in those of similar halides.

Earlier, the ionic mobilities were supposed to tend towards each other with increasing temperature, i.e. the transference number of salts tends to 0·5. The extended investigations of Gordon et al. [10], as well as Owen's measurements [8] have shown, however, that this is not correct, since even in solutions of KCl consisting of ions of approximately identical mobilities, the transference number of the Cl^- ion increases from 0·508 (18 °C) to 0·513 (45 °C), i.e. ionic mobilities do not change in the direction of equalization with increasing temperature.

The temperature dependence of the relaxation and electrophoretic effect in dilute solutions of strong electrolytes can be described by the Debye–Hückel–Onsager theory of electrostatic interactions of ions. Owing to the temperature dependence of the factors determining the ionic atmosphere, these effects are altered strongly by temperature. For example, for aqueous solutions of KCl, one obtains:

$$\text{at } 0 \text{ °C} \quad \Lambda = \Lambda_{0°}^0 - 4.74 \sqrt{c} \tag{4.3.8}$$

$$\text{at } 100 \text{ °C} \quad \Lambda = \Lambda_{100°}^0 - 313.2 \sqrt{c} \tag{4.3.9}$$

calculated according to the limiting law.

The temperature dependence of association and dissociation is rather vague. The degree of ion pair formation caused by purely coulombic forces decreases with increasing temperature, although this process also has a complicated mechanism (cf. Section 5.1.5). Electrolytic dissociation of covalently bound molecules is, in fact, a chemical reaction with the solvent similar to the process given in equation (4.3.1) which–unlike thermal dissociation–can be either endothermic or exothermic. Consequently, the degree of dissociation and the dissociation constant depend on temperature according to the van't Hoff equation as a function of the heat of dissociation determined by the nature of the electrolyte.

At high temperatures, conductivity changes in a more complicated way with temperature than it does below 100 °C, even in solutions of strong electrolytes. Studying the conductivity of aqueous solutions of alkali metal and alkali earth metal chlorides up to 300 °C (under pressure), Kondratiev and his co-workers [11] established that specific and equivalent conductivities have a maximum at a concentration of 0.5 mole l^{-1} and at a temperature of 250 °C. This maximum is shifted to lower temperatures with increasing concentration of the solution. The maximum in conductivity has been explained by the increased degree of association with rising temperature, since hydration decreases as a function of temperature and, due to this, the effective ionic diameter also decreases, so the ions can approach each other more closely. Since the dielectric constant also decreases simultaneously, the conditions can become more favourable for ion pair formation. According to Maximova and Yushkevich [12], there is also a maximum in the conductivity of the 20 per cent NaOH solution at 220 °C. It has been emphasized in interpreting this phenomenon that the hydrogen bonds in the water structure are broken by the joint effect of increasing temperature and the field of the dissolved ions, particularly in concentrated solutions. In solutions so concentrated as to contain only two or three water molecules per ion, the ions do not remain spherical. Breaking the hydrogen bonds, the anion (e.g. Cl^- ion) can enter the lattice-like structure of water and, thus, it can be inserted between the cation and the water molecules. Around the cation, $[\text{anion} - n\text{H}_2\text{O}]$ groups can be arranged, their number depending on its radius. In the vicinity of polyvalent cations, this type of bond is stronger.

The effect of hydroxide ions differs from that of monoatomic anions. In more concentrated solutions, OH····OH····OH bonds can develop.

Literature on page 338

Due to the increase in the kinetic energy of the species with increasing temperature, this arrangement decomposes progressively and the orientation effects decrease; so the individual interactions of the particles act more efficiently, which is indicated by the viscosity and conductivity of the solution. The distortion of the water structure alters the structure of the ionic atmosphere as well; the shielding of the electric field of ions by the ordered molecules in the hydrate sphere decreases. The binding of monomeric dipoles by ions decreases the mobility of water molecules less than that of the polymers comprising branching bonds, and the remaining hydrogen bonds are also weakened. All these effects act in the direction of decreasing the viscosity and increasing the ionic mobility. On the other hand, however, the conditions for interaction between oppositely charged ions are facilitated by the decrease in the order of the dipole molecules of water, and thus the possibility for formation of ion pairs and other local ionic groups is increased. This effect reduces more and more the mobility of ions and the conductivity of the electrolyte with increasing temperature. The more concentrated the solution, the more ordered it is under comparable conditions and the higher temperature is necessary for the loosening the water structure in the vicinity of the ions, i.e. the association supported by this becomes more efficient at higher temperatures.

On the basis of measurements on the conductivity of aqueous alkali chloride solutions, Mangold and Franck [13] have calculated the association constant up to 1000 °C (up to a pressure of 12 000 bar) according to the Fuoss theory. They have established that ion pair formation is significant even when the density of water is higher than $1 \cdot 0$ g cm^{-3}, but that ion triplet formation is negligible.

According to Renkert and Franck [14], the equivalent conductivity of aqueous KCl solutions increases rapidly with temperature and pressure (at normal temperature and pressure, this value is 140 Ω^{-1} cm^2 mole^{-1} and increases to 680 at 350 °C and 6 kbar). The increase in the product $\Lambda\eta$ with increasing pressure has been explained by the reduced effective diameter of the hydrated ions.

At higher temperatures, the conductivity of electrolyte solutions can be studied only under pressure, therefore, the effect of high temperature on conductivity is combined with that of pressure (see the next Section).

REFERENCES

to Section 4.3

[1]   R. A. Horne, B. R. Myers and G. R. Frysinger, *J. Chem. Phys.*, 1963, **39**, 2666.
[2]   R. A. Horne and R. A. Courant, *J. Phys. Chem.*, 1964, **68**, 1258.
[3]   A. V. Pamfilov and O. M. Dolgaya, *Zh. Fiz. Khim.*, 1963, **37**, 1800.
[4]   R. A. Horne and D. S. Johnson, *J. Chem. Phys.*, 1966, **45**, 21.
[5]   J. E. Smith and E. B. Dismukes, *J. Phys. Chem.*, 1963, **67**, 1160.
[6]   B. J. Steel, *J. Phys. Chem.*, 1965, **69**, 3208.
[7]   H. S. Harned and B. B. Owen, The Physical Chemistry of Electrolytic Solutions, p. 233 (New York, 1958).
[8]   B. B. Owen, *J. Chim. Phys.*, 1952, **49**, 672.

[9]   R. L. KAY and D. F. EVANS, *J. Phys. Chem.*, 1966, **70**, 2325.
[10]   R. W. ALLGOOD, D. J. LEROY and A. R. GORDON, *J. Chem. Phys.*, 1940, **8**, 418; W. R. ALLGOOD and A. R. GORDON, *J. Chem. Phys.*, 1942, **10**, 124; G. C. BENSON and A. R. GORDON, *J. Chem. Phys.*, 1945, **13**, 473; A. G. KEENAN, H. G. MCLEOD and A. R. GORDON, *J. Chem. Phys.*, 1945, **13**, 466.
[11]   V. P. KONDRATIEV and V. I. NILKIK, *Zh. Fiz. Khim.*, 1963, **37**, 100; S. V. GOVDAEV and V. P. KONDRATIEV, *Zh. Fiz. Khim.*, 1961, **35**, 1235; 1965, **39**, 2753.
[12]   I. N. MAXIMOVA and V. F. YUSHKEVICH, *Zh. Fiz. Khim.*, 1963, **37**, 903.
[13]   K. MANGOLD and E. U. FRANCK, *Ber. Bunsenges.*, 1969, **73**, 21.
[14]   H. RENKERT and E. U. FRANCK, *Ber. Bunsenges.*, 1970, **74**, 40.

4.4 PRESSURE DEPENDENCE OF ELECTROLYTIC CONDUCTIVITY

Conductivity of electrolyte solutions also depends on hydrostatic pressure. However, large pressure is required for an essential change in conductivity. It can be seen from Fig. 4.15, which gives rather general information, that the effect of pressure on conductivity depends strongly on the nature of the electrolyte and, to a lesser degree, on its concentration. The conductivity of weak electrolytes, e.g. $HgCl_2$, NH_4CN, shows a particularly large increase with pressure. In solutions of strong electrolytes, conductivity often exhibits a maximum as a function of pressure. The effect of pressure decreases with increasing concentration under comparable conditions.

There are less experimental observations available for the pressure dependence of conductivity than for the temperature dependence, since studies on the effect of pressure require measurements in the pressure range of some thousand bars, at various temperatures, which involves several experimental difficulties.

The pressure dependence of equivalent conductivity–like its temperature dependence–consists essentially of three factors: the change in the limiting value of mobility (which is connected with short-range interactions between ions and solvent molecules), the change in the electrostatic interaction between ions (with a rather long-range efficiency), and the change in the equilibrium of association or dissociation.

The pressure dependence of specific conductivity differs from that of the equivalent conductivity because of the compressibility of solutions. The

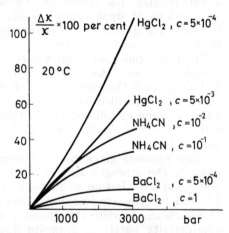

Fig. 4.15 The variation of specific conductivity with pressure in aqueous solutions of electrolytes at 20 °C (referred to the conductivity measured under a pressure of 1 atm.)

Literature on page 346

22*

relationship between the two kinds of conductivity (see e.g. equation (4.3.2)) is related to the compressibility of the solution:

$$\frac{\partial \varkappa}{\partial P} = \frac{1}{1000}\left\{c\left(\frac{\partial \varLambda}{\partial P}\right) + \varLambda\left(\frac{\partial c}{\partial P}\right)\right\},\tag{4.4.1}$$

where

$$\left(\frac{\partial c}{\partial P}\right) = -\frac{1}{V^2}\left(\frac{\partial V}{\partial P}\right).\tag{4.4.2}$$

The compressibility of electrolyte solutions appreciably differs from that of pure water, because the electric field of ions destroys the water structure, alters the packing of the molecules, and, particularly, the compressibility of water in the hydrate sphere. The increase in pressure increases the number of ions present in unit volume.

Of the factors resulting in the change of conductivity with pressure, the variation of the dissociation constant K can be regarded as relatively the most simple one. In this respect, the predominating factor is the change in volume $\varDelta V$ accompanying dissociation:

$$\frac{\partial \ln K}{\partial P} = -\frac{\varDelta V}{RT^2}.\tag{4.4.3}$$

In electrolyte solutions, $\varDelta V$ is always negative, since the electric field of the ions formed in the course of dissociation compresses the water molecules to a certain extent (electrostriction) which results in contraction. For example, in solutions of weak organic acids, $\varDelta V$ is about $-10 \text{ cm}^3 \text{ mole}^{-1}$ which increases the dissociation constant by about 23 per cent when the pressure rises from 1 atm to 500 atm.

The relaxation and electrophoretic effects depend on pressure in a more complicated way than does the dissociation constant. According to the Debye–Hückel–Onsager theory (cf. Section 4.2.3), in dilute solutions these effects are controlled by the viscosity, dielectric constant, and the limiting value of ionic mobilities, in addition to physical constants. All these properties change markedly under high pressure, because the water structure is also altered.

High pressure gives rise to several changes in the water structure. It changes the strength and number of the hydrogen bonds, modifies the statistical distribution of molecules, and the ratio of more or less lattice-like ordered clusters and disordered regions, decreases the average number of vacancies, hinders the vacancy formation, etc. These effects alter the conductivity partly in opposite directions, and the direction of the predominating effect depends on the nature of the electrolyte, on concentration, pressure and temperature. For example, the deformation and rupture of hydrogen bonds and the increase in the ratio of monomeric water molecules acts towards the increase in conductivity; on the other hand, the decrease in the average number of vacancies and the increase in the activation energy required for vacancy formation decreases conductivity. The complex nature of these effects of various magnitude, partly acting in opposite directions and depending specifically on the conditions, makes the pressure dependence of conductivity very complicated.

4.5 ACTIVATION ENERGY OF IONIC MIGRATION

According to the kinetic statistical theories of ionic migration (cf. Section 4.2.2), dissolved ions are in equilibrium positions corresponding to the average configuration formed with neighbouring solvent molecules, and jump from these to an adjacent equilibrium position after having gained the necessary activation energy by means of energy fluctuation. The electric field contributes to this disordered displacement by an ordered component, resulting in ionic migration in the direction of the field. The activation energy of the ionic migration process is determined by the height of the energy barrier separating the adjacent equilibrium positions (cf. Section 4.3), i.e. it is determined by the structure of the transition complex. According to the kinetic statistical theories, equivalent conductivities of ions (their relative mobilities) are described essentially by the following equation:

$$\lambda^0 = \frac{ze\mathrm{F}l^2}{6\mathrm{h}} \exp\left[-\frac{\Delta\mu_0^{\ddagger}}{\mathrm{R}T}\right] \qquad (4.5.1)$$

for the limiting case of infinitely dilute solutions, where z is the valence of the ion, l is the average distance between two adjacent equilibrium positions,* $\Delta\mu_0^{\ddagger}$ is the change in chemical potential in the course of the transition to the activated state referred to 1 mole. The various kinetic theories differ from one another only in the numerical factor.

The first development of the kinetic theory of ionic migration was then followed by a stagnation for about 25 years which, according to Hills [1], is related to the lack of detailed knowledge on the enthalpy of activation. In interpreting the change in conductivity under high pressure, the temperature coefficients or activation energies referred to constant pressure and constant volume (isobar and isochore) should be distinguished particularly clearly. Namely, in condensed phases, the relationship between isobar and isochore transitions is not as simple as for the state of perfect gases. It seems to be sufficient to refer to the fact that the relationship between enthalpy and internal energy at constant pressure and at constant volume for processes involving a change in the volume $\Delta V = V_2 - V_1$ under pressure P is:

$$\Delta H_P = \Delta U_P + P\Delta V =$$
$$= \Delta U_V + \int_{V_1}^{V_2} \left(\frac{\partial U}{\partial V}\right)_T \mathrm{d}V + P\Delta V. \qquad (4.5.2)$$

* Instead of the λ common in the kinetic theory of the absolute rate of processes, we denote this distance by l in order to avoid confusion with conductivity.

Literature on page 346

Newitt and Wassermann [2] and recently, Jobling and Lawrence [3], and Collins [4] have already emphasized that the parameters related to constant volume give more details on the mechanism of the process than those obtained at constant pressure.

According to Hills [1], the relationship between the isobar and isochore parameters of conductivity can be determined unequivocally as follows. The equivalent conductivity of a given ion in a given solution is a function of temperature and pressure: $\lambda^0 = f(T, P)$. Thus, the change in conductivity due to the effect of temperature and pressure is:

$$d \ln \lambda^0 = \left(\frac{\partial \ln \lambda^0}{\partial T} \right)_P dT + \left(\frac{\partial \ln \lambda^0}{\partial P} \right)_T dP; \qquad (4.5.3)$$

further

$$\left(\frac{\partial \ln \lambda^0}{\partial T} \right)_V = \left(\frac{\partial \ln \lambda^0}{\partial T} \right)_P + \left(\frac{\partial \ln \lambda^0}{\partial P} \right)_T \left(\frac{\partial P}{\partial T} \right)_V. \qquad (4.5.4)$$

With respect to the usual interpretation, the isobar energy (enthalpy) of activation that can be determined experimentally, is:

$$E_P = RT^2 \left(\frac{\partial \ln \lambda^0}{\partial T} \right)_P. \qquad (4.5.5)$$

While the isochore energy of activation is:

$$E_V = RT^2 \left(\frac{\partial \ln \lambda^0}{\partial T} \right)_V. \qquad (4.5.6)$$

The difference in the two activation energies will be:

$$E_P - E_V = -RT^2 \left(\frac{\partial \ln \lambda^0}{\partial P} \right)_T \left(\frac{\partial P}{\partial T} \right)_V. \qquad (4.5.7)$$

In this equation, $\partial \ln \lambda^0 / \partial P$ is closely related to the activation volume, $\Delta V^{\ddagger}$, and with the distance l between two adjacent equilibrium positions. According to the theory of transition state, reflected also in equation (4.4.4), one has:

$$\left(\frac{\partial \ln \lambda^0}{\partial P} \right)_T = - \Delta V^{\ddagger} + 2 \left(\frac{\partial \ln l}{\partial P} \right)_T. \qquad (4.5.8)$$

Utilizing the reasonable assumption that l is equal to V^{-3} and $d \ln l = 1/3\, d \ln V$, further, denoting the compressibility coefficient of the solvent by β, we obtain:

$$\left(\frac{\partial \ln \lambda^0}{\partial P} \right)_T = \frac{\Delta V^{\ddagger}}{RT} - \frac{2}{3} \beta.$$

If S is the entropy of the system, π the internal pressure of the solvent, and α the cubic heat expansion coefficient of the solvent, one can write:

$$\left(\frac{\partial P}{\partial T}\right)_V = \left(\frac{\partial S}{\partial V}\right)_T =$$

$$= \frac{1}{T}\left\{\left(\frac{\partial U}{\partial V}\right)_T + P\right\} =$$

$$= \frac{1}{T}(P + \pi) =$$

$$= \frac{\alpha}{\beta}. \tag{4.5.9}$$

Taking into account all these relationships:

$$E_P - E_V = (P + \pi)\Delta V^{\ddagger} + \frac{2}{3}(P + \pi)RT\beta. \tag{4.5.10}$$

On the other hand, it follows from equation (4.5.1):

$$E_P = (\Delta H_0^{\ddagger})_P =$$

$$= 2RT^2\left(\frac{\partial \ln l}{\partial T}\right)_P =$$

$$= \Delta H_0^{\ddagger} + 2RT^2\alpha \tag{4.5.11}$$

and

$$E_V = (\Delta U_0^{\ddagger})_V, \tag{4.5.12}$$

suppose that:

$$\left(\frac{\partial \ln l}{\partial T}\right)_V = 0.$$

On this basis, we obtain:

$$E_P - E_V = (\Delta H_0^{\ddagger})_P - (\Delta U_0^{\ddagger})_V + \frac{2}{3}RT\alpha =$$

$$= P\Delta V^{\ddagger} + \left(\frac{\partial U}{\partial V}\right)_T \Delta V_0^{\ddagger} + \frac{2}{3}RT^2\alpha =$$

$$= (P + \pi)\Delta V^{\ddagger} + \frac{2}{3}RT^2\alpha. \tag{4.5.13}$$

Literature on page 346

This equation is identical with equation (4.5.10); both of them imply the assumption that $d \ln l = 1/3 \ln V$ and $(P + \pi)$ is constant, which holds only in approximation.

The relationships mentioned above describe the equivalent conductivity of one of the ions, while, experimentally, the equivalent conductivity of an electrolyte containing at least two types of ions can be determined directly. For the calculation of ionic conductivity, the transference number should also be known. However, under high pressures for which the distinction between isobaric and isochoric conditions is particularly important, the transference numbers are not known, in general. Since, however, the transference numbers change slightly with pressure and temperature, according to experiments, the ionic mobilities under high pressures can be calculated in good approximation from the transference numbers measured under normal pressures. According to Howard [5], the E_P, E_V and $\Delta V^{\ddagger}$ values of electrolytes can be computed additively from the corresponding values of the ions forming them. Thus:

$$\frac{E_V}{RT^2} = \frac{1}{\Lambda^0}\left(\frac{\partial \Lambda^0}{\partial T}\right)_V =$$

$$= \frac{1}{\Lambda^0}\left(\frac{\partial(\lambda_c^0 + \lambda_a^0)}{\partial T}\right)_V =$$

$$= \frac{\partial}{\partial T}\left(t_c^0 \ln\lambda_c^0 + t_a^0 \ln \lambda_a^0\right); \qquad (4.5.14)$$

i.e.

$$E_V = t_c(E_c)_V + t_a(E_a)_V. \qquad (4.5.15)$$

In this way–although the individual ionic parameters cannot be determined–the effect of the type, size and charge of ions on these parameters can be well studied by comparing the properties of salts comprising a common cation or common anion, since these parameters are approximately additive.

There are only few experimental data on the energy and volume of activation of ionic migration. Hills [1] has established that, in non-aqueous solvents with no significant internal structure (e.g. nitrobenzene), the pressure coefficient of conductivity is negative, its activation volume is positive, and consequently, $E_P > E_V$. This indicates that most of the activation energy arises from the energy required for the vacancy formation (or for the expansion of the 'solvent cage'). With an increase in the specific volume (which can be realized at constant temperature by changing the pressure) E_V decreases, while $\Delta V^{\ddagger}$ increases. E_V is independent of the temperature and the ionic parameters, which shows that the structure of the activated state is similar to the ground state, and in the course of the activation, the rotational and vibrational degrees of freedom remain unchanged.

In aqueous solutions, the conditions are different, as it has been shown by Hills. On the basis of his own and others' measurements, he found that in dilute aqueous solutions E_V increases and $\Delta V^\ddagger$ decreases with increasing specific volume below 100 °C, and E_V is higher than E_P. A similar 'anomaly' has been revealed in the parameters of viscous flow, as well. In aqueous solutions, the effects arising from the liquid structure evidently overcompensate those occurring in structureless 'normal' liquids (which have not been interpreted satisfactorily so far). Pressure influences the different factors which determine the liquid structure variously; it partly destroys the lattice-like regions or clusters and alters the number of vacancies, as well as the polarization. The increase of E_V with specific volume is probably due to the increase in the number and size of ordered regions which makes water to be more rigid. The fact that the value $\Delta V^\ddagger$ becomes more negative with the increase of specific volume is connected with the structure of the ground state becoming more open.

The effect of pressure on the activation energy of ionic conductivity in aqueous solutions of KCl, KOH and HCl has been investigated by Horne, Myers and Frysinger [6] between 5 and 45 °C and up to a pressure of 6900 bar. The activation energy of the ionic conductivity of KCl solutions exhibits a change similar to that of viscosity. In HCl solutions, however, the changes in the activation energies of these two processes differ significantly. The deviations have been interpreted by supposing that at relatively low pressures the rotation of water molecules along the pathway of the migrating ion is the rate-determining process in the migration of hydrogen ions, but under pressures higher than about 1400 bar, the rotation is hindered and the jump of the proton from one water molecule to the other one becomes the rate-determining factor. The experimental discovery that KOH solutions are similar to KCl rather than HCl solutions with respect to changes in the activation energy should also be noted, though this has not been interpreted satisfactorily.

According to the studies of Pearson, Copeland and Benson [7], the conductivity of HCl solutions as a function of temperature shows a maximum at about 350 °C under high pressures, the conductivity falling rapidly after this maximum. A small maximum can also be observed in the conductivity of KCl solutions. This maximum in conductivity has been interpreted by considering changes in the dissociation constant (cf. Section 4.3).

The conductivity of aqueous solutions of some electrolytes has been investigated by Quist and Marshall [8] in a broad pressure and temperature range (1–4000 bar, 0–800 °C). It has been established by them that, under high pressures, there is a maximum in the conductivity of H_2SO_4, $KHSO_4$, K_2SO_4, HCl, HBr, as well as some alkali halide solutions, as a function of temperature (at 160 °C in H_2SO_4 solutions, at about 250 °C and 300 °C in HBr and KBr solutions, respectively, under a pressure of 500 bar, and at about 300 and 450 °C under 4000 bar). They have concluded from the measurements that the strength of acids decreases with increasing temperature and specific volume of the solution. Changes in the water structure

Literature on page 346

influence ionization significantly, in which hydration plays an important role, e.g.:

$$HBr(H_2O)_j + k\,H_2O \rightleftharpoons H(H_2O)_m^+ + Br(H_2O)_n^-,$$

and in this process, j, k, m, depend appreciably on temperature. Under high pressures and at high temperatures, NaBr is also a weak electrolyte. Above 400 °C, the equivalent conductivity of NaBr is independent of temperature calculated for constant density. Extrapolating to zero density, the equivalent conductivity of several 1–1 electrolytes (perhaps all of them) is approximately identical.

On the basis of the conductivity of tetra-alkyl ammonium halides measured in aqueous solutions in the 4–25 °C temperature range and under pressures up to 4000 bar, Horne and Young [9] confirmed the earlier opinion that the hydration of tetra-alkyl ammonium ions differs in nature from that of 'normal' (e.g. alkali metal) ions because of the hydrophobic nature of alkyl groups. In the vicinity of these large hydrophobic ions, coulombic forces play a less pronounced role in hydration than they do around simple ions, and the hydrostatic pressure influences hydration much less. Furthermore, it seems that in solutions of the large ions of this series, cation–cation interactions should also be taken into account. It should be noted that conductivity shows a maximum with increasing pressure in solutions of tetrasubstituted ammonium ions.

The investigation of changes in electrolytic conduction on the effect of pressure seems to be promising for further interpretation of the mechanism of ionic migration. However, owing to the low compressibility of liquids, applicable results for this purpose can be obtained only in experiments under very high pressures. Such experiments are accompanied by great technical difficulties, therefore, only the recent development in experimental methods has made it possible to perform them. The data available are not sufficient to outline a general theory on the effect of pressure.

REFERENCES

to Sections 4.4 and 4.5

[1]   G. J. HILLS, Chemical Physics of Ionic Solutions (ed. by B. E. CONWAY and R. G. BARRADAS), p. 521 (New York, 1966); S. B. BRUMMER and G. J. HILLS, Trans. Faraday Soc., 1961, 57, 1816.
[2]   D. M. NEWITT and A. WASSERMANN, J. Chem. Soc., 1940, 735.
[3]   A. JOBLING and A. S. C. LAWRENCE, Proc. Roy. Soc., 1951, A. 206, 257; J. Chem. Phys., 1952, 20, 1296.
[4]   F. C. COLLINS, J. Chem. Phys., 1957, 26, 398.
[5]   B. HOWARD, Dissert. (London, 1963) cited by HILLS [1].
[6]   R. A. HORNE, B. R. MYERS and G. R. FRYSINGER, J. Chem. Phys., 1963, 39, 2666.
[7]   D. PEARSON, C. S. COPELAND and S. W. BENSON, J. Amer. Chem. Soc., 1963, 85, 1044, 1047.
[8]   A. S. QUIST, E. U. FRANCK, H. R. JOLLEY and W. L. MARSHALL, J. Phys. Chem., 1963, 67, 2453; A. S. QUIST, W. L. MARSHALL and H. R. JOLLEY, J. Phys. Chem., 1965, 69, 2726; A. S. QUIST and W. L. MARSHALL, J. Phys. Chem., 1965, 69, 2984; 1966, 70, 3714; 1968, 72, 1545, 2100; 1969, 73, 978.
[9]   R. A. HORNE and R. P. YOUNG, J. Phys. Chem., 1968, 72, 1763.

4.6 CONDUCTIVITY OF WEAK ELECTROLYTE SOLUTIONS

Fundamentally, the mechanism of the transport of electricity in solutions of weak electrolytes does not differ in principle from that of strong electrolytes. However, the various factors determining electric conductivity influence the conductivity of weak electrolyte solutions in different ways than that of strong electrolytes. The main differences are in the ratio of the electrolyte present in ionic and non-ionic states and in the extent of electrostatic interaction between ions.

In solutions of weak electrolytes, the electrolyte is mostly present in the form of non-dissociated covalent molecules and only a small portion corresponding to the degree of dissociation exists as ions. The degree of dissociation is small, even in the most dilute solutions of weak electrolytes that can still be studied satisfactorily in experiments, and it decreases strongly with increasing concentration, (e.g. in 0·001 M acetic acid it is 0·12, while in 0·1 M solution it is 0·014 at 25 °C), unlike the dilute solutions of strong electrolytes in which dissociation is complete or, at least, the degree of dissociation does not differ too much from unity, so its relative change with dilution is small. On the other hand, since the concentration of ions is much lower because of the small degree of dissociation as compared with the solutions of strong electrolytes of the same concentration, the electrostatic interaction between ions and its change with concentration are much smaller in the solutions of weak than of strong electrolytes. Consequently, the concentration and temperature dependence of the conductivity of weak electrolytes arises mainly from the change in the degree of dissociation, and the electrostatic interaction of ions appears only as a small correction factor, unlike the conductivity of dilute solutions of strong electrolytes which depends on concentration mainly because of the electrostatic interaction of ions, while the change in the degree of dissociation of the small fraction of non-dissociated molecules incidentally present can be taken into account at most as a small correction. However, the formation of ion pairs kept together by electrostatic forces cannot be excluded in principle in solutions of weak electrolytes either, although this can take place only to a small extent in comparison to the formation of non-dissociated covalent molecules, and these two phenomena cannot be distinguished on the basis of conductivity.

The effect of the change in the degree of dissociation on the conductivity of weak electrolytes can be described in first approximation by the *Ostwald dilution law* deduced from the law of mass action. According to the law of mass action, the relationship between the degree of dissociation α and concentration c of the ideal solution of a binary electrolyte is:

$$\frac{\alpha^2 c}{1 - \alpha} = K_c \tag{4.6.1}$$

in which K_c is a concentration-independent dissociation constant (ioniza-

Literature on page 351

tion constant). If the electrostatic interaction of ions is negligibly small in the solution, so that ionic mobilities in a solution of concentration c do not differ appreciably from the limiting value of infinite dilution, and one has:

$$\frac{\Lambda}{\Lambda^0} = \alpha \tag{4.6.2}$$

and

$$\frac{\Lambda^2 c}{\Lambda^0(\Lambda^0 - \Lambda)} = \frac{\Lambda^2 c}{\Lambda^{0^2}\left(1 - \dfrac{\Lambda}{\Lambda^0}\right)} = K_c. \tag{4.6.3}$$

In order to emphasize the concentration dependence of conductivity, this relationship can be re-written in the following form:

$$1 - \frac{\Lambda}{\Lambda^0} = \frac{\Lambda^2 c}{\Lambda_0^2 K_c}, \tag{4.6.4}$$

and

$$\frac{1}{\Lambda} = \frac{1}{\Lambda^0} + \frac{\Lambda c}{\Lambda^{0^2} K_c}. \tag{4.6.5}$$

In this way, $1/\Lambda$ depends linearly on the product Λc in weak electrolyte solutions, in first approximation, and the slope of the straight line $1/\Lambda$ vs Λc is $1/\Lambda^{0^2} K_c$. If the limiting value of the conductivity of the ions of the weak electrolyte is known (from measurements in strong electrolytes), the dissociation constant of the weak electrolyte can be calculated from the Ostwald dilution law. When $K_c < 10^{-5}$, then $\alpha \ll 1$ and $\dfrac{\Lambda}{\Lambda^0} \ll 1$ in equations (4.6.1) and (4.6.4). Under such conditions, the concentration dependence of conductivity becomes:

$$\lg \Lambda = \text{const.} - \frac{1}{2}\lg c, \tag{4.6.6}$$

which shows a good agreement with experiments for dilute solutions.

For the dissociation equilibrium of weak electrolytes, the approximations given above are satisfactory only when, on one hand, the changes in equivalent conductivity are only due to the concentration dependence of the number of ions (i.e. their mobility in concentration c is the same as in infinite dilution), and on the other hand, when the activity coefficients of the ions and the non-dissociated molecules are unity: $\gamma_{ca} = \gamma_{mol} = 1$. Corresponding to the high accuracy requirements made possible by the methods of conductivity measurements, these criteria are met only in the most dilute solutions of weak electrolytes ($K \leq 10^{-5}$, $c < 10^{-3}$ mole l^{-1}). In a more accurate calculation valid for somewhat more concentrated solutions as well, it should be considered that, on one hand, the ionic mobility in a solution of concentration c is not identical with the limiting value of the conductivity of the ion–because of the electrostatic interaction–but its value corresponds to ionic concentration αc. If total dissociation is assumed, the equivalent conductivity of a solution of ionic concentration αc would be Λ_i, then:

$$\alpha = \frac{\Lambda}{\Lambda_i}. \tag{4.6.7}$$

However, except the most dilute solutions, activity coefficients differ from unity, thus, in the expression of the dissociation constant, activities should be taken into account instead of concentrations, i.e.:

$$\frac{\gamma_{ca}\,\alpha^2 c}{\gamma_{mol}\,(1-\alpha)} = K_a. \tag{4.6.8}$$

Taking these factors into account, a procedure has been outlined by MacInnes and Shedlovsky [1] for the exact calculation of the dissociation constant from conductivity data.

In a weak electrolyte solution of concentration c, the value of Λ_i corresponding to total dissociation in the above sense is:

$$\Lambda_i = \Lambda^0 - (A_1\Lambda^0 + A_2)\frac{\sqrt{\alpha c}}{1 + Ba\sqrt{\alpha c}} \tag{4.6.9}$$

on the basis of equation (4.2.68). Since ionic concentration is small ($\alpha c <$ $< 10^{-3}$ mole l^{-1}), the value of Λ_i does not depend too much on the ionic size a, hence the calculations can be performed by using an average value estimated to be $a = 4$ Å. The mean activity coefficient of the ions is, in good approximation:

$$\lg \gamma_{ca} = -\frac{A\sqrt{\alpha c}}{1 + Ba\sqrt{\alpha_c}} \tag{4.6.10}$$

on the basis of the Debye–Hückel theory of strong electrolytes (cf. Section 5.1.3). The activity of neutral molecules depends much less on concentration than that of ions, therefore, in the second approximation, it can be assumed that $\gamma_{mol} \approx 1$.

The dissociation constant of acetic acid calculated by the second approximation in the above sense is really constant within the experimental error in solutions of concentration lower than $c = 0\cdot006$ mole l^{-1} (Table 4.8). In solutions of higher concentration, however, the value of K_a decreases, indicating that even the second approximation is not satisfactory under such conditions. Probably, the deviations are caused mainly by the change in the viscosity of the solution (which has been disregarded above), and γ_{mol} is evidently not unity in such concentrations. According to Katchalsky, Eisenberg and Lifson [2], dimerization of the acid can also play a role in these deviations.

Literature on page 351

Table 4.8

The dissociation constant of acetic acid at 25 °C

$c,$ mole $\cdot$ l^{-1}	$\Lambda_{obs.}$	$\dfrac{\Lambda_{obs.}}{\Lambda^0}$	Λ_i	$\dfrac{\Lambda}{\Lambda_i} = \alpha$	$K_a \cdot 10^5$
0·000 028 01	210·38	0·5384	390·13	0·5393	1·753
0·000 111 35	127·75	0·3270	389·81	0·3277	1·754
0·000 218 4	96·493	0·2470	389·62	0·2477	1·752
0·001 028 3	48·146	0·1232	389·05	0·1237	1·751
0·002 414	32·217	0·0825	388·63	0·08290	1·752
0·005 912	20·962	0·0537	388·10	0·05401	1·750
0·020 0	11·566	0·0296	387·16	0·02987	1·740
0·050 0	7·358	0·0188	386·27	0·01905	1·726
0·100 0	5·201	0·0138	385·46	0·01349	1·700
0·200 0	3·651	0·0093	384·54	0·00949	1·653

For the determination of the dissociation constant of weak electrolytes, some other methods are also available: firstly, the measurement of the electromotive forces of corresponding Galvanic cells, as well as some spectrophotometric methods [4] in the case of coloured substances. Recently, Christensen, Oscarson and Izatt [5] have dealt with the thermodynamics of the dissociation of weak acids, giving accurate data on the dissociation constants of several acids.

The evaluation of the conductivity of weak acid solutions containing nonelectrolytes according to the Fuoss–Kraus procedure has been interpreted by Kortüm and Wenck [6].

Weak electrolyte solutions are generally regarded as solutions of free ions in a solvent consisting of neutral molecules in which the non-dissociated molecules (recombined ion pairs) of the electrolyte play the same role with respect to the ions as the solvent molecules do, and they have no special feature in conducting electricity. Recently, however, Onsager and Provencher [7] pointed out that this is an oversimplified picture of reality, at least in relation to the relaxation effect, since the theory of electrostatic interaction cannot be applied for solutions of weak electrolytes without restrictions, and the assumption that the relaxation and electrophoretic effects of the ionic atmosphere, as well as the chemical equilibrium, depend only on the concentration of free ions is not justified. In order to describe real conditions more reliably, the kinetics of the recombination of ions to molecules should also be taken into account, because this process does not take place with an infinitely high rate, either. The detailed analysis led to the consequence that, if the rate of recombination is at least commensurable with the rate permitted by diffusion, and further, if the average lifetime of the associates (molecules) is at least commensurable to the relaxation time of the ionic atmosphere, this gives rise to an appreciable decrease in the relaxation effect. The Debye–Falkenhagen theory on the dispersion of conductivity has been modified accordingly by Onsager and Provencher, and the effect of a weak stationary field on the state of binary electrolyte solutions of strong association has also been discussed in detail.

REFERENCES

to Section 4.6

[1]   D. A. MacInnes and T. Shedlovsky, Amer. Chem. Soc., 1932, **54,** 1429.
[2]   A. Katchalsky, H. Eisenberg and S. Lifson, J. Amer. Chem. Soc., 1951,
       73, 5889.
[3]   See e.g., H. S. Harned and R. W. Ehlers, J. Amer. Chem. Soc., 1932, **54,**
       1350; H. S. Harned and B. B. Owen, Chem. Rev., 1939, **25,** 31; R. G. Bates
       and G. D. Pinching, J. Amer. Chem. Soc., 1950, **72,** 1393.
[4]   See e.g. A. I. Biggs, Trans. Faraday Soc., 1954, **50,** 800; R. A. Robinson and
       A. I. Biggs, Trans. Faraday Soc., 1955, **51,** 901.
[5]   J. J. Christensen, J. L. Oscarson and R. M. Izatt, J. Amer. Chem. Soc.,
       1968, **90,** 5949.
[6]   G. Kortüm and H. Wenck, Ber. Bunsenges., 1966, **70,** 435.
[7]   L. Onsager and S. W. Provencher, J. Amer. Chem. Soc., 1968, **90,** 3134.

4.7 EFFECT OF NON-ELECTROLYTES ON THE CONDUCTIVITY OF AQUEOUS ELECTROLYTE SOLUTIONS

4.7.1 GENERAL

The effect of non-electrolytes on the conductivity of aqueous electrolyte solutions has been dealt with in several investigations. Organic substances (e.g. sugars), as well as organic liquids soluble in water or miscible with it can be added as non-electrolytes. More concentrated aqueous solutions of organic liquids can be regarded as solvent mixtures too, and their investigation gives some information on the effect of the nature of the solvent on the transport of electricity by ions.

An appreciable part of the investigation in this field is focussed on the effect of the systematic variation of some properties of the solvent on the conductivity of electrolyte solutions. Such a property is usually the dielectric constant or the viscosity which can be varied in a broad range at constant temperature by dissolving non-electrolytes in water.

Such experiments give valuable information on the properties of electrolyte solutions, however, the theoretical interpretation of their results is very difficult, since the presence of a non-electrolyte in the aqueous electrolyte solution changes not only one single property of the liquid (e.g. the dielectric constant), but almost all the other characteristics influencing the transport of electricity. The dissolved non-electrolyte alters the liquid structure (cf. Section 1.4.1), as well as the effect of the electric field of the ions. The first solvate sphere (primary solvation) around the ion is altered since the electric field of ionic charges attracts the dipole molecules of the non-electrolyte, in addition to those of water,* and some of the water molecules around the ion are replaced by non-electrolyte molecules.

---

* In the above sense, only organic substances consisting of dipole molecules can be applied, since those of non-polar molecules usually do not dissolve appreciably in water.

Literature on page 391

On the other hand, the effect of the field of ions on the liquid layers located somewhat farther from the ion (i.e. secondary hydration) is also significantly altered because of the modification of the liquid structure by the non-electrolyte. The non-electrolyte–depending on its properties and concentration–partly deforms, loosens, or breaks some hydrogen bonds between water molecules, and it can partly form hydrogen bonds with water molecules and some other connections of van der Waals or chemical nature. Thus, not only the lattice-like structure of water, but the structure and size of the more or less ordered regions and their statistical equilibrium with monomeric water molecules is altered, and various regions of different composition and arrangement can also develop which influence the statistical distribution. Thus, some conclusions can be drawn on the structure of liquids from the investigation of the effects of non-electrolytes. Owing to these many-sided interactions, detailed and extended experimental investigations are required to interpret the conditions of electrolyte solutions containing non-electrolytes.

With respect to the transport of electricity, the viscosity and dielectric constant of aqueous solutions are those well-measurable macroscopical quantities which are primarily altered by the non-electrolyte which is simultaneously present in the solution. Thus, the main part of these investigations has dealt with changes in conductivity caused by changes in viscosity and dielectric constant under the effect of the non-electrolyte. Of course, the influence of these two factors cannot be sharply separated, because changes in viscosity are accompanied by changes in the dielectric constant and *vice versa*, further, both changes are the consequence of a complicated transformation of the liquid structure. Hence, the separation of the effects of viscosity and dielectric constant implies a lot of assumptions, it is rather uncertain and only the first orientation is facilitated by it. The effect of these two factors can be studied also in pure non-aqueous solutions of electrolytes, but, their structure being essentially different from that of aqueous solutions, the results can be correlated to the properties of aqueous solutions in some respects though in a more difficult way than in the investigation of non-electrolytes dissolved in water. The studies on non-aqueous solutions are outside the scope of the present book.

With respect to the effect of non-electrolytes on aqueous electrolyte solutions, there is an essential difference between electrolytes containing ions that exhibit pure hydrodynamic migration and those comprising ions that also transport electricity by the prototropic mechanism. In the former solutions, the fundamental mechanism of migration is not altered by the non-electrolyte and only its conditions are modified. However, in solutions of ions involved in prototropic conduction, the presence of the non-electrolyte–particularly when in high concentration–essentially alters the transport mechanism of electricity; the prototropic mechanism can be suppressed by the hydrodynamic migration, and its nature can also be altered when proton exchange between the molecules of the dissolved non-electrolyte takes over the role of proton exchange between water molecules with increasing concentration of the non-electrolyte. However, the effect of the non-electrolyte on prototropic conduction cannot be easily separated from its effect on the hydrodynamic migration, partly because the oxonium ions participating in prototropic conduction also migrate by the hydrodynamic mechanism, and partly because the other ions of the electrolyte conduct by hydrodynamic migration.

The effect of non-electrolytes on the mobilities of ions conducting by the hydrodynamic mechanism appears in the conductivity in several respects as in diffusion (see Section 3.3.3); however, in the prototropic conduction the differences can be much greater.

With respect to the effect of non-electrolytes on the conductivity of aqueous solutions, some distinction should be made between the change in the limiting value of ionic conductivities in infinite dilution and the changes in conductivity in solutions of finite concentration relative to the former, which are caused mainly by the electrostatic interaction between ions. The latter effect can be studied by investigating the concentration dependence of conductivity in a given medium (water + a non-electrolyte present in a given concentration); for the change in the limiting value of mobility, valuable information can be obtained from the dependence of the conductivity of very dilute solutions on the nature of the non-electrolyte and its concentration. In other respects, however–according to the above statement–there is an essential difference between the effects exerted by non-electrolytes on the hydrodynamic migration of ions and on the conduction by the prototropic mechanism.

The effect of the change of the dielectric constant on the conductivity of electrolyte solutions is discussed in Section 4.2.3.4, whereas some other aspects will be dealt with below.

4.7.2 EFFECT OF CHANGES IN VISCOSITY DUE TO NON-ELECTROLYTES ON HYDRODYNAMIC MIGRATION

One of the determining factors in ionic conductivity (relative mobility) is viscosity. According to the simple hydrodynamic theory of ionic migration (see Section 4.2.1), the change in the viscosity of a solution modifies the ionic conductivity in such a way that the product $\lambda^0 \eta_0$ remains constant if the radius r of the ion is unchanged (see equation (4.2.6)). It has been established by Walden et al., [1] that the limiting value of the equivalent conductivity of tetra-alkyl ammonium picrate is:

$$\Lambda^0 \eta_0 = \text{const.} \tag{4.7.1}$$

in several solvents (at different temperatures, as well). This is the so-called Walden rule. This relationship–if indeed proved to be correct–would justify the assumption that the radii of the relatively large ions of this electrolyte are the same at every temperature and in each of the solvents investigated, i.e. these large ions are not appreciably solvated or hydrated. Furthermore, this would justify the assumption that microviscosity determining the translational movement of ions is identical with the viscosity of the solutions measured macroscopically. On the basis of these assumptions, Walden concluded, taking into account the known equivalent conductivity of the picrate ion in the aqueous solution, that for this ion $\lambda^0 \eta_0 = 0.270$ cm^2 Ω^{-1}eq^{-1} poise, and that the relationship $\lambda^0 \eta_0 =$ constant is also valid for the limiting value of the equivalent conductivity of large ions separately. According to the recent investigation of Kraus [2a], however, this relationship does not hold for the picrate ion either, e.g. $\lambda^0 \eta^0 = 0.30$ in pyridine, and $\lambda^0 \eta_0 = 0.24$ in ethylene dichloride, and–in the lack of direct data on the transference number–ionic conductivity can be estimated on the basis of the assumption that the mobilities of the tetra-n-butyl ammonium ion and the tris-phenyl borofluoride ion are identical in every solvent (within about 5 per cent). Calculating on this basis, however, the product $\lambda^0 \eta_0$ for both the picrate ion and the tetraethyl ammonium ion varies with the solvent. Moreover, as Kraus' accurate investigations showed, even the product $\Lambda^0 \eta_0$ of tetraethyl ammonium picrate is not constant. The Walden rule is even less valid in solutions of small ions.

The approximative nature of the Walden rule has been proved by the measurements of Kay, Evans and Cunningham [2b] on electrolytes containing large 'hydrophobic' cations, too. According to them, in aqueous solutions of the halides of i-Am$_3$BuN$^+$ and n-Am$_4$N$^+$, the product $\Lambda^0 \eta_0 \approx$ ≈ 0.160, while in methanol and acetonitrile it is ≈ 0.194. This variation has been explained by the assumption that the hydrocarbon groups form hydrogen bonds with the neighbouring water molecules to a higher extent than with the molecules of the non-aqueous solvents investigated. According to them, the degree of ion pair formation is appreciably higher in solvents containing hydroxyl groups than could be expected on the basis of the simple electrostatic theory, which can be attributed to the possibility of association between ion pairs in immediate contact with each other and separated by solvent molecules as well.

Although the Walden rule is not rigorously valid, it is not doubted that the product $\Lambda^0 \eta_0$ depends much less on the nature of the solvent than equivalent conductivity and viscosity do separately. This verifies that Stokes' Law, giving the basis of the Walden rule, reflects the essential feature of the translational movement of the ions correctly, but some other significant interactions should also be accounted for which are disregarded in Stokes' Law. In studying these latter important factors, the investigation of the change in the product $\Lambda \eta$ is indispensable.

The deviations from Stokes' Law can evidently be attributed partly to the alteration of the size of the ions as kinetic entities with the composition of the solvent, due to the primary solvation (see Section 5.2.2), and partly to the fact that the liquid structure, and so the microviscosity, are altered differently and to a different extent by the electric field of the various ions, depending on their nature. In various salts, the conditions are very different, therefore, the experiments regarding the changes in conductivity of aqueous electrolyte solutions due to the presence of non-electrolytes give valuable information in certain respects on the mechanism of ionic migration. The theoretical evaluation of such experiments, however, is hindered by the fact that direct interaction of the ions with the non-electrolyte molecules (primary solvation) also alters the migrating kinetic entity and viscosity can also be changed by various factors.

In order to describe the relationship between ionic mobility and the viscosity of the solution, Stokes et al., [3] have determined the limiting conductivity of several ions in sugar, mannitol and glycerol solutions (Table 4.9). The limiting value of the equivalent conductivity of the ions is not influenced directly by the dielectric constant, since the electrostatic interaction of the ionic charges, as well as ion association, is eliminated from these data. Nevertheless, the factors which caused the change in the limiting conductivity and viscosity are not simple, since these solutions cannot be considered as uniform media. The change in viscosity caused by the non-electrolyte is also influenced by the field of the ions, further, the size of the ion as a migrating kinetic entity and its other properties, as well as the effect of the ion surrounded by the primary solvate sphere on the adjacent liquid layers (i.e. the secondary hydration, cf. Section 5.2.3) are also altered by the solvation competing with the hydration of the ions. Since, however, these effects are combined, and no method is known to separate them unequivocally in experimental investigations, one has to be satisfied with enlightening these complex effects by changing the conditions in various ways and with trying to gain some information about the influence of the single

Table 4.9

The limiting conductivity of some ions in aqueous solutions containing saccharose, mannitol and glycerol at 25 °C

Ion	Non-electrolyte concentration	Saccharose				Mannitol 10 per cent		Glycerol			
		10 per cent		20 per cent				10 per cent		20 per cent	
	relative fluidity $\dfrac{\eta_{water}}{\eta_{ne}}$	0·756		0·525		0·747		0·775		0·579	
		λ_{ne}^0	$\dfrac{\lambda_{ne}^0}{\lambda_{water}^0}$	λ_{ne}^0	$\dfrac{\lambda_{ne}^0}{\lambda_{water}^0}$	λ_{ne}^0	$\dfrac{\lambda_{ne}^0}{\lambda_{water}^0}$	λ_{ne}^0	$\dfrac{\lambda_{ne}^0}{\lambda_{water}^0}$	λ_{ne}^0	$\dfrac{\lambda_{ne}^0}{\lambda_{water}^0}$
H^+		294·3	0·841	230·2	0·658	292·8	0·837	—	—	—	—
K^+		59·7	0·812	46·1	0·627	58·6	0·797	60·0	0·817	47·6	0·648
Na^+		40·6	0·810	31·1	0·621	39·6	0·790	40·8	0·819	32·4	0·647
Li^+		31·0	0·802	23·6	0·616	30·1	0·778	—	—	—	—
Ag^+		49·5	0·800	37·6	0·607	48·3	0·780	49·6	0·801	39·1	0·632
Ca^{2+}		46·8	0·787	34·8	0·589	—	—	—	—	—	—
Mg^{2+}		41·8	0·788	30·9	0·582	—	—	—	—	—	—
La^{3+}		54·2	0·778	39·5	0·567	—	—	—	—	—	—
$N(n\text{-}Am)_4^+$		13·3	0·761	9·6	0·550	—	—	—	—	—	—
Cl^-		62·2	0·815	48·2	0·631	61·1	0·800	62·1	0·813	49·2	0·644
Br^-		63·1	0·807	48·4	0·619	62·3	0·797	63·0	0·806	49·4	0·632
I^-		61·2	0·796	46·4	0·604	60·9	0·792	61·4	0·799	47·4	0·617
NO_3^-		57·9	0·810	44·6	0·624	57·4	0·803	58·4	0·817	46·0	0·644
ClO_4^-		54·1	0·803	41·2	0·612	—	—	—	—	—	—

Literature on page 391

factors. The problem is far from solved, but several observations are available already.

Stokes *et al.* [3, 4] have measured the concentration dependence of electrolytes in aqueous solutions containing 10 and 20 per cent of saccharose and glycerol, respectively, and 20 per cent of mannitol. They have calculated the limiting values of the equivalent conductivity of the ions investigated in aqueous solutions containing these non-electrolytes by means of extrapolation (λ^0_{ne}, see Table 4.9). If the Walden rule were valid ($\lambda^0\eta = $ constant) the ratio $\lambda^0_{ne}/\lambda^0_{water}$ (λ^0_{water} is the limiting value of the conductivity of a given ion in pure aqueous solution) should be equal to the ratio of the fluidities, i.e. with the ratio η_{water}/η_{ne}. It can be seen in the Table that this relationship does not hold for any ion investigated. The relative mobility ($\lambda^0_{ne}/\lambda^0_{water}$) is higher for each ion than the relative fluidity of the solution, i.e. the decrease in viscosity. The values for the large tetra-*n*-amyl ammonium ion show the best approach to one another, but, in a sugar solution of 20 per cent concentration, the relative mobility is higher by about 5 per cent than the relative fluidity. The relative mobility of the cations investigated increases in the order $N(Am)_4^+ < La^{3+} < Mg^{2+} < Ca^{2+} < Ag^+ < Li^+ < Na^+ < K^+ (< H^+)$ which (may be with the exception of Ag^+) corresponds to the order of ionic sizes. A similar relationship is valid for the anions, too. The decrease in mobility depends somewhat on the nature of the non-electrolyte as well, since in solutions of various non-electrolytes of identical fluidity, the relative mobilities of the same ion differ somewhat from each other, although viscosity has the predominating effect. Thus, the increase in viscosity decreases conductivity the more, the larger the ion, but the ratio of the decrease is smaller than that expected according to the Walden rule, only the largest ions approaching this value.

The decrease in ionic mobilities under the effect of dissolved non-electrolytes has been qualitatively explained by Stokes *et al.* [4] by supposing that the ions, large in comparison with the molecules of the solvent and the non-electrolyte, migrate in such a way—in accordance with the original Stokes' Law—that they push away the molecules of both water and the non-electrolyte. For small ions, however, the non-electrolyte molecules are obstacles to go round and for them the medium in which they move consists essentially of water molecules. In this way, the increased frictional resistance with respect to small ions can be regarded as a consequence of the lengthening of the pathway they move along in the course of their migration due to going round these obstacles ('obstruction theory'). A quantitative treatment of this phenomenon is not available yet—mainly because there is no reliable theory of the structure of liquids.

It has been shown, however, by Stokes *el al.* [4] that the validity of the Walden rule cannot be expected theoretically for the change in viscosity on the effect of non-electrolytes, either. Namely, in first approximation the aqueous solution of a non-electrolyte can be considered to be a material system in which isolated spheres (large in comparison with the water molecules and the ions) are suspended in a conducting medium. As for the conductivity changes caused by dielectric particles suspended in liquids, it can be established, on the basis of the investigations of Fricke [5a] and others, that if the suspended particles are large in comparison with the molecules of the liquid and their concentration is low, the following relation is valid for the specific conductivity $\varkappa'$ of the suspension:

$$\frac{\varkappa'}{\varkappa} = 1 - 1 \cdot 5 \, \Phi,\qquad\qquad(4.7.2)$$

where $\varkappa$ is the conductivity of the pure liquid, and Φ is the volume fraction of the dielectric spheres. In order to compare this with the relative mobility, it should be transformed to equivalent conductivity, since the number of equivalents is related to the total volume of the solution in these latter calculations, while, in equation (4.7.2) above, $\varkappa$ refers to the volume outside the dielectric particles. Performing this transformation according to Stokes, the limiting value of the equivalent conductivity of small ions in the presence of dielectric spheres of volume fraction Φ is:

$$\frac{\lambda^0_{ne}}{\lambda^0_{water}} = \frac{1 - 1 \cdot 5 \, \Phi}{1 - \Phi} \approx 1 - 0 \cdot 5 \, \Phi. \qquad\qquad(4.7.3)$$

In contrast with this, the macroscopic relative viscosity is:

$$\frac{\eta_{ne}}{\eta_{water}} = 1 + 2 \cdot 5 \, \Phi \qquad\qquad(4.7.4)$$

according to the Einstein relationship (see Section 2.4.2) under identical conditions (when Φ is small). Thus, it is seen that non-electrolyte molecules, if they can be considered dielectric spheres in first approximation, influence ionic mobility and viscosity to different degrees. For example, isolated spheres present in such a concentration that the viscosity of water is increased by 5 per cent, decrease the equivalent conductivity by only 1 per cent.

This relationship was not confirmed quantitatively by experiments either, because in each case investigated, conductivity decreased to a higher degree than could be expected on the basis of the above deduction (e.g. 5 per cent increase in viscosity decreases the mobility of small ions by up to 3–3·5 per cent). The real conditions are, thus, somewhere in between the behaviour corresponding to Stokes' Law (expected to hold when the ions are very large in comparison with the molecules of water and the non-electrolyte) and the behaviour corresponding to the obstruction theory (which can be expected when the molecules of the non-electrolyte are very large in comparison with the water molecules and the ions). This can be attributed to several reasons. One is that the size of the non-electrolyte molecules investigated is not large enough with respect to the size of the water molecules and the ions and further, the concentration of the non-electrolyte is also higher than that justifying the approximation given above. These interpretations thus imply only qualitatively one of the reasons

Literature on page 391

giving rise to the deviations from the Walden rule or the simple Stokes' Law.

It is also seen in Table 4.9 that the change in mobility caused by the non-electrolyte depends not only on viscosity, but on some other factors as well. In solutions of different non-electrolytes with identical viscosities, a given ion has not the same mobility. The change in viscosity due to glycerol decreases ionic mobility to a somewhat higher degree than that due to mannitol and saccharose. This can be explained in at least two ways, according to Stokes:

(1) ions are somewhat solvated by glycerol molecules,

(2) the glycerol molecules are smaller than those of saccharose and mannitol, thus the ionic mobilities are closer to those corresponding to Stokes' Law.

Moulik [5b] studied the conductivities of aqueous solutions of electrolytes containing urea, saccharose, mannitol and glycerol. According to him, Fricke's obstruction theory is unsatisfactory for the description of the variations in conductivity; apart from other factors, the change in the association of the electrolyte in the presence of a dissolved non-electrolyte must also be taken into account.

In addition to these factors, the change in the mobility of the hydrogen ion is evidently connected with the prototropic conduction mechanism (see Section 4.7.3).

There is a complex relationship between the concentration of the non-electrolyte present in the solution and the conductivity caused by the electrolyte present. The theory of this relationship has been studied by Ebeling, Falkenhagen and Kraeft [6].

Summarizing, it can be stated that no accurate and general relationship is known between the viscosity of a solution and the mobility of the ions. It can hardly be doubted, however, that one of the factors influencing this relationship is the ratio of the sizes of the molecules and ions.

The role of the change in viscosity in the alteration of ionic mobilities can be elucidated from another side, too. The effect of the change in viscosity due to dissolved non-electrolytes has been compared by Robinson and Stokes [7] with the effect of the change in viscosity due to changes in temperature. It can be seen in Fig. 4.16 that the two types of viscosity effects are not identical. The Walden rule holds for the temperature-induced change in

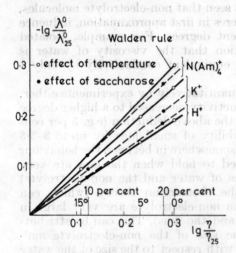

Fig. 4.16 Relationship between the variations of ionic conductivity and viscosity, when viscosity is altered by temperature (○) or by saccharose (•). The viscosities and conductivities are referred to the values measured in pure water at 25 °C

mobility only in the case of the largest tetra-amyl ammonium ion, while the change in mobility due to saccharose is already somewhat lower. For other ions–whose typical representative is the K^+ ion–the relationship $\lambda^0 \eta^p = $ constant approximately describes this change where the power $p < 1$, instead of the former value. The mobility is usually decreased to a higher degree by the increase in viscosity due to the presence of a non-electrolyte than that due to the decrease in temperature. Thus, the effect of non-electrolytes cannot be attributed simply to the change in the water structure (to the change of 'structural temperature'), but it also has some specific effects on conductivity. It is to be noted that the change in the mobility of the hydrogen ion with viscosity deviates appreciably from the Walden rule ($p \approx 0 \cdot 6$), still the effects of temperature and saccharose hardly differ and the small difference has a sign opposite to that observed for other ions. This is evidently a consequence of the prototropic conduction mechanism, and, perhaps, it is connected with the rotation of water molecules, which is probably the rate-determining step in the prototropic conduction mechanism under common conditions: this becomes hindered on decreasing the temperature, as does the effect of the presence of saccharose molecules in the solution.

Several research workers have dealt with the conductivity of aqueous electrolyte solutions containing non-electrolytes, including also concentrated solutions of non-electrolytes (mainly alcohols and dioxan) in which the non-electrolyte can be regarded rather as a component of the solvent mixture [8a]. In most of these investigations, the dependence of conductivity on the concentration of the electrolyte has mainly been studied in non-electrolyte solutions of given composition, aiming primarily at the interpretation of the electrostatic interaction of ions and, secondly, at the investigation of association and the determination of the limiting value of conductivity in the given medium. (Recently, e.g. Amis and Castel [8b] have found that the dependence of the conductivity of $MgSO_4$, dissolved in a water–ethanol mixture, on the concentration of the electrolyte slightly deviates from that expected on the basis of the Onsager equation.) However, experimental investigations focussed on the conductivity of a given electrolyte in a medium of systematically varied composition, i.e. in aqueous solutions containing various amounts of different non-electrolytes, have also contributed appreciably to explaining the extremely complicated conditions in solutions. Since the conduction mechanism of hydrogen and hydroxide ions differs from that of other ions, the results of investigations on the variation of the conductivity of these ions should be treated separately from those of electrolytes conducting only by hydrodynamic migration.

The conductivity and transference numbers of KF and KCl have been investigated extensively as a function of temperature in aqueous solutions of various non-electrolytes (or in solvent mixtures of water and a non-electrolyte) by Erdey-Grúz et al. Of the properties of the medium, the effect of change in viscosity on the conductivity has been studied by them.

Literature on page 391

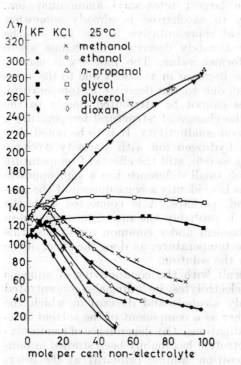

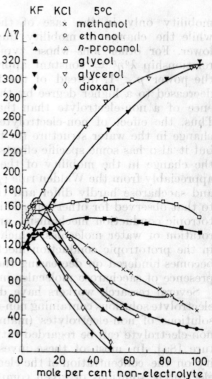

Fig. 4.17 The product $\Lambda\eta$ of KF and KCl in aqueous solutions of non-electrolytes at 5 °C

Fig. 4.18 The product $\Lambda\eta$ of KF and KCl in aqueous solutions of non-electrolytes at 25 °C

Measuring the conductivity of very dilute solutions of KCl and KF in the presence of non-electrolytes soluble in water, Erdey-Grúz *et al.* have established that the product $\Lambda\eta$ is increased by methanol [9], ethanol [10], propanol [11], and dioxan [12] present in low concentrations, while on increasing the concentration of the non-electrolyte to higher than 4–10 mole per cent, the product $\Lambda\eta$ decreases after passing a maximum (see Figs 4.17 and 4.18: the concentration of the electrolyte is 0·01 mole l^{-1}). In aqueous solutions of each of these non-electrolytes, viscosity also has a maximum as a function of concentration (Fig. 4.19), and this maximum is the higher and sharper, the lower the temperature. The increase in viscosity and its maximum indicate that the liquid structure is strengthened by the non-electrolyte (which is connected with the successive occupation of the structural cavities of water, see below); complexes less mobile than the components are formed which undergo progressive decomposition with increasing electrolyte concentration. It is also in accordance with this that a maximum appears in the activation energy of the viscous flow of these non-electrolyte solutions, too (Fig. 4.20), though it is found at a somewhat lower concentration than the maximum of η. The maximum in the value of

$\Lambda\eta$ is probably related to the same phenomenon. The fact that in KF and KCl solutions the maximum of the $\Lambda\eta$ value appears at lower non-electrolyte concentration than the maximum of viscosity can probably be attributed to the structure-breaking effect of the field of ions. It is noteworthy that the activation energy of conductivity also has a maximum as a function of the concentration of the non-electrolyte in these aqueous electrolyte solutions (Fig. 4.38). This maximum is at a concentration somewhat higher than that corresponding to the maximum in the activation energy of viscous flow. The activation energy of conductivity is generally somewhat lower than that of viscous flow which indicates that the structure of the non-electrolyte solution, which can be regarded as a liquid mixture, is loosened by the electric field of the ions.

Aqueous solutions containing non-electrolytes in identical concentration have different dielectric constants (Fig. 4.21); this also influences the change of the product $\Lambda\eta$ under the conditions described above.

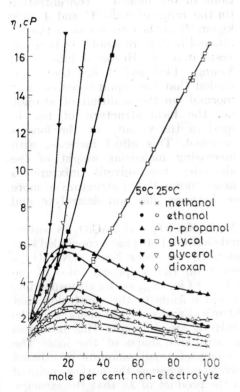

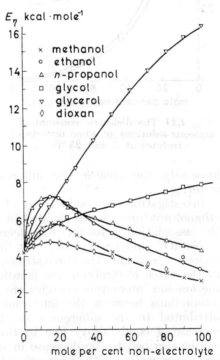

Fig. 4.19 The viscosity of aqueous solutions of some non-electrolytes at 5 and 25 °C

Fig. 4.20 The activation energy of viscous flow in aqueous solutions of some non-electrolytes

Literature on page 391

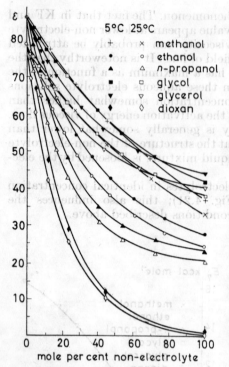

Fig. 4.21 The dielectric constant of aqueous solutions of some non-electrolytes at 5 and 25 °C

According to Pedersen and Amis [13], the conductivity of aqueous solutions of CsCl changes similarly under the effect of ethanol. A smooth minimum of equivalent conductivity is observed in solutions containing 70 weight per cent alcohol at 25 °C, while the product $\Lambda\eta$ has a significant maximum in about 22 weight per cent ethanolic solutions. In dilute solutions, the equivalent conductivity changes linearly with $\sqrt{c}$ also in the presence of ethanol. The temperature and pressure dependence of the conductivity of aqueous KCl solutions of 0·1 mole l^{-1} concentration (in the range of 0–20 °C and 1–4500 kg cm^{-2}) in the presence of methanol, ethanol and n-propanol has been investigated by Horne, Johnson and Young [14a] and it has been concluded that the liquid becomes more 'normal' on the addition of alcohol, i.e. the local structure of the liquid in the vicinity of the ions is loosened. This effect increases with increasing molecular weight of the alcohols; the solvent mixture is more 'normal' (its structure is more loosened), the greater the difference between the non-electrolyte and water.

Investigating the conductivity of NaCl, NaClO$_3$ and NaClO$_4$ in water–ethanol mixtures, Accascina, de Lisi and Goffredi [14b] have concluded that the association constant K_{ass} decreases in the order NaCl $>$ NaClO$_3$ $>$ $>$ NaClO$_4$, and log K_{ass} depends linearly on the reciprocal value of the dielectric constant. The three straight lines of log K_{ass} *vs.* 1/ε corresponding to the three electrolytes are parallel, which leads to the conclusion that the ion–ion interaction energies are identical and only the energies of the interactions between the ions and solvent are different. This has been attributed to the differences in the size and shape of the ions. The increase of conductivity due to the effect of small amounts of ethanol can be ascribed to the decrease in solvation, whereas the authors pointed out in connection with the increase in the product of $\lambda\eta$ that the stronger the structure of the liquid, the more reduced the efficiency of the structure-making effect of the ions; this is why the local viscosity decreases and mobility increases. The liquid structure is loosened with decreasing dielectric constant, and the structure-making effect of the ions increases progressively.

The conductivity of aqueous solutions of polyvalent electrolytes changes similarly with the effect of alcohols. Measuring the change in the conductivity of $CoCl_2$, $CoSO_4$, $MgSO_4$ and $Zn(ClO_4)_2$ solutions caused by ethanol, Davies and Thomas [15] have stated that in dilute solutions Λ depends linearly on $\sqrt{c}$, but the product $\Lambda\eta$ is not constant. In aqueous solutions of $MgCl_2$, the minimum of equivalent conductivity can be found in a solution containing about 70 weight per cent of alcohol, according to Than and Amis [16].

The conductivity of solutions of large ions also shows a similar variation. From the measurements of Foster and Amis [17] and Whorton and Amis [18] it can be seen that the conductivity of aqueous solutions of tetraethyl ammonium picrate also passes a minimum with increasing concentration of methanol or ethanol, while the product $\Lambda^0\eta$ exhibits a maximum.

The conditions are somewhat altered in the presence of polyvalent alcohols. In aqueous KF and KCl solutions containing ethylene glycol, Erdey-Grúz, Kugler and Hidvégi [19] have shown that the product $\Lambda\eta$ is almost independent of the concentration of the non-electrolyte (Figs 4.17 and 4.18), although the increase in viscosity is 44-fold at 5 °C and 17-fold at 25 °C within the concentration range investigated. In glycerol-containing aqueous solutions of KF and KCl, the product $\Lambda\eta$ increases monotonically up to three times the value of that measured in pure water, as shown by Erdey-Grúz and Kugler [20], while viscosity becomes 4300 times as high as its original value at 5 °C, and the increase is 1000-fold at 25 °C. The activation energy of viscous flow in ethylene glycol and glycerol is hardly higher than that of the conductivity. Accascina and Goffredi [21] have measured the conductivity of LiCl in mixtures of water and glycol. The conductivity of glycerol solutions of NaCl as a function of dielectric relaxation time and viscosity has been investigated in detail by Bartoli, Birch, Toan and McDuffie [22a]. They have stated that specific conductivity depends on temperature not according to the Arrhenius law. They concluded from their measurements that the movement of the ions is hindered by the dynamic-structural rearrangement of the solvent rather than by the retardation force arising from the continuum-viscosity.

The different properties of mixtures of water and glycol or glycerol present in low concentration and those of mixtures of monovalent alcohols and water are also probably connected with the fact that the rather large molecules of the former do not fit into the structural cavities of water (see below), therefore, the water structure is already altered by them in low concentration.

The particular effect of polyvalent alcohols on the ionic conduction of electricity has been indicated by the investigations of DeSieno, Greco and Mamajek [22b] on tetra-alkyl ammonium halides dissolved in glycol. Analysing their experimental data on the basis of the Fuoss–Onsager theory, they have not detected ionic association; however, the sizes of the ions in solution calculated from conductivity have been found to be smaller than

Literature on page 391

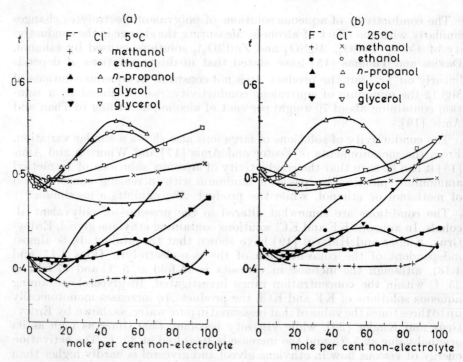

Fig. 4.22 The transference number of fluoride and chloride ions in KF and KCl
    solutions in the presence of non-electrolytes (a) at 5 and (b) at 25 °C

the crystallographic ones. On the other hand, the Walden product is strikingly high.

According to the studies of Erdey-Grúz and Majthényi [23–26a], the
transference number of fluoride and chloride ions in aqueous solutions of KF
and KCl is decreased equally by methanol, ethanol, propanol, ethylene
glycol and glycerol added in low concentrations (Figs 4.22(a) and (b)),
then, after passing a minimum in solutions of about 10–20 mole per cent
non-electrolyte content, it increases again with increasing concentration.
In solutions containing methanol [23], the minimum is very smooth, while
in ethanolic solutions [24], the curve of the transference number passes
a maximum at a non-electrolyte concentration of about 50 mole per cent.
In methanolic and ethanolic solutions, the transference number of the fluoride ion is smaller at 5 °C than at 25 °C within the whole concentration range
of the non-electrolyte; the relative temperature coefficient of the transference
number of chloride ions reverses its sign in methanolic and ethanolic solutions at a concentration of about 5 mole per cent and 10 mole per cent,
respectively, which is followed by another change of sign in the methanolic
solution. In glycol solutions, the relative temperature coefficient of the
transference number of the Cl⁻ ion and (in a higher concentration of the
non-electrolyte) that of the fluoride ion also reverses its sign. In glycerol

solutions, however, the transference number of the chloride ion is higher at 25 °C than at 5 °C, within the entire concentration range.

Carmo and Spiro [26d] carried out measurements in solutions of KCl in glycol, and established that the transference number of the Cl⁻ ion decreases on the effect of small amounts of water, when starting from pure glycol solutions.

Thus, it can be established that hydroxyl-containing non-electrolytes (mono- and polyalcohols) alter the mobilities of F^-, Cl^- and K^+ ions differently. A relatively low concentration of the non-electrolyte decreases the mobility of the F^- and Cl^- ions more than that of K^+ ions, while the conditions are reversed in solutions of higher concentration. This change (the extreme value of the transference number) occurs in solutions of alcohols at a concentration of about 5–20 mole per cent, thus, in this concentration range the non-electrolyte alters the water structure in such a way that, from this, the structural change favours the motion of the fluoride and chloride ions more than that of K^+ ions. This change is probably connected with the fact that in low concentration the molecules of the non-electrolyte are located in the cavities of the quasi-crystalline lattice of water (if their size is suitable to fit into them) and the water structure is only deformed and strengthened but not destroyed by them. In other respects, it has been pointed out already by Mitchell and Wynne-Jones [26b] and Mikhailov [26c] that, in water–alcohol mixtures, the knee or the extreme value of the isotherms of several properties appears at an alcohol content of about 0·1 mole fraction, which can be explained by complete occupation of the structural cavities of water. In solutions of higher non-electrolyte concentration, already no space is available for the dissolved molecules in the structural cavities of water, which results in the development of a liquid structure different from that of water, and this influences the properties in a different way from that of water.

Recently, Oakes [26e] measured the chemical shift of the magnetic resonance of the proton in the hydroxyl group and came to the conclusion that–in contrast with widely accepted opinions–the formation of hydrogen bonds between the water and alcohol molecules takes place to a considerable extent in dilute alcohol–water mixtures. In such mixtures, the mobility of water molecules is restricted. This is caused primarily by the hydrogen bonds formed between the OH groups of the alcohol and the water molecules, but, in a smaller degree, the mobility of the water molecules surrounding the alkyl chain is also limited. This latter effect increases with the length of the alkyl chain.

The extreme value of the transference number is not directly related with macroscopic viscosity: its maximum is at higher non-electrolyte concentrations in aqueous solutions of methanol and ethanol while no extreme values can be observed in aqueous glycol and glycerol solutions. On the other hand, from the temperature dependence of the transference number, it follows that the activation energy of the migration of F^-, Cl^- and K^+

Literature on page 392

ions is altered by these non-electrolytes to different degrees, depending on their concentration.

Effect of dioxan on the conductivity of aqueous solutions. 1-4-dioxan (which has no hydroxyl groups, although its oxygen atoms can form hydrogen bonds with water molecules) behaves somewhat differently from alcohols. According to the measurements of Erdey-Grúz and Nagy-Czakó [27], the transference numbers of K^+ and Li^+ ions, as well as those of Cl^- and F^- are not altered too much by dioxan in the very dilute KCl, KF and LiCl solutions investigated by them (the change is not higher than 0·05), but these small changes still have some characteristic features (Fig. 4.23). In KF solutions, the transference number of K^+ ions shows only a small decrease with increasing dioxan concentration in the range investigated, the curves being parallel at 5 and 25 °C. The transference number of K^+ ions in KCl solutions increases slightly with the effect of dioxan in low concentration (which is probably connected with the occupation of the structural cavities of water) then, starting from a dioxan content of about 5 mole per cent, it decreases appreciably. After passing a minimum at a dioxan content of 20 mole per cent, it increases again.

The transference number of Li^+ ions in LiCl solutions gives an initial increase with increasing dioxan content higher than that of K^+ ions in KCl solution, then, after passing a smooth maximum at a dioxan concentration of about 25 mole per cent, it decreases. In these electrolytes the temperature coefficient of the transference number reverses its sign at 15 and 5 mole per cent dioxan content, respectively. These observations indicate that ionic radii have a significant role in the alteration of ionic mobility due to the presence of dioxan in the cases investigated. The mobilities of K^+ and F^- ions, which have the same radii in crystals ($r_{K^+} = r_{F^-} = 1·33$ Å), are altered approximately in the same ratio on the addition of dioxan: the transference number is almost independent of the non-electrolyte concentration in the range investigated. In KCl and LiCl, the cation is smaller than the anion ($r_{Li^+} = 0·68$ Å, $r_{Cl^-} = 1·81$ Å), and in these electrolyte solutions the increase in dioxan concentration results initially in a higher relative increase in the mobility of the cation than in that of the larger anion, then this effect is reversed; this occurs at a higher dioxan concentration in the solutions of the very small bare Li^+ ions than in those of K^+ ions.

Bard, Wear, Griffin and Amis [28a] also investigated the transference number of Cl^- ions in more concentrated solutions of LiCl containing dioxan. They found the transference number curves rather similar to those obtained by Erdey-Grúz and Nagy-Czakó in dilute solutions; the deviations appearing (local maximum) can perhaps be attributed to the difference in the concentration or to the different methods of measurement.

The above changes of the transference number indicate that dioxan–depending on its concentration–alters ionic mobility by means of at least two opposite effects. It can be assumed that dioxan decomposes the hydrogen bonds between water molecules–and this effect increases with increasing concentration. Thus, the binding of water molecules located in the primary hydrate sphere to the farther ones (i.e. secondary hydration) decreases and mobility is enhanced in this way. The smaller the ions, the higher is this effect. Since the dipole moment of dioxan molecules is much less than that of water (that of dioxan is $\mu_d = 0.40$ Debye unit in solution, while that of water is $\mu_{H_2O} = 1.84$), the replacement of water molecules in the primary hydrate spheres of ions by dioxan molecules becomes significant only in solutions of higher concentration. This process results in an increase in the size of the solvated ion, and, as experiments indicate, this is also higher, the smaller the ion. An appreciably higher dioxan concentration is required to make this effect predominating in solutions of strongly hydrated Li^+ ions, than in those of the less hydrated K^+ ions.

Studying the mobility of alkali metal and halide ions in aqueous solutions containing 0–15 mole per cent dioxan, Kay and Broadwater [28b] have concluded that the increase in mobility observed up to a concentration of about 10 mole per cent is not a consequence of the stronger structure-making effect of the ions, but arises from the dehydration caused by dioxan which is not compensated by solvation with dioxan. The reorientation rate of the water molecules increases, owing to dehydration. Since this is also the result of the structure-breaking effect of ions, these two effects are

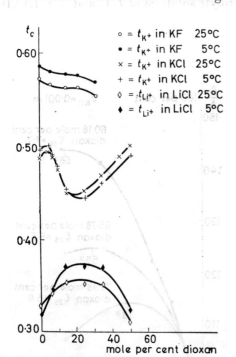

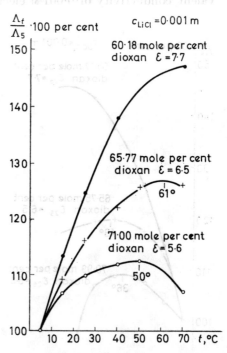

Fig. 4.23 The variation of the transference numbers of K^+ and Li^+ ions in 0.01 M KF, KCl and LiCl solutions at 5 and 25 °C on the effect of dioxan

Fig. 4.24 Relative changes in the conductivity of LiCl with temperature in solutions of various dioxan contents

Literature on page 392

similar from many viewpoints, and they cannot be distinguished by, for example, NMR measurements.

The concentration dependence of the conductivity of $NaNO_3$ dissolved in mixtures of water and dioxan was investigated in detail by Ramana-Murti and Yadav [28c]. By the use of a computer and the Fuoss–Onsager equation, they calculated the values of Λ°, the association constant K_a, and the 'ionic size' a. The product $\Lambda^\circ\eta$ and the value of a are dependent on the composition of the solvent mixture in the solutions of this electrolyte, too, and the $\lg K_a{-}1/\varepsilon$ correlation is approximately linear.

The changes in the structure of aqueous electrolyte solutions due to dioxan has been elucidated from another side by Erdey-Grúz, Kugler, Nagy-Czakó and Balthazár-Vass [29] on the basis of the temperature dependence of conductivity. According to their measurements, the equivalent conductivity of LiCl, $LiClO_4$ and KCl in solutions containing dioxan in high concentrations (and, in this way, having a low dielectric constant) increases initially with increasing temperature, then, after passing a maximum, it decreases (see Figs 4.24–26 in which the ratio of the equivalent conductivity of $0{\cdot}001$ M electrolyte solutions at t °C and 5 °C (Λ_t/Λ)

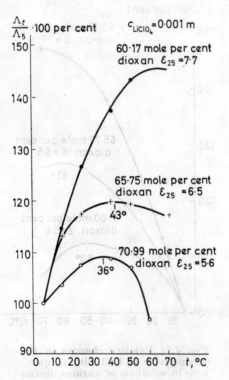

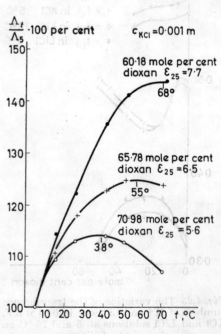

Fig. 4.25 Relative changes in the conductivity of $LiClO_4$ with temperature in solutions containing dioxan

Fig. 4.26 Relative changes in the conductivity of KCl with temperature in dioxan solutions of various concentrations

are given for solutions containing dioxan in different concentrations). In the presence of dioxan, conductivity shows a smaller increase with increasing temperature than in pure aqueous solutions, and, after passing a maximum in solutions of dioxan content higher than about 60 mole per cent, it decreases at higher temperatures. The maximum conductivity and the temperature corresponding to this maximum is the lower, the higher the dioxan concentration. The temperature of the maximum increases with the dielectric constant of the solution (Fig. 4.27). In solutions containing 60 mole per cent dioxan, the conductivity maximum of LiCl is higher and appears at a higher temperature than that of KCl, but this difference decreases with increasing dioxan concentration, and it diminishes or takes an opposite sign in 71 mole per cent solutions of dioxan. This indicates that the temperature coefficient of the conductivity of the lithium salt also decreases with the decrease in the concentration of water as compared with that of the potassium salt, which is connected, perhaps, with the fact that in the presence of sufficient amounts of water the size of the hydrated Li^+ ion decreases more rapidly with the increase in temperature (so its mobility increases) than that of (the less hydrated) K^+ ion. However, when solvation by dioxan becomes predominating, this difference disappears. Although there had been an earlier view that ions were solvated almost exclusively by the strongly polar water molecules in dioxan–water mixtures, it has been pointed out in the detailed investigations of Grunwald [30] that the solvation of cations by dioxan is also significant, since cations form a bond with the oxygen-containing part of the dipole molecules of the solvent, and the energy of the ionic charge–solvent dipole interaction is not much more negative for water than for dioxan. In the solvation of anions an appreciable role is attributed to the hydrogen bonds, and, in this respect, water molecules are preferred to dioxan.

The maximum, appearing in the temperature dependence of the conductivity of dioxan-containing salt solutions, is probably a consequence of

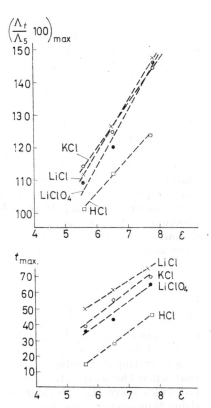

Fig. 4.27 Changes in the temperature of the maximum in conductivity of some electrolytes with the dielectric constant of aqueous dioxan solutions

Literature on page 392

24

several types of interactions. The initial increase can evidently be attributed to the predominating effect of factors similar to those resulting in an increase in conductivity in pure aqueous solutions. In addition to these, the dioxan present strongly decreases the dielectric constant of the solution which enhances ionic association. Since the dielectric constant of a solution of given composition decreases with temperature [31], this factor results in an increase in association with rising temperature. On the other hand, an increase in the thermal energy decreases association.

The maximum in the temperature dependence of conductivity, however, can hardly be attributed purely to changes in the degree of association, since it appears in both LiCl and $LiClO_4$ solutions, though LiCl associates markedly in aqueous dioxan solutions, as shown by Fabry and Fuoss [32], while the association of $LiClO_4$ is negligibly small according to Accascina and Schiavo [33]. It is to be noted, however, that conductivity as a function of temperature has a maximum only in solutions containing more dioxan than 55–60 mole per cent. In these solution, as shown by Erdey-Grúz and Nagy-Czakó [34], the water structure is destroyed by dioxan so much that prototropic conduction of hydrogen ions cannot take place. Thus, in these solutions, mostly monomeric or at least oligomeric water molecules are present, or such ones that are bound to dioxan molecules by hydrogen bonds. Since, according to the conclusions discussed above, the predominating effect giving rise to the appearance of the conductivity maximum can hardly be attributed to changes in the degree of association, it should be concluded that it arises from the increase in the size of the ions migrating as kinetic entities. Maybe this can be explained by supposing that when the water structure is very much loosened by dioxan, the water molecules leave the secondary hydration sphere of ions progressively with rising temperature, and they are replaced by the much larger dioxan molecules. According to this assumption, the water molecules located in the immediate vicinity of ions are deformed in such a way that they form a stronger connection with the oxygen atoms of the dioxan molecules under the given conditions than with water molecules. In other words, increasing temperature loosens or destroys the water-to-water bonds much more than the water-to-dioxan ones. This is also supported by the fact that the initial increase in conductivity due to elevated temperatures is the lower and the maximum appears at the lower temperature, the higher the dioxan concentration. In any case these processes must be in overall exethermic.

The effect of dioxan present in aqueous electrolyte solutions on conductivity is very complex, according to the experiments. In order to interpret this effect unequivocally, the structure of the aqueous solutions of dioxan should be known. Rather little information has been collected, however, on the structure of aqueous solutions of dioxan by means of direct measurements. The dielectric constant of water decreases rapidly with the effect of dissolved dioxan (see Fig. 4.21), and, according to Hasted, Haggis and Hutton [35], the relaxation time increases in the presence of dioxan which is explained by the development of a hydration sphere around dioxan molecules by means of $>O\cdots H-O$ hydrogen bonds. On the basis of the measurements of Williams [36] and Tourky, Rizk and Girgis [37], the dipole moment of water molecules dissolved in dioxan calculated from the dielectric constant of the solution is 1·93 Debye unit, which is higher than the dipole moment of the molecules in water vapour (1·84). This indicates that in dioxan solution, dielectric polarization is higher than in the gaseous phase, and this can be ascribed to hydrogen bonds between water and dioxan molecules. According to the latter authors, the electric anisotropy of water dissolved in dioxan decreases rapidly with increasing water concentration up to a water-content of about 60 mole per cent, then it approaches the value corresponding to pure water smoothly. From this, they concluded that the rotation of the molecules in water is hindered by six neighbouring molecules (the retardation of rotation shows a rapid increase at water-contents higher than 60 mole per cent). Thus—in accordance with the observations of some other authors—the coordination number in water is higher than the four generally assumed (see e.g. Kirkwood's papers [38]). It should be noted that the activation energy of macroscopic viscosity is lower in solutions containing a small amount of water than in pure water, it is equal to that of water in a solution of about 60 mole per cent concentration, while in more concentrated solutions, it is higher than that of water.

The rather strong connection between water and dioxan molecules is also indicated by the fact that the viscosity of the liquid (see Fig. 4.19), as well as the activation energy of viscous flow, increases initially on the effect dioxan dissolved in water, then after passing a rather sharp maximum at about 0·2 mole fraction of dioxan, it decreases. The maximum of viscosity becomes smoother with increasing temperature. According to Shkodin, Flevitskaia and Lozhnikov [39a], this indicates that there is an initial structure-making effect of dioxan on water evidently due to hydrogen bonds. Complex formation of four water molecules and one dioxan molecule has been postulated by them. This complex forms hydrogen bonds with the excess water molecules, which results in strengthening of the structure. In solutions of higher dioxan concentrations, there is not enough water for the formation of such complexes, which results in a decrease in viscosity.

On the other hand, Hammes and Knoche [39b] have concluded on the basis of their investigations on thermodynamic properties and ultrasonic absorption that relatively stable complexes are formed between water and dioxan molecules in 2 : 1 and 2 : 2 (W_2D and W_2D_2) ratios by means of intermolecular hydrogen bonds.

Our inadequate knowledge of the structure of aqueous solutions of dioxan makes the interpretation of the transport processes taking place in them difficult. However, on the other hand, experiments on transport processes contribute to the elucidation of the structure of these solutions. The investigation of aqueous solutions containing dioxan or mixtures of water and dioxan can also facilitate the clearing up of the extremely complicated structure of water as a liquid since, in dioxan containing a small amount of water, there is no doubt that water molecules exist in monomeric state; the polymers and the characteristic liquid structure of water develop prog-

Literature on page 392

ressively with increasing water concentration. In this way, the development of the structure of liquid water can be investigated. The conditions are complicated by the interaction of both monomeric and polymeric water molecules with dioxan molecules, as well, which leads to the formation of various structural units. Evidently, this is the main reason why the experiments could not be explained unequivocally in this way, either.

4.7.3 EFFECT OF NON-ELECTROLYTES ON PROTOTROPIC CONDUCTION

The prototropic conduction mechanism of hydrogen and hydroxide ions (cf. Section 4.2.2.2; a new mechanism for proton mobility in water–including proton transfer step shifting the oxonium ion across two hydrogen bonds in water–recently described by Fang, Godzik and Hofacker [39c]) is closely related to the structure of water. The binding of water molecules by hydrogen bonds contributes to the relay-like proton transport between water molecules and oxonium ions, because proton exchange probably takes place along the hydrogen bonds. All the effects altering the liquid structure caused by these connections probably also alter the ratio and nature of the conduction taking place by prototropic mechanism in the conductivity of acid and base solutions. The majority of non-electrolytes that are very soluble in water contain hydroxyl groups, being liquids which have a significant structure in their pure state and are themselves also suitable for transport of electricity by the prototropic mechanism, although to a different extent than water.

Several authors have dealt with the conductivity of acid and base solutions containing non-electrolytes, mainly with respect to the change in conductivity with the concentration of the electrolyte. A smaller number of investigations have been carried out on the conductivity of acid and base solutions containing different non-electrolytes in various concentrations.

The conductivity of aqueous hydrochloric acid solutions containing methanol and ethanol has been investigated earlier, mainly by Goldschmidt and Dahl [40], Walden [41], Thomas and Marum [42], Bezman and Verhoek [43], Kortüm and Wilski [44], Dnieprov [45], el Aggan, Bradley and Wardlaw [46] and Tourky and Mikhail [47]. The change in the conductivity of aqueous hydrochloric acid solutions on the addition of ethylene glycol has been studied by Dnieprov [45], and Kirby and Maass [48], while the effect of glycerol has been measured by Dnieprov [45], Conway, Bockris and Linton [49], Woolf [50], and Accascina, D'Aprano, and Goffredi [51a]; further, the effect of butanol has been investigated by Accascina, de Lisi and Goffredi [51b]. It has been stated that the equivalent conductivity of aqueous hydrochloric acid solutions decreases on the addition of alcohols, passes a minimum with increasing alcohol concentration, then increases again in solutions containing very small amounts of water. According to Tourky, Mikhail and Abdel-Hamid [54], the conductivity of hydrochloric acid varies similarly in mixtures of t-butanol and water.

In solutions containing non-electrolytes, the relationship between the prototropic conduction mechanism of hydrogen (oxonium) and hydroxide ions and the nature and concentration of the non-electrolyte, and the temperature has been studied extensively by Erdey-Grúz *et al.* Since it can be supposed that oxonium (H_3O^+) and hydroxide ions transport electricity not only by proton transfer, but, partly, by common hydrodynamic migration as well, the conductivity due to the two types of mechanism should be separated in order to facilitate the interpretation of experiments. These two types of conductivity cannot be measured separately; however, they can be calculated in approximation, assuming that the hydrodynamic mobilities of H_3O^+ and OH^- ions do not differ significantly from those of ions which are identical with them in size and charge in the given medium but cannot conduct by prototropic mechanism. In the crystalline lattice, the radius of the oxonium ion is $r_{H_3O^+} = 1.38-1.40$ Å and that of the hydroxide ion is $r_{OH^-} = 1.32-1.40$ Å. The radii of potassium and fluoride ions are nearly equal to these ionic sizes ($r_{K^+} = 1.33$ Å, and $r_{F^-} = 1.33$ Å), therefore, in first approximation, it can be assumed that the difference between the conductivities of HCl and KCl solutions under the same conditions corresponds to the conductivity of the hydrogen ion due to the prototropic mechanism, while the difference in the conductivities of KOH and KF solutions under the same conditions represents the prototropic conductivity of hydroxide ions. On the other hand, the comparison of the conductivities of KF and KCl solutions gives information on the change in the conditions when the ionic radius changes from the value $r_{F^-} = 1.33$ Å to $r_{Cl^-} = 1.81$ Å.

The comparison on the basis of ionic radii measured in crystalline lattices is strongly hypothetical, of course, with respect to hydration and solvation phenomena. Owing to the complicated nature of the group of phenomena summarized under the term hydration or solvation (cf. Section 5.2), it is questionable what radius should be introduced in the calculations regarding solutions. For example, according to the investigation of Kapustinsky [52a], in aqueous solutions or in the hydrated state the radii of positive atom ions are about 0.28 Å larger while those of negative ones are smaller by about the same value than those measured in ionic crystals, when taking into account the layer of adjacent water molecules. Assuming that this rule is valid approximately for oxonium and hydroxide ions, too, the radii of the ions compared above is: $r^h_{H_3O^+} = 1.66-1.68$ Å, $r^h_{K^+} = 1.61$ Å, $r^h_{OH^-} = 1.04-1.14$ Å, $r^h_{F^-} = 1.05$ Å, and $r^h_{Cl^-} = 1.53$ Å in hydrated state (taking into account the adjacent water molecules). The comparability is also reduced by considering that the effect of polyatomic ions on the solvent in their environment is hardly identical with that of the atom ions of identical average size and charge. Nevertheless, the differences in the conductivities of electrolytes conducting by both hydrodynamic migration and the prototropic mechanism and of those conducting only by hydrodynamic migration give satisfactorily reliable information on the change in the prototropic mechanism due to the presence of non-electrolytes, all the more so, since the effect of non-electrolytes on the conductivity of solutions conducting in these two ways is very different.

Literature on page 392

If the new assumptions of Zatsepina [52b] were confirmed, according to which the hydroxide ion is strongly bound to one molecule of water by a hydrogen bond and is present as a $H_3O_2^-$ ion in which the distance between the two oxygen atoms is 2·7 Å, the conclusions in connection with the size of the hydroxide ion should be modified to a certain degree.

Effect of non-electrolytes on the conductivity of hydrogen ions. The equivalent conductivity of dilute HCl solutions containing methanol, ethanol, *n*-propanol, ethylene glycol, and glycerol varies with the concentration of the non-electrolyte according to Figs 4.28 and 4.29, as measured by Erdey-Grúz and Kugler [12, 19, 20]. With increasing concentration of non-electrolyte the equivalent conductivity decreases rapidly, and, after passing a minimum in alcoholic solutions containing small amounts of water, it increases again. The conductivity minimum is at about 90 mole per cent alcohol content and it is sharpest in methanolic solutions, while it diminishes in *n*-propanol and glycerol solutions. The equivalent conductivity of KOH solutions also fall rapidly with the concentration of the non-electrolyte (Figs 4.30 and 4.31), but the minimum appears only in methanolic solutions. This minimum found at a methanol content of about 75 mole per cent is rather smooth.

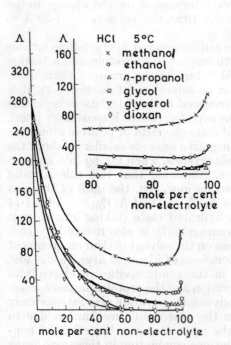

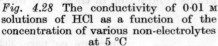

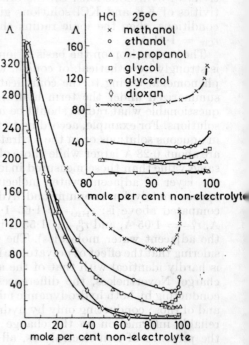

Fig. 4.28 The conductivity of 0·01 M solutions of HCl as a function of the concentration of various non-electrolytes at 5 °C

Fig. 4.29 The conductivity of 0·01 M solutions of HCl as a function of the concentration of various non-electrolytes at 25 °C

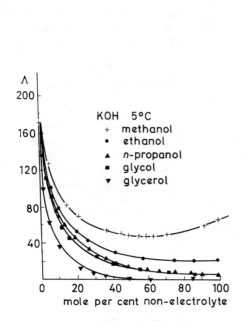

Fig. 4.30 The conductivity of 0·01 M solutions of KOH as a function of the concentration of various non-electrolytes at 5 °C

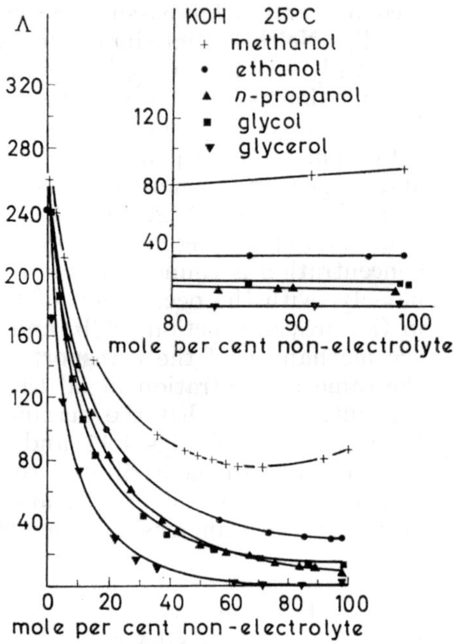

Fig. 4.31 The conductivity of 0·01 M solutions of KOH as a function of the concentration of various non-electrolytes at 25 °C

The variation of the product $\Lambda\eta$ caused by the effect of non-electrolytes is noteworthy (Figs 4.32 and 4.33). In solutions of monovalent alcohols, the initial part of the curve describing the dependence of the product $\Lambda\eta$ on the non-electrolyte concentration is similar to those observed by Erdey-Grúz and Kugler in KF and KCl solutions (Figs 4.17 and 4.18). The product $\Lambda\eta$ has a maximum in solutions containing approximately 8–10 mole per cent of alcohol. The product $\Lambda\eta$ decreases markedly with the effect of monovalent alcohols, then, after passing a smooth minimum in solutions of about 90–92 mole per cent alcohol content, it increases slightly. The decrease in the value of $\Lambda\eta$ in HCl solutions under the effect of monovalent alcohols is essentially higher than in KCl solutions, while KOH solutions show no essential difference from KCl and KF solutions. The effect of glycol and glycerol, however, differs significantly from that of monovalent alcohols. In solutions containing glycol, the maximum in the value of $\Lambda\eta$ is smoother than in the presence of monovalent alcohols and it appears at a higher concentration (12–14 mole per cent). On the other hand, the minimum is at a lower non-electrolyte concentration (about 85 mole per cent) but it is much sharper than that due to monovalent

Literature on page 392

alcohols, and, after passing the minimum, the product $\varLambda\eta$ increases more rapidly. Neither a maximum nor a minimum can be found in the value of $\varLambda\eta$ in glycerol solutions, but it increases monotonically, and, after passing an inflexion point at a glycerol concentration of about 65 mole per cent, it increases rapidly in solutions containing small amounts of water.

Comparing the dependence of the product $\varLambda\eta$ determined by Erdey-Grúz and Kugler on the non-electrolyte concentration in HCl, as well as in KCl and KF solutions (e.g. Figs 4.17 and 4.33), it can be supposed that the maximum appearing in solutions of about 6–8 mole per cent non-electrolyte concentration is connected mainly with the change in the liquid structure, namely, with the occupation of the structural cavities of water described in the previous section, [26b] and [26c], and it has only a small effect on the mechanism of the transport of electricity. This maximum is at about the same concentration in all the three electrolyte solutions. Taking into account, further, that the maximum of $\varLambda\eta$ is higher and sharper at 5 °C than at 25 °C (cf. Figs 4.17 and 4.18, as well as 4.32 and 4.33, respectively), this can be attributed to formation of complexes between water and non-electrolyte molecules and so to the structure-making effect which decreases with increasing temperature. The maximum value of $\varLambda\eta$ is

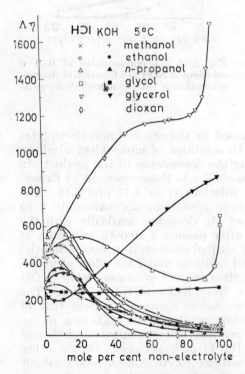

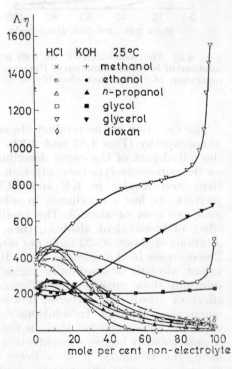

Fig. 4.32 The product $\varLambda\eta$ in HCl and KOH solutions *vs.* non-electrolyte concentration at 5°C

Fig. 4.33 The product $\varLambda\eta$ in HCl and KOH solutions vs. non-electrolyte concentration at 25 °C

evidently related to the maximum in the viscosity of electrolyte solutions (Fig. 4.19), though it appears at a much lower non-electrolyte concentration.

Accepting the concept of Ageno and Frontall [53], according to which the rather low viscosity of water is related to the fact that the members of linear polymers are in continuous exchange with the monomeric molecules in it, it can be supposed that the increase in viscosity caused by alcohols can be attributed to the saturation of hydrogen bonds by the alcohol which hinders the exchange with monomeric water molecules and, thus, decreases its rate.

Investigating the variation of conductivity of aqueous perchloric acid solutions on the addition of methanol and ethanol, Goldenberg and Amis [56] concluded that, according to the theory of Wicke, Eigen and Ackermann [57], three water molecules are bound by the H_3O^+ ion in the primary hydration process, which leads to the development of a pyramidal complex of composition $H_9O_4^+$. In addition to this, secondary hydration also takes place. The transport of electricity takes place partly by the hydrodynamic migration of $H_9O_4^+$ ions, and partly by protonic jumps caused by the fluctuation in the distribution of electrons in the $H_9O_4^+$ ion. According to this view, the rate-determining step is the formation of hydrogen bonds between this ion and the neighbouring water molecules. On the effect of alcohols, the water structure is gradually destroyed which retards the rate-determining step. With decreasing concentration of water molecules suitable for hydrogen bond formation at the periphery of the complex $H_9O_4^+$, these ions become stabilized, their contribution to the transport of electricity by means of hydrodynamic migration increases, thus, conductivity decreases. Another cause of this decrease is the progressive solvation of H_3O^+ ions by alcohol instead of water, which increases their size. In water-free alcoholic solutions or at low water concentrations, however, protonic jumps take place along the hydrogen-bonded chains of alcohol molecules.

From the fact that not only viscosity but also the product $\Lambda\eta$ has a maximum, it follows that in this concentration range the conductivity decreases to a lower degree with the effect of non-electrolytes than macroscopic viscosity increases. This can probably be attributed to the fact that the liquid structure formed by participation of the non-electrolyte is perturbed more extensively by the electric field of the ions and, in this way, the microviscosity is decreased to a higher degree in comparison with macroscopic viscosity than in pure water or in non-electrolyte solutions of higher concentration. Reduction of the size of the hydrate sphere moving along with the ion can hardly be expected in this range, because, even if the replacement of the water molecules in the hydrate sphere by alcohol molecules decreases the structure-making effects exerted by hydrogen bonds on the farther water layers, this replacement would give rise to the growth in size of the migrating particles and, thus, to a decrease in their mobility, since the molecular weight of alcohol molecules is higher than that of water. It should be noted that the viscosity of mixtures of glycol and water has no maximum (Fig. 4.19), but a maximum appears in the product

Literature on page 393

$\Lambda\eta$ in hydrochloric acid solutions. This supports the view that the liquid structure altered by the presence of a non-electrolyte in low concentration is loosened more by the field of the ions than that of pure water. On the other hand, the viscosity of dioxan–water mixtures exhibits a maximum, but the maximum in the product $\Lambda\eta$ of KF and KCl dissolved in these mixtures is still very small; in hydrochloric solutions there is no maximum, or 'a degenerated maximum' appears, since the $\Lambda\eta$ values are independent of the dioxan content up to about 5 mole per cent. This also indicates that opposite effects predominate in this concentration range which act by increasing partly viscosity and partly mobility. Possibly this is connected with the presence of two oxygen atoms in the dioxan molecule or with the hydrogen bonds formed by them.

The decrease in the value of $\Lambda\eta$ as a function of the non-electrolyte concentration after passing the maximum can probably be attributed to several interactions. On one hand, the ratio of non-electrolyte molecules replacing water molecules in the solvation process increases, resulting in increased size of the solvated ions forming a kinetic entity: this reduces mobility, in addition to the increase in viscosity. On the other hand, the structural cavities of water are occupied by alcohol molecules and the liquid structure developing gradually will correspond to that of the non-electrolyte which is, perhaps, less loosened by the field of the ions than that existing in the range of the maximum in the product $\Lambda\eta$. The increase in the degree of association of ions may become appreciable, too, as a result of the lowered dielectric constant with the increase in the non-electrolyte content of the solution (cf. Fig. 4.22). For solutions containing monovalent alcohols in concentrations higher than about 30 mole per cent, the value of $\Lambda\eta$ is the lower, the lower the dielectric constant of the solution, which indicates the effect of association. In the presence of glycol and glycerol, however, it is not this effect that predominates, since, e.g. the dielectric constants of methanol–water and glycol–water mixtures of 40 mole per cent concentration are approximately identical (at 25 °C it is 54·5), while that of the glycerol–water mixture of the same concentration is smaller (52). However, the order of the product $\Lambda\eta$ in KCl and HCl solutions of 40 mole per cent non-electrolyte content is glycerol > glycol > methanol (cf. Figs 4.18, 4.21 and 4.33). In the effect of these latter non-electrolytes on conductivity, the increase in the degree of association has hardly a prevailing influence. The relatively small decrease in conductivity in comparison with the increase in macroscopic viscosity can probably be attributed to the fact that the structure formed on binding the molecules of the liquid to one another (which results in high macroscopic viscosity) is greatly loosened by the electric field of ions. Consequently, viscosity significantly decreases in the vicinity of ions, and their mobility does not decrease to such an extent that would correspond to the increase in macroscopic viscosity. Evidently, this effect acts in both KCl and HCl solutions. In the latter ones, however, an additional factor also contributes to this common effect, that is the transport of electricity by the prototropic mechanism is also altered under the effect of the non-electrolyte. Comparing the dependence of the product $\Lambda\eta$ in KCl and HCl solutions on the glycol and glycerol concentration (Figs 4.18 and 4.33), it is seen that, in HCl solutions of medium glycol concentration (about 20–28 mole per cent), the $\Delta\eta$ value decreases with increasing concentration, while in glycerol solutions of similar concentration, it shows a smaller increase than in the corresponding KCl solutions. This can evidently be ascribed to the fact that, in this concentration range, the ratio of the prototropic conduction mechanism decreases or is overshadowed. From 80 mole per cent glycol or glycerol concentration, the product $\Delta\eta$ increases rapidly in HCl solutions unlike the glycerol and glycol solutions of KCl (the value of $\Delta\eta$ shows a small decrease in them). This increase of $\Delta\eta$ is evidently caused by the increasing ratio of conduction taking place by means of the prototropic mechanism occurring again in solutions containing small amounts of water, because of the development of the liquid structure of glycol or glycerol.

From the phenomena discussed above, it can be concluded that the transport of electricity by the prototropic mechanism of hydrogen ions requires a rather extended. more or less ordered connection of identical molecules containing hydroxyl groups (cf. Section 4.2.2.2) which, of course, is a short-range order as compared with macroscopic sizes. If the more or less ordered regions present in water are destroyed on interacting with non-electrolyte molecules, the possibility of proton transfer is reduced and a progressively increasing portion of electricity is transported by the hydrodynamic migration of H_3O^+ ions. According to the above observations, the percentage of the transport of electricity by means of the prototropic mechanism is lowest in solutions containing glycol and glycerol in a concentration of about 80 mole per cent. Under such conditions, the proton transfer through water molecules is already hindered very much by the extended destruction of the water structure, while the proton transfer between alcohol molecules, schematically shown as:

$$ROH_2^+ + R\overset{\frown}{O}H \rightarrow ROH + ROH_2^+$$

is still not favoured enough by the development of a liquid structure corresponding to that of glycol or glycerol. With further decreases in the water content, the liquid structure of glycol or glycerol is built up rapidly, i.e. more and more molecules will be bound together by hydrogen bonds which enhances the ratio of the prototropic mechanism in the transport of electricity. The mechanism of proton transfer between water and glycol or glycerol molecules is evidently not the same in details, for the very reason that the sizes and shapes of the molecules are very different. The presence of two or three hydroxyl groups per molecule significantly facilitates prototropic conduction through glycol and glycerol molecules. The rapid increase in the value of $\Delta\eta$ in solutions containing a small amount of water shows that this acts appreciably in a direction opposite to that of the lower translational and rotational mobility caused by the larger size of glycerol molecules as compared with water.

The change in the product $\Delta\eta$ can be explained similarly in hydrochloric acid solutions containing monovalent alcohols. The continued decrease

Literature on page 393

following the decrease after the maximum observed in corresponding KCl solutions arises from the retardation of the prototropic mechanism. In this concentration range, the increasing concentration of the non-electrolyte progressively reduces the regions of the water molecules bound together by hydrogen bonds, so proton transfer is more and more restricted and the transport of electricity is limited to the hydrodynamic migration of H_3O^+ ions. In this range, alkyl-oxonium ions are still not formed in a significant amount, since the proton has a higher affinity to water than to alcohols. For example, according to Wicke, Eigen and Ackermann [57], in methanolic solution the equilibrium constant of the process:

$$H_3O^+ + CH_3OH \rightleftharpoons H_2O + CH_3OH_2^+$$

is $K = 0 \cdot 23$. In propanolic solution, K is equal to $0 \cdot 002$ as measured by Murgulescu, Barbulescu and Greff [58a]. Thus, the majority of the protons of the dissolved acid are in the form of H_3O^+ ions even in solutions containing a relatively small amount of water. In solutions exhibiting a minimum in either the product $\Lambda\eta$ or the conductivity, electricity is transported mostly by H_3O^+ ions migrating hydrodynamically; proton transfer through water is already hindered and that through alcohol molecules does not predominate yet. After passing the minimum, however, the concentration of the alcohol is already so high that, on one hand, the formation of alkyl-oxonium ions becomes significant:

$$H_3O^+ + ROH \rightarrow H_2O + ROH_2^+,$$

on the other hand, the development of a liquid structure corresponding to that of the alcohol (the formation of polymers by means of hydrogen bonds) takes place to such an extent that proton transfer through alcohol molecules becomes progressively possible, schematically:

$$ROH_2^+ + R\overset{\frown}{O}H \rightarrow ROH + ROH_2^+.$$

This results in an increase in conductivity in solutions containing water in very low concentrations.

Recently, similar results have been obtained by Tourky and Abdel-Hamid [58b] measuring the conductivity of HCl dissolved in mixtures of methanol, ethanol and n- and i-propanol with water. De Lisi and Goffredi [58c] have also established by studying the conductivity of HCl dissolved in n-and i-butanol, that the conductivity decreases with the effect of water present in low concentrations, then it increases again with increasing water content.

According to de Lisi and Goffredi [55], the equilibrium constant of the process

$$C_3H_7OH_2^+ + H_2O \rightleftharpoons C_3H_7OH + H_3O^+$$

is $K = 1038 \pm 100$ at 25 °C, and its standard free enthalpy is $\Delta G^° = 4,11 \pm \pm 0.06$ kcal/mole, calculated on the basis of the conductivity of hydrochloric acid dissolved in 2-propanol containing a small amount of water. They also surveyed the results of other measurements, and pointed out that in pure alcohols of the $C_nH_{2n+1}OH$ homologous series the products $\Lambda^\circ\eta$ and $\lambda_H^+\eta$ plotted as a function of the dielectric constant show a minimum at about $\varepsilon = 25$. On the basis of conductivity values measured in different alcohols, they inferred that the mobility of the proton depends not only on the size and shape of the alkyl groups, but other factors also play a part. Their statements are at variance with the theory of Conway, Bockris and Linton (cf. Section 4.2.2.2.), which requires a monotonic decrease in the rotational velocity of the alcohol molecules in the process of prototropic conduction and, consequently, a similar decrease in the product $\lambda_H^+\eta$, with increasing number of carbon atoms; according to the experiments, this product decreases with increasing number of carbon atoms in the series BuOH, PrOH, EtOH, and only the change from EtOH to MeOH gives rise to an increase.

According to Erdey-Grúz and Kugler [29], in HCl solutions containing dioxan, the value of $\Lambda\eta$ decreases monotonically after an initial constant period. This is probably the consequence of the relatively high ionic association due to the low dielectric constant of dioxan solutions of higher concentration, as well as to the fact that dioxan molecules do not transfer protons, which is also revealed by the transference number measurements (see below).

The phenomenon observed in salt solutions (cf. Section 4.7.2) that an increase in temperature reduces conductivity under certain conditions is shown to an increased degree in the temperature dependence of the conductivity of HCl solutions containing dioxan. According to the measurements of Erdey-Grúz, Kugler, Nagy-Czakó and Balthazár-Vass [29], the relative equivalent conductivity of HCl solutions at t °C related to that of a solution at 5 °C varies with the temperature t, as shown in Fig. 4.34. In HCl solutions containing dioxan in sufficiently high concentrations, the conductivity ratio $\Lambda_{t\,°C}/\Lambda_{5\,°C}$ changes with temperature similarly to that of the corresponding KCl and LiCl solutions (cf. Figs 4.24–4.26); with the difference that in the case of HCl the maximum in the value of $\Lambda_{t\,°C}/\Lambda_{5\,°C}$ already appears in aqueous solutions of lower dioxan content than in the

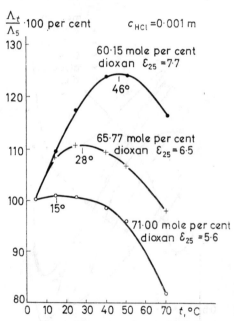

Fig. 4.34 The ratio of the equivalent conductivities of 0·001 M HCl solutions measured at t°C and 5 °C, as a function of temperature in solutions of various dioxan content

Literature on page 393

case of the salts. Further, in HCl solutions, the maximum of conductivity is lower and appears at lower temperatures, while the decrease in conductivity after the maximum is much more rapid than in salt solutions. In solutions containing more than 64 mole per cent of dioxan, the conductivity is lower at 70 °C than at 5 °C, and at 71 mole per cent dioxan content, the decrease is almost 20 per cent.

The temperature dependence of the conductivity of HCl solutions containing dioxan can probably be attributed to two groups of consequences of the interactions. In one of them, those effects of the field of ions on the liquid structure are involved which influence the hydrodynamic migration of ions and are hardly different in hydrochloric acid solutions and in the corresponding salt solutions. The effects of the interactions in the other group modify the prototropic conduction mechanism. From the temperature dependence of the ratio $\Lambda_{t\,°C}/\Lambda_{5\,°C}$, it can be concluded that the prototropic conduction mechanism is hindered to a rapidly increasing degree with increasing temperature in solutions of appropriate dioxan concentration. This can probably be ascribed to the fact that dioxan destroys the water structure progressively as its concentration increases, it reduces the size of the clusters or polymers bound together by hydrogen bonds, it separates the water molecules from one another, and, thus, the proton transfer between H_2O molecules is restricted, its pathway is shortened, and finally its possibility is excluded. Further, the water structure already loosened by dioxan is destroyed more rapidly by the increase in temperature. Therefore, in solutions containing dioxan in sufficient amounts, the ratio of the prototropic mechanism in conductivity decreases considerably under the effect of increasing temperature. A loosened structure can promote association too.

Variation of the transference number of the hydrogen ion under the effect of non-electrolytes. The change in the transference number of the hydrogen ion with the effect of non-electrolytes added is in accordance with the conclusions obtained from the variation in conductivity with the effect of non-electrolytes. In aqueous hydrochloric acid solutions, the variation of the transference number of the hydrogen ion with non-electrolyte concentration is given in Figs 4.35 and 4.36 based

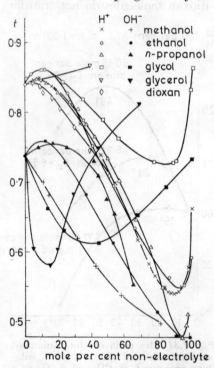

Fig. 4.35 The transference number of hydrogen and hydroxide ions in HCl and KOH solutions as a function of the concentration of non-electrolytes at 5 °C

on the investigations of Erdey-Grúz, Majthényi and Nagy-Czakó [23–26, 59a]. The non-electrolyte present in a low concentration somewhat increases the transference number of the hydrogen ion. This is probably connected with the strengthening of the liquid structure under such conditions (which can evidently be attributed to the occupation of the structural cavities of water by alcohol molecules); this has also been indicated by some other phenomena discussed above. According to experiments, the structure-making effects hinder conduction by means of hydrogen ions less than that by an ion migrating only hydrodynamically. This is comprehensible considering that the H_3O^+ ions with a relatively large mass are not displaced when conducting by the prototropic mechanism, only the proton with a much smaller mass is transferred from one molecule to the next.

After the maximum of the transference number, a further increase in the concentration of monovalent alcohols results in a marked decrease in the transference number of the hydrogen ion, and it has a minimum at about 90–95 mole per cent concentration, and after passing this point, it increases rapidly. The minimum is the lowest in propanolic solution at 25 °C: $t_{H^+} = 0.510$. In this way, conduction by the prototropic mechanism diminishes under such conditions. In methanolic and ethanolic solutions, the minimum values are $t_{H^+} = 0.558$ and 0.567, respectively, thus in these

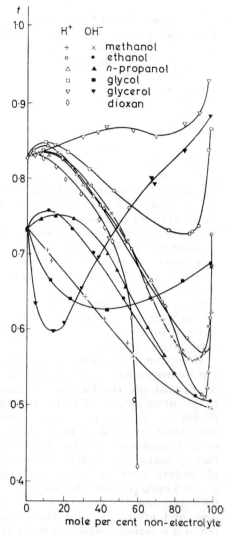

Fig. 4.36 The transference number of hydrogen and hydroxide ions in HCl and KOH solutions as a function of the concentration of non-electrolytes at 25 °C

solutions the ratio of the prototropic mechanism in the transport of electricity must be small. It should be noted that the order of the minima of the transference numbers does not follow the order of the molecular weight of the alcohols. Some other phenomena (e.g., the change in the transference number of the hydroxide ion and

Literature on page 393

of salts, see Fig. 4.22) also indicate that the effect of methanol on the transport processes of aqueous solutions shows a certain difference from that of other monovalent alcohols. The increase in the transference number after the minimum is probably caused by a prototropic mechanism through alcohol molecules. The transference number of the hydrogen ion, however, does not reach the value of pure water in pure monovalent alcohols either.

The transference number of the hydrogen ion varies to a certain degree similarly under the effect of glycol and glycerol (Figs 4.35 and 4.36), but the minimum is appreciably smaller; moreover, in the case of glycerol at 25 °C, the transference number in the relative minimum is higher than in pure water due to the rather high initial increase. In addition to this, the minimum of the transference number appears at a lower concentration of the non-electrolyte solution than in the case of monovalent alcohols. It should also be noted that the transference number of the hydrogen ion in almost water-free glycol and particularly in glycerol, is higher than that in pure water. The differences in the properties of glycol and glycerol in this respect as compared with monovalent alcohols can be ascribed to the presence of two or three hydroxyl groups per molecule which can mediate proton transfer. In both glycol and water there are two OH groups, but these have a common oxygen atom in water. The transference number of the hydrogen ion in glycol is higher than that in water, perhaps the independence of the oxygens of the two hydroxyl groups facilitates the prototropic mechanism.

In low concentrations, dioxan also increases the transference number of the hydrogen ion in water, but the decrease is monotonic indicating that no prototropic conduction mechanism can occur via dioxan. According to the investigations of Erdey-Grúz and Nagy-Czakó [59a], the transference number of hydrogen ions in hydrochloric acid solution varies with the concentration of dioxan as shown in Figs 4.35 and 4.36. The increase of the transference number due to the effect of dioxan in low concentrations can, perhaps, be explained by the structure-making effect of the non-electrolyte mentioned above in connection with several phenomena. From about 5 mole per cent dioxan content onwards, the transference number of hydrogen ions decreases moderately to a dioxan content of about 52 mole per cent, and, after this point, the decrease is rapid, and it becomes lower than 0·5 in a solution containing 60 mole per cent of dioxan. In such solutions, the ratio of the prototropic mechanism in conduction is thus negligible at 25 °C*. In this way, dioxan is unsuitable for proton transfer by the relayed mechanism.

Effect of non-electrolytes on the conductivity of hydroxide ions. The high mobility of hydroxide ions in aqueous solutions–in a certain degree similarly to that of the hydrogen ion–can also be attributed to the fact that they transport electricity partly by the prototropic mechanism, schematically:

* In solutions containing dioxan in high concentration, the transference number could not be measured because of technical difficulties at 5 °C.

$$O^{-} + H\!\!-\!\!O \rightarrow O\!\!-\!\!H + O^{-}$$

$$\begin{array}{cccc} | & | & | & | \\ H & H & H & H \end{array}$$

There is, however, a significant difference between the prototropic conduction mechanism of hydrogen and hydroxide ions. According to the theory of Conway, Bockris and Linton [49], the rate of proton transfer is determined by the rotation of the water molecules as dipoles in the inner hydrate sphere of the ions in aqueous solutions, necessary to make possible the next (i.e. the running) proton transfer (as compared with this, the proton transfer itself along the hydrogen bond connecting neighbouring water molecules is a fast process), since immediately after the proton transfer, the neighbouring water molecules have an unfavourable orientation, and they reach a position that makes the next proton transfer possible only after the required rotation. However, the interaction resulting in this rotation of the solvent molecules is not identical in the vicinity of the oxonium ion and hydroxide ion. After the proton transfer of the oxonium ion, two effects act toward rotation in the configuration $H_2O\!-\!H_3O^+$: partly the effect of the ion on the dipole molecules of the solvent, and partly the repulsion between the two OH groups facing each other as a result of the protonic jump. According to the theory mentioned above, three quarters of the force resulting in rotation arises from this latter effect at normal temperatures and in aqueous solutions. After the transfer of a proton to a hydroxide ion, however, two hydrogen bonds are not facing each other; thus, under such conditions, the rotation of solvent molecules is produced only by the effect of the electric field of OH^- ions on the dipole molecules of the solvent. In accordance with this, the rate of rotation given by the theory mentioned above is lower, and the ratio of the prototropic mechanism in the transport of electricity is smaller than in solutions of hydrogen ions.

The difference in the prototropic conduction mechanisms of hydrogen and hydroxide ions is also revealed by the fact that the conductivity and transference number of acids and bases are altered differently by non-electrolytes, as indicated by the experiments. From the investigations of Erdey-Grúz and Kugler [12, 19, 20], it follows that the product $\Delta\eta$ shows an initial increase under the effect of methanol, ethanol and propanol of increasing concentration in aqueous KOH solutions, then, after passing a maximum at an alcohol content of about 6–8 mole per cent, it decreases appreciably, and, in solutions containing a small amount of water or in water-free solutions, it hardly differs from the corresponding values of KCl or KF solutions (Figs 4.32 and 4.33). The effect of glycol and glycerol differs markedly from this. The product $\Delta\eta$ in KOH solutions shows a smooth decrease up to about 15 mole per cent glycol content, then it remains almost constant in a broad concentration range, and, finally, it shows a small increase in solutions of low water concentration. In the presence of glycerol, the minimum of $\Delta\eta$ is at about 8 mole per cent, and, from this

Literature on page 393

point on, it increases appreciably. The structural changes of water caused by the electrolyte present in a relatively low concentration have an opposite influence on the product $\Lambda\eta$ in glycolic and glycerolic KOH solutions as compared with KF, KCl and HCl solutions investigated in this way. In rather low concentrations, glycol and glycerol decrease the conductivity of hydroxide ions to a higher extent than that expected on the basis of the increase in viscosity. Unlike other ions studied, the structure-breaking effect of the electrostatic field of hydroxide ions is, thus, overcompensated by another effect. Possibly this is in accordance with the higher asymmetry (higher dipole moment) of the OH^- ion than that of H_3O^+. The rapid increase in the value of $\Lambda\eta$ is particularly noteworthy in solutions containing a higher concentration of glycerol, in which it becomes much higher than in any of the electrolyte solutions mentioned above (except hydrochloric acid solutions containing glycerol). This effect of polyvalent alcohols is probably related to the possibility of proton transfer in the presence of two or three OH groups in the molecule without the rotation of the molecules transmitting it; thus the rate of proton transfer rises appreciably together with the ratio of the prototropic mechanism in the transport of electricity.

Effect of non-electrolytes on the transference number of the hydroxide ion. The conclusions discussed above are supported by the studies of Erdey-Grúz et al. [23–26a] on the variation of the transference number of hydroxide ions due to the effect of non-electrolytes in KOH solutions (Figs 4.35 and 4.36). In solutions of monovalent alcohols of high concentration, the transference number of hydroxide ions decreases monotonically and rapidly, and in pure alcohols it reaches the value of about 0·5, moreover, at 5 °C, it is lower than that. It follows from this, that the proton transfer through hydroxide ions is not mediated by monovalent alcohols, and the decrease in the transference number or in the product $\Lambda\eta$ is caused mainly by the progressive decrease in the size of the water clusters or polymers bound together by hydrogen bonds due to the decomposition of the water structure, and the range within which relay-like proton transfer can take place becomes more and more narrow. The unsuitability of monovalent alcohols to mediate the proton transfer of hydroxide ions is probably related to the hindrance of the rotation of their molecules required for prototropic conduction because of their expanded shape due to the presence of the alkyl group. In the case of hydroxide ions, the force acting in the direction of the rotation–being much smaller than in the case of hydrogen ions–is unable to bring about the rotation of the alcohol molecules with any significant probability.

It should be noted that in ethanol and propanol solutions of low concentration, the transference number of hydroxide ions shows an initial increase with increasing concentration of the non-electrolyte, then, after passing a maximum at 12 and 18 mole per cent, respectively, it decreases. On the other hand, the transference number decreases monotonically from the beginning under the effect of methanol. Although in several respects methanol does not fit into the order of the properties of monovalent aliphatic alcohols as a function of their molecular weight, the above deviation is still striking, which indicates that methanol alters the water structure in a somewhat different way than other monovalent alcohols. Perhaps, methanol molecules break or deform more hydrogen bonds on forming a connection with water molecules when present in a small ratio as compared with ethanol or propanol molecules, and the rotation of water molecules is hindered by their stronger binding. It is not excluded either that the complexes of water molecules with ethanol and propanol molecules present in a low concentration increase the proton transfer involving no rotation, owing to their structure which is hindered on decreasing the amount of water as a result of some factors unknown so far.

The change in the transference number of the hydroxide ion under the effect of glycol and glycerol is to be particularly noted. These non-electrolytes, especially the latter, decrease the transference number of hydroxide ions rapidly in low concentration (Figs 4.35 and 4.36), then, with a further increase in concentration, they increase it rapidly. The minimum in the transference number appears in solutions of 40 mole per cent and 15 mole per cent glycol and glycerol content, respectively (the latter is close to the concentration corresponding to the maximum of the transference number in ethanolic solutions).

In glycerol solutions containing a small amount of water, the transference number of hydroxide ions is higher not only than that corresponding to pure aqueous solutions, but the transference number measured in pure aqueous solutions for hydrogen ions as well. This shows that glycol and glycerol have at least two opposite effects on the transference number of hydroxide ion in aqueous potassium hydroxide solutions. In low concentrations, these electrolytes decrease the ratio of the prototropic mechanism in the transport of electricity. In solutions containing glycol and glycerol in higher concentrations, the ratio of the prototropic mechanism increases rapidly which indicates that, under such conditions, proton transfer takes place not through water molecules, but *via* glycerol. The prototropic conduction mechanism through glycol is almost as effective as that through water, while that through glycerol is much more efficient.

Considering the very high viscosity of concentrated aqueous solutions of glycol and glycerol, it cannot be assumed that glycol or glycerol molecules should turn after proton transfer to make the next protonic jump possible. It seems to be likely that a rearrangement takes place within the $CH_2OHCH_2OH_2^+$ or $CH_2OHCHOHCH_2OH_2^+$ ions which permits the departure of a proton from another hydroxyl group in the molecule. These processes are probably facilitated by the extended network of hydrogen bonds, which results simultaneously in high viscosity. The decrease in the transference number of hydroxide ions in solutions of glycol and glycerol present in a low concentration can be attributed to the fact that these non-electrolytes hinder the rotation of water molecules required for prototropic conduction by means of the hydrogen bonds formed between their two or three hydroxyl groups and water molecules. In solutions of low concentration,

Literature on page 393

however, such a network of hydrogen bonds between glycol or glycerol molecules and water, which would mediate the continuous proton transfer, cannot be formed.

4.7.4 CHANGES IN THE ACTIVATION ENERGY OF THE TRANSPORT OF ELECTRICITY DUE TO THE EFFECT OF NON-ELECTROLYTES

In aqueous solutions of HCl and KOH, the activation energy of transport of electricity increases because of the effect of non-electrolytes present in a low concentration. This can be seen in Fig. 4.37, which summarizes the data obtained in the studies of Erdey-Grúz and Kugler [12, 19, 20]. In water and in most of the solutions containing non-electrolytes, the activation energy of the transport of electricity in KOH is higher than that in HCl, and this difference is increased by non-electrolytes. A non-electrolyte present in low concentrations increases the activation energy in each case investigated, while in higher concentrations, it decreases it- disregarding two exceptions. The decrease is greatest under the effect

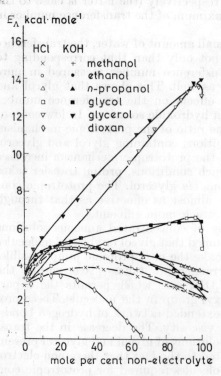

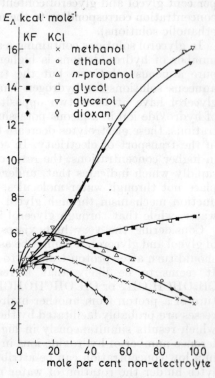

Fig. 4.37 The activation energy of conductivity in HCl and KOH solutions as a function of the non-electrolyte concentration, calculated from the conductivities measured at 5 and 25 °C

Fig. 4.38 The activation energy of conductivity in KF and KCl solutions as a function of the non-electrolyte concentration

of dioxan, while the activation energy of the transport of electricity by HCl is decreased much less by glycol and glycerol in solutions containing a small amount of water. By comparing these data with the activation energies of the transport of electricity in KF and KCl solutions conducting only by hydrodynamic migration, it can be stated (Fig. 4.38) that there are several similarities in the dependence on the concentration of the non-electrolyte. The activation energy of the transport of electricity by HCl and KOH is generally somewhat smaller than that by KF and KCl; further, the maximum in the activation energy corresponds to a somewhat lower concentration in HCl and KOH solutions. A characteristic difference appears in the rapid decrease in the activation energy of the hydrogen ion in ethanolic, glycolic and glycerolic solutions of low water content, which is evidently a consequence of the prototropic conduction mechanism of hydrogen ions predominating again. The activation energy of the transport of electricity by KOH does not decrease even in the most concentrated glycolic and glycerolic solutions, which indicates that the prototropic conduction mechanism of the hydroxide ion made possible by glycol and glycerol has an appreciably dissimilar nature as compared with that of hydrogen ions.

The above observations on the activation energy of the transport of electricity show that the prototropic conduction mechanism usually decreases the activation energy, i.e. the rate of transport by the prototropic mechanism shows a smaller increase with temperature than that taking place according to the hydrodynamic mechanism. This is probably connected with the fact that the prototropic mechanism does not require the translational motion of the whole ion with a rather large mass. The fact, however, that the activation energy of the transport of electricity by KOH is increased to a greater extent by a non-electrolyte present in a low concentration than for HCl, indicates that the structure-making effect on the structure of the liquid in this concentration range hinders the prototropic conduction by hydroxide ions to a greater degree than in the case of hydrogen ions. The reason for this is, perhaps, that the change in the liquid structure in this range influences the ion–dipole interaction more strongly than the repulsion between two OH groups coming up against each other in the course of the proton transfer by hydrogen ions.

The activation energies above have been calculated from measurements at 5° and 25 °C, thus they are average values corresponding to this temperature range. Owing to the complex structure of aqueous electrolyte solutions containing non-electrolytes, it is not excluded that the activation energy of the transport of electricity depends markedly on temperature. Maybe the activation energies calculated on the basis of measurements carried out in a narrow temperature range (1–2 °C) would deviate from the average values given above. However, no experimental data are available with respect to this.

The physical factors determining the activation energy of the transport of electricity have been investigated recently in the theoretical studies of Laforgue-Kantzer, Laforgue and Khanh [59b].

* * *

Literature on page 393

The matter discussed above regarding the prototropic conduction mechanism of hydrogen and hydroxide ions also proves that this phenomenon is influenced by several factors in a complicated way, and that these are different in the case of the two ions. For a satisfactory interpretation of these phenomena, considerable theoretical and experimental work seems to be needed in the future.

4.8 ELECTRONIC CONDUCTION IN AQUEOUS SOLUTIONS

In solutions, charge transport due to the electric potential gradient takes place generally by the migration of ions. Levich [60] recognized, however, on the basis of Frumkin's note that charge transport by means of electrons (i.e. electronic conduction) can also take place in solutions containing an oxidation–reduction system, the phenomenon being similar to electronic diffusion (cf. Section 2.3.5). The usually non-orientated electron transfer resulting in the oxidation-reduction process is directed by the potential gradient in space, and thus it can, in principle, give rise to a macroscopic charge displacement, which contributes to the conduction of electricity. The overall electric current density i is the sum of the ionic and electronic currents (i_i and i_e, respectively); under such conditions:

$$i = i_i + i_e.$$

If electron transfer takes place within a distance λ in the solutions of an oxidation-reduction pair:

$$Ox_1 + Red_2 \underset{k_2}{\overset{k_1}{\rightleftharpoons}} Ox_2 + Red_1$$
$$\quad c_1 \qquad c_2 \qquad\qquad c_3 \qquad c_4$$

and the rate constants in the opposite directions are k_1 and k_2; the current density of the electronic conduction caused by an electric potential gradient ψ is:

$$i_e = \frac{F^2 \lambda^2}{RT} (k_1 c_1 c_4 + k_2 c_2 c_3) \, \text{grad} \, \psi, \qquad (4.8.1)$$

as deduced by Dahms [61].

The density of the ionic current flowing together with this is:

$$i_i = \Sigma \, \lambda_i c_i z_i \cdot \text{grad} \, \psi, \qquad (4.8.2)$$

where λ_i is the equivalent conductivity of the ith type of ions present in concentration c_i and z_i is its valence. The ratio of the electronic and ionic currents is:

$$\frac{i_e}{i_i} = \frac{F^2 \lambda^2 (k_1 c_1 c_4 + k_2 c_2 c_3)}{RT \, \Sigma \, \lambda_i c_i z_i}. \qquad (4.8.3)$$

Since λ is in the order of magnitude of 10^{-8}–10^{-7} cm, the electronic current is significant, in addition to the ionic one, only in the case where the reaction rate constant of the oxidation–reduction system is higher by several orders of magnitude than 10^8. Such oxidation–reduction pairs are rare, but possible. Diffusion measurements in solutions of the oxidation–reduction system ferrocene-ferricen carried out by Ruff [62] indicated an observable contribution of electronic conduction to the charge transport process in these solutions. A real electronic conduction, however, takes place only within a distance of a few Ångstroms, as shown above.

REFERENCES

to Sections 4.7 and 4.8

[1]   P. WALDEN, H. ULICH and G. BUSCH, Z. phys. Chem., 1926, **123**, 429; P. WAL-
       DEN and E. J. BIRR, Z. Phys. Chem., 1931, **153**, 1.
[2a]   C. A. KRAUS, Ann. N. Y. Acad. Sci., 1949, **51**, 789.
[2b]   R. L. KAY, D. F. EVANS and G. P. CUNNINGHAM, J. Phys. Chem., 1969, **73**,
       3322.
[3]   B. J. STEEL and R. H. STOKES, J. Phys. Chem., 1958, **62**, 450; J. M. STOKES
       and R. H. STOKES J. Phys. Chem., 1958, **62**, 497.
[4]   B. J. STEEL, J. M. STOKES and R. H. STOKES, J. Phys. Chem., 1958, **62**, 1514;
       F. ACCASCINA, J. Amer. Chem. Soc., 1959, **81**, 4995.
[5a]   H. FRICKE, J. Phys. Chem., 1953, **57**, 934.
[5b]   S. P. MOULIK, Electrochim. Acta, 1972, **17**, 1491.
[6]   W. EBELING, H. FALKENHAGEN and W. D. KRAEFT, Ann. Phys. (Leipzig)
       1966, **18**, 15.
[7]   R. A. ROBINSON and R. H. STOKES, Electrolyte Solutions p. 309 (London,
       1959).
[8a]   See e.g. G. KORTÜM, S. D. GOKHALE and H. WILSKI, Z. Phys. Chem., NF.,
       1955, **4**, 286; G. KORTÜM and A. WELLER, Z. Naturforsch., 1950, **5a**, 451, 590;
       TH. SHEDLOVSKY, The Structure of Electrolytic Solutions (ed. by W. J. HAMER)
       p. 268 (New York, 1959).
[8b]   E. S. AMIS and J. F. CASTEL, J. Electrochem. Soc., 1970, **117**, 213.
[9]   T. ERDEY-GRÚZ, E. KUGLER and A. REICH, Acta Chim. Acad. Sci. Hung.,
       1958, **13**, 429; cf. N. G. FOSTER and E. S. AMIS, Z. Phys. Chem. NF., 1955,
       3, 365; S. PETRUCCI, Acta Chem. Scand., 1962, **16**, 760.
[10]   T. ERDEY-GRÚZ, E. KUGLER and J. HIDVÉGI, Acta Chim. Acad. Sci. Hung.,
       1959, **19**, 89.
[11]   T. ERDEY-GRÚZ, E. KUGLER and L. MAJTHÉNYI, Electrochim. Acta, 1968,
       13, 947.
[12]   T. ERDEY-GRÚZ and E. KUGLER, Acta Chim. Acad. Sci. Hung. In press.
[13]   L. G. PEDERSEN and E. S. AMIS, Z. Phys. Chem. NF., 1963, **36**, 199.
[14a] R. A. HORNE, D. S. JOHNSON and R. P. YOUNG, J. Phys. Chem., 1968, **72**, 866.
[14b] F. ACCASCINA, R. DE LISI and M. GOFFREDI, Electrochim. Acta, 1971, **16**, 101.
[15]   C. W. DAVIES and G. O. THOMAS, J. Chem. Soc., **1958**, 3660.
[16]   A. THAN and E. S. AMIS, Z. Phys. Chem. NF., 1968, **58**, 196.
[17]   N. G. FOSTER and E. S. AMIS, Z. Phys. Chem. NF., 1956, **7**, 360.
[18]   R. WHORTON and E. S. AMIS, Z. Phys. Chem. NF., 1958, **17**, 300.
[19]   T. ERDEY-GRÚZ, E. KUGLER and J. HIDVÉGI, Acta Chim. Acad. Sci. Hung.,
       1959, **19**, 363.
[20]   T. ERDEY-GRÚZ and E. KUGLER, Acta Chim. Acad. Sci. Hung., 1968, **57**,
       301 cf. L. A. WOOLF, J. Phys. Chem., 1960, **64**, 500.

[21]   F. ACCASCINA and M. GOFFREDI, *Ricerca Sci.*, 1968, **37**, 1126.
[22a] F. J. BARTOLI, J. N. BIRCH, N. H. TOAN and G. E. MCDUFFIE, *J. Chem. Phys.*, 1968, **49**, 1916.
[22b] R. P. DE SIENO, P. W. GRECO and R. C. MAMAJEK, *J. Phys. Chem.*, 1971, **75**, 1722.
[23]   T. ERDEY-GRÚZ and L. MAJTHÉNYI, *Acta Chim. Acad. Sci. Hung.*, 1958, **16**, 417.
[24]   T. ERDEY-GRÚZ and L. MAJTHÉNYI, *Acta Chim. Acad. Sci. Hung.* 1959, **20**, 73; cf. J. O. WEAR, C. V. MCNULLY and E. S. AMIS, *J. Inorg. Nucl. Chem.*, 1961, **19**, 278; 1961, **20**, 100.
[25]   T. ERDEY-GRÚZ and L. MAJTHÉNYI, *J. Inorg. Nucl. Chem.*, 1959, **20**, 175.
[26a] T. ERDEY-GRÚZ, L. MAJTHÉNYI and I. NAGY-CZAKÓ, *J. Inorg. Nucl. Chem.*, 1967, **53**, 28.
[26b] A. G. MITCHELL and W. F. K. WYNNE-JONES, *Disc. Faraday Soc.*, 1953, **15**, 161.
[26c] V. A. MIKHAILOV, *Zh. Strukt. Khim.*, 1961, **2**, 677.
[26d] M. CARMO and M. SPIRO, *J. Phys. Chem.*, 1972, **76**, 712.
[26e] J. OAKES, *Faraday Transact. II.*, 1973, 1311.
[27]   T. ERDEY-GRÚZ and I. NAGY-CZAKÓ, *Acta Chim. Acad. Sci. Hung.*, 1971, **67**, 283.
[28a] J. R. BARD, J. O. WEAR, R. G. GRIFFIN and E. S. AMIS, *J. Electroanal. Chem.*, 1964, **8**, 419.
[28b] R. L. KAY and T. L. BROADWATER, *Electrochim. Acta*, 1971, **16**, 667.
[28c] M. V. RAMANA-MURTI and C. C. YADAV, *Electrochim. Acta*, 1972, **17**, 643.
[29]   T. ERDEY-GRÚZ, E. KUGLER, I. NAGY-CZAKÓ and K. BALTHAZÁR-VASS, *Acta Chim. Acad. Sci. Hung.*, 1972, **71**, 353.
[30]   E. GRUNWALD, Electrolytes (ed. by B. PESCE) p. 62, (Oxford, 1962).
[31]   G. ÅKERLÖF and O. A. SHORT, *J. Amer. Chem. Soc.*, 1936, **58**, 1241; J. E. LIND and R. M. FUOSS, *J. Phys. Chem.*, 1962, **66**, 1727.
[32]   T. L. FABRY and R. M. FUOSS, *J. Phys. Chem.*, 1964, **68**, 971, 974.
[33]   F. ACCASCINA and S. SCHIAVO, Chemical Physics of Ionic Solutions (ed. by B. E. CONWAY and R. G. BARRADAS) p. 515 (New York, 1966).
[34]   T. ERDEY-GRÚZ and I. NAGY-CZAKÓ, *Acta Chim. Acad. Sci. Hung.* 1971, **67**, 283; cf. T. L. FABRY and R. M. FUOSS, *J. Phys. Chem.*, 1964, **68**, 971.
[35]   J. B. HASTED, G. H. HAGGIS and P. HUTTON, *Trans. Faraday Soc.*, 1951, **47**, 577.
[36]   J. W. WILLIAMS, *J. Amer. Chem. Soc.*, 1930, **52**, 1838.
[37]   A. R. TOURKY, H. A. RIZK and Y. M. GIRGIS, *J. Phys. Chem.*, 1961, **65**, 40.
[38]   J. G. KIRKWOOD, *J. Chem. Phys.*, 1939, **7**, 911.
[39a] A. M. SHKODIN, N. FLEVITSKAIA and V. A. LOZHNIKOV, *Zh. Obsch. Khim.*, 1968, **38**, 1006.
[39b] G. G. HAMMES and W. KNOCHE, *J. Chem. Phys.*, 1966, **45**, 4041.
[39c] J. R. FANG, K. GODZIK and G. L. HOFACKER, *Ber. Bunsen Ges.*, 1973, **77**, 980.
[40]   H. GOLDSCHMIDT and P. DAHL, *Z. Phys. Chem.*, 1924, **108**, 121; 1925, **114**, 1.
[41]   P. WALDEN, *Z. Phys. Chem.*, 1924, **108**, 341.
[42]   L. THOMAS and E. MARUM, *Z. Phys. Chem.*, 1929, **143**, 191.
[43]   I. I. BEZMAN and F. H. VERHOEK, *J. Amer. Chem. Soc.*, 1945, **67**, 1330.
[44]   G. KORTÜM and H. WILSKI, *Z. Phys. Chem. N. F.*, 1954, **2**, 256.
[45]   G. F. DNIEPROV, *Uchenie Zapiski LGU*, 1939, **40**, 19.
[46]   A. M. EL AGGAN, D. C. BRADLEY and W. WARDLAW, *J. Chem. Soc.*, **1958**, 2092.
[47]   A. R. TOURKY and S. Z. MIKHAIL, *Egypt. J. Chem.*, 1958, **1**, 1, 13, 187.
[48]   P. KIRBY and O. MAASS, *Canad. J. Chem.*, 1958, **36**, 456.
[49]   B. E. CONWAY, J. O'M. BOCKRIS and H. LINTON, *J. Chem. Phys.*, 1956, **24**, 834.
[50]   L. A. WOOLF, *J. Phys. Chem.*, 1960, **64**, 500.
[51a] F. ACCASCINA, A. D'APRANO and M. GOFFREDI, *Ricerca Sci.*, 1964, **34** (II.a° 443).
[51b] F. ACCASCINA, R. DE LISI and M. GOFFREDI, *Electrochim. Acta*, 1970, **15**, 1209.

[52a] A. F. KAPUSTINSKY, *Acta Chim. Acad. Sci. Hung.*, 1958, **15**, 351.
[52b] T. N. ZATSEPINA, *Zh. Strukt. Khim.*, 1971, **12**, 969.
[53] M. AGENO and C. FRONTALL, *Bull. Soc. Ital. Fis.*, 1966, **50**, 19.
[54] H. R. TOURKY, S. Z. MIKHAIL and A. A. ABDEL-HAMID, *Z. phys. Chem.*, 1973, **252**, 289.
[55] R. DE LISI and M. GOFFREDI, *Electrochim. Acta*, 1972, **17**, 2001.
[56] N. GOLDENBERG and E. S. AMIS, *Z. Phys. Chem. N. F.*, 1961, **30**, 65; 1962, **31**, 145.
[57] E. WICKE, M. EIGEN and TH. ACKERMANN, *Z. Phys. Chem. N. F.* 1954, **1**, 340.
[58a] I. G. MURGULESCU, FL. BARBULESCU and A. GREFF, *Revue Roumaine de Chimie*, 1965, **10**, 387.
[58b] A. R. TOURKY and A. A. ABDEL-HAMID, *Z. Phys. Chem.*, 1971, **248**, 9.
[58c] R. DE LISI and M. GOFFREDI, *Electrochim. Acta*, 1971, **16**, 2181.
[59a] T. ERDEY-GRÚZ and I. NAGY-CZAKÓ, *Acta Chim. Acad. Sci. Hung.*, 1971, **67**, 283.
[59b] D. LAFORGUE-KANTZER, A. LAFORGUE and T. C. KHANH, *Electrochim. Acta*, 1972, **17**, 151.
[60] V. G. LEVICH, Advances of Electrochemistry (ed. by P. DELAHAY) 1966, **4**, 314.
[61] H. DAHMS, *J. Phys. Chem.*, 1968, **72**, 362.
[62] I. RUFF, *Electrochim. Acta*, 1970, **15**, 1059.

5. APPENDIX

Transport processes in electrolyte solutions are closely related to the inter-actions of the molecules and ions present. These relationships are complex and various, they interpenetrate complicatedly, and can be expressed by means of various forces. Nevertheless, the interactions between the species of electrolyte solutions and the phenomena caused by them can be classified into two main groups with respect to their nature. In one of the groups, the predominating forces are electrostatic attraction and repulsion between the electric charges of ions, i.e. coulombic forces; they mainly modify the spatial distribution of dissolved ions and decrease ionic mobility. The theories discussing this effect are usually summarized under the term: *the electrostatic theory of strong electrolytes*. In the other group, the interactions between ions and solvent molecules are considered; on one hand, the electric or some other atomic field of ions disturbs or destroys the structure of water (or, in general, that of the solvent); on the other hand, it binds solvent mole-cules to the ion in a more or less ordered arrangement (but, usually, not by covalent bonds). These phenomena, called *solvation* or–in the case of water *hydration*, are very complicated in themselves. Their common feature is, however, that several properties of the solvent, mainly its structure and, hence, its enthalpy, entropy, molar volume, compressibility and the mobility of the molecules are altered in the vicinity of ions. The mobility of water molecules has great importance in transport processes and is influenced by the ionic field in two opposite directions: the mobility of water molecules is increased by destroying the lattice-like structure of water, while it is decreased by the ordering and fixing nature of the ion-dipole interaction, as well as of other van der Waals-type forces. If the resultant force depending on the relative magnitude of these two types of effects results in a decrease in the mobility of water molecules, *positive hydration* (or in short: hydration) takes place, while if the resultant effect increases the mobility of water molecules as compared with that of pure water, it is usually quoted as *negative hydration*. In addition to the interaction between the permanent dipole moments of polar water molecules and ions, the polarization (induced dipole moment) of molecules and the dispersion effect also influence hydration.

In fact, the electrostatic interaction between ions and the solvation effects between ions and the solvent cannot be clearly separated from each other. The attraction of ions of opposite charge and the repulsion of those of identical charge does not leave the water molecules located between them unaltered, whereas the interaction between ions and solvent molecules

Literature on page 439

influences the coulombic forces, and locally alters the dielectric constant of the medium. A relative separation of these two groups of phenomena according to the predominating effects is still justified, since it makes possible a better survey and theoretical discussion of the complicated properties of electrolyte solutions.

5.1 INTERACTION OF DISSOLVED IONS (THEORY OF STRONG ELECTROLYTES)

It has been known since the end of the last century that the formation of the ions of electrolyte solutions takes place simultaneously with the dissolution of a pure solid or liquid 'electrolyte'. The theory of Arrhenius has been the starting point of modern electrochemistry and states that ions are formed on dissociation of the 'molecules' of an electrolyte; for binary electrolytes, in general:

$$KA \rightleftharpoons K^{z+} + A^{z-}$$

where KA is the electrolyte, while the cation and anion formed from it are K^{z+} and A^{z-}. However, by widening and sophisticating our knowledge of electrolyte solutions, it has been proved in this century that electrolyte molecules do not dissociate spontaneously, and that the ions moving with a relative freedom in solution are the products of interactions between the solute and the solvent.

Considering either substances consisting of covalent molecules in pure state *(potential electrolytes)*, or ionic crystals containing ions fixed in the crystal lattice *(true electrolytes)*, the interaction of the solvent with the solute, or with the ions formed from it, is a decisive factor in ionization. The work required for the decomposition of covalent molecules or crystal lattices is provided by the energy of solvation (or hydration), mostly with a small contribution from the heat of dissolution as well. Due to the rather complicated interactions between dissolved ions and solvent molecules, as well as between the ions themselves, various species (hydrated and complex ions, monomeric and associated molecules, and clusters of a lattice-like structure modified by the electric field of ions etc.) are usually present in solution. A theory reflecting the properties of electrolyte solutions really reliably should take into account all types of species and all interactions. However, no such theory has been successfully elaborated so far owing to the complicated nature of the conditions, and it cannot be expected in the near future, either.

The various theories on electrolyte solutions explain certain groups of properties and the characteristics of some processes which predominate under actual conditions in an approximation permitted by the development of the theoretical treatment, usually on the basis of greatly simplifying assumptions. Consequently, even the most complex up-to-date theories show only approximate agreement with the observations in a limited range of conditions. For example, the electrostatic theory of strong electrolytes investigates the electrostatic interaction of the electric charges of ions, treating the solvent as a continuum, taking into account only some changes in its dielectric constant. The theories on solvation or hydration, on the other hand, deal with the direct interaction of ions with solvent molecules, almost neglecting the interaction between the ions themselves. Much evidence should still be collected about the properties of electrolyte solutions, and the processes occurring in them. Extended and accurate experiments which will also reveal small effects are required; furthermore, the theoretical methods should be developed considerably before the elaboration of a theory on the properties of electrolyte solutions, reflecting the real conditions satisfactorily reliably and completely, could be expected. Nevertheless, the current theories–which can be considered as first approximations–also give some valuable information for the interpretation of several phenomena, and give direction to future experimental and theoretical research.

5.1.1 RATIO OF FREE IONS IN SOLUTION

According to the Arrhenius theory, the ions of an electrolyte solution are formed in the course of the *dissociation* of electrolyte molecules; this process can be described similarly to thermal dissociation by the law of mass action. According to the classification developed at that time, electrolytes with a dissociation constant lower than about 10^{-3} are weak electrolytes, those having a dissociation constant from about 10^{-3} to one are medium strong electrolytes, and those having a dissociation constant higher than one are strong electrolytes. However, with the improvement of our knowledge on electrolyte solutions, this classification has not proved to be satisfactory, all the more so because some experiments indicated that, for solutions of strong electrolytes and for the changes of their conductivities, the law of mass action does not hold in its original form. It became clear that the nature of the process in the course of which free ions (i.e. those having a relatively high mobility) are formed from solutes consisting of covalent molecules in pure state (e.g. hydrochloric acid) is entirely different from that producing ions from ionic crystals. Further, it has also been established in Bjerrum's studies that two ions can form an ion pair or, incidentally, three of them form an ion triplet by means of electrostatic attraction without forming any covalent bonds between them. This phenomenon is ionic *association* and leads to a decrease in the ionic concentration of the solution like the suppression of dissociation. These two processes, however, differ from each other in their nature, because the products of suppressed dissociation are covalent molecules, while ion pairs or triplets kept together by electrostatic attraction are mainly formed in ionic association, within which ions retain their relative independence. In spite of their different structures, ion pairs or triplets formed from oppositely charged ions decrease the conductivity of the solution in a similar way to non-dissociated molecules.

Literature on page 439

As a consequence of all the facts mentioned above, the term 'strong electrolyte' is no longer clear cut. *Strong electrolytes* may be interpreted as being those present as almost 100 per cent free ions in very dilute solutions, irrespective of whether the additional part of the solute forms non-dissociated molecules or ion pairs. In this sense, both HCl and KCl are strong electrolytes, although in their pure state and in their 'non-dissociated form' their bonding type is different. However, the concept of strong electrolytes can also be defined as referring to compounds that form no covalent molecules either in the pure state or in solution, but ionic lattices and ion pairs (or triplets), respectively (they may be called *true electrolytes* too); in this case, *weak electrolytes* have ions capable of forming covalent molecules in solution. According to this definition, for example, both hydrochloric acid and nitric acid are weak electrolytes, in addition to the common weak organic acids and bases [1].

The more concentrated the solution, the higher the ratio of ion pair formation under comparable conditions. In solutions of identical concentration, it also depends on the nature of the electrolyte. NaCl, KCl, the halides and perchlorates of other alkali and alkali earth metals, as well as perchloric acid, showing no trace of marked ion pair formation in dilute aqueous solutions, are typical examples of *non-associating electrolytes*. In more concentrated solutions, however, ions of these electrolytes do associate. *Associating electrolytes* are those already forming ion pairs in an appreciable ratio in dilute solutions as, for example, $MgSO_4$ in aqueous solutions and as most of the salts in non-aqueous solutions do.

Even the classification of electrolytes in the categories of associating, non-associating, and weak is not entirely unequivocal, partly because no sharp distinction can be made between dilute and more concentrated solutions, in the limiting cases it depends to a certain degree on subjective judgements which category a given electrolyte (e.g. $LiNO_3$, $Mg(NO_3)_2$, rare earth halides) belongs to. On the other hand, there are some electrolytes consisting of ions which show no or only limited association in dilute solutions, while covalent complex ions are formed in more concentrated ones which, in addition, can also associate with the counter-ion. For example, the ions of zinc iodide do not associate in dilute solutions, while in more concentrated ones ZnI_4^{2-} is formed by covalent bonds and shows appreciable association with Zn^{2+} ions.

Electrically neutral ion pairs cannot always be clearly distinguished from non-dissociated molecules. The latter can be reliably identified when, for example, the vibrational band of the molecule can be found in the infrared or Raman spectra. The absence of covalent molecules is not, however, proved if no corresponding bands have yet been observed in the spectra.

The concept of association itself is not entirely clear either, since ions of opposite charge evidently meet each other (collide) occasionally in every solution, and they remain together for a longer time than colliding neutral molecules do. On the other hand, ion pairs are also in a kinetic equilibrium with free ions and sooner or later they decompose. The limiting case of association can perhaps be that two ions are considered an ion pair when the average duration of their being together is longer than the vibrational period of the pair of the species.

Several properties of dilute electrolyte solutions can be described in first approximation by taking into account only the relatively long-range electrostatic forces between ions and neglecting all the others which are regarded as short-range interactions in comparison with them. In this respect, it seems to be permissible to consider all the substances that are present almost completely (95–100 per cent) as free ions in dilute solution as strong electrolytes, irrespective of whether the small additional part forms ion pairs or covalent molecules.

5.1.2 DISTRIBUTION OF IONS IN DILUTE SOLUTIONS OF STRONG ELECTROLYTES

In electrolyte solutions, the predominating interaction *between ions* is coulombic attraction and repulsion due to their electric charges, and these forces are inversely proportional to the square of the distance. In addition, van der Waals forces independent of the free electric charges are also operative; their effect is, however, inversely proportional to a higher power of the mutual separation of the particles, thus their efficient range is much shorter than that of coulombic forces. Van der Waals forces are significant only between adjacent particles, while the coulombic forces can reach to distances of several atomic diameters. Hence, several properties of the dilute solutions of strong electrolytes can be described in a rather good approximation by taking into account only electrostatic attraction and repulsion. The laws of electrostatics, however, are related to macroscopic phenomena, thus the conditions existing at the order of magnitude of atomic sizes can only approximately be described by them by introducing simplifying assumptions.

Our model for dilute solutions of strong electrolytes is based on the theory of Debye and Hückel [2]. This theory considers the solvent to be a continuum of dielectric constant ε in which the distribution of ions is determined by the thermal energy and the energy of their electrostatic interaction. Of course, the distribution always varies due to fluctuations and only its average value remains the same. Owing to the coulombic forces, every positively charged ion attracts the negative ones and repels the others of positive charge, and *vice versa*. In this way, in the vicinity of each ion, oppositely charged ions are in excess on the statistical average, i.e. every ion is surrounded by an 'ionic atmosphere' with an average charge opposite to that of the 'central ion'.

Thermal energy, acting in the direction of uniform distribution, and electrostatic energy, acting simultaneously towards an ordering according to the sign of the charge, yield a peculiar distribution in electrolyte solutions. Results indicate that any small but still macroscopic volume of the solution is electrically neutral, i.e. ions are in a uniform overall distribution

Literature on page 439

(electroneutrality condition). If, however, one considers a coordinate system *fixed* with respect to any of *the ions*, there is an excess of oppositely charged ions on the statistical average in the vicinity of the ion, i.e. there is an excess of positive or negative charges which decreases asymptotically to zero with increasing distance from the ion. This coordinate system is in permanent movement, and, since such a coordinate system can be assigned to every ion, the ionic atmospheres described by means of them penetrate into one another.

The Debye–Hückel theory starts from the fact that, around an arbitrarily chosen ion, the ions are arranged according to the Boltzmann distribution law on the statistical average. According to this, in a system containing a sufficiently high number of particles, the number of particles in unit volume (n') or their concentration c' can be given as:

$$n' = n \exp\left[-\frac{E}{kT}\right],$$

and   (5.1.1)

$$c' = c \exp\left[-\frac{E}{kT}\right]$$

at a point where the potential energy of the particles is E, if in the 'bulk' of the solution 'far' from the ions, the total number and concentration of the particles in unit volume are n and c, respectively. Thus, the first step in the theory is the calculation of the electrostatic potential as a function of the distance from the ion, to enable the application of the distribution law of statistical mechanics mentioned above. In the ionic atmosphere formed around each ion considered a centre, the oppositely charged ions present in excess result in a charge density ϱ that can be assumed to be 'smeared' in first approximation. Charge density is the difference of positive and negative electric charges at a given point in unit volume. The potential corresponding to coulombic forces acting between electric charges is described by the Poisson equation of general validity. Expressed in Cartesian coordinates, it is:

$$\Delta^2\psi = \left(\frac{\partial^2\psi}{\partial x^2} + \frac{\partial^2\psi}{\partial y^2} + \frac{\partial^2\psi}{\partial z^2}\right) =$$

$$= -\frac{4\pi\varrho}{\varepsilon},$$ (5.1.2)

if ψ is the electric potential at the point where the charge density is ϱ and ε is the dielectric constant of the medium.

For spherically symmetrical charge distribution, at a given point the potential is the function of the distance r only, measured from the centre of the central ion. Under such conditions, the Poisson equation becomes:

$$\frac{1}{r^2}\frac{d}{dr}\left(r^2\frac{d\psi}{dr}\right) = -\frac{4\pi\varrho}{\varepsilon}.$$ (5.1.3)

For the dielectric constant ε, the Debye–Hückel theory takes the macroscopic dielectric constant of the solvent which, however, is only a first approximation acceptable at most for very dilute solutions. Since the liquid present between the ions is the decisive factor with respect to their electrostatic interaction, and as the structure and state of the liquid is evidently altered by ions, it is not in fact justified to apply the dielectric constant measurable macroscopically, being the average of various local values. So far, however, it has not been determined reliably and satisfactorily what dielectric constant should be taken. In this respect, neither have unequivocal results been yielded by the investigations of Glueckauf [3], who calculated the change of the dielectric constant in the vicinity of ions on the basis of a continuum model. However, his theory makes it probable that water structure is changed discontinuously in the strong electric field of ions, perhaps as a result of the temporary formation of 'ice-bergs', proposed by Frank and Evans [4].

Another deficiency of these starting assumptions in the theory of electrolytes is that the Poisson equation is valid rigorously only for continuous charge distribution, whereas the charge of the ionic atmosphere has a discrete distribution. Furthermore, the size of the ionic atmosphere is, in fact, not large enough to eliminate the individuality of charge carriers in comparison with it and to permit consideration of the charge excess as a continuum. However, this deficiency could not be successfully eliminated whilst retaining the requirement for the theory to yield results comparable to experiments.

In order to describe the potential conditions in the ionic atmosphere in a more detailed way, let us examine an arbitrary ion, say, the jth positive one whose charge is $+z_j e$. There are excess negative ions in its environment, i.e. the probability of finding negative ions in volume elements in its vicinity is higher than that of finding positive ions. Since electrolyte solutions as a whole are electrically neutral, the charge of the total ionic atmosphere surrounding the central jth ion is equal to that of the central ion with an opposite sign. If a is *the smallest distance* to which the central ion can be approached by another ion, and ϱ_j is the charge density at a distance r from the jth ion measured in the coordinate system fixed to the jth ion, one has:

$$\int_a^\infty 4\pi r^2 \varrho_j \, dr = -z_j e. \qquad (5.1.4)$$

It should be emphasized that the coordinate system is not fixed in space (e.g., to the vessel of the solution), since in such a system, the time average of the charge density is zero everywhere. The subscript j indicates that the respective relationships are valid only in the coordinate system fixed to the jth ion, i.e. to a system always moving with respect to the vessel.

For the ionic atmosphere one can state without additional assumptions that the probability of finding any ith ion in any elementary volume dV is higher, the lower its potential energy there, and the higher its concentra-

Literature on page 439

tion in the bulk of the solution. At a sufficiently large distance from the central ion, the electric forces produced by it are negligibly small, thus the probability of finding an ion there is proportional to $n_i \, dV$ (n_i is the number of the ith kind of ions in unit volume). Beyond these two statements, no other *sound* basis is available for the distribution of ions in the ionic atmosphere. It has been assumed by Debye and Hückel that, on the statistical average, the other ions are distributed according to the Boltzmann distribution law around the individual ion considered as the central one. In this way, at a point related to the central ion, where the potential is ψ_j, the local concentration of ions is:

$$n_i' = n_i \exp - \left[- \frac{z_i e \psi_j}{kT} \right]. \tag{5.1.5}$$

The net charge density at this point is:

$$\varrho_j = \Sigma \, n_i z_i e \exp - \left[- \frac{z_i e \psi_j}{kT} \right], \tag{5.1.6}$$

since each ion has a charge of $z_i e$. According to the charge distribution calculated from the Boltzmann theorem, charge density would be an exponential function of potential. However, this is in contradiction with that theorem of electrostatics which states that potentials arising from two or more charges or from a system of charges add up linearly. If, for example, the charge of each ion (and in this way, the charge density) is increased to twice its original value, the potential should also increase by a factor of two at every point, according to the laws of electrostatics. On the basis of equation (5.1.6) deduced from the Boltzmann theory, however, the relationship is exponential. Thus, the Boltzmann distribution, in general, contradicts the Poisson theorem, i.e., rigorously speaking, they cannot be applied simultaneously for the reliable description of real conditions. Under certain conditions, however, the Boltzmann theorem can still be used to calculate the charge density, in approximation. Namely, equation (5.1.6) can be expanded into a series*:

$$\varrho_j = \Sigma \, n_i z_i e - \Sigma \, n_i z_i e \left(\frac{z_i e \psi_j}{kT} \right) + \Sigma \frac{n_i z_i e}{2!} \left(\frac{z_i e \psi_j}{kT} \right)^2 - \Sigma \frac{n_i z_i e}{3!} \left(\frac{z_i e \psi_j}{kT} \right)^3 + \dots \tag{5.1.7}$$

Since an electrolyte solution is macroscopically neutral, we have:

$$\Sigma \, n_i z_i e = 0, \tag{5.1.8}$$

* The expansion of the function e^{-x} into a series is, generally:

$$e^{-x} = 1 - x + \frac{x^2}{2!} - \frac{x^3}{3!} + \frac{x^4}{4!} - \dots$$

and so the first term in equation (5.1.7) can be omitted. If $z_i e \psi_j \ll kT$, i.e., the electric potential energy of the ith kind of ions is much lower than the average thermal energy, the terms of second and third power are negligibly small and retaining the term of first power is satisfactory in first approximation. With such a restriction, the Boltzmann distribution also yields a linear relationship between ϱ_j and ψ_j:

$$\varrho_j = - \sum \frac{n_i z_i^2 e^2 \psi_j}{kT}, \tag{5.1.9}$$

and

$$n_i' = n_i \left(1 - \frac{z_i e \psi_j}{kT} \right). \tag{5.1.10}$$

In very dilute solutions, the criterion $z_i e \psi_j \ll kT$ evidently gives a good approximation for the majority of the ith kind of ions which are rather far from the central jth ion, but it cannot be correct for ions close to the central one. In connection with this, however, it should be taken into account that the properties of dilute solutions of strong electrolytes differ significantly from those of ideal mixtures, which indicates that electric interactions between ions cannot be neglected in comparison with thermal energy. In general, this approximation is entirely justified only in the limiting case of infinitely dilute solutions.

The Boltzmann distribution is in much better accordance with the laws of electrostatics, as compared with the general case above, if the solution contains only one single electrolyte of symmetrical valence type. In this case, $z_1 = -z_2$ and $n_1 = n_2$, consequently, the terms of even powers in equation (5.1.7) are zero, and under such conditions one obtains:

$$\varrho_j = - 2n_1 z_1 e \left(\frac{z_1 e \psi_j}{kT} \right) - \frac{1}{3} n_1 z_1 e \left(\frac{z_1 e \psi_j}{kT} \right)^3 - \ldots \tag{5.1.11}$$

In this expression, the term of third power can be neglected in a much better approximation than the term of second power in the expression corresponding to the general case.

The relationship between the potential of the ionic atmosphere and the distance measured from the central ion can be derived from the Poisson law given in equation (5.1.3). Taking into account equation (5.1.11), we can write:

$$\frac{1}{r^2} \frac{d}{dr} \left(r^2 \frac{d\psi_j}{dr} \right) = \frac{4\pi e^2}{\varepsilon kT} \sum n_i z_i^2 \psi_j =$$

$$= \varkappa^2 \psi_j, \tag{5.1.12}$$

where $\varkappa$ is given by:

$$\varkappa^2 = \frac{4\pi e^2 \sum n_i z_i^2}{\varepsilon kT}. \tag{5.1.13}$$

Literature on page 439

$\varkappa$ has a dimension of reciprocal length and its value depends on the concentration and valence of the ions, as well as on the dielectric constant of the solution and on temperature. $1/\varkappa$ is the so-called *characteristic distance* (the 'radius' or 'thickness' of the ionic atmosphere).

Equation (5.1.12) is a second-order linear differential equation. Its general solution is:

$$\psi_j = A\, \frac{e^{-\varkappa r}}{r} + B\, \frac{e^{\varkappa r}}{r}, \tag{5.1.14}$$

where A and B are integration constants. Since the potential has a finite value even at a large distance from the central ion, $B = 0$, i.e.:

$$\psi_j = A\, \frac{e^{\varkappa r}}{r}. \tag{5.1.15}$$

In order to calculate the constant A, one can start with the other limiting case, since, if r tends to zero, ψ_j becomes the potential of a point charge. This is:

$$\psi_j = \frac{z_i e}{\varepsilon r}. \tag{5.1.16}$$

If $r \to 0$, then $e^{-\varkappa r} \to 1$, thus:

$$\frac{z_i e}{\varepsilon r} = \frac{A}{r}, \tag{5.1.17}$$

and

$$A = \frac{z_i e}{\varepsilon}, \tag{5.1.18}$$

while the potential will be:

$$\psi_j = \frac{z_i e}{\varepsilon r}\, e^{-\varkappa r}. \tag{5.1.19}$$

On the basis of this expression, the length $1/\varkappa$ can gain a certain figurative physical meaning. Namely, expanding the power of e into series and retaining only the first term, one has:

$$\psi_j = \frac{z_i e}{\varepsilon r}\,(1 - \varkappa r) = \frac{z_i e}{\varepsilon r} - \frac{z_i e}{\varepsilon}\cdot\varkappa. \tag{5.1.20}$$

In this expression:

$$\frac{z_i e}{\varepsilon r} = \psi' \tag{5.1.21}$$

being the potential of the point-like central ionic charge at a distance r measured from it; whereas:

$$\frac{z_i e}{\varepsilon\, \dfrac{1}{\varkappa}} = \psi_j'' \tag{5.1.22}$$

can be regarded as the potential at the position of the central ion which would be caused by a spherical shell located at a distance $1/\varkappa$, if the charge of the ionic atmosphere were distributed uniformly all over it. In this sense, $1/\varkappa$ can be called as the *'thickness'* or *'radius'* of the ionic atmosphere. In this way:

$$\psi_j = \psi'_j - \psi''_j. \tag{5.1.23}$$

That is, at point r in the ionic atmosphere, the potential can be regarded as the algebraic sum of the potentials arising from the lone central ion and from the ionic atmosphere alone.

The above deduction yields an acceptable approximation for the calculation of A only when the solution is very dilute and the ions can be considered as point charges in comparison with the size of the ionic atmosphere (i.e., the radii of ions are negligibly small as compared with the thickness of ionic atmosphere).

If, however, the thickness of the ionic atmosphere is not large (the solution is not dilute enough), one can obtain a better approximation when applying equation (5.1.4) in calculating A, which takes into account the fact that ions can approach one another only to a distance a, i.e., they are not point-like, but have a radius of $a/2$, on the average. Introducing the expression of the potential given in equation (5.1.15) into equation (5.1.19) describing charge density, one has:

$$\varrho_j = -A\,\frac{e^{-\varkappa r}}{r}\sum\frac{n_i z_i^2 e^2}{kT} = -A\,\frac{\varkappa^2 \varepsilon\, e^{-\varkappa r}}{4\,\pi r}. \tag{5.1.24}$$

On the other hand, this value of the charge density can be inserted into equation (5.1.4) (which is, in fact, the expression of the electroneutrality of the solution as a whole) and one obtains:

$$A\varkappa^2\,\varepsilon \int\limits_a^\infty re^{-\varkappa r}\,\mathrm{d}r = z_j\,\mathrm{e}. \tag{5.1.25}$$

Under such conditions, this yields:

$$A = \frac{z_j\,\mathrm{e}}{\varepsilon}\cdot\frac{e^{\varkappa a}}{1+\varkappa a}, \tag{5.1.26}$$

while the potential is:

$$\psi_j = \frac{z_j\,\mathrm{e}}{\varepsilon}\cdot\frac{e^{\varkappa a}}{1+\varkappa a}\cdot\frac{e^{-\varkappa r}}{r}. \tag{5.1.27}$$

This relationship is the fundamental equation of the Debye–Hückel theory of strong electrolytes from which phenomena arising from the interactions of ions can be deduced. Of course, only approximations valid within the limits mentioned above can be expected from these deductions even in the optimum case (namely, if other forces have no influence). As a further

Literature on page 439

restriction of its validity, it should also be pointed out that a represents the minimum distance between the centres of ions, i.e., the sum of their effective radii in solution. This implies the assumption that every ion is a sphere of the same radius a, and this is a very rough approximation in several cases, (as indicated by, e.g., ionic mobilities).

In calculating the potential of the ionic atmosphere, Müller [5] and Gronwall, LaMer and Sandved [6] increased the accuracy by taking into account the higher terms in the series expansion. However, no essentially new results have been yielded by these calculations. Wicke and Eigen [7] criticized the correctness of the application of the Boltzmann distribution function. Although the concentration of ions of opposite charge is higher than the average around each ion, this increasing concentration is limited by the finite dimensions of ions. Consequently, there cannot be an arbitrary number of counter-ions in the environment of the central one, but the number of immediate neighbours has a 'saturation' limiting value depending on their size and packing. Wicke and Eigen calculated with a distribution function accounting for this condition. However, it has been shown by Robinson and Stokes [2] that the Debye–Hückel equation does not yield unreasonably high ionic concentrations in the vicinity of the central ion, even in rather concentrated solutions of small ions. Thus, the improvement of the theory suggested in this respect has not resulted in any appreciable advancement.

One of the simplifying assumptions of the Debye–Hückel theory has been the replacement of the potential which fluctuates from point to point owing to the corpuscular character of the ionic atmosphere by an average value. This assumption has been eliminated recently by Outhwaite [8] in a further development of the theory. According to his treatment, the potential of the ionic atmosphere oscillates from a value of $\varkappa a = 1 \cdot 49$ onwards in solutions of symmetrical electrolytes.

Concentration dependence of the characteristic distance $1/\varkappa$. In the theory of strong electrolytes, $\varkappa$ plays an important role in the description of various properties, and in a certain sense, it gives information on the size of the ionic atmosphere. In accordance with the definition in equation (5.1.3), it is:

$$\varkappa = \sqrt{\frac{4 \pi e^2 \Sigma n_i z_i^2}{\varepsilon kT}}. \tag{5.1.28}$$

Since n_i is the number of ions in 1 ml volume, the concentration expressed in the usual unit of mole l^{-1} is:

$$c_i = \frac{1000 \, n_i}{N}. \tag{5.1.29}$$

Introducing the concept of *ionic strength* (I), used in the theory of strong electrolytes, one has:

$$I = \frac{1}{2} \Sigma c_i z_i^2, \tag{5.1.30}$$

which leads to the following formulation of $\varkappa$:

$$\varkappa = \sqrt{\frac{8\,\pi\,e^2\,N}{1000\,\varepsilon\,kT}} \cdot \sqrt{I}\,. \qquad (5.1.31)$$

Thus, $\varkappa$ is proportional to the *square root* of ionic strength (and to the square root of ionic concentration in solutions of 1–1 electrolytes).

Inserting the values of the constants into equation (5.1.31), the thickness of the ionic atmosphere in solutions of 1–1 electrolytes will be:

$$\frac{1}{\varkappa} = 1 \cdot 4 \times 10^{-10} \sqrt{\frac{\varepsilon T}{c_i}}\,, \qquad (5.1.32)$$

while in aqueous solutions at 298 K, one obtains:

$$\frac{1}{\varkappa} = \frac{2 \times 10^{-8}}{\sqrt{c_i}}\,. \qquad (5.1.33)$$

In this way, $1/\varkappa \approx 10^{-7}$ cm in 0·01 M solutions of 1–1 electrolytes. Since the radius of ions is of the order of magnitude of 10^{-8} cm, the thickness of the ionic atmosphere is about ten times as great as the size of the ions, while in 0·0001 M solutions it is hundred times as great as that ($1/\varkappa \approx 10^{-6}$ cm). Thus, in dilute solutions, the thickness of the ionic atmosphere is significant in comparison with the size of the individual ions. However, in 1 M solutions, $1/\varkappa \approx 10^{-8}$ cm, i.e. it has the same order of magnitude as the size of ions. Under such conditions, ions cannot be regarded as point-like charges, even in a rough approximation, and their size should also be taken into account. Ions can be treated as point-like charges, in first approximation, only in solutions more dilute than 0·01 mole l^{-1}.

In solutions of a given concentration of ions of different valence, the thickness of the ionic atmosphere is the smaller, the larger the charge (valence) of ions. For example, in solutions of electrolytes with ions of 1–1 valence present in a concentration of 0·0001 M, $1/\varkappa = 96 \times 10^{-8}$ cm, in solutions of 1–2 electrolytes, this value is 56×10^{-8} cm, in those of 2–2 and 1–3 electrolytes, it is 48×10^{-8} cm, and 39×10^{-8} cm, respectively, whereas in those of 1–4 electrolytes it is 30×10^{-8} cm. This is the consequence of the fact that the higher the ionic charge, the stronger the electrostatic interaction.

5.1.3 CHEMICAL POTENTIALS AND ACTIVITY COEFFICIENTS OF DISSOLVED ELECTROLYTES

One of the important thermodynamic data of solutions is the chemical potential (partial molar free enthalpy) of the solute. If x_j is the molar fraction of the jth ion in a solution, and its ('rational') activity coefficient is γ_j related to the mole fraction, its chemical potential in this solution becomes:

Literature on page 439

$$\mu_j = \mu_j^0 + RT \ln \gamma_j \, x_j =$$
$$= \mu_j^0 + RT \ln x_j + RT \ln \gamma_j \qquad (5.1.34)$$

where μ_j^0 is the chemical normal potential of the jth kind of ions at temperature T. If the solution were ideal, the chemical potential of the jth kind of ions would be:

$$\mu_j^{id} = \mu_j^0 + RT \ln x_j. \qquad (5.1.35)$$

Comparing this with equation (5.1.34), the chemical potential in real solutions is:

$$\mu_j = \mu_j^{id} + RT \ln \gamma_j, \qquad (5.1.36)$$

i.e.

$$\mu_j - \mu_j^{id} = \Delta\mu_j = RT \ln \gamma_j. \qquad (5.1.37)$$

Thus, the measure of the deviation of real solutions from ideal ones is essentially the activity coefficient. One of the fundamental tasks of the theory of real solutions of electrolytes is, thus, the calculation of the activity coefficient of dissolved ions.

With respect to the solute, a solution is rigorously ideal only in the case when the interaction between the molecules of the solute and the solvent does not appreciably differ from the interaction between the solvent molecules themselves.* Also in non-electrolyte solutions, this occurs only when the molecules of the solute and the solvent are chemically very similar. Otherwise, solutions of non-electrolytes are not ideal either. Here, the interaction of the solute with the solvent appears in short-range forces which increase the value of $\ln \gamma_j$ linearly with increasing concentration, i.e. they increase the chemical potential in comparison with ideal solutions (Fig. 5.1). In electrolyte solutions, however, the logarithm of the mean activity coefficient (and, at the same time, the chemical potential) of ions decreases with increasing concentration in the most dilute solutions, according to experiments, then, after passing a minimum, it increases. The decrease in the chemical potential of dissolved ions is a consequence of attraction and repulsion acting between them. These electrostatic forces have a long-range efficiency, thus, these are the predominating factors in dilute solutions. In addition to the mutual interaction between ions, some interactions also exist between ions and

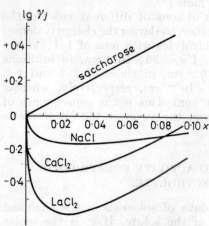

Fig. 5.1 Concentration dependence of the activity coefficient of a non-electrolyte and electrolytes

* Sufficiently dilute solutions are always ideal with respect to the solvent.

solvent molecules (ion-dipole effects, structure-making or structure-break-ing effects, etc.), however, they have a much shorter-range efficiency than electrostatic interactions. Consequently, interactions between ions and solvent molecules become significant only in more concentrated solu-tions; these are responsible for the upward deviation in the plot $\ln \gamma_j$ vs. x (Fig. 5.1).

The theory of electrolyte solutions should take into account all these effects. The electrostatic theory of strong electrolytes, however, calculates only with coulombic forces of long-range efficiency, thus, its validity is limited only to the most dilute solutions, all the more so since some other simplifying assumptions are also involved in this theory.

5.1.3.1 CALCULATION OF THE ACTIVITY COEFFICIENT FROM THE INTERACTION BETWEEN IONS

Thermodynamically, the deviation of the chemical potential of real solu-tions from that of ideal ones can be unequivocally described by the activity coefficient, as shown in equation (5.1.37). Therefore the theory of activity coefficients is usually considered satisfactory, since by knowing the activity coefficient, the change in free enthalpy can simply be calculated. The Debye–Hückel theory of strong electrolytes takes into account only the change in the chemical potential of dissolved ions which arises from the long-range coulombic forces of electric charges, hence it can be valid, in principle, only under particular conditions–very dilute solutions–where the short-range interactions are not significant.

In first approximation, very dilute electrolyte solutions differ from solu-tions of non-electrolytes–otherwise similar in composition–in the creation of the ionic atmosphere. The central ion possesses an electrostatic potential energy due to the ionic atmosphere, and this value is responsible for the difference between the chemical potential of ions and that of neutral molecules, under identical conditions. The potential caused by the ionic atmosphere at the position of the central ion is described by equation (5.1.22) and the change in the chemical potential of the ion brought about by the potential energy arising from this is:

$$\Delta \mu_j = - \frac{N z_j^2 e^2}{2\varepsilon} \varkappa , \qquad (5.1.38)$$

related to 1 mole. The factor of two in the denominator comes from the fact that each ion in the solution is simultaneously a central ion and a member of an ionic atmosphere, thus, if potential energy were calculated as $z_j e \psi''$, each ion would be taken twice in the calculation. Inserting the value of $\varkappa$ from equation (5.1.31), we have:

$$\Delta \mu_j = RT \ln \gamma_j = - \frac{N z_j^2 e^2}{2\varepsilon} \sqrt{\frac{8 \pi e^2 N}{1000 \, \varepsilon \, kT}} \sqrt{I} . \qquad (5.1.39)$$

Literature on page 439

Combining the constant quantities for a given solvent at a given temperature in factor A and performing the transformation to Briggsian logarithm, one gets:

$$\lg \gamma_i = -Az_j^2 \sqrt{I},$$ (5.1.40)

where

$$A = \frac{1}{(\varepsilon T)^{3/2}} \cdot \frac{e^3}{2 \cdot 303 \, k^{3/2}} \sqrt{\frac{2\pi \, N}{1000}} =$$

$$= \frac{1 \cdot 8246 \times 10^6}{(\varepsilon T)^{3/2}} \, \text{mole}^{-1/2} \, l^{1/2} \, \text{degree}^{3/2}.$$ (5.1.41)

However, the individual activity coefficients of ions cannot be determined experimentally, since the measurements yield the mean activity coefficient γ_{ca}; its general definition in binary electrolytes is:

$$\gamma_{ca} = \sqrt[\nu]{\gamma_c^{\nu_c} \gamma_a^{\nu_a}},$$ (5.1.42)

where ν_c and ν_a denote the number of cations and anions, respectively, formed on the dissociation of one 'molecule', while z_c and z_a are their valences $(\nu = \nu_a + \nu_c)$. In a solution containing a single electrolyte, one obtains for the mean activity coefficient:

$$\lg \gamma_{ca} = -A|z_c z_a| \sqrt{I}$$ (5.1.43)

on the basis of equation (5.1.40), taking into account that the electroneutrality condition predicts $\nu_c z_c = \nu_a z_a$. This is *the Debye–Hückel limiting law of the activity coefficient* (the so-called square-root law), the validity of which—owing to the simplifying assumptions applied in the deduction—can be expected, in fact, only in the limiting case of infinitely dilute solutions. According to experiments, this relationship gives a good agreement with observed values from the most dilute solutions to about $0 \cdot 001$ mole l^{-1} concentration: the value of $\lg \gamma_{ca}$ varies linearly with the square-root of concentration, and the slope of the straight line (the value of $A|z_c z_a|$) also corresponds to the theory. This is a rather notable result of the theory, since the expression for A only involves some universal constants, the temperature, and the dielectric parameter of the solvent, without empirical constants to be fitted to observed results. With further increase in concentration, however, increasing deviations appear: the measured values are usually higher than the calculated ones (cf. Section 5.1.4), which, of course, markedly limits the practical applicability of the limiting law.

The second approximation. The validity range of the limiting law regarding the activity coefficient of dissolved electrolytes is somewhat extended if ions are considered as rigid spheres of diameter a in second approximation (more accurately, a is the distance of closest approach of two ions). In this case, the potential caused by the ionic atmosphere at the position of the central ion is:

$$\psi_j'' = \frac{z_j \, e}{\varepsilon r} \left(\frac{e^{\varkappa r}}{1 + \varkappa a} \cdot e^{-\varkappa r} - 1 \right), \qquad (5.1.44)$$

taking into account equations (5.1.21), (5.1.22) and (5.1.27). This expression, valid in second approximation, determines the potential caused by the ionic atmosphere at all points for which $r \geq a$. Since in this approximation ions are considered to be rigid spheres of diameter a, no other ion can appear at a distance $r < a$ from the central ion. Due to the spherical symmetry of the ionic atmosphere, at all points where $r < a$ the potential is the same as at points where $r = a$, i.e.:

$$\psi_j'' = \frac{z_j \, e}{\varepsilon} \cdot \frac{\varkappa}{1 + \varkappa a}, \qquad (5.1.45)$$

on the basis of equation (5.1.44). In this approximation, the effect of ions in the ionic atmosphere on the potential of the central ion is the same as if the charge of the ionic atmosphere were uniformly distributed around it over a spherical surface of radius $a + 1/\varkappa$. If the value of $1/\varkappa$ is taken as the measure of the *thickness of the ionic atmosphere* in this case too, it must be taken into account that this 'thickness' is measured from the distance $r = a$. In very dilute solutions, $1/\varkappa$ is large in comparison with the ionic diameter a, and, in first approximation, $a + 1/\varkappa \approx 1/\varkappa$, i.e. the thickness of the ionic atmosphere can be measured from the centre of the central ion. However, for example, in 1 M solution of 1–1 electrolytes, $1/\varkappa$ is about 3 Å, i.e. it is less than the closest approach distance of two ions. Thus, it seems as if in more concentrated solutions the ionic atmosphere would penetrate into the inside of the central ion. However, this is not correct, because the first approximation can by no means be accepted in such solutions, while, in the second approximation, the thickness of the ionic atmosphere is to be measured from the distance $r = a$.

Taking into account the above calculations, the mean activity coefficient of dissolved electrolytes is given by the following expression, in second approximation:

$$\ln \gamma_{ca} = - \frac{|z_c z_a| \, e^2}{2 \, \varepsilon \, kT} \cdot \frac{\varkappa}{1 + \varkappa a}. \qquad (5.1.46)$$

Introducing the value of $\varkappa$, one has:

$$\ln \gamma_{ca} = - \frac{A |z_c z_a| \sqrt{I}}{1 + Ba \sqrt{I}}, \qquad (5.1.47)$$

where A is the quantity defined in equation (5.1.41), and:

$$B = \left(\frac{8\pi \, Ne^2}{1000 \, k} \right)^{1/2} \cdot \frac{1}{(\varepsilon T)^{1/2}} =$$

$$= \frac{50 \cdot 29 \times 10^8}{(\varepsilon T)^{1/2}} \, cm^{-1} \cdot mole^{-1/2} \, l^{1/2} degree^{1/2}. \qquad (5.1.48)$$

Literature on page 439

Numerical values of A and B depend on the dielectric constant of the liquid, the temperature, and the smallest possible distance a between the ions, in addition to the universal constants. Although this distance can formally be taken as the *diameter of the ions*, this, however, is not identical with the value calculated from the lattice structure of corresponding ionic crystals measured by means of X-ray diffraction, i.e. with the sum of the crystallographic radii of the cation and anion. No method is available for directly determining the 'ionic diameter' a implied in the Debye–Hückel theory, thus the mean activity coefficient of dissolved electrolytes cannot be calculated in second approximation from data determined independently.

The value of the 'ionic diameter' a in the relationships given above is, in fact, obtained empirically in such a way that the value of the mean activity coefficient calculated according to equation (5.1.47) would give the best agreement with the experimental ones (to as high ionic strengths as possible). According to experience, the values of $a \approx (3-6) \times 10^{-8}$ cm are the most appropriate ones. For example, in NaCl solutions, the theoretically calculated activity coefficients agree with the experimentally observed ones relatively well if 4·0–4·8 Å is substituted for a, although the sum of the crystallographic radii is $0·95 + 1·8 = 2·8$ Å. For $CaCl_2$ solutions, a should be 5 Å instead of the crystallographic sum 2·8 Å, while for $LaCl_3$, it is 6–7 Å instead of the crystallographic value 3·0 Å. This difference is probably connected with the fact that ions are hydrated in solution, though a is not necessarily equal to the sum of the radii of the two hydrate spheres, since hydrate spheres can be deformed in the collisions between ions. Nevertheless, the difference in the 'ionic diameter' a and the sum of the crystallographic radii can yield some information on hydration.

The values of activity coefficients calculated from equation (5.1.47) can be brought into a better agreement with experiment when the numerical value of a is not taken as a constant, but somewhat different values are applied in the various concentration ranges. The condition that a is not entirely constant is reasonable to a certain extent according to the original interpretation as well, not only because the extent of hydration can also vary with changes in concentration, but because hydrated ions are not, in fact, rigid spheres and they can be deformed or 'penetrate into each other', depending on the conditions of the collision, altering the smallest possible separation distance. In addition to this real change in the diameter, however, the introduction of a concentration-dependent value of a into the calculation facilitates agreement with the experiments, since, implicitly, some other short-range interactions will also be taken into account by this parameter, in addition to the effect of the ionic sizes, which also depends on concentration.

For a more detailed discussion of the problems related to parameter a in the Debye–Hückel theory, see the paper of Fernández-Prini and Prue [9].

The second approximation–even when a constant ionic diameter a is assumed–describes the dependence of the activity coefficient of ions on the ionic strength or the ionic concentration more reliably than the limiting

law in first approximation (cf. the schematic representation in Fig. 5.2). However, deviations have already been observed
in dilute solutions. Since B is of the
order of magnitude of $0 \cdot 3 \times 10^8$ in aqueous
solutions, the value of Ba in equation
(5.1.47) is of the order of magnitude of
unity (in $0 \cdot 001$ M solutions of 1–1 electrolytes $(1 + Ba\sqrt{I}) \approx 1 \cdot 03$), i.e. the value
of $\lg \gamma_{ca}$ calculated on the basis of the second approximation differs by about 3 per
cent from that of the corresponding first
approximation limiting law even in such
dilute solutions. Nevertheless, the second
approximation also describes the real conditions acceptably only in dilute solutions.
By choosing the value of a appropriately
and regarding it as independent of concen-

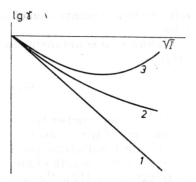

Fig. 5.2 Variation of the activity
coefficient of dissolved ions with
ionic strength; *1*, the first approximation by Debye and Hückel, *2*, the
second approximation, *3*, empirical

tration, the agreement between the calculated and observed values of the activity coefficient is acceptable up to an ionic strength of about $I = 0 \cdot 1$ in
aqueous solutions. The value of a required for this has a plausible physical
meaning as well. In a solution of ionic strength $I \approx 0 \cdot 1$, the average distance
between ions is not greater than 20 Å, and their mutual electric energy is in
the same order of magnitude as their average thermal energy (kT). This
indicates that the distribution function applied in the deduction of equation
(5.1.47) agrees satisfactorily with the experiments in such approximation.

Consideration of the short-range interactions. The second approximation in
the theory of dissolved strong electrolytes considers the ions as rigid non-
deformed spheres. The long-range coulombic forces are taken into account
in the term $-A|z_c z_a|\sqrt{I}$, while the other interactions are represented in
the term $(1 + Ba\sqrt{I})$ in equation (5.1.47). It is evident, however, that
this is also a rough approximation to reality, not only because ions can be
deformed so that their short-range interactions cannot be described by a
model of rigid spheres, but because, in fact, the short-range interactions between solvent molecules and ions should also be considered in a more extended
way, too. According to Robinson and Stokes [10], it can be assumed that
these long-range interactions influence the chemical potential as a linear
function of ionic strength and they can be represented by an empirical
term–linear in concentration–in the expression for the activity coefficient:

$$\lg \gamma_{ca} = -\frac{A|z_c z_a|\sqrt{I}}{1 + Ba\sqrt{I}} + DI, \qquad (5.1.49)$$

where D is a fitting parameter, similar to a. This expression yields values
in agreement with the observed ones within the limits of experimental

Literature on page 439

error up to a concentration of about 1 M in aqueous solutions of 1–1 electrolytes.

A more simple approximate formula has been suggested by Güntelberg [11]:

$$\lg \gamma_{ca} = - \frac{A |z_c z_a| \sqrt{I}}{1 + \sqrt{I}},$$ (5.1.50)

which, in fact, applies the ionic diameter $a = 3.04$ Å in equation (5.1.47), irrespective of the nature of the electrolyte. This equation does not involve any fitting parameters, yet it gives a good agreement with experiments up to an ionic strength of about $I \approx 0.1$ for several electrolytes. According to Guggenheim [12], the agreement can be improved by adding a term linear in concentration (and, at the same time, introducing a parameter D that should be fitted to experimental data):

$$\lg \gamma_{ca} = - \frac{A |z_c z_a| \sqrt{I}}{1 + \sqrt{I}} + DI.$$ (5.1.51)

Table 5.1 shows the deviation in the activity coefficients of NaCl calculated in various ways.

Table 5.1

The mean activity coefficient of aqueous solutions of NaCl at 25 °C

c, mole · 1^{-1}	$-\lg \gamma_{ca}$ according to (5.1.43)	$-\lg \gamma_{ca}$ according to (5.1.50)	$-\lg \gamma_{ca}$ according to (5.1.51)	$-\lg \gamma_{ca}$ according to experiments
0·001	0·0162	0·0157	0·0155	0·0155
0·005	0·0362	0·0338	0·0330	0·0327
0·01	0·0511	0·0465	0·0449	0·0446
0·05	0·1162	0·0933	0·0853	0·0859
0·10	0·1614	0·1227	0·1063	0·1072

Recently, Kremp, Ulbricht and Kelbg [13] calculated the distribution function up to the order $\varkappa^2 a^2 b$ by taking into account the short-range interactions, in addition to the coulombic ones.

As has been already discussed, the Debye–Hückel theory treats the solvent as a continuum, and considers only the mutual interactions of ions; furthermore, some simplifying assumptions form the basis of the deduction. Owing to the model and methods it is a very much simplified description of conditions more complicated in reality. Since the time of its elaboration, considerable work has been devoted to the development of theories reflecting reality in a more detailed way and taking into consideration the complicated interactions of molecules and ions in solutions by means of more detailed calculations. Of these, the detailed thermodynamic theory of Fowler and Guggenheim [14] also considers the fact that the electric free enthalpy of electrolyte solutions does not consist exclusively of the partial molar free enthalpy of the electrolyte, but the energy of the solvent arising from the electric field of ions, caused mainly by the interaction between ions and dipole molecules, also contributes to it. A more accurate elimination of the mathematical difficulties in the Debye–Hückel theory has been tried in the papers of Gronwall, La Mer and Sandved [15] and LaMer, Gronwall and Greif [16]. On the basis of earlier suggestions, Bogoliubov [17] and Mayer [18] deduced the fundamental equations on dilute solutions of strong electrolytes starting with the principles of statistical mechanics. The theory has been developed further by Falkenhagen and Kelbg [19] who account for the short-range forces in a more detailed way. Gurikov [20] has pointed out on the basis of some earlier papers by Hertz and Fuoss that the theory can be transformed by using the corresponding distribution functions to agree with the experiments up to significantly higher concentrations. The general limitations in the distribution of ions have been emphasized by Stillinger [21a], extending the theory of ionic atmosphere to a mixture of electrolytes of arbitrary ionic charges.

Starting from the deficiencies in the fundamental principles of the Debye–Hückel theory, it has been pointed out by Kirkwood and Poirier [21b] that in the limiting case the rigorous statistical mechanical calculations, taking into account both the coulombic and the short-range forces, also lead to the Debye–Hückel limiting law. Thus the validity of this limiting law is not influenced by the difficulties connected with the application of the Boltzmann and Poisson law, as well as with their coordination with the precise principles of statistical mechanics.

Frank [22] questioned the acceptability of the results of the Debye–Hückel theory even for the most dilute electrolyte solutions. One of his starting points was that, according to the most accurate measurements, in the most dilute solutions the value of lg γ_{ca} is lower than that calculated from the Debye–Hückel limiting law, even in electrolyte solutions in which association cannot be assumed (Fig. 5.3). On the other hand, the change in the activity coefficient can also be described by the cube-root law:

$$\lg \gamma_{ca} = a - b\sqrt[3]{c}, \quad (5.1.52)$$

in which a and b are constants (Fig. 5.3). Though the extrapolation of this relationship to zero concentration leads to lg $\gamma_{ca} = 0$ only when assuming that the value of a changes in solutions already too dilute for accurate measurements, its agreement with experiments in the most dilute and

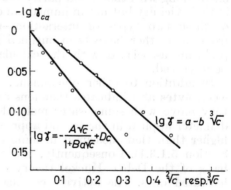

Fig. 5.3 Mean activity coefficients of NH$_4$Cl in aqueous solution. The solid lines are the calculated values on the basis of the $\sqrt{c}$ and $\sqrt[3]{c}$ relationship respectively

Literature on page 439

easily measurable solutions is better than that of the square-root limiting law. In other words, the Debye–Hückel limiting law can be brought into agreement with empirical values only by making one of its 'constants' concentration-dependent, moreover, in an unreasonable direction. As an interpretation of the cube-root law, Frank suggested that the structure of the electrolyte solution is altered in the concentration range of 0–0·001 M, and the ionic atmosphere is transformed to have a pseudo-lattice-like structure. The specific value of b characteristic of the electrolyte can be explained by the fact that the interaction of ions with the structure of water is specific. This novel model of Frank on the structure of dilute electrolyte solutions, also supported by thermodynamic arguments, is interesting, but it still cannot be considered to be accepted.

For the performance of the statistical mechanical theories of electrolyte solutions, see the review of Poirier [23].

These papers, as well as several other works have shed light on the properties of the solutions of strong electrolytes from new points of view, and have clarified several minor details [24], but, in spite of the detailed mathematical treatment applied, they do not provide a satisfactory theory of electrolyte solutions. This is why the Debye–Hückel theory is usually the most suitable one to start with in a first survey of the theory on the properties of dilute solutions of strong electrolytes.

5.1.3.2 VARIATION OF THE ACTIVITY COEFFICIENT DUE TO THE INTERACTION BETWEEN IONS AND SOLVENT MOLECULES

Experience with electrolyte solutions made it doubtless that dissolved ions interact with the solvent, too. Partly, ions bind solvent molecules and then move along with them, and partly they alter the structure of solvent layers in their vicinity but not in immediate contact with them. The main features of these two types of interactions–which cannot be separated sharply from each other–have been outlined in Section 5.2, and here only some problems associated with the calculation of the activity coefficient will be discussed.

In addition to other phenomena, the Debye–Hückel theory of strong electrolytes also indicates that ions are generally in a hydrated state, i.e. they move with some water molecules bound tightly to them. The ionic diameter applied in the second approximation of the theory is significantly higher than that measured in ionic crystals of the given electrolytes (cf. Section 5.1.3.1). Consequently, the activity coefficient calculated on the basis of the Debye–Hückel theory corresponds not to bare ions, but to hydrated ones. The activity coefficient and the concentration connected with it, however, is generally referred to the non-hydrated substance, since, in order to characterize the composition of the solution, experimental data are used to calculate the number of moles of the water-free solute that are in a given amount of the solvent (molality, m) or in a given volume of the solutions (molarity, c) or the number of its moles calculated for the sum of the moles of the solvent and the solute (mole fraction, x). These different concentration data are not uniformly influenced by hydration.

The molarity of a given solution is defined by the same numerical value referred either to the water-free or to the hydrated solute, since hydration does not alter the number of particles dissolved in a given volume (both the water-free and the hydrated ion represents one particle). The disadvantage of the molarity scale is, however, that the concentration of a given solution is represented by different numerical values at different temperatures. On the other hand, the molality and mole fraction values of a given solution depend on whether they are referred to water-free or to hydrated solute, because water molecules bound to a hydrate sphere do not add to the 'free' solvent.

Also interpretation of the partial molar free enthalpy (chemical potential) of the solute depends on whether it is referred to the water-free or to the hydrated solute. The free enthalpy of a given amount (e.g., 1 mole) of a solution, ΔG, is a definite value determined by objective properties, which is, of course independent of the subjective choice of units applied for the solute to express its concentration. The partial molar free enthalpy of the solvent is:

$$\bar{G}_A = \left(\frac{\partial G}{\partial n_A}\right)_{n_B, T, P} \tag{5.1.53}$$

where n_A and n_B are the moles of the solvent and the solute, respectively. This quantity is also independent of the state of solute B that the concentration is referred to, since the quantity $\bar{G}_A$ denotes the change in the free enthalpy when 1 mole of the solvent is added to a very large amount of the solution, irrespective of whether the solvent added interacts with the solute or not. Starting with this, Stokes and Robinson [25] have elaborated a treatment to take into account hydration in the calculation of activity coefficients, as well as for the correlation of the activity coefficient referred to mole fraction ('rational' $^x\gamma$) and that referred to molality ('conventional', $^m\gamma$).*

Consider a given amount of the solution containing 1 mole of the water-free substance B dissolved in S moles of the solvent, and 1 mole of substance B dissociates into ν_c moles of cations and ν_a moles of anions. Let us calculate the objectively given free enthalpy G of this amount of the solution in two ways. When considering first the solute as non-hydrated, we have:

$$G = S\bar{G}_A + \nu_c\mu_c + \nu_a\mu_a, \tag{5.1.54}$$

if μ_c and μ_a are the chemical potentials of the cation and anion referred to the non-hydrated state, respectively. In the other case, h moles of the solvent are assumed to be bound to $\nu = \nu_c + \nu_a$ moles of the ions of the

---

* In the present work the activity coefficient with no additional indices is referred generally to the concentration unit in which the composition is given. Here, the use of superscripts is required for the discussion of the correlation between the two types of activity coefficients.

Literature on page 439

solute. If μ_c' and μ_a' denote the corresponding chemical potentials calculated on the basis of the hydrated state of ions, one obtains:

$$G = (S - h)\bar{G}_A + v_c\mu_c' + v_a\mu_a'. \tag{5.1.55}$$

Expressing the chemical potentials by means of mole fractions and the corresponding activity coefficients $^x\gamma'$, the latter two equations give:

$$\frac{v_c(\mu_c^0 - \mu_c'^0)}{RT} + \frac{v_a(\mu_a^0 - \mu_a'^0)}{RT} + \frac{h\bar{G}_a^0}{RT} + h \ln a_A + v \ln \frac{S + v - h}{S + v} +$$

$$+ v_c \ln {}^x\gamma_c + v_a \ln {}^x\gamma_a = v_c \ln {}^x\gamma_c' + v_a \ln {}^x\gamma_a'. \tag{5.1.56}$$

For the limiting case of infinite dilution, $S \to \infty$ and all the activity coefficients γ tend to 1; the activity of the solvent a_A also tends to unity. Thus, all the logarithmic terms tend to 0 and consequently the first three terms in equation (5.1.56) containing the chemical potentials give a sum of zero as well. Introducing the mean activity coefficient instead of the individual ones of the ions, the following expression holds:

$$\ln {}^x\gamma_{ca}' = \ln {}^x\gamma_{ca} + \frac{h}{v} \ln a_A + \ln \frac{S + v - h}{S + v}, \tag{5.1.57}$$

assuming that h is the same in the given solution and in infinite dilution. This relationship can be expressed with respect to molality and the conventional mean activity coefficient ($^m\gamma_{ca}$), taking into account that:

$$S = \frac{1000}{M_A m} \quad \text{and} \quad {}^x\gamma_{ca} = {}^m\gamma_{ca} (1 + 0{\cdot}001\, v\, M_A\, m),$$

if M_A is the molecular weight of the solvent. From equation (5.1.57), one obtains, in this way:

$$\ln {}^x\gamma_{ca}' = \ln {}^m\gamma_{ca} + \frac{h}{v} \ln a_A + \ln [1 + 0{\cdot}001\, M_A\, (v - h)m]. \tag{5.1.58}$$

If the conventional activity coefficient $^m\gamma_{ca}$ is known up to the concentration of the given solution, the value of a_A can be calculated (or, alternatively, $^m\gamma_{ca}$ can be calculated if a_A is known). On the basis of this equation, the rational mean activity coefficient of the electrolyte can be calculated from the conventional activity coefficient under such conditions, assuming that one mole of the electrolyte is hydrated by h moles of water.

In the deduction of equation (5.1.58), the only non-thermodynamic assumption is that the hydration number h is independent of concentration. Thus, this relationship holds only for solutions where there is a large number of water molecules unbound in hydrate spheres and in this sense free, and further, the forces acting between the solute and the solvent can be saturated, i.e. dissolved species bind only a certain number of water molecules, even if the latter are present in a high excess. Several facts support the idea that the constancy of h can be accepted approximately up to that concentration of the solution at which about a quarter of the solvent molecules are bound in the hydrate spheres of ions.

The relationship between equation (5.1.58) given by Stokes and Robinson and equation (5.1.49) by Debye and Hückel can be explained by pointing out that the latter concerns the forces between ions, while the former considers the forces between ions and solvent molecules. Debye and Hückel regarded the properties of solutions of non-solvated ions as different from those of ideal solutions because of electrostatic interactions, whereas, according to the description of Stokes and Robinson, solutions of hydrated ions deviate from ideal ones as a result of the electrostatic interactions. This model is closer to reality, and it partly supports the use of the dielectric constant of the solvent in the theory. Namely, it has been pointed out by Hasted, Ritson and Collie [26] that the decrease in the dielectric constant of electrolyte solutions in comparison with that of pure water can be attributed almost completely to the effect of the first layer of water molecules around the ions. According to them, the dielectric constant of water that is outside the first layer is almost uninfluenced by the ions. Thus, if hydrated ions are regarded as the dissolved species, it is justifiable to calculate their mutual interactions by using the dielectric constant of the pure solvent. A similar result has also been obtained by Buckingham [27] in calculating the energies of dipole and quadrupole interactions of water molecules in the first hydrated layer with one another and with ions. In this way, the dielectric saturation effect of water molecules outside the first hydration sphere is negligibly small.

Nevertheless, it has not been established unequivocally whether the Stokes–Robinson theory indeed reflects the interactions between ions and solvent molecules reliably. The main difficulty in this respect is the determination of the number of water molecules h bound by one mole of the solute, since the methods applied to obtain the hydration number are extremely uncertain and the various methods lead to results very different from one another. Stokes and Robinson [25] tried to draw some conclusions on the hydration number of electrolytes by comparing thermodynamic conditions in electrolyte and non-electrolyte solutions, and they have calculated the hydration number of several electrolytes on this basis. However, their results are questionable and not even self-consistent (e.g. their hydration numbers are not additive, namely, the hydration number of a given cation shows an apparent dependence on the anion in dilute solutions, too). Thus, further investigations are required to elucidate the situation. This has been attempted in the theory of Glueckauf [28] in which statistics based on the volume fraction have been applied instead of those based on mole fractions used in the calculations before. However, these results cannot be regarded as convincing either, though the hydration numbers are additive in this theory.

Literature on page 440

5.1.4 ACTIVITY COEFFICIENTS OF MORE CONCENTRATED
ELECTROLYTE SOLUTIONS

In the limiting case of very dilute solutions, the Debye–Hückel limiting law describes the relationship between the activity coefficient of strong electrolytes and the ionic strength or, in the case of solutions containing one single electrolyte, the ionic concentration, with satisfactory precision. Its validity can be extended somewhat by applying a second approximation (equation (5.1.46)), but this already implies an arbitrary constant, the 'ionic diameter' a, which cannot be determined by other measurements but must be fitted to the measured data of the activity coefficient; thus, in principle, its value is arbitrary. This deficiency is even more characteristic of equation (5.1.49) which involves two empirical constants that cannot be determined independently. No theory holding satisfactorily for solutions more concentrated than about $0 \cdot 001$ mole l^{-1} and based on a sound theoretical basis (with no empirical constants applied) has been elaborated yet. This is evidently due to the very complicated nature of the interactions prevailing in electrolyte solutions. Nevertheless, efforts have been made to clarify the conditions in more concentrated electrolyte solutions theoretically.

According to experiments, the logarithm of the activity coefficient decreases linearly with the square root of concentration in the most dilute solutions, in accordance with the Debye–Hückel theory. In a marked concentration range, above a concentration of about $0 \cdot 001$ mole l^{-1}, however, the decrease is linear with the cube root of concentration, as experiments show, i.e. the following empirical relationship is valid in good approximation:

$$\lg \gamma_{ca} = a - b \sqrt[3]{c}, \tag{5.1.59}$$

where a and b are empirical constants (this is the so-called *cube root law*). In connection with the theory of more concentrated solutions of strong electrolytes, firstly it should be clarified where the difference between the ranges of the square and cube root dependence on concentration comes from. This is all the more important, since the cube root law also occurs in the earlier theory of Bjerrum [29] for the limiting case of infinitely dilute solutions. This, however, is in contradiction with experiments, thus, the Bjerrum theory is not correct; the validity of the square root law in the most dilute solutions is explained satisfactorily by the later Debye–Hückel theory. It should be shown, however, why the Debye–Hückel limiting law is not valid for more concentrated solutions; whether because the theory is, in fact, not a reliable model of the real conditions even in infinitely dilute solutions, or because the conditions significant with respect to the assumptions in the theory are altered objectively at the border of the validity ranges of the square root and cube root laws. Several aspects of this problem have been studied by many authors; the following analysis is based mainly on the work of Frank and Thompson [30], with reference to one of the plausible possibilities of the theoretical advancement.

The basic assumption of the Debye–Hückel theory is the Poisson equation given in equation (5.1.2) assuming a uniform charge distribution, i.e. assuming conditions under which the discrete nature of ionic charges can be neglected and electric charges can be regarded as a continuum, on the statistical average. Thus, the theoretical model is a structureless charge distribution in a structureless liquid of uniform dielectric constant. The charge distribution was calculated by the Boltzmann theorem in equation (5.1.1) supposing that the energy $E = z_i\,e\,\psi$ produced by the ionic atmosphere can be calculated as a statistical mean, eliminating the fluctuations. It can be shown, however, that this assumption can be accepted only for the most dilute solutions. According to the theory, for example, in 0·01 mole l^{-1} solutions of 1–1 electrolytes at 25 °C, 50 per cent of the effect of the ionic atmosphere arises from 0·8 of the inner ion (i.e. from a single ion residing there for 80 per cent of the time). In order to realize this, 0·8 part of the charge of this ion should bring about the spherically symmetrical charge distribution acting as a continuum around the central ion. Further, this 0·8 part of the ion, together with the central one, should give rise to a spherically symmetrical potential distribution in the next layer (at a distance of 25–46 Å) in which there are 4·2 ions on the statistical average under such conditions, and this would yield a further 25 per cent of the effect. This model and the calculation based on it are evidently not satisfactory. It can be understood in this way that in solutions of 0·01 mole l^{-1} concentration the Debye–Hückel theory does not give results in satisfactory agreement with experiments.

The case is different in, for example, 10^{-8} M solutions in which theoretical difficulties similar to the above ones arise with respect to only 5–10 per cent of the effect. Under such conditions, the theory yields results in satisfactory agreement with experiments. The limit of the rigorous validity of the Debye–Hückel theory is between 10^{-8} and 10^{-2} mole l^{-1} concentrations. The estimation of the limit is enhanced by the separation of the long-range interactions from the short-range ones, extended only to the immediate neighbours. From this point of view, the validity of the Debye–Hückel theory in the most dilute solutions is due to the fact that under such conditions even the immediately neighbouring ions are quite far from one another causing only a small perturbation of the long-range interactions by their short-range ones. Thus, mainly the farther parts of the ionic atmosphere influence the central ion containing a sufficiently high number of ions to regard their charge distribution as a continuous one, on the statistical average. With increasing concentration, i.e. with decreasing average separation of ions, however, the forces of the short-range interaction between the immediate neighbours increase more rapidly than the long-range ones. From a certain concentration on, ions are under the effect of the immediate neighbours rather than that of the farther ones. Beyond this point, the Debye–Hückel theory is already not a satisfactorily reliable description of the interactions between ions. In order to estimate the concentration limit mentioned above, one can start with a comparison of the thickness of the

Literature on page 440

ionic atmosphere $1/\varkappa$ with the average separation of ions $\bar{l}$ in solution. If n moles are dissolved in a solution of volume V and it dissociates into v ions, one obtains:

$$\bar{l} \approx \sqrt[3]{\frac{V}{v\mathrm{N}}}, \qquad (5.1.60)$$

and in aqueous solutions of 1–1 electrolytes at 25 °C:

$$\frac{1}{\varkappa} = 3 \cdot 0 \sqrt[2]{c} \ \text{Å}, \quad \text{and} \quad \bar{l} = 9 \cdot 4 \sqrt[3]{c} . \qquad (5.1.61)$$

The actual value of $\bar{l}$ depends, in fact, on the arrangement of ions, but it influences neither the order of magnitude of $\bar{l}$ nor its $\sqrt[3]{c}$-dependence; it can alter only the numerical factor. Owing to the different powers in the concentration dependence, $1/\varkappa$ is much higher than $\bar{l}$ in very dilute solutions, i.e. the spherical shell representing the ionic atmosphere is farther from the central ion than the closest neighbouring ion, on the average, which is the result of the partial compensation of the effect of positive and negative ions 'smeared' by the thermal motion. In a concentration of about $c' = 0 \cdot 001$ mole 1^{-1}, $1/\varkappa \approx \bar{l} \approx 90$ Å, while in higher concentrations, $1/\varkappa$ is lower than $\bar{l}$. Under such conditions, the ionic atmosphere corresponding to the Debye–Hückel theory has a smaller thickness than the real average distance of the closest neighbouring ions, i.e. a 'negative compensation' of the average distance of the negative and positive ions would occur; that is physically impossible. Thus, it is evident that reality is not reliably reflected by the theory under such conditions. The correctness of the description based on the statistics applied in the Debye–Hückel theory cannot be expected, because it can be shown that, in solutions more concentrated than the value of c', there are less than five positive and negative ions within the distance $1/\varkappa$. A similar conclusion has also been drawn by Fuoss and Onsager [31].

In the development of the theory of electrolyte solutions more concentrated than the limiting concentration c', Frank and Thompson started with the fact that the logarithm of the activity coefficient and the equivalent conductivity, as well as the diffusion coefficient, depend linearly on the cube root of concentration in good approximation within the concentration range of about $0 \cdot 001$–$0 \cdot 1$ mole 1^{-1} (Figs 5.4, 5.5 and 5.6). Since the average distance $\bar{l}$ between ions is also proportional to the cube root

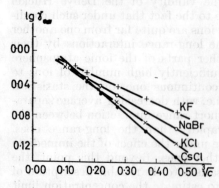

Fig. 5.4 Variation of the activity coefficients of some 1–1 electrolytes with the cube-root of concentration at 25 °C

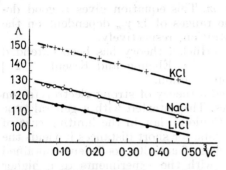

Fig. 5.5 Variation of the equivalent conductivities of some 1–1 electrolytes with the cube-root of concentration at 25 °C

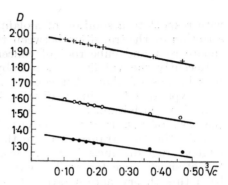

Fig. 5.6 Variation of the differential diffusion coefficient of some electrolytes with the cube-root of concentration at 25 °C

of concentration, the empirical correlation in equation (5.1.9) makes plausible the assumption that value $\bar{l}$ is one of the quantities characteristic of the properties in solutions more concentrated than the limiting concentration c'.

In the description of electrolyte solutions for which $1/\varkappa < \bar{l}$, it has been assumed by Frank and Thompson that in solution the ions have a lattice-like arrangement–expanded appreciably and smeared by the thermal motion–somewhat similar to that in NaCl crystals. Within short distances, the positive and negative ions also alternate more or less regularly in solutions, however, long-range order cannot be formed because of the thermal motion. Although thermal motion destroys the lattice-like arrangement within longer distances in solution, it also contributes to its relative stability. The role of the thermal motion is in some respects similar to the short-range repulsion forces acting in the crystal lattice that are not efficient in solution owing to the large average distance between ions. On the basis of the disturbed lattice-like arrangement of ions, the electrostatic potential energy has been calculated, which corresponds to the change in free enthalpy caused by electric charges as compared with the corresponding state of ideal solutions. This calculation leads to a complicated expression of the activity coefficient which yields the cube root relationship given in equation (5.1.59) after some acceptable simplifications.

The diffuse lattice theory of Frank and Thompson gives a reasonable interpretation of some properties of electrolyte solutions, however, it cannot be regarded as confirmed because of some questionable assumptions and the limited agreement with experiments. The same can be stated about the theory of Glueckauf [32a] also based on volume fraction statistics, which extends the Debye–Hückel equation by the following term:

$$X_m = \frac{-0.06 \, m^2}{\left(1 + \dfrac{m}{2}\right)^2} \tag{5.1.62}$$

Literature on page 440

with respect to a solution of molality m. This equation gives a good description of the transition between the ranges of $\lg \gamma_{ca}$ dependent on the square root and cube root of concentration, respectively.

Recently the validity of the Debye–Hückel theory has been extended to solutions of about 1 M concentration by Ebeling and Krienke [32b] who expanded the binary distribution function.

Levich and Kirianov [33a] elaborated a theory of strong electrolytes on the basis of statistical thermodynamics. They started with a general type of potential that transforms into the Coulomb law in the limiting case of large average distances between ions, while for short distances it yields the law of short-range forces. The expression of activity coefficients obtained from this theory agrees satisfactorily with the experiments to a higher concentration range than does the Debye–Hückel equation.

5.1.5 ASSOCIATION OF IONS

5.1.5.1 ASSOCIATION AND DISSOCIATION

From the dependence of the conductivity and other properties of electrolyte solutions on state parameters, it should be supposed that a smaller or greater portion of ions is involved in groupings with one another. The groupings of the species are electrically neutral in solutions of symmetrical electrolytes, whereas they may have free charges in those of asymmetrical ones (e.g. $CaCl^+$). This phenomenon can formally be called both association and dissociation, when it is considered from the point of view of the completely ionized state or from the point of view of the completely 'non-dissociated' state (that combined into completely neutral species), respectively. If the major part of the electrolyte is present in the form of ions in the solution, the term association seems to be more justified, while, if the majority of the species is neutral, the term dissociation should be used.

Beyond this formal view, ions can form connections with one another in two different ways in nature with respect to the essential feature of the phenomenon: they can form *ion pairs* held together by electrostatic attraction and repulsion or they can combine into *molecules* kept together by covalent bonds. For the sake of unanimity, it seems to be useful to call the formation of ion pairs (eventually ion triplets) bound together by coulombic forces *association*, whereas to consider the formation of covalent molecules as the suppression of *dissociation*.

In the course of ionic association, oppositely charged ions approach each other by means of electrostatic attraction of their free charges to a distance where short-range repulsion forces become equal to the attraction effect. At this distance, the potential energy of their interactions exhibits a minimum and this corresponds to the equilibrium position which, of course, can be easily perturbed by thermal energy. In the case of ion pair formation, the electron shells of the two ions do not become combined in the state belonging to the minimum of potential energy, but they remain separate, eventually deforming each other. When removing the ions from each other, their potential energy increases monotonically (its absolute value decreases), approaching zero asymptotically (Fig. 5.7a).

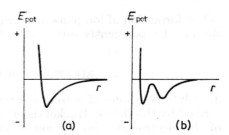

Fig. 5.7 The change in the potential energy with distance r between the components; (a) associated ion pair and (b) covalent bonded molecule

The formation of a non-dissociated molecule also starts with the electrostatic attraction of oppositely charged ions, because this force acts in a longer range than the quantum mechanical forces in covalent bonds. After approaching each other sufficiently, two oppositely charged ions form a covalent bond by means of shared electrons reaching a distance where the repulsion between their closed electron shells forms an equilibrium with the attraction force of the covalent bond. The minimum of the potential energy of interaction corresponds to the equilibrium state in this case too, but on separating the atoms involved in the covalent bond from each other, their potential energy does not increase monotonically to zero, but it passes a local maximum which is the energy of activation required for the decomposition of the molecule (Fig. 5.7b). The formation of covalent molecules can usually be detected by spectroscopical methods.

Covalent molecules are usually present in solutions of weak electrolytes where the degree of dissociation is low, so the concentration of ions is also so low that ion pairs cannot be formed in any noticeable amount. The dissociation of weak electrolytes is actually a chemical process which will not be dealt with in detail here.

Ion pairs are formed mainly in solutions of strong electrolytes by the association of a small portion of ions present. This association is the limiting case of the general electrostatic interaction taking place in solutions of strong electrolytes. In addition to ionic association, covalent molecules can also be formed in solutions of strong electrolytes in some cases; moreover, ions of identical charge can eventually form covalent bonds with each other. This interaction is, however, not characteristic of the solutions of strong electrolytes. The contribution of non-electrostatic forces to ionic association has been recently investigated in detail by Schwarzenbach [33b].

Covalent molecules or ion pairs of symmetric electrolytes do not migrate under the effect of an electric field (they are only oriented as dipoles towards the electric lines of forces in a degree influenced by the thermal energy), thus their formation has the consequence of complete elimination of two ions from the transport of electricity. From the thermodynamical point of view, however, one dipole species (molecule or ion pair) replaces two ions, i.e. the number of species is reduced to a half. Recently, thermodynamic and chemical kinetic aspects of ion pair formation have been analysed in detail by Nancollas [34a], while the problems of ion pairs have been reviewed by Schwarz [34b].

Literature on page 440

The formation of ion pairs in solution also alters the solvation conditions; this effect was recently surveyed by Smid [34c].

5.1.5.2 THEORY OF ION PAIR FORMATION

In dilute solutions of strong electrolytes, the statistical distribution described by the Debye–Hückel theory develops, in general, as a consequence of the electrostatic interaction of ions (see Section 5.1.2). It may occur, however, that ions approach one another to a distance shorter than that corresponding to the average distribution, and their mutual attraction energy becomes higher than the average energy of the thermal motion $3/2\ kT$ per ion or $3/2\ RT$ per mole. Although such an ion pair formation is only temporary, since the energy fluctuation destroys them sooner or later, their decomposition, however, takes place only after several collisions with other molecules. Thus, such ion pairs behave temporarily as kinetic units. No unequivocal principle can be found to define the cases when two ions can be regarded as an ion pair in this sense of the word. In this respect, reasonable assumptions must be applied in the theory of ionic association.

The approximate theory of ion pair formation has first been suggested by Semenchenko, then–independently of him–it has been outlined by Bjerrum [35]. As a starting point of his theory, two ions are regarded as an ion pair with respect to association if they approach each other to such a short distance that the energy of their electrostatic interaction is not lower than 2kT. The electrostatic potential energy of two ions with charges z_ce and z_ae located at a distance r and in a medium with dielectric constant ε is:

$$\frac{|z_c||z_a|\,e^2}{\varepsilon r}.$$

According to the above assumption, those ions can be considered to form ion pairs if they obey the following relationship:

$$\frac{|z_c||z_a|\,e^2}{\varepsilon r} \geq 2\,kT. \tag{5.1.63}$$

In order to calculate the average ratio of ion pairs, Bjerrum regarded two ions as an ion pair when being at a maximum distance q from each other, for which one obtains:

$$\frac{|z_c||z_a|\,e^2}{\varepsilon q} = 2\,kT. \tag{5.1.64}$$

According to this definition, ions that approach each other to a distance shorter than the critical q value in the course of their rapid movement without forming a temporary kinetic unit by remaining together will also be taken into account as ion pairs. The fraction of such ion pairs, however, is small in comparison with those remaining together temporarily.

In order to calculate the relative concentration of ion pairs, the number of the ith kind of ions within a spherical shell of volume:

$$dV = 4 \pi r^2 dr \qquad (5.1.65)$$

of thickness dr located at a distance r from the jth ion studied should be determined. If the average density of the ith kind of ions (their number per cm³) is n_i in the solution, their density at a position of potential ψ_j is, according to the Boltzmann theorem:

$$n_i' = n_i \exp\left[- \frac{z_i e \psi_j}{kT} \right]. \qquad (5.1.66)$$

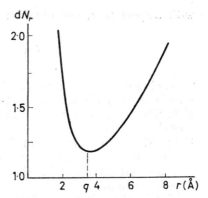

Fig. 5.8 The relative amount of oppositely charged ions in a spherical shell of thickness 0·1 Å at a distance r from the central ion in aqueous solutions of 1–1 electrolytes

The ions considered with respect to ion pair formation are so close to each other that the Debye–Hückel forces acting between the ions can be neglected, and the potential at a distance r from the jth ion is:

$$\psi_j = \frac{z_j e}{\varepsilon r}. \qquad (5.1.67)$$

Taking into account all the relationships given above, the number of ions within the spherical shell of thickness dr at a distance r from the jth ion studied is:

$$dN_r = 4 \pi n_i \exp\left[- \frac{z_i z_j e^2}{\varepsilon kT r} \right] r^2 dr. \qquad (5.1.68)$$

On the basis of this theory, for example, the number of oppositely charged ions within a spherical shell of thickness 0·1 Å at various distances r from any central ion can be calculated. The number of these ions passes a minimum, as shown in Fig. 5.8: the potential caused by the central ion decreases with increasing distance measured from it, together with the number of oppositely charged ions per unit volume, but the volume of the spherical shell of thickness 0·1 Å increases. The resultant of these two opposite effects is the minimum of dN_r found at a distance of 3·57 Å in aqueous solutions of 1–1 electrolytes at 25 °C. In the vicinity of the jth central ion studied the number of ions of identical charge is small in the spherical shell and it increases monotonically with the distance measured from it. By computing the distance corresponding to the minimum from the differential quotient of the term:

$$r^2 \exp\left[- \frac{z_i z_j e^2}{\varepsilon kT r} \right]$$

Literature on page 440

with respect to r, one obtains:

$$r_{\min} = q = \frac{|z_i z_j|\,\mathrm{e}^2}{2\varepsilon kT}. \tag{5.1.69}$$

The spherical shell containing the minimum quantity of oppositely charged ions is at that distance from the central ion where the electric energy is $2kT$. Thus, according to Bjerrum's deduction, those oppositely charged ions should be regarded as associated ion pairs that are closer to one another than 3·75 Å in solutions of 1–1 electrolytes (for electrolyte solutions of other valence types, this value is 3·75 $|z_c z_a|$ Å). For the distribution of ions located farther than this value, the Debye–Hückel theory holds in approximation.

If a is the minimum distance to which oppositely charged ions can approach each other (they are in 'contact' with each other), then, according to the Bjerrum theory, those ions form associated ion pairs that are at a distance of at least a and not more than q from each other. Consequently, the *degree of association* δ can be calculated by integrating equation (5.1.68) within these limits:

$$\delta = 4\pi n_c \int_a^q \exp\left[-\frac{z_c z_a\,\mathrm{e}^2}{\varepsilon kT r}\right] r^2\,\mathrm{d}r. \tag{5.1.70}$$

Introducing the following notations:

$$\frac{z_c z_a\,\mathrm{e}^2}{\varepsilon kT r} = x \tag{5.1.71}$$

and

$$\frac{z_c z_a\,\mathrm{e}^2}{\varepsilon kT a} = b; \tag{5.1.72}$$

further, taking into account that:

$$\frac{z_c z_a\,\mathrm{e}^2}{\varepsilon kT q} = 2, \tag{5.1.73}$$

the integration gives the formula:

$$-\left(\frac{|z_c z_a|\,\mathrm{e}^2}{\varepsilon kT}\right)^3 \int_b^2 \frac{\mathrm{e}^x}{x^4}\,\mathrm{d}x. \tag{5.1.74}$$

The value of the integral:

$$\int_2^b x^{-4}\,\mathrm{e}^x\,\mathrm{d}x = Q(b) \tag{5.1.75}$$

can be found in corresponding tables. With reference to all the above equations, one obtains for the degree of association:

$$\delta = 4\pi n_c \left(\frac{|z_c z_a| e^2}{\varepsilon\, kT}\right)^3 Q(b). \tag{5.1.76}$$

When calculating the concentration in mole l^{-1} units $(c = 1000\, n/N)$, one has:

$$\delta = \frac{4\pi N c}{1000} \left(\frac{|z_c z_a| e^2}{\varepsilon\, kT}\right)^3 Q(b). \tag{5.1.77}$$

The law of mass action is valid for the equilibrium of ionic association and of the dissociation of ion pairs, since, in the deduction, the nature of the forces holding together the species in the 'non-dissociated' state has no specific importance. Thus, if the activity coefficient of ions is γ, and the activity coefficient of ion pairs does not deviate too much from unity, the dissociation or association constant becomes, according to the law of mass action, (when $\alpha = 1 - \delta$ is the degree of dissociation of ion pairs):

$$K_{\text{dissoc.}} = \frac{\alpha^2 \gamma^2 c}{(1-\alpha)} =$$

$$= \frac{(1-\delta)^2 \gamma^2 c}{\delta} = \frac{1}{K_{\text{assoc.}}}. \tag{5.1.78}$$

In very dilute solutions, $\alpha \approx 1$ and $\gamma \approx 1$, consequently:

$$K_{\text{assoc.}} \approx \frac{1-\alpha}{c} = \frac{4\pi N}{1000} \cdot \left(\frac{|z_c z_a| e^2}{\varepsilon\, kT}\right)^3 Q(b). \tag{5.1.79}$$

Since definite b and $Q(b)$ values correspond to each value of a $(< q)$, the association constant depends on the minimum distance to which two ions can approach each other. On the basis of the dependence on the activity coefficient and equation (5.1.74), the dependence of the degree of association on the value of a, i.e. essentially, on the size of the ions, can be calculated by successive approximation. It is seen in Fig. 5.9, demonstrating the degree of association calculated for aqueous solutions of 1–1 electrolytes of 0·1 mole l^{-1} concentration at 18 °C, that, in the cases of $a = 2$ and 1·4 Å, δ is equal to 2·5 per cent and 10 per cent, respectively, and association becomes predominating only for values lower

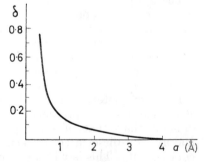

Fig. 5.9 The dependence of the degree of association on the minimum separation distance a of ions in aqueous solutions of 1–1 electrolytes in concentration 0·1 mole l^{-1} at 18 °C

Literature on page 440

than 0·6 Å. Since such small ionic sizes do not occur, the extent of association can only be small in aqueous solutions of 1–1 electrolytes. In solvents of low dielectric constant, however, the degree of association is much
higher.

For the determination of the association constant in dilute solutions,
Fuoss [36] suggested a method, combining the Fuoss–Onsager conductivity
equation with the law of mass action (see Section 4.2.3.4.). From the equation describing the concentration dependence of the conductivity of dilute
electrolyte solutions, the limiting value of the equivalent conductivity,
the ionic size, and the association constant can be calculated.

It has not been unequivocally clarified whether water molecules remain
between the ions when ion pairs are formed, or are pushed away and this,
of course, influences the value of a. Since the Bjerrum theory is based on
a continuous distribution function, ions can approach each other to distances equal to a fraction of the size of the solvent molecule as well. According
to Fuoss [37], however, the probability that two ions approach each other
to a distance equal to a fraction of the size of a solvent molecule is very
small, and two ions can only be regarded as an associated ion pair, when no
solvent molecules are inserted between them. On this basis, the association
constant of 1–1 electrolytes is:

$$K_{assoc.} = \frac{4\pi N a^3 e^b}{3000}. \qquad (5.1.80)$$

The lower the dielectric constant of the medium, the higher the extent
of ionic association, under comparable conditions. In the expression of
the association constant in equation (5.1.79), the value of ε occurs not
only explicity, but implicitly in the values of $Q(b)$, too. Accounting also for
this dependence, $\lg K_{assoc.}$ varies approximately linearly with the reciprocal
value of the dielectric constant as predicted by the Bjerrum theory.

The ion pair formation reduces the electric conductivity of the solution,
therefore, the association constant can be calculated, in principle, from
the concentration dependence of conductivity [38]. In dilute aqueous
solutions, however, the degree of association is low, therefore, it is very
complicated to determine what part of the observed concentration dependence of conductivity arises from ion pair formation, and what part comes
from other factors (cf. Section 4.2.3.4), since, in a solution containing an
almost completely ionized electrolyte, the detection of some percentage
of ion pairs in the presence of almost 100 per cent of ions is much more
difficult than that of some percentage of ions in a solution containing
mainly non-dissociated molecules observable directly by means of their
conductivity. This is why an abundant amount of reliable data is available
on dissociation constants of weak electrolyte solutions, in contrast with the
case of the association of strong electrolytes in aqueous solutions. The
determination of the association constant of strong electrolytes is also
hindered by the fact that the degree of association is higher, the more
concentrated the solution, while the theoretical calculations on conductivity
are valid in the best approximation and yield the more reliable conclusions,

the more dilute the solution. Nevertheless, it can be stated that the degree of association is higher in solutions of polyvalent ions than in those of monovalent ones under comparable conditions, therefore it can be detected more easily as well.

The limitations of the Bjerrum theory. Bjerrum's electrostatic theory of ionic association is in agreement with experiments in the main, however, it also involves questionable conclusions, and it can be accepted only as a first approximation. In this theory ions are considered as rigid spheres which can approach each other to a distance a and the decrease in the coulombic forces is calculated on the basis of the macroscopic dielectric constant. In solutions containing polyvalent ions, the use of the macroscopic dielectric constant can hardly be condemned as shown by Robinson and Stokes [39]. For example, in solutions of 3–3 electrolytes (e.g., $La[Fe(CN_6)]$) the ions cannot approach each other to a distance shorter than 7·2 Å, and the limit of ion pair formation is 32·1 Å. Between the spherical shells of radii 7·2 and 32·1 Å around the ions, there are about 5000 water molecules, if the volume of water molecules is assumed to be the same (30 Å^3) as in pure water. Around the ions of 2–2 electrolytes, the critical distance of pair formation is 14·3 Å, and within this sphere there are approximately 400 water molecules. Although there are some ions within this spherical shell that approach each other temporarily to a very short distance, the majority are, however, separated by several water molecules only a few of which are in a state polarized by the ion in its immediate vicinity. Under such conditions, probably no appreciable error is caused when calculating with the macroscopic dielectric constant.

The situation is different in aqueous solutions of 1–1 electrolytes. At room temperature, the critical distance given by the Bjerrum theory is 3·57 Å, and the whole volume of the site of ion pair formation is 190 Å^3. It provides space enough for only about four water molecules–in addition to the two ions forming the pair–adjacent to the ions and strongly polarized by them. Under such conditions, a calculation using the macroscopic dielectric constant is evidently not justified. No reliable information is available, however, for the degree of the polarization of water molecules. On the basis of studies on the dielectric constant in concentrated electrolyte solutions, Ritson and Hasted [40] concluded that the orientation polarization is complete in the immediate vicinity of positive ions (they are in a dielectrically saturated state) and they give rise to a dielectric constant corresponding to deformation polarization only. The dielectric constant, however, increases rapidly with the distance from the ion, and it reaches the value of pure water at a distance of about 4–5 Å. According to Glueckauf [41], however, even the adjacent water molecules are not in the state of dielectric saturation; he made some calculations on the decrease of the dielectric constant of water as a function of the distance from the ion on the basis of a detailed molecular model. Detailed calculations have also been carried out by Rosseinsky [42a] on the variation of the dielectric

Literature on page 440

constant of water in the vicinity of ions and on the distance a. Though these studies led to different results, it still seems that novel theories on dielectric saturation will not essentially alter the picture obtained from the macroscopic dielectric constant of the solvent.

With respect to the minimum distance a between the ions, the Bjerrum theory does not always yield satisfactory results, either. For example, TlCl is a rather strongly associating electrolyte in aqueous solution ($K_{assoc.} \approx 3$ as obtained in conductivity measurements), and the Bjerrum theory gives a value of $a \approx 1$ Å, though the sum of the crystallographic radii of Tl$^+$ and Cl$^-$ ions is 3.26 Å. Since no spectroscopic evidence indicates that TlCl forms covalent molecules, it cannot be explained how the ions assumed to be rigid spheres can approach each other to a much shorter distance than the sum of their radii. The explanation of the contradictions here and in some other cases is that ions are not rigid spheres, but they can be deformed when coming close to each other; anions of a rather loose electron shell can be deformed strongly under the effect of cations.

Of 1–1 electrolytes, the properties of the solutions of those indicate pair formation which contain a polyatomic anion (nitrate, chlorate, perchlorate, bromate, etc.). In the case of the coplanar triangular NO_3^- ion, the value of a being lower than that corresponding to the crystallographic radius can perhaps be explained by the fact that, in the direction perpendicular to the plane of the triangle, the cation can approach it to a much shorter distance than the sum of the crystallographic radii; however, this can hardly be acceptable in connection with anions of non-coplanar structure. It is to be noted with respect to this, that the properties of salts mainly comprising a slightly hydrated or non-hydrated cation indicate association. For example, in dilute solutions of the salts of the strongly hydrated Li$^+$ ion, no trace of association can be observed. Properties indicating association appear only in solutions of cations the electric field of which is not 'saturated' by the water molecules in the hydrate sphere, and in this way, their polarization effect on anions is strong. Owing to the deformation accompanying polarization, ions can approach each other to a distance shorter than the sum of their crystallographic radii, as well. Therefore, the doubts of Robinson and Stokes, as to whether the explanation of these phenomena by ion pair formation is not a too much simplified model of the real conditions, are justified. Perhaps the assumption that these phenomena, or their appreciable part at least, can be attributed to the general interaction between the cation and the anion, becoming a dipole due to polarization is a better approximation to reality.

The association of large complex ions can hardly be ascribed to electrostatic interaction only. This is indicated e.g. by the studies of Masterton and Bierly [42b] which show that, in 10^{-4}–10^{-3} N aqueous solutions of $[Co(NH_3)_5NO_2]SO_4$ the association constant calculated on the basis of conductivity by the Shedlowsky method is markedly higher than that observed in solutions of 2–2 electrolytes containing small ions, under comparable conditions. Since on the effect of purely electrostatic forces, association should decrease with increasing ionic size, it has been supposed that dispersion forces acting between the cation and anion contribute considerably to association in solutions containing complex ions, because the dispersion forces are approximately independent of electric charge.

The theory of the interaction energy of ion pairs has recently been improved by Levine and Rozenthal [43], while Mozumder [44a] has studied theoretically the rate of the pair formation of ions.

Taking into account the electrostatic interactions and the static linear dielectric response function, the Bjerrum ion pair concept has been extended by Stillinger and White [44b] to solutions of symmetrical electrolytes so concentrated that all the ions are already paired.

In solutions of *asymmetric electrolytes*, the detection of ion pair formation is even more difficult than in those of symmetrical electrolytes, since ion pairs of asymmetric electrolytes have free charge (e.g. $NaSO_4^-$, $BaCl^+$), and therefore they take part in the transport of electricity. Thus the decrease in conductivity caused by them is smaller than that brought about by the formation of uncharged symmetrical ion pairs. On the other hand, the correctness of the simplifying assumptions introduced in the Debye–Hückel–Onsager theory is questionable even for lower concentrations of the solutions of polyvalent ions than in the case of monovalent ones. Hence, it is difficult to judge to what extent the deviation of the experimentally determined conductivity from the calculated one arises from the approximate nature of the theory and its unsatisfactory precision, and to what extent it is due to ion pair formation. On the basis of measurements on the solubility of asymmetrical electrolytes, Monk *et al.* [45] have given convincing evidence on ion pair formation ($NaSO_4^-$, $LaSO_4^+$).

Ion pair formation can also be detected by absorption spectrometry, when ion pairs absorb at other wavelengths than simple ions. For example, the ultra-violet absorption band of Pb^{2+} and $PbCl^+$ differ from each other, so the formation of this ion pair can be detected [46]. Similarly, the formation of the ion pair $PbNO_3^+$ could also be detected optically [47].

The role of hydrogen bonds in ion pair formation. The deviation of the electric conductivity of large ions from the Debye–Hückel–Onsager limiting law, the dependence of the activity and osmotic coefficients on concentration, and several other phenomena indicate that large ions have no primary hydrate sphere, and their secondary hydration effect is mainly a structure-making one on the water-layers in their immediate vicinity. These ions are so to say hydrophobic and their structure-making effect on water is higher, the larger they are. It has been pointed out by Diamond [48a] that, in solutions of large ions with no primary hydration, ion pair formation is also facilitated by the fact that water in their vicinity exhibits an effect keeping together the hydrophobic ions by means of its structure, in addition to the electrostatic attraction between their charges, since the hydrogen bonds between water molecules act in the direction of increased interaction of water molecules with one another and decreased distortion of water structure. The effect of the liquid structure facilitating ionic association differs from the Bjerrum ion pair formation brought about by electrostatic forces:

Literature on page 440

it appears only in liquids of extended hydrogen bonding, and the larger the ionic size and the smaller the charge, (e.g. tetra-alkylammonium iodide), the larger this effect.

Hydrogen bonds have a role not only in the association of ions, but also in the association of acid molecules, as stated by Kampschulte-Scheuing and Zundel [48b] on the basis of the infrared spectra of acids and their solutions. According to them, when the solvent molecules do not possess acceptor properties in respect of hydrogen bonding, p-toluene-sulphonic acid studied by them is present in non-dissociated and associated states. When, however, the solvent molecules have an acceptor group, the fraction of the acid existing in non-associated and dissociated states is the higher, the higher the basicity of the solvent molecules.

5.1.5.3 ASSOCIATION OF THREE AND FOUR PARTICLES

In addition to ion pairs, ion triplets can also be formed in electrolyte solutions, since, according to electrostatics, when placing an additional charged sphere symmetrically to two oppositely charged ones, i.e. forming a triplet with a configuration $+ - +$ or $- + -$, the binding energy of a system built up in this way is higher by about 50 per cent than that of the pair $+ -$. Thus, in principle, the energetical precondition of the formation of ion triplets in addition to ion pairs is satisfied; these, of course, have free positive or negative charges unlike ion pairs. However, this is appreciable only in media of low dielectric constant. According to Fuoss and Kraus [49], in addition to the equilibrium:

$$MX \rightleftharpoons M^+ + X^-$$

corresponding to ion pair formation in solutions of a binary electrolyte MX, the equilibria:

$$(MXM)^+ \rightleftharpoons MX + M^+,$$

and

$$(XMX)^- \rightleftharpoons MX + X^-$$

also take place. The equilibrium constant of ion pair formation is:

$$K_p = \frac{[M^+][X^-]}{[MX]} \approx \alpha_p^2 c, \qquad (5.1.81)$$

when the dielectric constant of the solution is so low that the degree of dissociation of ion pairs (α_p) is low and $1 - \alpha_p \approx 1$. If ions M^+ and X^- have identical size and the formation of ions $(MXM)^+$ has the same probability as that of ions $(XMX)^-$, the equilibrium constant of triplet formation is:

$$K_{tr} = \frac{[MX][M^+]}{[MXM^+]} =$$

$$= \frac{[MX][X^-]}{[XMX^-]}. \qquad (5.1.82)$$

The total concentration of the solution can be written as:

$$c = [\text{MX}] + \frac{1}{2}[\text{M}^+] + \frac{1}{2}[\text{X}^-] + \frac{3}{2}[\text{MXM}^+] + \frac{3}{2}[\text{XMX}^-]. \quad (5.1.83)$$

The degree of dissociation of ion triplets is:

$$\alpha_{tr} = \frac{[\text{MXM}^+]}{c} =$$

$$= \frac{[\text{XMX}^-]}{c}. \quad (5.1.84)$$

If the value of α_{tr} is small, one has:

$$K_{tr} \approx \frac{\alpha_p}{\alpha_{tr}} c$$

and $\quad (5.1.85)$

$$\alpha_{tr} \approx \frac{\sqrt{K_p c}}{K_{tr}}.$$

Further, if the solution is so dilute that the limiting value of equivalent conductivity can be applied, the conductivity corresponding to the simple ions is:

$$\Lambda^0 = \lambda^0_{\text{M}^+} + \lambda^0_{\text{X}^-}. \quad (5.1.86)$$

The equivalent conductivity of triplet ions takes the form:

$$\Lambda^0_{tr} = \lambda^0_{\text{MXM}^+} + \lambda^0_{\text{XMX}^-}. \quad (5.1.87)$$

In this way, the conductivity measurable directly in a solution of concentration c is:

$$\Lambda = \alpha_p \Lambda^0 + \alpha_{tr} \Lambda^0_{tr} =$$

$$= \sqrt{\frac{K_p}{c}} \Lambda^0 + \frac{\sqrt{K_p c}}{K_{tr}} \Lambda^0_{tr}. \quad (5.1.88)$$

This expression has the following form:

$$\Lambda = A c^{-1/2} + B c^{1/2} \quad (5.1.89)$$

according to which conductivity passes a minimum as a function of concentration.

Thus, in the case of marked ion triplet formation, the curve of equivalent conductivity *vs.* concentration exhibits a minimum; the concentration

Literature on page 440

and conductivity corresponding to it can be calculated by differentiating equation (5.1.88) (the differential quotient is zero at the minimum):

$$c_{\min} = \frac{A}{B} = \frac{K_{tr}\,\Lambda^0}{\Lambda_{tr}^0}$$

and

$$\Lambda_{\min} = 2\sqrt{AB}\,.$$

(5.1.90)

On the basis of these relationships, the two equilibrium constants can be calculated from the conductivity and concentration corresponding to the minimum:

$$K_{tr} = c_{\min}\left(\frac{\Lambda_{\min}}{2\,\Lambda^0}\right)^2,$$

and

$$K_p := c_{\min}\frac{\Lambda_{tr}^0}{\Lambda^0},$$

(5.1.91)

further:

$$\Lambda_{\min} = 2\,\alpha_{p,\,\min}\,\Lambda^0 = 2\,\alpha_{tr,\,\min}\,\Lambda_{tr}^0.$$

(5.1.92)

In this way, the minimum in conductivity depends on both the single and triplet ions. The ion pairs being uncharged do not influence conductivity directly.

The above equations regarding the formation of triplet ions are valid only for solutions of low dielectric constant in which α_p and α_{tr} are small, and the Debye–Hückel electrostatic interaction of ions can also be neglected. In aqueous solutions, the probability of triplet formation is very low. If, however, the solution also contains a non-electrolyte (e.g., dioxan) in a sufficient concentration which decreases the dielectric constant to a suitably low value, the formation of triplet ions can become noticeable even in aqueous solutions.

In more concentrated solutions of low dielectric constant, ion quadrupoles can also be formed by the addition of an ion to a triplet or by the combination of two ion pairs. These are probably responsible in many cases for the complex dependence of the conductivity of concentrated solutions on concentration. From the freezing point depression of solutions in benzene, the existence of such ions can be concluded with a rather good certainty [50], but they are seldom formed in aqueous solution.

5.1.5.4 CORRELATION BETWEEN ASSOCIATION AND COMPLEX ION FORMATION; LOCALIZED HYDROLYSIS

The Bjerrum theory of ionic association discussed in Section 5.1.5.2 treats the formation of ion pairs on an electrostatic basis and–to a certain degree–in a formal way. In reality, however, cations and anions can be bound together by several means, and such differences can be somewhat revealed by the detailed analysis of phenomena.

Ion pairs can be formed so that the cation and anion remain separated from each other by one or more water molecules in their hydrate spheres (*outer sphere association*, or ion pair formation: $[M^+(H_2O)_nX^-]$). It may occur that water molecules are pushed away from the space between the ions and the latters get in close contact with each other (*inner-sphere association*, or ion pair formation: $[M^+X^-]$). The difference between these two types of pair formation has been emphasized by Smithson and Williams [51] who point out that the outer-sphere ion pair formation of transition metals hardly influences the low-intensity absorption bands in the visible spectrum, while the inner-sphere association has an appreciable effect on it. These two types of pair formation are not necessarily accompanied by large differences in the thermodynamic equilibrium constant of association. For example, according to spectroscopical measurements, outer-sphere association takes place in $CoSO_4$ solutions, whereas in cobalt(II) thiocyanate solutions, it is of inner-sphere type although their association constants ($2 \cdot 3 \times 10^2$ and $1 \cdot 2 \times 10^2$ l mole^{-1}) do not differ too much.

Not all the methods of determining association constants are suitable to distinguish between these two types of association. Still, they influence several phenomena differently, which explains the appreciable deviation observed in some cases between the results obtained by various methods. For example, the conductivity and the electromotive force of corresponding Galvanic cells are altered in the same way by outer- and inner-sphere ion pairs, while only inner-sphere association has a marked effect on the absorption of light in the visible spectral region. Ultraviolet absorption, however, can be influenced by both types of association. The different effects make it possible within certain limits to distinguish between outer- and inner-sphere association.

Inner-sphere association is quoted by several authors as *complex ion formation* unlike the outer-sphere interaction which is called *ion pair formation*. Although the concept of complex ion can be extended, it is still justi-fied–as emphasized recently by Nancollas [52]–to retain the concept of complex formation for cases in which short-range forces giving rise to covalent bonds prevail, and association caused mainly by long-range elec-trostatic forces should be regarded as ion pair formation–namely, an inner-sphere one–when there is no water molecule between the associated ions.

Outer- and inner-sphere association can occur simultaneously. For example, Posey and Taube [53] have shown by spectrophotometric methods that rapid formation of $Co(NH_3)_5(H_2O)^{3+} \cdot SO_4^{2-}$ outer-sphere ion pairs takes place as indicated by the change in the ultraviolet spectrum, when mixing solutions containing $Co(NH_3)_5(H_2O)^{3+}$ and SO_4^{2-} ions. This is ac-companied by a slow change in the visible spectrum caused by the transfer of the SO_4^{2-} ions into the inner sphere, when $Co(NH_3)_3SO_4^+ \cdot H_2O$ ions are formed.

The two types of ion pair formation are also supported by the ultra-sonic absorption measurements of Eigen and Tamm [54] and Atkinson and Kor [55]. The absorption of ultra-sound depends on the compressibility

Literature on page 440

of the solution which, however, is influenced by the number of water molecules with altered compressibility due to electrostriction produced by ions. In this way, the following ion pairs can be distinguished in solutions of sulphates of bivalent metals: $Me^{2+}(H_2O)_3 \cdot SO_4^{2-}$, $Me^{2+}(H_2O) \cdot SO_4^{2-}$, and $Me^{2+}(SO_4^{2-})$.

On the basis of the variation of the conductivity of electrolyte solutions under the effect of high pressures, outer- and inner-sphere ion pair formation can also be distinguished, since the equilibrium of the outer-sphere interaction is much less sensitive to pressure than the inner-sphere one, because the former alters hydration to a much lesser extent. On this basis, it has been concluded by Horne, Myers and Frysinger [56] that $FeCl^{2+}$ is an inner-sphere ion pair, while $Fe(H_2O)_x \cdot NO_3^{2+}$ is an outer-sphere one in which the cation retains its hydrate sphere.

Localized hydrolysis. The properties of some strong electrolyte solutions cannot be interpreted by the general relationships on hydration and ion pair formation. For example, in solutions of chlorides, bromides, iodides, chlorates and perchlorates of the same concentration, the order of the activity coefficients is $Li^+ > Na^+ > K^+ > Rb^+ > Cs^+$ which is in accordance with the increasing hydration from Cs^+ to Li^+ ions. However, in solutions of hydroxides, the order of the activity coefficients is the opposite: $Li^+ < Na^+ < K^+ < Rb^+ < Cs^+$. This order cannot be explained by ion pair formation: for the interpretation of the low conductivity of LiOH solutions, an extended ion pair formation should be assumed which is very unlikely for strongly hydrated Li^+ ions. To explain this phenomenon, Robinson and Harned [57] have assumed 'localized hydrolysis' in the solutions of hydroxides: according to this, water molecules in the hydrate spheres of cations are strongly polarized, and the positive part of their dipoles points outwards from the cation, e.g.:

$$Na^+ \cdots \overset{-}{O}H - \overset{+}{H}.$$

The 'bound hydrogen ion' in the strongly polarized water molecules exerts sufficiently large force on relatively small ions such as OH^- to form a

$$Na^+ \cdots \overset{-}{O}H - \overset{+}{H} \cdots \overset{-}{O}H$$

group by means of a bond of short lifetime. This can be regarded as being formed by the mediation of a polarized water molecule. The smaller the cation, the more polarized the water molecules are, due to which this effect increases from Cs^+ to Li^+ ions. Such an interaction acts in the direction of decreased activity coefficient, thus, it can be explained on the basis of this assumption that the activity coefficient of LiOH is small in comparison with that of CsOH. A similar relationship holds for other proton acceptor anions (e.g. formates, acetates), and further, according to Stokes [58], for magnesium and barium acetate in dilute solution, as well. The Mg^{2+} ion is much more strongly hydrated than the Ba^{2+} ion, therefore it gives rise to localized hydrolysis to a greater extent. Diamond [59] has extended this hypothesis to alkali halides, too: the effect is large in solutions of lithium halides and it is negligibly small in those of Cs halides. In solutions of rubidium and caesium halides, the order of activity coefficients is: $Cl^- > Br^- > I^-$. This can be regarded as the normal order, if no localised hydrolysis occurs. The opposite order of lithium, sodium and potassium halides is to be attributed to localized hydrolysis. In this respect, the effect of hydrated cations is larger than that of anions, but, in the case of a given cation, the degree of hydrolysis is higher in the presence of small anions than in that of large ones.

REFERENCES

to Section 5.1

[1]   For solvation resulting in ionization of acids and some other effects, see e.g. D. J. G. IVES and P. MARSDEN, *J. Chem. Soc.*, **1965**, 649.
[2]   P. DEBYE and E. HÜCKEL, *Phys. Z.*, 1923, **24**, 185. As a review see e.g. N. A. IZMAILOV, Elektrokhimiya rastvorov (Moscow, 1966); R. A. ROBINSON and R. H. STOKES, Electrolyte Solutions (London, 1959); P. M. V. RÉSIBOIS, Electrolyte Theory (New York, 1968); H. FALKENHAGEN, Theorie der Elektrolyte (Leipzig, 1970); S. PETRUCCI (ed.) Ionic Interactions (from Dilute Solutions to Fused Salts). Vol. 1—2 (London, 1971).
[3]   E. GLUECKAUF, Chemical Physics of Ionic Solutions (ed. by B. E. CONWAY and R. G. BARRADAS), p. 67 (New York, 1966).
[4]   H. S. FRANK and M. W. EVANS, *J. Chem. Phys.*, 1945, **13**, 507.
[5]   H. MÜLLER, *Phys. Z.*, 1927, **28**, 324; 1928, **29**, 78.
[6]   T. H. GRONWALL, V. K. LAMER and K. SANDVED, *Phys. Z.*, 1928, **29**, 358.
[7]   E. WICKE and M. EIGEN, *Z. Elektrochem.*, 1952, **56**, 551; 1953, **57**, 319.
[8]   C. W. OUTHWAITE, *J. Chem. Phys.*, 1969, **50**, 2277.
[9]   R. FERNÁNDEZ-PRINI and J. E. PRUE, *Z. Phys. Chem.*, 1965, 228, 373.
[10]   R. A. ROBINSON and R. H. STOKES, Electrolyte Solutions, p. 231 (London, 1959).
[11]   E. GÜNTELBERG, *Z. Phys. Chem.*, 1926, **123**, 199.
[12]   E. A. GUGGENHEIM, *Phil. Mag.*, 1935, **19**, 588.
[13]   D. KREMP, H. ULBRICHT and G. KELBG, *Z. Phys. Chem.*, 1969, **240**, 65.
[14]   R. H. FOWLER and E. A. GUGGENHEIM, Statistical Thermodynamics, Chapter IX (Cambridge, 1949).
[15]   T. H. GRONWALL, V. K. LAMER and K. SANDVED, *Phys. Z.*, 1928, **29**, 358.
[16]   V. K. LAMER, T. H. GRONWALL and L. J. GREIF, *J. Phys. Chem.*, 1931, **35**, 2245.
[17]   N. N. BOGOLIUBOV, Problemy dinamicheskoi teorii v statisticheskoi khimii (Moskva 1946).
[18]   J. E. MAYER, *J. Chem. Phys.*, 1950, **18**, 1426.
[19]   H. FALKENHAGEN and G. KELBG, *Ann. Phys.*, 1953, **11**, 60; 1954, **14**, 391; *Z. Phys. Chem.*, 1955, **204**, 111; *Naturwiss.*, 1955, **42**, 10; G. KELBG, Chemical Physics of Ionic Solutions (ed. by B. E. CONWAY and R. G. BARRADAS) p. 29 (New York, 1966).
[20]   YU. V. GURIKOV, *Zh. Strukt. Khim.*, 1962, **3**, 10.
[21a] F. H. STILLINGER, *J. Chem. Phys.*, 1968, **49**, 1991.
[21b] J. K. KIRKWOOD and J. C. POIRIER, *J. Phys. Chem.*, 1954, **58**, 591.
[22]   As a review see H. S. FRANK, Chemical Physics of Ionic Solutions (ed. by B. E. CONWAY and R. G. BARRADAS) p. 53 (New York, 1966).
[23]   J. C. POIRIER, Chemical Physics of Ionic Solutions (ed. by B. E. CONWAY and R. G. BARRADAS), p. 9 (New York, 1966).

[24]   As a review see, e.g., Chemical Physics of Ionic Solutions (ed. by B. E. CONWAY and R. G. BARRADAS, New York, 1966); G. H. NAN-COLLAS, Interactions in Electrolyte Solutions (Amsterdam, 1966).

[25]   R. H. STOKES and R. A. ROBINSON, *J. Amer. Chem. Soc.*, 1948, **70**, 1870; further development of this method: Electrolyte Solutions, p. 289 (London, 1959).

[26]   J. B. HASTED, D. M. RITSON and C. H. COLLIE, *J. Chem. Phys.*, 1948, **16**, 1.

[27]   A. D. BUCKINGHAM, *Disc. Faraday Soc.*, 1957, **24**, 151.

[28]   E. GLUECKAUF, *Trans. Faraday Soc.*, 1955, **51**, 1235; in The Structure of Electrolytic Solutions (ed. by W. J. HAMER), p. 97 (New York, 1959).

[29]   N. BJERRUM, *Z. Elektrochem.*, 1918, **24**, 321.

[30]   H. S. FRANK and P. T. THOMPSON, in The Structure of Electrolytic Solutions (ed. by W. J. HAMER), p. 113 (New York, 1959).

[31]   R. M. FUOSS and L. ONSAGER, *J. Phys. Chem.*, 1957, **61**, 668.

[32a]   E. GLUECKAUF, in The Structure of Electrolyte Solutions (ed. by W. HAMER) p. 97 (New York, 1959).

[32b]   W. EBELING and H. KRIENKE, *Z. Phys. Chem.*, (1971) 248, 274.

[33a]   V. G. LEVICH and V. A. KIRIANOV, *Zh. Fiz. Khim.*, 1962, **36**, 1646.

[33b]   G. SCHWARZENBACH, *Pure and Applied Chem.*, 1970, **24**, 307.

[34a]   G. H. NANCOLLAS, Chemical Physics of Ionic Solutions (ed. by B. E. CONWAY and R. G. BARRADAS) p. 197 (New York, 1966).

[34b]   M. SCHWARZ, *Accounts Chem. Research*, 1969, **2**, 87; *Usp. Khim.* 1970, **39**, 1260.

[34c]   I. SMID, *Usp. Khim.*, 1973, **42**, 799.

[35]   N. BJERRUM, *K. danske vidensk. Selsk.*, 1926, **7**, No. 9.

[36]   R. M. FUOSS, *J. Amer. Chem. Soc.*, 1957, **79**, 3301.

[37]   R. M. FUOSS, *J. Amer. Chem. Soc.*, 1958, **80**, 5059.

[38]   G. H. NANCOLLAS, Interaction in Electrolyte Solutions p. 24 (Amsterdam, 1966).

[39]   R. A. ROBINSON and R. H. STOKES, Electrolyte Solutions, p. 421 (London, 1959).

[40]   D. M. RITSON and J. B. HASTED, *J. Chem. Phys.*, 1948, **16**, 11.

[41]   E. GLUECKAUF, *Trans. Faraday Soc.*, 1964, **60**, 1637.

[42a]   D. R. ROSSEINSKY, *J. Chem. Soc.*, **1962**, 785.

[42b]   L. MASTERTON and T. BIERLY, *J. Phys. Chem.*, 1970, **74**, 139.

[43]   S. LEVINE and D. K. ROZENTHAL, Chemical Physics of Ionic Solutions (ed. by B. E. CONWAY and R. G. BARRADAS), p. 119 (New York, 1966).

[44a]   A. MOZUMDER, *J. Chem. Phys.*, 1969, **50**, 3153, 3162.

[44b]   F. H. STILLINGER and R. J. WHITE, *J. Chem. Phys.*, 1971, **54**, 3395, 3405.

[45]   I. L. JENKINS and C. B. MONK, *J. Amer. Chem. Soc.*, 1950, **72**, 2695; T. O. DENNEY and C. B. MONK, *Trans. Faraday Soc.*, 1951, **47**, 992.

[46]   W. C. VOSBURGH and G. R. COOPER, *J. Amer. Chem.*, *Soc.*, 1941, **63**, 437; R. T. FOLEY and R. C. ANDERSON, *J. Amer. Chem. Soc.*, 1949, **71**, 909.

[47]   H. M. HERSHENSON, M. E. SMITH and D. N. HUME, *J. Amer. Chem. Soc.*, 1953, **75**, 507.

[48a]   R. M. DIAMOND, *J. Phys. Chem.*, 1963, **67**, 2513.

[48b]   I. KAMPSCHULTE-SCHEUING and C. ZUNDEL, *J. Phys. Chem.*, 1970, **74**, 2363

[49]   R. M. FUOSS and C. A. KRAUS, *J. Amer. Chem. Soc.*, 1933, **55**, 2387.

[50]   See F. M. BATSON and C. A. KRAUS, *J. Amer. Chem. Soc.*, 1934, **56**, 1027.

[51]   J. M. SMITHSON and R. J. P. WILLIAMS, *J. Chem. Soc.*, **1958**, 457.

[52]   G. H. NANCOLLAS, Interactions in Electrolyte Solutions, p. 94 (Amsterdam, 1966).

[53]   F. A. POSEY and H. TAUBE, *J. Amer. Chem. Soc.*, 1953, **75**, 1463; 1956, **78**, 15.

[54]   M. EIGEN and K. TAMM, *Z. Elektrochem.*, 1962, **66**, 107.

[55]   G. ATKINSON and S. K. KOR, *J. Phys. Chem.*, 1963, **69**, 128.

[56]   R. A. HORNE, B. R. MYERS and G. R. FRYSINGER, *Inorg. Chem.*, 1964, **3**, 452.

[57]   R. A. ROBINSON and H. S. HARNED, *Chem. Rev.*, 1941, **28**, 419.

[58]   R. H. STOKES, *J. Amer. Chem. Soc.*, 1953, **75**, 3856.

[59]   R. M. DIAMOND, *J. Amer. Chem. Soc.*, 1958, **80**, 4808.

5.2 INTERACTION OF DISSOLVED IONS WITH THE SOLVENT (THE FUNDAMENTALS OF THE THEORY OF HYDRATION)

The formation of electrolyte solutions from ionic crystals or covalent substances is a consequence of the interactions between the ions or atoms of the solute and the molecules of the solvent. A crystal or a pure liquid dissolves when the energy of this interaction is sufficient for the decomposition of ionic crystals or covalent molecules and for the dispersion of the species, in addition to the thermal energy. It is already evident from this, that electrolyte solutions cannot be regarded as material systems in which ions dispersed by the solvent act only on one another by coulombic forces disturbed by the thermal energy, the solvent being only a neutral medium. On the contrary, the interaction between the ions and solvent molecules essentially influence the properties of the solutions, too, and its result is usually called *solvation* or,–in aqueous solutions–*hydration* [1a]. In several sections of the previous chapters, the problems related to this have necessarily been mentioned; here only the most important aspects of hydration will be summarized.

The interactions between dissolved ions and the solvent are so many-sided and complicated that no unequivocal definition of general validity could have been given for the concept of hydration. According to the original interpretation, it means the binding of water molecules by ions which will move along with the ions forming a kinetic entity with them. However, ions also affect the water molecules located farther from them by means of their electric field and some other effects (e.g. their size). They more or less alter the lattice-like arrangement, as well as the statistical distribution or equilibrium between the ordered regions and the disordered monomeric water molecules. All these interactions appear differently in the various phenomena, depending on the actual conditions. Since the various interactions of ions with the solvent cannot be sharply separated from one another, the interpretation of Mishchenko [1a] seems to be justified, according to which the totality of all types of interactions between solute particles and the solvent should be understood under the term hydration or solvation.

5.2.1 CLASSIFICATION OF HYDRATION PHENOMENA

In accordance with the general interpretation given above, hydration (or solvation, in general) takes place in both electrolyte and non-electrolyte solutions. In the narrower sense, hydration is usually quoted for electrolyte solutions.

Water structure is usually also altered by dissolved non-electrolytes (cf. Section 1.4.1). Solute molecules can form covalent or hydrogen bonds with water molecules, thus, on one hand, their size increases, but on the other hand, they modify the water structure or the statistical distribution of the

Literature on page 489

equilibrium of various structural units of water in their vicinity. But even without any chemical bonds, they modify the water structure in their vicinity, since in several cases they show a structure-making effect, building up and expanding the lattice-like regions (cf. the 'iceberg' effect discussed in Section 1.4.1), only by their size and some other short-range interactions (mainly dipole–dipole and dispersion forces). However, in non-electrolyte solutions, no coulombic forces of relatively long range are in operation, therefore the phenomena of hydration are of less importance here than in electrolyte solutions.

One of the general and chemically non-specific causes of ionic hydration in electrolyte solutions is the effect of their electric field on the dipole molecules of water, which can be called *physical hydration* in short. It has a general importance not only because it takes place around each ion irrespective of its chemical nature, but because it influences the water molecules not only in the immediate vicinity of the ion, but also acts on the farther ones owing to the relatively long range of the coulombic forces. In addition to physical hydration, in some cases, ions bind water molecules by specific chemical interactions as well. This phenomenon, called *chemical hydration*, often occurs in solutions of transition metal ions with incomplete electron shells. The products of the chemical interaction are complex ions of stoichiometric composition.

Although the electric field of ionic charges is extended to a distance of several molecules, there is still an appreciable difference between the effect exerted on water molecules in the immediate neighbourhood of ions and on those farther from them—in accordance with the nature of electrostatic forces. The effect on the immediate neighbours can be called (according to Bockris) *primary* or (according to Samoilov) *nearby* hydration, while the effect on the farther water molecules can be called *secondary* or *remote* *hydration*. This latter effect can be interpreted, in a certain sense, as the influence of the ion together with its primary hydrate sphere on their surroundings. It can be assumed that mainly the water molecules bound in primary hydration move together with the ions in transport phenomena, while the effect of secondary hydration can be ascribed rather to the change in the medium. However, these phenomena cannot be sharply separated from each other, since secondary hydration also modifies the mobility of the solute species.

Discussing hydration, one often thinks implicitly only of primary hydration or considers phenomena related to hydration as though they, as a whole, were connected with water molecules moving together with ions. This interpretation can be excused by the difficulty of separating the two types of hydration, however, it does not help a proper explanation of the mechanism of transport phenomena. Since primary and secondary hydration influence various phenomena that are observable experimentally in many ways, it is not surprising that the results of various methods regarding the extent of hydration differ significantly from each other.

The number of solvent molecules in the immediate neighbourhood of the ions can be regarded as the measure of primary hydration. In a narrower sense, this is the *coordination number of hydration* (or, generally, that of solvation), or in short, the *hydration number*. Usually, however, in a broader- or looser–sense of the word, hydration number means the assumed number of water molecules moving together with the ion required for the formal interpretation of the phenomena observed experimentally, when hydration is supposed to alter merely the size of ions as kinetic entities, leaving the medium unaltered.

With respect to the many-sided and complicated nature of the interaction of ions with solvent molecules, the concept of hydration number is questionable. In certain opinions, the introduction of this concept is considered to lead to confusion rather than to promote the development of a model reliably reflecting real conditions. Nevertheless, it is a fact that hydration numbers of a particular ion believed to be determined by various methods differ very much–sometimes by several hundred per cent–from one another (see e.g. Table 5.2). In spite of this, distinction must still be made between a complex moving as a kinetic entity in solution and the effect brought about by this complex on the bulk of the liquid, even if the composition,

Table 5.2

Hydration numbers of some ions calculated by various methods

Method \ Ion	H^+	Li^+	Na^+	K^+	Rb^+	Cs^+	F^-	Cl^-	Br^-	I^-
Water transport [a]	5	62	44	29	—	—	—	27	29	31
Water transport [b]	—	120	66	16	14	13	—	16	15	15
Water transport [c]	—	22	13	7	—	6	7	5	5	—
Diffusion [d]	—	24	17	10	6	4	16	10	7	4
Ionic mobility [e]	0·2	160	71	22	—	—	—	21	20	20
Ionic mobility [f]	—	5	4	4	—	—	—	4	—	1
Ionic activity [g]	8	6	—	0	—	—	—	3	—	—
Entropy [h]	4	5	2·5	2	2	—	5	2	1·5	0·5
Compressibility [i]	—	4·0	4·5	3·5	3·0	2·5	4·5	2·0	1·8	1·5
Ultrasonic absorption [j]	—	3·6	4·8	4·1	—	—	—	2·3	1·7	—
Partial mole fraction [k]	0·3	2	1	0	0	0	0·4	0	0	0
Density [l]	—	6	4	4	0	0	4	0	0	0

[a]   G. BABOROSKY, *Z. Phys. Chem.*, 1927, **129**, 129; *Collect. Czechosl. Chem. Comm.* 1938, **11**, 542.
[b]   H. REMY, *Z. phys. Chem.*, 1925, **118**, 161; 1926, **124**, 397; 1927, **126**, 161; *Trans. Faraday Soc.*, 1927, **23**, 381.
[c]   A. J. RUTGERS and Y. HENDRIKX, *Trans. Faraday Soc.*, 1962, **58**, 2184.
[d]   G. JANDER, *Z. phys. Chem.*, 1930, **149**, 97; 1942, **190**, 81; *J. Amer. Chem. Soc.*, 1915, **37**, 722.
[e]   E. H. RIESENFELD, *Z. phys. Chem.*, 1909, **66**, 672.
[f]   H. ULICH, *Trans. Faraday Soc.*, 1927, **23**, 392.
[g]   N. BJERRUM, *Z. anorg. allg. Chem.*, 1920, **109**, 275.
[h]   H. ULICH, *Z. Elektrochem.*, 1930, **36**, 457; *Z. phys. Chem.*, 1934, **168**, 141.
[i]   J. O'M BOCKRIS and P. SALUJA, *J. Chem. Phys.*, 1972, **76**, 2140
[j]   D. S. ALLAM, W. H. LEE, *J. Chem. Soc.*, **A 1966**, 426.
[k]   E. DARMOIS, *J. Phys. Radium*, 1941, **2**, 2.
[l]   J. D. BERNAL and R. H. FOWLER, *J. Chem. Phys.*, 1933, **1**, 515.

Literature on page 489

size or structure of this kinetic entity depends very much on the conditions. Practically, however, it is very difficult to distinguish between these two groups of phenomena that are different in principle, because a certain property of the solution as a whole observed directly consists usually of both types of interactions.

On the basis of NMR measurements, van Geet [1d] came to the conclusion that $Na(H_2O)_4^+$ ions exist in dilute solutions.

Recently Hinton and Amis [1b], as well as Strehlow, Knoche and Schneider [1c] gave a comparative review of our knowledge on various aspects of hydration and solvation, and showed the reasons for the confusions in this field. The problems of solvation in solvent mixtures has been reviewed by Strehlow and Schneider [1c].

On the basis of the results of recent research, the reliability can hardly be doubted of Samoilov's view [2] that it is not justified to speak of hydration in such an absolute sense whether there are water molecules strongly bound to the ions–i.e. adsorbed to them–in solution or not. With respect to the interaction between ions and water molecules, an important point is the strength of the bond between the ion and the water molecules compared with that existing between water molecules themselves. Moreover, in this respect it is not the energy required to break the bonds between adjacent water molecules (as required for evaporation) that is the determining factor, but the energy (Δw) necessary to stretch the bond between neighbouring water molecules a short distance Δr to enable the exchange of this molecule with another one. This distance is approximately $\Delta r = 0.8$ Å in aqueous solutions; the energy Δw required for such a stretching governs nearby (primary) hydration. Thus, primary hydration can, in fact, be regarded as the adsorption of water molecules on ions.

The molecules of 'hydrate water' adsorbed on ions are in permanent exchange with those in the 'bulk' of the liquid (see Section 5.2.2). It can be assumed that the exchange requires a stretching of the bond between the ion and the water molecule adsorbed on it by about 0.8 Å. The ratio of the energy Δh required to stretch this bond and the energy Δw controls primary hydration. If $\Delta h > \Delta w$, the water molecule adsorbed on the ion forms a stronger bond with the ion than with the neighbouring water molecule; this can be called *positive hydration*, according to Samoilov. Under such conditions, the adjacent water molecules spend a longer time at the ion–on the average–than bound together. If $\Delta h < \Delta w$, water molecules are bound more weakly to the ion than to the neighbouring water molecules (the ion is hydrated more weakly than water) and this can be called *negative hydration*. If, finally, it happens that $\Delta h = \Delta w$, water molecules are bound to the ion with the same bond strength as to each other and the ion fits with almost no perturbation into the structure of water. In this way, negative hydration does not mean that no water molecules are bound to the ions, it but means only that the ion–water molecule bond is weaker than the water molecule–water molecule one. Under such conditions, the neighbouring water molecules spend a longer average time bound together than adsorbed on the ion. The adsorption of water molecules on ions is negligibly small only in the case when $\Delta h \ll \Delta w$.

5.2.2 PRIMARY HYDRATION

Among the phenomena involved in the concept of hydration, the effect of the electric field of ions on the immediate neighbouring dipole molecules of water is often a predominating factor. The smaller the size and the higher the charge of the ion, the more important this effect is, because the electric force at the periphery ('surface') of the ion is greater. This interaction orientates the dipole molecules more or less in the direction of the lines of force against the disordered thermal motion, it decreases their mobility, and bringing about a partial (or total) dielectric saturation, it increases their dielectric relaxation time with respect to outer electric effects. One might expect that the magnitude of this ion–dipole force or the ratio of it and the thermal energy can easily be estimated, since, according to electrostatics, the potential energy of electric interaction between a point-like electric charge e and an electric dipole of momentum μ at a distance r from the former is $(e\mu \cos \vartheta)/\varepsilon r^2$, if ϑ is the angle between the axis of the dipole and the direction of r. Calculating in this way, and assuming a vacuum ($\varepsilon = 1$) between the ion and the water molecule, the potential energy would be $(124 \cos \vartheta)/r^2$ kcal mole^{-1} for monovalent ions which– supposing complete orientation ($\cos \vartheta = 1$)–is higher than the average thermal energy ($\mathrm{R}T \approx 0.6$ kcal mole^{-1}) at room temperature, up to a distance of about 14 Å. This calculation is, however, evidently not correct, because there is no vacuum between ions and neighbouring water molecules. Calculating with the macroscopic dielectric constant of water, the potential energy between the ion and the dipole is in the order of magnitude of $(1.5 \cos \vartheta)/r^2$ kcal mole^{-1} which is much lower than the average thermal energy, even for a distance r acceptable as the smallest possible one. However, this treatment is not correct either, since, evidently, it is not justified to calculate the interaction between the ion and adjacent water molecules using the macroscopic dielectric constant. The uncertainty is increased when considering that the dipole moment of water molecules is even increased by the polarization effects of ions, in comparison with its permanent value.

With respect to other properties of electrolyte solutions, in addition to hydration, there is some uncertainty in the question how a large dielectric constant should be applied in the calculation of the interaction of dissolved ions with one another and with the solvent molecules, i.e. how the dielectric constant of the liquid is altered by ions in their immediate vicinity [3].

For the dielectric constant of water adjacent to dissolved ions, a reasonably good approximation to reality has been given in the detailed calculations of Ritson and Hasted [4] (cf. also Section 5.1.2). Calculating the dielectric constant of water as a function of the distance from a point-like electric charge, two different models have led to the same conclusion that dielectric saturation is complete up to a distance of about 2 Å from the point-like elementary charge (the field of this charge practically fixes the axes of the permanent dipoles). In this range, the dielectric constant of

Literature on page 489

water arises only from the electronic and atomic polarization of water (no orientation polarization contributes to it), and its value can be estimated to be about 4–5. From this distance on, the dielectric constant increases rapidly up to a distance of about 4 Å from the point charge, and beyond this distance the use of the macroscopic dielectric constant is justified. In this way–since the radii of simple cations are about 0·5–1·6 Å, while the diameter of one water molecule is 2·8 Å–the first layer of water molecules around monovalent cations can be considered to be dielectrically saturated (thus this is responsible primarily for the decrease in the dielectric constant caused by dissolved electrolytes). At the periphery ('surface') of polyvalent positive atom ions, the field strength is higher (all the more so, since their radii are smaller than those of comparable monovalent ones), the bond can probably be strong even in the second layer of water molecules around them, and the dielectric saturation is extended. The effect of anions binding water molecules or resulting in dielectric saturation can be estimated to be lower; this is partly because they are usually larger than cations (the majority are polyatomic) and partly that the rotational freedom of water molecules bound to anions is higher than of those bound to cations under comparable conditions because of the structure of the water molecule.

Even if one accepts that on the basis of the above consideration, calculations should be performed using a value of $\varepsilon = 4$–5 in the immediate vicinity of ions, no reliable model for the energy of this part of hydration can be obtained, because it is not known how the polarization and orientation of water molecules in the vicinity of ions are influenced by the other molecules, and how it is affected by the other ions. One can get a detailed model of the number and state of water molecules belonging to the first hydrate sphere of the ions only by applying simplifying assumptions. The reality of this picture, however, is rather doubtful, all the more so, since the values calculated from assumptions, apparently reasonable on the basis of various empirical facts, are very different even under comparable conditions. Nevertheless, it can be supposed at least that the binding energy of water molecules in the first layer around monoatomic cations is high as compared with the average thermal energy, thus these molecules usually move together with the ions. In the second layer, however, the bond is already much weaker and its energy reaches the thermal energy probably only around small polyvalent ions.

The distribution of solvent molecules around ions has been investigated in detail by Azzam [5a]. He came to the theoretical conclusion that primary hydration around ions can be extended only to a monomolecular layer of water. According to him, primary hydration takes place only around cations of radii smaller than 1·7 Å and around anions of radii smaller than 2·1 Å. Discussing the orientation of water molecules around ions, he concluded that the orientation of water molecules around cations is identical with that supposed by Bernal and Fowler, whereas the orientation around anions is different (see Fig. 5.10).

Electrostatic forces–unlike quantum-mechanical valence forces–cannot be saturated in the chemical sense, i.e. the laws of stoichiometry do not hold for them. For sterical reasons, however, the number of water molecules in the hydrate sphere adjacent to the ion is still limited. If, for example, the size of the ion is approximately the same as that of water molecules, it can be supposed that the ion occupies the place of one water molecule in the water structure, and thus it has four nearest neighbours and eight in the second sphere, i.e. hydration occurs in integ-

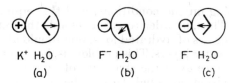

K^+ H_2O F^- H_2O F^- H_2O
  (a)   (b)   (c)

Fig. 5.10 The orientation of the water molecule (a) around a cation, according to Bernal and Fowler and Azzam; (b) around an anion, according to Bernal and Fowler; (c) around an anion, according to Azzam

ral steps. The justification of this assumption, however, may be limited by various conditions.

Water molecules can be bound to ions not only by electrostatic interactions, but by chemical bonds as well. This is also revealed by the fact that stoichiometric complexes e.g. $Cu(H_2O)_4^{2+}$, $Co(H_2O)_6^{2+}$, can be identified reliably even in crystals, not only in solutions. In these, the water molecules form covalent bonds with ions, similarly to other complex compounds. In some cases, the connections between them can be analogous to hydrogen bonds.

Ions usually move together with water molecules bound tightly to them, i.e. the ion hydrate, in this sense, forms a kinetic entity in several respects. It does not follow from this, however, that the hydrated ion is a permanent formation. The water molecules bound by hydration sooner or later exchange with farther water molecules due to energy fluctuation. The rate of this exchange can sometimes be observed even in the case when the hydrate is, in fact, a covalent complex. For example, according to the measurements of Hunt and Taube [6] on ^{18}O exchange, the average half-life of a water molecule staying in the complex ion $[Cr(H_2O)_6]^{3+}$ is about 40 hours, while in the ion $[Co(NH_3)_5H_2O]^{3+}$ it is 24 hours. In several other ions, however, the average dwelling time of water molecules is much shorter. Investigating the exchange of water molecules in the hydrate spheres of transition metals, Luz and Schulman [7] and McDonald and Phillips [8a] have established that its rate is usually high except for Cr^{3+} and Fe^{3+} ions. It is still justified to speak of hydration in each case, when the rate of the exchange of water molecules located around ions is low in comparison with the rate of the Brownian motion.

With respect to the transport processes in aqueous electrolyte solutions, it should also be taken into account that knowledge of the number of water molecules moving together with the ion is not sufficient, since the hydrated ion regarded as a kinetic entity moves in a medium of non-uniform structure. In water, present as a solvent, there are probably monomeric water molecules in addition to the lattice-like regions and more or less ordered regions of different types may simultaneously be present. These various states of water are in dynamic equilibrium with one another depend-

Literature on page 489

ing on the conditions, and this equilibrium is altered by the field of ions. Thus, in order to get a reliable theoretical description of the transport processes of hydrated ions, one should also know the effect of ions on the farther water layers. This problem is, however, far from being solved, all the more so, since even the structure of pure water is not known with satisfactory precision, although this is the precondition of the unequivocal description of the alteration of this structure by ions. The influence of hydration on transport processes can be discussed theoretically only by applying marked simplifying assumptions, and the theoretical treatments based on various starting assumptions yield results contradicting one another in several cases.

Nearby hydration also affects the association of ions. This has been pointed out recently by Samoilov, Yastremskii and Tarasov [8b] and by Tereshkovich, Valenchuk, Kuprik, Volkova and Skidan [8c] on the basis of their studies in solutions of alkali metal chlorides and iodides, and of cobalt and nickel nitrates, respectively. A detailed survey has been given by Smid [8d] concerning the structure of solvated ion pairs.

The role of primary hydration in the translational movement of ions. In most of the theories on hydration, it is supposed that ions transport the water molecules bound in their primary hydrate sphere in the course of their translational displacement. In this way, the ion and its primary hydrate sphere form a kinetic entity with respect to translation. Though the water molecules in the hydrate sphere exchange from time to time with those arriving from the farther environment, according to these models, however, this does not influence the fact that the hydrated ion as a whole is the kinetic entity of the translational motion. However, this model is not convincing and phenomena indicating hydration can also be interpreted in another way.

According to the theory of Samoilov [2], hydration should be explained as the effect of ions on the translational movement of water molecules in their immediate vicinity and not by supposing their displacement together with them. In accordance with the problems discussed above, experiments indicate that certain ions decrease the mobility of water molecules adjacent to them *(positive hydration)*, while the mobility of the water molecules is increased around other ions as compared with pure water *(negative hydration)*. There is a certain reason for retaining the coordination number of hydration in this theory too, although this does not mean the number of water molecules moving together with ions. In Samoilov's interpretation, the coordination number of ions means the average number of water molecules, constantly in motion, located adjacent to ions in dilute electrolyte solutions. These water molecules have properties different from those of the others in some respects, so their number can also be obtained from thermochemical measurements. According to conclusions drawn from several points of view, ions develop their environment consisting of water molecules in such a way that it differs from the structure of pure water as little as possible.

The coordination number of hydration is not constant even in a given solution, as it also depends on the translational motion of ions. The local loosening of the liquid structure in the course of translation decreases the coordination number. In connection with hydration it must also be taken into account that the quantitative description of the structure of liquids can be carried out only by means of the corresponding distribution function that also reflects the short-range order of molecules. For a description of the short-range order of molecules and their thermal movement, the model corresponding to the lattice-like structure proved to be useful in several cases but must, however, be applied carefully. Sometimes the liquid is treated as a 'damaged' solid, i.e. a crystal containing many vacancies. However, expressions referring to such a quasi-crystalline structure of liquids can lead to confusions as has also been pointed out by Hildebrand [9a], since, in addition to the fact that no long-range order exists in liquids, the translation of their molecules is also followed by alterations in the short-range order when compared with that existing in crystals. The essential feature of the short-range order is that certain distances between atoms or molecules are more frequent than others and the same is valid for certain orientations too, if–as in water–the intermolecular forces are orientated.

As supposed by Samoilov, the continuous exchange of water molecules around ions is not excluded when considering that the energy of interaction between ions and adjacent water molecules is higher than the energy of interaction between water molecules themselves, since the exchange of a water molecule adjacent to an ion with another in the bulk does not require the removal of a water molecule from the primary hydrate sphere, but only its exchange with another (see Section 5.2.1). The energy barrier (i.e. activation energy) of this latter process is much lower than that of the departure of a water molecule, which explains the high exchange rate. In other words, the exchange rate does not depend on the total energy of interaction of the ion with water molecules, only on the variation of the energy of interaction at small distances from the ion.

The effect of ions on the translational motion of water molecules can be very different depending on their size, charge, shape and, incidentally, some other properties. If the exchange rate is low, the mobility of water molecules close to ions decreases and (positive) hydration is large. If, however, the exchange is accelerated as compared with the exchange between neighbouring molecules in pure water, negative hydration occurs. In this way, primary hydration is governed by kinetic factors and the effect of the electrostatic field of ions is only a secondary influence on hydration, i.e. it predominates only with respect to the state of the somewhat distant water molecules.

On the basis of studies on the dielectric constant of alkali and alkali earth halide solutions, Kokovina, Samoilov and Yastremskii [5b] have concluded that the increase in the average dwelling time of adjacent water molecules in the vicinity of cations, i.e. the strengthening in nearby hydration, is followed by a decrease in the average dwelling time of water molecules around anions. Water molecules partly replace anions in the vicinity of cations, because the potential barrier to be overcome by water molecules

Literature on page 489

to leave the immediate vicinity of a cation is higher than the corresponding potential barrier for anions. The stronger the hydration of the cations, the higher the portion of anions displaced from the environment of cations.

In order to estimate the mobility of water molecules adjacent to ions, one can start–following Samoilov's treatment–with the fact that the average dwelling time of a water molecule τ in its stationary state in pure water is determined by the height of the potential barrier E separating two adjacent positions. In the immediate vicinity of ions, the energy barrier separating two neighbouring water molecules from each other is $E + \Delta E$ to which an average dwelling time τ' will correspond. ΔE is the change in the potential energy barrier between neighbouring water molecules caused by the effect of an ion; this and the corresponding ratio τ'/τ are decisive in hydration. The average dwelling time of a molecule in its equilibrium position is:

$$\tau = \tau_0 \exp\left[\frac{E}{RT}\right], \tag{5.2.1}$$

and

$$\tau' = \tau_0 \exp\left[\frac{E + \Delta E}{RT}\right],$$

where τ_0 is a constant.

The ratio of the average dwelling times, on the other hand, is:

$$\frac{\tau'}{\tau} = \exp\left[\frac{\Delta E}{RT}\right]. \tag{5.2.2}$$

ΔE governing hydration can be calculated from the change of ionic conductivity with temperature. According to the Pisarzhevskii–Walden rule, the product of the limiting value of ionic conductivity and the viscosity of the solution is approximately constant, independent of temperature:

$$\lambda_i^0 \eta^0 = \text{const.} \tag{5.2.3}$$

from which:

$$\frac{1}{\lambda^0} \frac{\partial \lambda^0}{\partial T} = -\frac{1}{\eta^0} \frac{\partial \eta^0}{\partial T}. \tag{5.2.4}$$

Viscosity also depends on the activation energy of the displacement of molecules. The viscosity of pure water is:

$$\eta_w = A \exp\left[\frac{E}{RT}\right] \tag{5.2.5}$$

and that of a solution extrapolated to infinite dilution is:

$$\eta^0 = A' \exp\left[\frac{E + \Delta E}{RT}\right], \tag{5.2.6}$$

where A and A' are constants. By differentiating these two equations, and eliminating the ratio E/RT one has:

$$\frac{1}{\eta^0}\frac{\partial\eta^0}{\partial T} = \frac{1}{\eta_w}\frac{\partial\eta_w}{\partial T} - \frac{\Delta E}{RT}.$$

(5.2.7)

Substituting this value into equation (5.2.4), one obtains:

$$\frac{1}{\lambda^0}\frac{\partial\lambda^0}{\partial T} = -\frac{1}{\eta_w}\frac{\partial\eta_w}{\partial T} + \frac{\Delta E}{RT}.$$

(5.2.8)

The change in the energy barrier ΔE controlling hydration can be calculated in this way from the relative temperature coefficient of ionic conductivity and the viscosity of water. Table 5.3 shows values of ΔE calculated for some ions and the crystallographic radii r_i of the ions. It is seen in the Table that $\Delta E > 0$ for Li^+, Na^+, Mg^{2+} and Ca^{2+} ions, thus τ'/τ is higher than unity; the mobility of water molecules located in the immediate vicinity of these ions is lower in their solutions than in pure water, i.e. positive hydration takes place. In solutions of K^+, Cs^+, Cl^-, Br^- and I^- ions, $\Delta E < 0$ and $\tau'/\tau < 1$; the mobility of water molecules increases in the vicinity of these ions, thus negative hydration occurs. The sign changes between the Na^+ ion of radius 0.98 Å and the K^+ ion of radius 1.33 Å; the ionic size causing neither positive nor negative hydration is evidently between these two values. In this way, such a monovalent ion would not change the mobility of water molecules in its vicinity and its size would be approximately identical with that of water molecules (the radius of a water molecule is 1.39 Å as given by Bernal and Fowler [57]). From measurements on the magnetic susceptibility of electrolyte solutions, Yergin and Kostrova [9b] have concluded that the transition between positive and negative hydration is at an ionic radius of about 1.1 Å at room temperature.

Table 5.3

Change in the energy of nearby water molecules in the vicinity of some ions

(r_i is the ionic radius in crystals, according to Goldschmidt)

Ion	ΔE, kcal · mole^{-1}	r_i Å	Ion	ΔE, kcal · mole^{-1}	r_i Å
Li$^+$	0.73	0.78	Mg^{2+}	0.01	0.78
Na$^+$	0.25	0.98	Ca^{2+}	0.45	1.06
K$^+$	—0.25	1.33	Cl$^-$	—0.27	1.81
Cs$^+$	—0.33	1.65	Br$^-$	—0.29	1.96
			I$^-$	—0.32	2.20

Studying the sizes of the structural cavities in water and the activation energy barrier to be overcome for the displacement of water molecules

Literature on page 491

located in the hydrate sphere of the ions, Lyashchenko [9c] found that the Li^+ ion is positively hydrated, while the K^+ ion gives rise to negative hydration within the temperature range 0–40 °C, whereas the negative hydration of the Na^+ ion observed at low temperatures becomes positive at 26·5 °C. In general, the increase in temperature facilitates positive hydration. By extrapolation, Lyashchenko has supposed that the negative hydration of the K^+ ion is transformed into a positive one at about 80–100 °C, while the positive hydration of Li^+ ions becomes negative at about -10 to -15 °C.

Emelyanov, Nikoforov and Popov [9d] measured the self-diffusion of water and, in accordance with the above facts, they have drawn the conclusion that F^- ions are hydrated positively; Emelyanov and Yagodarov [9e] considered the negative hydration of the Rb^+ ion to be stronger than that of K^+, but weaker than that of the Cs^+ ion. Solovkin [9f] carried out calculations concerning the hydration of alkali metal halides in the presence of negatively hydrated cations.

It has been emphasized by Samoilov that the exchange of water molecules in the primary hydrate sphere, including negative hydration, is not in contradiction with the fact that hydration is a strongly exothermal process. According to his opinion, the main part of the energy effect is caused by secondary hydration connected with farther water molecules.

Even if it is questionable whether the difference in the temperature coefficients of ionic conductivity and viscosity arises completely from the factors mentioned above, the various effects of ions on the translational mobility of water molecules follows from other phenomena as well (hydration entropy, the effect of ions on the self-diffusion of water of various isotopic compositions).

The exchange of the water molecules in the primary hydrate sphere of ions can be directly detected. Nevertheless, it cannot be considered to be an unequivocally verified fact that hydration is based only on the kinetic effect given in Samoilov's interpretation. It has not been disproved that, in addition to the change in the mobility of water molecules, those moving together with the ions–ion hydrates in the classical sense of the word– also have a role in the group of phenomena summarized under the term hydration.

From a study of the dielectric relaxation in solutions of alkali metal halides, Giese [9g] has drawn conclusions regarding the dwelling time of water molecules. According to him, the variation of the relaxation time with the concentration of the electrolyte can be attributed to the fact that the average dwelling time of water molecules in the hydrate spheres of ions is of the same order of magnitude as the average reorientation period of the molecules. In view of the frequency dependence of the dielectric relaxation of electrolyte solutions, Kaatze [9h] assumes that the orientation polarization is far weaker in the hydrate spheres of certain ions than in pure water; that is, the dynamic behaviour of water is altered in the vicinity of ions. According to these calculations, the average number of water molecules being apparently totally prevented in rotation is 3.9, 2.2 and 11 around the OH^-, NH_4^+ and H^+ ions, respectively. The reorientation period of water molecules in the proton hydrate is far longer than in pure water.

In a study of the hydration of organic cations and anions, Uedaira and Uedaira [10a] have started with the calculation of the activation energy required by a water molecule to leave the immediate environment of an ion. By comparing this with the activation energy of self-diffusion of the ion, they have concluded that hydrophobic groups and the carboxyl group result in positive hydration, while sulpho groups give rise to a negative one.

Nearby hydration of ions is noticeably influenced by the isotopic composition of water. It has been stated by Samoilov, Rabinovich, Volosova and Borina [10b] on the basis of the density of aqueous and non-aqueous solutions of LiCl, NaCl and KCl–in agreement with other phenomena–that nearby hydration is reduced by the replacement of H_2O molecules by D_2O ones.

Nearby hydration (in general, solvation) is also dependent on *temperature*. Krumgalz [10c] pointed out in a comprehensive paper that the hydration of most ions increases with increasing temperature, owing to the fact that higher temperature causes the rupture of a certain amount of hydrogen bonds in water, thus the ratio of free water molecules available for bonding by the ions also increases. At the same time, the thermal motion is also enhanced, lowering the dwelling time of the solvent molecules in the hydrate spheres of the ions. Therefore, the average number of water molecules moving together with the ions will be lower, decreasing the hydration number.

As the net result of the opposing effects, the radii of hydrated ions decrease with increasing temperature only in solutions of strongly hydrated ions (Li^+, Mg^{2+}). The apparent radii of hydrated hydrogen and hydroxid ions increase with increasing temperature, which is clearly in connection with the suppression of the prototropic conduction mechanism. The negative hydration of larger ions decreases with increasing temperature, and that of K^+, Rb^+, Cs^+ and NH^+ ions becomes positive at 50, 100 and 150 °C, respectively.

Several facts indicate that tetra-alkylammonium ions comprising a large ($> C_3H_7$) alkyl group are not hydrated by monomeric H_2O molecules, owing to their large size and low 'surface' charge density. The alkyl groups of these ions can be assumed to be situated in the structural cavities of water thus forming a lattice-like structure. The ordered regions in liquid water become more and more destroyed with increasing temperature, thus the number of clusters available for the alkyl groups is also reduced, resulting in a decrease in the apparent size of the hydrated ion. In liquids not possessing a strong structure, solvation is only slightly or not at all dependent on temperature.

The effect of pressure on primary hydration. It is a consequence of the loose structure of water, which contains a lot of cavities, that the liquid structure is significantly altered by the effect of pressure, namely, its order decreases. The rupture of the water structure under pressure–which is revealed in the pressure dependence of viscosity (cf. Section 2.2)–probably takes place because the tetrahedral structural cavities in the lattice-like regions are

Literature on page 490

successively occupied with increasing pressure by water molecules that had originally been members of the lattice-like structure forming hydrogen bonds with other water molecules (Samoilov and Sokolov [11], Vdovenko, Gurikov and Legin [12]). Thus the occupation of the cavities by water molecules is accompanied by the rupture of a part of the hydrogen bonds, resulting in a reduced average binding of molecules. (In this way, the increase in pressure influences the structure of water in a similar way to that of temperature to a certain limit—up to the value where all structural cavities are occupied.) On the other hand, changes in water structure influence hydration. According to Yastremskii and Samoilov [13], the destruction of the liquid structure causes strengthening of hydration, while the loosening of the structure decreases hydration. Thus, in electrolyte solutions, primary hydration is also altered by changes in the pressure (and, similarly, by changes in temperature), in addition to the alterations of the structure of the free solvent. The resultant of these two effects is revealed macroscopically.

The structure-breaking effect of pressure on water is confirmed by the pressure dependence of the activation energy of transport processes, as shown by Horne, Courant and Johnson [14] which also indicates indirectly a change in hydration.

The effect of pressure on hydration also depends on how much the interactions between the species are orientated in the liquid. According to Nevolina, Samoilov and Seifer [15], the increase in pressure has presumably a smaller direct effect on the hydration of approximately spherically symmetrical ions than that exerted on the state of the 'free' water outside the hydrate spheres. Since the increase in pressure decreases the average binding of water molecules, the hydration of spherically symmetrical ions, e.g. Na^+, K^+ should increase with pressure, according to this suggestion. On the other hand, the hydration of non-spherically symmetrical ions is expected to decrease. In this respect, the NH_4^+ ion cannot be regarded as a spherically symmetrical one forming four hydrogen bonds with adjacent water molecules. The consequence of this is that this ion has a structure-breaking effect on water, as shown in the X-ray diffraction studies of Radchenko and Ryss [16a]. Thus it can be expected that the hydration of the NH_4^+ ion decreases with the increase in pressure. This view is supported by the accurate density measurements of Nevolina, Samoilov and Seifer [15] in NaCl, KCl and NH_4Cl solutions, up to a pressure of 1000 bar. The size of the hydrate sphere calculated from the experiments increases with pressure below 45 °C around Na^+ and K^+ ions, whereas it decreases around NH_4^+ ions. This effect diminishes above 45 °C, because the water structure is so destroyed at this temperature due to the effect of the thermal motion that further destruction due to the increase in pressure is already not significant.

Some investigations using the molecular orbital theory also help to clarify the hydration conditions. Burton and Daly [16b] have applied the semi-empirical CNDO method for the hydration of ions. Calculations have been made on the structure, energy of hydration, and force constants of Li^+, Na^+, Be^{2+} and Mg^{2+} ions. Studying the effect of ions on the hydrogen bonds of water molecules adjacent to and farther from them, it has been established that the effect of substituted ammonium ions on the water molecules in their vicinity is somewhat different from the effect of alkali metal ions. Their calculations on the hydration of H_3O^+ ions show that the energy of hydrogen bonds is not additive.

5.2.3 SECONDARY HYDRATION

In the macroscopic phenomena induced by dissolved ions in water, the modification of the structure of water layers not in immediate contact with ions also has an appreciable role in addition to primary hydration. This effect, called secondary hydration, is even more complicated and complex than primary hydration is, and it cannot be sharply separated from the latter. The alteration of the water structure does not originate only from the electric field of ions, because it can be established on the basis of several experiments that uncharged molecules alter the liquid structure in their vicinity as well (cf. Section 1.4.1). Here only some aspects of the secondary hydration will be dealt with.

Starting with the Samoilov model and limiting our review only to the main characteristics of the interactions for the sake of a better survey, secondary hydration will mean the ensemble of the effects brought about collectively by the ion and its primary hydrate sphere on the neighbouring water molecules and the more or less ordered regions formed by them. Recently, some efforts have been made by Nightingale [17] to review these interactions from a uniform point of view. According to him, four main types of hydration effects of ions can be distinguished.

Hydration of Type I is caused by ions (Li^+, Na^+, Be^{2+}, Mg^{2+}, Fe^{2+}, Ce^{3+}, F^-) having an electric field strong enough to orientate the dipoles of water molecules in the direction of the lines of forces at their periphery (at their 'surface'). This primary effect taking place at the immediate periphery of ions can be called *peripheric hydration*. The effect of the complex formed by the ion and its peripheric hydrate sphere on the structure of the farther water layers in its vicinity is *aperipheric hydration*. If the water molecules in the peripheric hydrate sphere are bound to the ion more strongly than to one another, positive hydration takes place, in the Samoilov sense. Under such conditions, the common effect of peripheric and aperipheric hydration, i.e. the net hydration effect of the ion, is usually a structure-making one when compared to the bulk of the liquid. Such ions are hydrated in the classical sense, and the electric field is strong enough at their periphery to adsorb one (or more) layers of water molecules. In the first layer, the binding of water molecules is so strong that their vibrational degree of freedom becomes reduced. The bi- and trivalent ions in this group reduce or even stop the rotational degree of freedom of the water molecules in the first layer (e.g., Eu^{3+} ion). At the periphery of the ions of this type, the electric field strength is high, these ions have a large effective hydrodynamic volume, and the B coefficient of their viscosity is positive (cf. Section 2.4). In solutions of such ions, the activation energy of the

Literature on page 490

viscous flow is higher than in pure water, which indicates that the ion hydrate moving as a kinetic entity hinders viscous flow not only as a dissolved species in the Einstein sense; the hindrance decreases relatively with increasing temperature. The probable reason for this is that the 'icebergs' around the ions are 'molten' to a larger extent than the suppression of the ice-like structure which takes place in pure water.

The ions classified into Type II by Nightingale are those which–although hydrated peripherically and having a structure-making effect–result in a negative activation energy of the viscous flow of their solutions. Ba^{2+}, IO_3^- and SO_4^{2-} ions belonging to this type are the largest ones among those of the same charge. At their periphery, the field strength is rather low which limits peripherical hydration. The orientation of water molecules around ions is facilitated by increasing temperature and the B coefficient of viscosity increases. Tetramethyl ammonium ion is also hydrated peripherically (unlike other tetra-alkylammonium ions), and this hydration increases somewhat with increasing temperature.

The ions belonging to Type III (at 25 °C, these are K^+, Rb^+, Cs^+, NH_4^+, Cl^-, Br^-, I^-, NO_3^-, ClO_3^-, ClO_4^- and IO_3^-) exhibit a weak bond towards the water molecules at their periphery, they are hydrated negatively, and the primarily hydrated complex has a structure-breaking effect on its environment in the liquid; its aperipheric effect destroys or loosens the ice-like structure of water in its vicinity. Due to the effect increasing the disorder, the B coefficient of the viscosity of their solutions is negative, as well as the activation energy of the viscous flow, indicating that the weak electric field of ions orientates water molecules more and more easily with increasing temperature. The classification into this type, however, depends markedly on temperature, e.g. at 40 °C, K^+, NH_4^+, Cl^- and BrO_3^- ions already have a positive B coefficient and a positive value of dB/dT. Although data are hardly available near the boiling point, it seems that only strong electrolytes contribute to the viscosity of water with a positive increment at elevated temperatures, which does not depend too much on the size and structure of ions.

Large ions belong to Type IV in Nightingale's classification (e.g. tetraalkylammonium ions) at the periphery of which the electric field strength is so weak that no appreciable peripherical hydration takes place. These ions behave as non-hydrated, Einstein-type dissolved species in good approximation (cf. Section 2.4.2), having only an aperipheric hydration effect. They strengthen the ice-like structure of water, because they do not participate in the vibration of the structural clusters of the solvent owing to the hydrophobic nature of their periphery. The activation energy of their viscous flow is positive, which can be attributed to the structure-making effect of such ions, and is connected with the easier melting of the icebergs grown by the ions (similarly to the melting of peripheric icebergs of Type I).

Negative hydration and the structure-breaking effect of some ions have been interpreted by Engel and Hertz [18] in a way other than that of the Samoilov theory. In the interpretation of their NMR relaxation experiments and of some earlier empirical data, they started with the fact that the electron population of water molecules is altered and their hydrogen bonds

are modified in the vicinity of neutral atoms or molecules in such a way
that it results in a structure-making effect on water (secondary hydration
or 'iceberg' formation, cf. Section 1.4.1). An important consequence of
this is that molecular motion is slowed down and the mobility of water
molecules around the dissolved neutral species decreases. Although the
arrangement of water molecules around the neutral solute species can be
different, their calculations showed that only those configurations are realized with appreciable probability which do not deviate too much from
the configuration of the highest probability. They have given a theoretical
analysis of the case when a solute particle gets charged. If the charge
is small or the particle is large and thus the electric field strength at its
periphery is low, various types of configuration of water molecules showing
a greater deviation from the maximum probability are formed in a greater
number in the vicinity of solute species. This results in an increase in
entropy and enthalpy, and simultaneously it enhances the rate of the
molecular motion, as well, that appears in the group of phenomena called
negative hydration. However, if the electric field strength is high at the
periphery of the ion (the ionic charge is large or the size is small), the ordering effect is so strong that the configuration of the neighbouring water
molecules around the ions can only deviate a little from the maximum probability (of course, this differs from that existing around neutral molecules)
which results in the decrease of enthalpy, entropy and mobility. According
to this model, negative hydration is caused by ions which, on one hand,
give rise to 'iceberg' formation to a considerable extent owing to their
specific properties but independently of their charge; on the other hand,
the electric field strength is so weak at their periphery that their effect
on increasing the mobility of neighbouring water molecules predominates
due to the 'melting of icebergs', and the electric field is not strong enough
to bring about a tight order and enable the effect which decreases the
mobility of molecules to function.

The fact that the presence of organic substances usually diminishes
negative hydration in solutions has been interpreted by Engel and Hertz
as negative hydration, being a cooperative effect of the ensemble of several
water molecules. If the presence of non-electrolyte molecules disturbs this
ensemble, negative hydration is also reduced or even eliminated. The
Engel–Hertz theory gives an explanation of several details of the phenomena of negative hydration as well, however, verification of its correctness
needs further study.

Negative hydration is much more rare in non-aqueous solutions than in
aqueous ones, so in some ways this phenomenon is a speciality, following
from the particular structure of water. However, it has recently been pointed out by Nikiforov [19a] using a spin-echo relaxometer in glycerolic KI
solutions that the mobility of glycerol molecules is increased by electrolytes,
i.e. they have a negative solvation effect. Magnetochemical measurements
by Ergin and Kostrova [19b], made in glycol and glycerol solutions of alkali
metal halides, also indicate negative hydration.

Literature on page 490

5.2.4 EXPERIMENTS INDICATING HYDRATION

Several physico-chemical properties of electrolyte solutions indicate interactions between dissolved ions and the solvent, which have been summarized under the term solvation or hydration. The relationship between the mobility of ions dissolved in water (equivalent conductivity) and atomic weight, as well as the transference phenomena observed in the presence of dissolved non-electrolytes were discovered first as having such properties. Hydration can be indicated by the concentration dependence of diffusion, conductivity, and activity coefficient of electrolytes in light of the Debye–Hückel theory, and by the influence of electrolytes on the solubility of non-electrolytes and on their distribution between two solvents, by the compressibility and ultrasonic absorption as well as by light and NMR absorption, thermodynamic properties and some other phenomena of electrolyte solutions. The model of hydration constructed on the basis of these experiments and the estimation of the hydration number, however, should be accepted with reservation, because the results are rather contradictory owing to the reasons discussed above.

Although, strictly speaking, the hydration number cannot be defined unambiguously, it is helpful as a classification parameter in the comparative structure investigations on electrolyte solutions, as has been pointed out recently in detail particularly by Berecz and Achs-Balla [19c]. On the basis of simplified geometrical considerations and neglecting changes in the dielectric properties due to hydration, they interpreted and systematized the hydration numbers of ions with various sizes and changes, using available experimental data. They arrived at the conclusion that–in accordance with earlier views–in aqueous solutions those ions, whose crystallographic radii are smaller than the radius of the water molecule give rise to positive hydration. Further, they emphasized that positive hydration is also brought about by ions having a coordination number higher than 6, related to water. They have discussed, in agreement with earlier opinions that, as shown by several physical–chemical properties, the character of the hydrated ion also affects the water molecules situated beyond the hydrate sphere of the ion, that is, secondary hydration must also be considered. The number of water molecules being in the regions disturbed by the ions also depends on the size, charge and electron structure of the ion, and it may be $1.2-2.4$ times higher than the hydration number. These conclusions made possible the definition and determination of the maximal hydration number (h_{max}), that is, the number of all water molecules that suffer changes in their structural conditions owing to the presence of the ion, as compared with their original condition in the bulk water. The maximal hydration number is shown as a function of the atomic number in Fig. 5.11. It can be seen that, on one hand, the ions can be divided into categories according to their valence. On the other hand, within the categories, h_{max} decreases with increasing atomic number (thus with an increasing number of electrons, too) of the ion, down to the argon configuration, that is, the action to polarize the solvent becomes weaker. Further increase in the atomic number within an actual category does not alter the hydration of ions significantly. In respect of

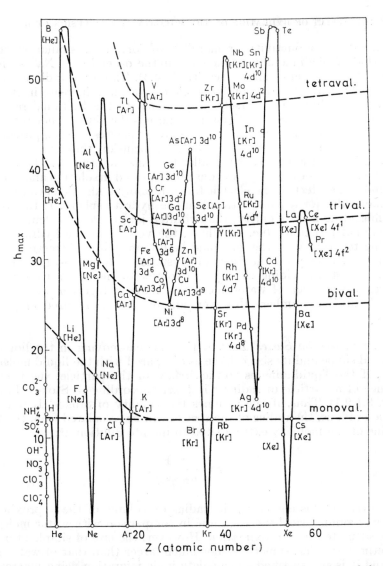

Fig. 5.11 The maximum hydration number of some ions as a function of the atomic number

the maximum hydration number, the limit of the destroying and strengthening effect is at about $h_{max} \approx 12$. According to Berecz and Achs-Balla, a proton disturbs the original attachment of 12 water molecules. Their calculations have been confirmed by experimental data on the viscosities and tensions of electrolyte solutions.

Literature on page 490

5.2.4.1. EFFECT OF HYDRATION ON THE MOBILITY AND DIFFUSION OF IONS

The equivalent conductivity or mobility of ions with similar structure, mainly that of alkali metal ions, increases in the order $Li^+ < Na^+ < K^+ < Rb^+$ as experiments show (see Table 4.4). The ionic radii calculated from the structure of corresponding ionic crystals also increase in the same order. Since mobility is inversely proportional to the frictional resistance (see equation (4.2.6)) and frictional resistance is the higher–under otherwise comparable conditions–the larger the moving species, the empirical facts regarding ionic mobilities lead us to the conclusion that the size of the migrating ions is the larger, the smaller their size in crystals. The cause of this phenomenon is evidently that ions are hydrated in solution and they migrate under the effect of the electric force, forming a kinetic entity with the water molecules (or a part of them) in the hydrate sphere. If the volume of the migrating kinetic entity (the hydrated ion) could be calculated from the frictional resistance (i.e. mobility), the volume of the hydrate sphere migrating together with the ion could be obtained by subtracting that of the bare ion, known on the basis of crystallographic investigations, from the total volume. Dividing this by the volume of the water molecule, one could get the number of water molecules moving together with the ion, i.e. the hydration number. However, all this can be realized only hypothetically.

Conclusions based on Stokes' Law. If the particle moving in the liquid has a spherical shape and its size is large as compared with the liquid molecules, further, if the liquid adheres to the surface of the particle, the frictional resistance of a particle of radius r is $1/6\pi\eta r$ according to Stokes' Law (cf. Section 3.1.2.1). Thus, if one assumes the validity of Stokes' Law for the migration of ions, the radius of the species can be calculated from the limiting value of the mobility extrapolated to infinite dilution λ^0:

$$r = \frac{|z|\mathbf{F}^2}{6\pi\eta^0\lambda^0}. \tag{5.2.9}$$

On the basis of this equation, the radius or volume of the approximately spherically symmetrical species, large in comparison with water molecules, can be calculated in approximation. However, the size of simple atom ions is sometimes smaller, sometimes not much larger than that of water molecules and it is not justified to calculate ionic migration using macroscopic viscosities (cf. Section 4.2), thus their frictional coefficient differs appreciably from that calculated from Stokes' Law. Under such conditions, Stokes' Law is a rather semiquantitative estimation of the trend in the mobility with the change in radius, but it is not suitable for calculating the volume of the hydrated ion.

Nevertheless, it can be established from the mobilities that similar hydrated ions (belonging to the same column in the periodic system) have a larger size, the smaller their crystallographic radii. According to the general models of hydration, this can be ascribed to the fact that the higher the field strength, the greater the effect resulting in hydration, and the smaller the size of the ion, the greater the strength of the electric field at the 'periphery' of ions of large charge. In this way, the decrease in the size of the ion is overcompensated by the increase in volume accompanying hydration.

It is to be noted that in solutions, the mobilities of large ions vary much less in proportion than their crystallographic sizes. For example, at 25 °C, the mobility of Li^+ ion is $\lambda^0_{Li^+} = 38.6$, that of Rb^+ ion is $\lambda^0_{Rb^+} = 66.2$, while their crystallographic radii are $r_{Li^+} = 0.60$ Å and $r_{Rb^+} = 1.48$ Å. Similarly, $\lambda^0_{Mg^{2+}} = 53.0$ and $\lambda^0_{Ba^{2+}} = 63.6$, while $r_{Mg^{2+}} = 0.65$ Å and $r_{Ba^{2+}} = 1.35$ Å. In addition, the mobility tends to a limiting value with the increase in the crystallographic ionic size, e.g. the mobility of a Cs^+ is no greater than that of a Rb^+ ion, that of a I^- ion is even slightly less than that of a Br^- ion. It seems that there is an optimum size as a result of all interactions that tend to develop on hydration. It seems to be reasonable that ions which have a size close to or exceeding this optimum value even in their bare state do not become hydrated (e.g. Cs^+ and I^- ions).

In order to determine the volume of hydrated ions and hence their hydration number, one should know the accurate relationship between the radius and mobility of small species as well. No reliable relationship of this kind is available yet. In order to eliminate this deficiency, Robinson and Stokes [20] tried to estimate the correction by which Stokes' Law could be transformed into a relationship approximately valid for small ions. Their deduction started with the mobility of tetra-alkyl ammonium ions for which some detailed studies by Kraus et al. [21a] have been published. From the theoretical point of view, these ions are also important because their shape is approximately spherically symmetric, they have a small charge, and their size is large, and it can be enlarged systematically by changing the substituent alkyl groups. From the bond lengths $N-C$ and $C-C-H$ and from the van der Waals radii, as well as from the molar volume, the effective radii r of tetra-alkylammonium ions can be calculated rather reliably (see Table 5.4). On the other hand, from the limiting value of mobility, the

Table 5.4

Radii of tetra-alkylammonium ions calculated from their molecular structure (r) and from Stokes' Law (r_{St})

Ion	r, Å	r_{St}, Å	$\dfrac{r}{r_{St}}$
$(CH_3)_4N^+$	3.47	2.04	1.70
$(C_2H_5)_4N^+$	4.00	2.84	1.42
$(C_3H_7)_4N^+$	4.52	3.92	1.15
$(C_4H_9)_4N^+$	4.94	4.71	1.05
$(C_5H_{11})_4N^+$	5.29	5.25	1.01

Literature on page 490

radius r_{St} corresponding to Stokes' Law can be calculated on the basis of equation (5.2.9):

$$r_{St} = \frac{0.820 \, |z|}{\lambda^0 \eta^0}. \tag{5.2.10}$$

It is seen in Table 5.4 that Stokes' Law is valid in a satisfactory approximation for ions of radius larger tha nabout 5 Å, while for ions smaller than this, an empirical correction factor of r/r_{St} can be determined by which the probable value of the real radius can be calculated from the Stokes' one. On the basis of Stokes' Law corrected in this way, the radius of the hydrated ion can be calculated and its volume estimated from mobilities. Since the volume of the bare ion is small as compared with that of the hydrated ion obtained in this way, the average number of water molecules migrating together with the ion (i.e. the hydration number h), can be obtained in first approximation, when the volume of the hydrated ion is divided by the volume of one water molecule (that is 30 Å^3, disregarding the change in the volume of water molecules occurring in the hydrate sphere owing to electrostriction). This treatment, of course, can be applied even in approximation only for ions with Stokes radii not markedly smaller than 2 Å, since for ions that have a smaller value of r_{St}, no conclusions can be drawn for the correction factor obtained in investigations on tetra-alkylammonium ions. It can be stated that this factor rapidly decreases with decreasing r_{St}. The reliability of this method of calculation is also reduced by the fact that the agreement between r and r_{St} is not unequivocal evidence for the validity of Stokes' Law, because it can occur by the almost incidental equality of opposite effects. The hydration numbers calculated by the Robinson–Stokes method can be found in Table 5.5 in which:

$$r = \frac{0.820 \, |z|}{\lambda^0 \eta^0} \left(\frac{r}{r_{St}}\right) \text{Å},$$

Table 5.5

Crystallographic radii (r_{cr}), the Stokes (r_{St}) and the corrected radii (r) corresponding to the hydrated state, volume, and hydration number of some ions

Ion	r_{cr}, Å	λ^0, (25 °C)	r_{St}, Å	r Å	v	h
Na$^+$	0.97	50.10	1.83	3.3	150	5
Li$^+$	0.60	38.68	2.37	3.7	210	7
Mg^{2+}	0.65	53.05	3.46	4.4	360	12
Ca^{2+}	0.99	59.50	3.09	4.2	130	10
Sr^{2+}	1.13	59.45	3.09	4.2	310	10
Ba^{2+}	1.35	63.63	2.88	4.1	290	9—10
Zn^{2+}	0.74	53.0	3.46	4.4	360	12
La^{3+}	1.15	69.75	3.95	4.6	410	13—14

while the volume of the hydrated ion is:

$$v = \frac{4}{3}\pi\, r^3 \text{ Å}^3.$$

The reliability of the results of these calculations is also limited by the statement by Burton and Daly [16b] that the effect of substituted ammonium ions on the water molecules in their vicinity is somewhat different from that of alkali metal ions.

On the basis of a review on the conductivity of non-aqueous electrolyte solutions, Monica and Senatore [21b] have concluded that the Stokes radii of tetra-alkylammonium ions (except the methyl derivative) calculated on the basis of equation (5.2.10) vary linearly with the crystallographic radii. On the other hand, the product $\lambda^0\eta^0$ depends linearly on the viscosity of the solvent. It has been assumed by them on this basis that the majority of tetra-alkylammonium ions are partly solvated on solution.

The extrapolation from the experiments with solutions of tetra-alkylammonium salts to the properties of alkali metal ions is rather uncertain (cf. e.g. Burton and Daly [16b]). It has been pointed out recently by Schwabe [21c] that the activity coefficient, molar volume, conductivity, and some other properties of acids show a concentration dependence different from those arising due to the effect of alkali metal ions. The deviations have been explained by the fact that the quaternary ammonium salts–unlike alkali metal ions–give rise to cavity formation in the liquid because of their hydrophobic nature, these cavities being surrounded by a dense water shell.

The conclusions drawn from the comparison of ionic radii, deduced empirically from experiments on electrolyte solutions with crystallographic radii, should be handled with caution because of the fact that the sizes of the ions in gaseous phase are larger than the crystallographic and electrochemical ones, as shown by Stokes [22]. This is evidently a result of the self-compression taking place in the condensed state. It has not been clarified whether the deviations of the electrochemical radii (or a part of them) from the crystallographic ones can be attributed to differences in the cohesional conditions between crystalline and solution phases. This is why Friedman's query [23] whether the application of the Born equation in the calculations on hydration are reasonable.

Results very different from the above have been obtained by Rutgers and Hendrikx [24] determining the number of H_2O molecules transported from an aqueous solution into a D_2O one and that of D_2O molecules transported into H_2O from an electrolyte solution prepared with D_2O, across a cellophane membrane under current, taking into account diffusion, as well as the H_2O–D_2O exchange in the hydrate spheres (see Table 5.6).

On the basis of the Stokes–Einstein law in equation (3.1.33), some conclusions regarding the size of the diffusing species can also be obtained from the diffusion coefficients, but only in the case when its size is large

Literature on page 489

in comparison with solvent molecules. For small ions, the Stokes–Einstein Law is unvalid and the hydration of the diffusing ions can be obtained from the diffusion coefficient in this way at most with an approximation like that obtained from ionic mobilities deduced from conductivity. Some other, rather indirect, relationships can also be established between diffusion coefficients and the hydration number of electrolytes, for which see Section 3.2.

Table 5.6

Hydration numbers of some ions on the basis of measurements on the transport of H_2O into D_2O

Ion	h	Ion	h
Li+	22	F⁻	7
Na+	13	Cl⁻	5
K+	7	Br⁻	5
Cs+	6	NO_3^-	6
Mg^{2+}	36	acetate	11
Ca^{2+}	29	SO_4^{2-}	12
Ba^{2+}	28		
Cu^{2-}	34		
Zn^{2+}	44		
Cd^{2+}	39		

Conclusions overriding Stokes' Law. Conclusions rather different from the conventional ones have been obtained by Horne and Birkett [25] on the basis of conductivity measurements in normal and heavy water, as well as in solutions containing the isotopes of some metal ions. According to them, the number of water molecules in the hydrate sphere of the Na+ ion is approximately identical with the average number of water molecules in the Frank–Wen type molecular clusters at the same temperature (cf. Section 1.3.3.8). The cation replaces one water molecule in the cluster. The complicated effect of high pressures on the hydration number has been explained by the formation of a hydrate sphere consisting of three layers.

Although several facts indicate that the larger the ionic size, the smaller the extent of hydration under comparable conditions, the shape of ions also influences hydration. For example, from conductivity measurements on water and water–dioxan mixtures, Accascina, D'Aprano and Triolo [26a] have concluded that ClO_4^- ions are hydrated by more water molecules than ClO_3^- ions, though their size is approximately the same. This can be attributed to the fact that the ClO_4^- ion has a regular tetrahedral shape with the chlorine atom in its centre, and with the oxygen atoms at the vertices. This configuration fits in well with the tetrahedral lattice-like structure of water. According to them, six water molecules form hydrogen bonds with ClO_4^- ions. In this way, a hydrated ion of large volume and approximately spherical shape is formed that associates to a much smaller extent than ClO_3^- ions with a pyramid shape. The tetrahedron of ClO_4^- has four planes with three oxygen atoms on each, while the ClO_3^- ion has only one plane of such type.

Gusev [26b] examined the hydration conditions of the perchlorate ion on the basis of measurements of conductivity, and has drawn the conclusion that in aqueous solutions of $HClO_4$ the hydrogen ion binds no more than 10 water molecules. In studies where metal perchlorates were dissolved in aqueous solution of perchloric acid, the changes caused by dissolved cations were examined, and the following correlation could be established between the concentrations of free water, c'_{H_2O}, the water orientated by the cations, c''_{H_2O}, the perchloric acid concentration, c_{HClO_4}, the metal salt concentration $c_{Me(ClO_4)_x}$, and the number of water molecules, n_{Me}, orientated by one cation:

$$c'_{H_2O} + c''_{H_2O} = 9.95\, c_{HClO_4} + n_{Me}\, c_{Me(ClO_4)_x}$$

In solutions of 0.5 mole/l concentration, the n_{Me} values are 6.5, 5.9 and 5.4 for $Zn(ClO_4)_2$, $Cd(ClO_4)_2$ and $Hg(ClO_4)_2$, respectively. The value of n_{Me} increases at higher temperatures indicating an increased ratio of monomeric water molecules appearing as a result of the destruction of the water structure; this facilitates the orientation by the field of the cation.

The Stokes equation generally used on ionic mobility (equation (5.2.9)) is, in fact, only a first approximation which takes into account only two specific properties: the radius characteristic of the ion and the viscosity characteristic of the solvent. It does not take into account the dielectric properties of the solvent, although ionic migration is hindered by a contributing effect of these factors as compared with the motion of non-electrolyte molecules, since ions orientate the dipole molecules of water in their vicinity to a certain degree in the direction of the electric field when migrating. After an ion has passed by, the water molecules rearrange into the disordered state corresponding to thermal equilibrium. The theory of dielectric frictional effects arising from this has been outlined by Boyd [28] and Zwanzig [29] on the basis of Fuoss' suggestion [27]. The dielectric contribution to viscosity depends on the Debye-type relaxation time constant of the dielectrics τ, on the static dielectric constant ε_s, and on the dielectric constant extrapolated to high frequencies ε_∞. The ionic mobility is, according to this theory:

$$\lambda^0 = \frac{|z|\, F^2}{6\,\pi\eta\left(r + \dfrac{C}{r^3}\right)}, \tag{5.2.11}$$

where

$$C = \frac{2}{3}\,\frac{\tau}{6\,\pi\eta^0}\cdot\frac{z^2 e^2}{\varepsilon_s}\cdot\frac{\varepsilon_s - \varepsilon_\infty}{\varepsilon_s}. \tag{5.2.12}$$

In aqueous solutions at 25 °C, C is about 10 Å⁴. As pointed out by Frank [30], these relationships lead to conclusions that appreciably limit the

Literature on page 490

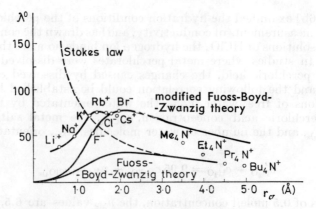

Fig. 5.12 The limiting value of equivalent conductivity of some ions *vs.* the Pauling crystallographic radii

reliability of conclusions on hydration obtained from ionic mobilities, since equation (5.2.11) can be transformed as follows:

$$\lambda^0 = \frac{Ar^3}{C + r^4},\tag{5.2.13}$$

where, in aqueous solutions at 25 °C, $A = 91 \cdot 7\,|\,z\,|$, if r is measured in ångstrøms. This relation differs markedly from equation (5.2.10) following from the simple Stokes' Law, and the difference diminishes only in the limiting case of large values of r. According to equation (5.2.13), λ^0 has a maximum as a function of the crystallographic radius at:

$$\lambda_m^0 = A\,\frac{3^{3/4}}{4}\,C^{-1/4}$$

in conductivity units which corresponds to a radius of $r_m = (3C)^{1/4}$ Å. Fig. 5.12 shows the values of λ^0 *vs.* the crystallographic ionic radii for some monovalent ions. It can be seen that the maximum in the conductivity of alkali metal ions appears at about $1 \cdot 7$ Å, while in the case of halide ions it is at $2 \cdot 1$ Å; however, the position of the maximum differs from the value of $r_m = 2 \cdot 34$ Å predicted theoretically. The deviation from the theoretical value is even larger for the maximum conductivity: instead of the predicted value of $\lambda_m^0 = 29 \cdot 4$, the maximum is at about 74 in conductivity units, and no reasonable value of r can be assumed to which a value of $\lambda_m^0 = 75$ would correspond. (Such a problem does not appear in connection with the simple Stokes' Law, because however high the observed ionic conductivity, it can be explained by assuming a small effective radius that still cannot be considered to be unreal.) From this, Frank [30] concluded that the model for ionic migration should be investigated further. On discussing the various possibilities, the Fuoss–Boyd–Zwanzig theory should be modified with respect to viscosity. In ionic migration, the viscosity of the solvent predominating in regions farther from the ions is not the determining factor (yielding most of the viscosity measured macroscopically), but the microviscosity of the immediate neighbourhood of the ion which is appreciably smaller. The electric field of ions destroys the structure of water, and, thus, it increases the mobility of the molecules. This structure-breaking effect–as one of the factors accompanying friction in ionic migration–also appears in the translational movement of strongly hydrated ions with a structure-making effect, since where the hydrate sphere, forming a kinetic entity with the ion, meets the region of water of unaltered structure the fitting becomes distorted, resulting in the destruction of the water structure at these places. Thus, with respect to migration, structure-making ions also have a structure-breaking effect. The curve calculated according to the relationship modified with respect to this fits well with the experimental points (see Fig. 5.12).

The mechanism of the change in viscosity had also been discussed by Frank [30]. It is hardly doubted that ions migrate *via* jumps between neighbouring equilibrium positions, however, it is not clear whether the displacements take place through a small number of jumps covering large distances, as has been suggested by Eyring [31], or by many small jumps. According to Frank, the relationship implying $1/r^3$ indicates that the dielectric effect on ionic mobility arises almost completely from water molecules adjacent to the ion, and it is connected with the cooperative rotation of the neighbouring molecules about the migrating ion. It is probable that some jumps are mainly translational and the ion takes part in them together with its immediate neighbours, while others are mainly cooperative movements in which the ion is displaced between neighbouring molecules helped by the rotation of the latter. If the former process predominates, the effective kinetic entity of migration is large and the resistance to its motion has a rather 'viscous' nature, and the value of r corresponding to λ^0 belongs to the right-hand branch of the curve in Fig. 5.12. The conditions are probably such for the Li^+ ion, being strongly hydrated, i.e. having a structure-making effect, but the distortion between the hydrate sphere and the 'unaltered' water results in disorder. If, however, the cooperative process predominates, the kinetic unit displaced is the bare ion, the resistance to its movement is rather of the electrostatic nature and arises from the fact that the ion should 'wait' for the rotation of the neighbouring water molecules to make the jump able to proceed with a low activation energy. Under such conditions, the correlation between λ^0 and r corresponds to the left-hand branch of the curve in Fig. 5.12. F^- and the other halide ions probably migrate by such a mechanism. The difference in the migration mechanism of alkali metal ions and halide ions is also indicated by the fact that the heat of transfer accompanying migration approximately is higher, the smaller the ionic radii of alkali ions, while the heat of transfer of halide ions decreases with the increase in the radius, as has been shown

Literature on page 490

30*

by Agar and Turner [32a]. This also confirms that the migration mechanism of the two groups of ions is different.

The recent investigations discussed above–even if no definitive interpretation could be obtained–make one cautious about conclusions on hydration drawn from ionic mobilities. It is likely that the mobility of ions depends on several factors, more numerous than believed earlier, and several experimental investigations are required to clarify the mechanism of ionic migration; that is, in fact, a precondition of obtaining reliable conclusions on hydration from mobilities. The introduction of the dielectric relaxation retardation makes it questionable whether the radii of tetraalkylammonium ions can be calculated from the simple Stokes' Law.

Investigating the concentration dependence of the dielectric constant in solutions of $0 \cdot 2$–$1 \cdot 0$ mole l^{-1} concentration, Samoilov and Yastremskii [32b] have concluded that the hydration of Na^+ ions is lower in D_2O than in H_2O, and the nearby hydration decreases in a higher proportion than the distant one. On the other hand, the hydration of ions increases more rapidly with the increase in temperature in D_2O than in H_2O. This can be attributed to the fact that the liquid structure of D_2O is stronger than that of H_2O. The increase in temperature alters the structure of D_2O to a higher extent than that of H_2O. On the basis of magnetochemical studies, Ergin and Kostrova [32c] have also concluded that the nearby hydration of alkali metal ions is smaller in D_2O than in H_2O under the same conditions. According to their investigations, the ionic radius at which positive hydration transforms into negative hydration is lower in D_2O ($0 \cdot 95$ Å at 25 °C) than in H_2O ($1 \cdot 1$ Å).

5.2.4.2 CORRELATION BETWEEN TRANSFERENCE NUMBER AND HYDRATION

Transference number measurement–carried out by the Hittorf procedure modified by Buchböck [33] and Washburn [34]–has been regarded for a long time as the most important method for determining the relative hydration numbers of ions. The essential feature of this method is the measurement of transference numbers in electrolyte solutions in which a non-electrolyte (e.g. sugar) is dissolved in an initially uniform concentration. It is an empirical fact that the non-electrolyte concentration is changed in the solution around the electrodes in the course of electrolysis. Provided that the non-electrolyte is neutral with respect to the experiment, i.e. it does not migrate directly under the effect of the electric field strength or bind to ions, ionic hydration can be obtained from the change in concentration brought about by electrolysis. Namely, if ions are hydrated, they also transport water in the course of their migration under the effect of the electric field, altering the non-electrolyte concentration. For example, cations migrating into the area of the cathode bring water there, while the anions leaving it take away some water depending on their transference number and hydration number. The net change in the amount of water can be determined from the change in the non-electrolyte concentration in the cathode area. From this data, and from the transference number, the difference in the hydration number of cations and anions can be simply calculated in the case of a binary electrolyte. Taking the more or less arbitrarily assumed hydration number of an ion as a reference, the hydration numbers of other ions, referred to this, can be calculated on the basis of such measurements. The hydration numbers determined by this apparently direct method had been accepted until it was discovered that the seemingly reasonable fundamental assumption in the calculation is not reliable.

In order to determine the hydration number of the cation and anion separately from the difference in hydration determined by the measurement of the transference numbers of the cation and anion without arbitrary assumptions, Erdey-Grúz, Hunyár, Pogány and Váli [35] have studied the diffusion of electrolytes in aqueous solutions containing a non-electrolyte in initially uniform concentration, since, if ions transport no non-electrolyte molecules but only water in their hydrate sphere, the sum of the hydration numbers of the cation and anion could be calculated from the change in the non-electrolyte concentration developed in the course of diffusion. From this and from the difference of the hydration numbers obtained from transference experiments, the transference number of the cation and anion could be calculated separately with no arbitrary assumption in this respect. Erdey-Grúz *et al.* found, however, that the assumption that ions do not transport non-electrolyte molecules but only water is not acceptable. Measuring the change in concentration caused by the diffusion of electrolytes in various non-electrolyte solutions, it has been discovered that the ions do transport non-electrolyte molecules in a significant amount (for details see Section 3.3.3). Recently, it has also been observed by Longsworth [36] and Hale and De Vries [37a] that ions also transport non-electrolytes. On the basis of our current knowledge on the structure of solutions this is not surprising, because non-electrolytes, soluble in water, contain polar groups or atoms, and are dipoles like water molecules; further they can be suitable for hydrogen bonding. In this way they can be bound to ions in a similar way to water molecules.

Khoo (37b] measured the transference number of hydrochloric acid in a mixture of water and glycerol, and established that the amount of water transported toward the cathode by 1 Faraday (related to glycerol) increases with increasing glycerol concentration up to 30% glycerol content, then it decreases. In his opinion, this should be attributed not to the dependence of the transference number of ions on the composition of the solvent, but to the alteration of the structure of the solvent, and to the fact that the ratio of water and glycerol molecules is changed in the solvate spheres of the ions; the molecules of the non-electrolyte interact with both the ions and the water molecules.

Since it has been ascertained that ions, dissolved in water, also transport non-electrolyte molecules, no reliable conclusions regarding the difference of hydration numbers can be obtained from transference number measurements.

Literature on page 490

5.2.4.3 EFFECT OF HYDRATION ON THE CONDUCTIVITIES AND ACTIVITY
COEFFICIENTS OF STRONG ELECTROLYTES, AND ON THE SOLUBILITIES
OF NON-ELECTROLYTES

The Debye–Hückel theory on the conductivity of strong electrolyte solutions (cf. Section 4.2.3) and its second approximation on the activity coefficient (cf. Section 5.1.3) treats the ions as spheres of diameter a (more accurately, as spheres that can approach each other to a minimum distance a
between their centres). For determining ionic diameters, however, it yields
no sound method in principle since the value of a is regarded as a dispos-
able parameter that should be determined empirically so as to make the
theoretically deduced concentration dependence of conductivity and the
activity coefficient approach the relationship determined experimentally
as well as possible. The ionic diameters suitable for this (3·5–6 Å) are appreciably larger than those measured in ionic crystals, which indicates that in
solutions the kinetic entities are hydrated ions. The minimum distance
to which two ions can approach each other is not the sum of the radii of the
two ions, but is much larger, indicating that ions are separated from each
other by a hydrate sphere even when in 'contact' with each other under
normal conditions.

Although a fairly reliable conclusion can be drawn from the Debye–
Hückel theory stating that the size of ions moving as kinetic entities in
electrolyte solutions is larger than that of bare ones, no quantitative results
can be obtained for the real diameter of the hydrated ion, and, thus, for its
volume, since in the theory the differences in the conductivity or activity
coefficient calculated in the first approximation and those determined
experimentally are taken into account by fitting the diameter a to the
empirical correlation, i.e. calculating as though all the deviations arose
from primary hydration. Although primary hydration is evidently an
important factor in these differences, the effects of ions on the farther regions
of the solution cannot be neglected. These, however, are not known sufficiently well to be taken into account quantitatively. Thus, the 'diameter' a
of the theory is, in fact, a quantity introduced conventionally and representing many types of interactions.

Some conclusions regarding hydration have been drawn by Gusev [37c]
from the concentration dependence of the specific conductivity ($\varkappa$) of aqueous acid solutions. It has been established that the maximum in specific
conductivity corresponds to a solution in which the concentration of the
hydrogen ion is one tenth that of water, $[H_2O] : [H^+] \approx 10$. Under the
effect of salts and non-electrolytes the maximum of the curve $\varkappa$ vs. $[HA]$
decreases and is shifted toward lower acid concentrations.

To explain these phenomena, Gusev assumed that the water molecules
bound in the hydrate spheres of the ions do not participate in prototropic
conduction, which only takes place via 'free' water molecules forming the
essentially unaltered water structure. On this basis, he concluded from the
position of the maximum that H_3O^+ ions hinder the participation of four
other water molecules in proton exchange, i.e. the hydration number of
H_3O^+ is four. The position of the maximum in specific conductivity is independent of the anions, according to the experiments, which leads to the

conclusion that anions do not hydrate, i.e. water molecules do not stay in the vicinity of the anion for a time long enough to become hindered in their participation in proton transfers. From the changes in the conductivity maximum under the effect of salts, Gusev has concluded that the hydration number of monovalent metal ions is three or four, on the average, while that of the most of the bivalent ones is about eight.

Ionic hydration is also indicated by the fact that the solubility of non-electrolytes in water is markedly decreased by electrolytes. This phenomenon, known as 'salting out', arises essentially from the decrease in the activity of water due to the presence of the electrolyte. Qualitatively, the phenomenon of salting out also proves unequivocally the hydration of ions, however, quantitative calculation of the hydration numbers can be carried out only by introducing some simplifying assumptions. If one assumes that the decrease in the activity of water under the effect of dissolved electrolytes is only due to the binding of a certain number of molecules by ions in primary hydration, while the water outside the primary hydrate spheres remains unaltered (i.e. secondary hydration is disregarded), further, if dissolved ions are supposed to have no influence on the non-electrolyte present in the same solution (which is very doubtful on the basis of the facts discussed in the previous section), the sum of the hydration numbers of the two ions in the electrolyte can be calculated from the decrease of the solubility. Thus, for example, from the fact that 55·5 moles of pure water dissolve 3 moles of phenol, but the 55·5 moles of water present in an NaCl solution of 1 mole l^{-1} concentration dissolve only 2 moles of phenol, it can be concluded that 37 moles of free water are present in this NaCl solution and 18·5 moles of water do not act as a solvent, but are bound in the hydrate spheres of Na^+ and Cl^- ions. The sum of the hydration numbers of these two ions is thus eighteen. The hydration numbers calculated in this way decrease with the salt concentration (e.g. in NaCl solutions of 3 mole l^{-1} concentration, there are only 8 moles of water per mole of salt). However, calculating the hydration number from the decrease of the solubility of different non-electrolytes by the same electrolyte, can produce very different results–apparently under the same conditions–indicating that the preconditions of the calculation are not acceptable.

The change in the distribution quotient of a non-electrolyte in water and another immiscible solvent under the effect of an electrolyte dissolved in water is a phenomenon similar to the decrease in solubility. This method has been used widely, for example, by Sugden [38a] who investigated the change in the distribution of acetic acid between water and amyl alcohol. Experiments with various non-electrolytes did not give the same result with this method either.

To a certain degree, the effect of electrolytes on the surface tension and adsorption of non-electrolyte solutions can be interpreted similarly. Surface tension decreases, in general, under the effect of the electrolyte in non-electrolyte solutions and the adsorbed amount increases under otherwise identical conditions. The decrease in surface tension is caused by the in-

Literature on page 491

crease in the amount of the non-electrolyte adsorbed on the free surface of the solution.

All these phenomena connected with the decrease in the activity of water due to the presence of electrolytes reliably confirm the existence of hydration qualitatively, but they lead to very different results when performing the same type of experiments using different substances. For example, in a solution of 1 mole l^{-1} concentration, the hydration number of LiCl is found to be 8–19, that of NaCl is 8–23, while that of KCl is 3–17 indicating that, in addition to the primary hydration of ions, secondary hydration and solvation by the non-electrolyte present in the solution also play an important role.

Recently, several researchers have investigated the effect of non-electrolytes dissolved in water on the hydration and solvation of ions. The results of measurements on the free enthalpy of the transfer of alkali metal chlorides from water into a methanol–water mixture have been interpreted by Feakins and Voice [38b] by means of interactions similar to those playing a part in the 'hardness' and 'softness' of acids and bases. According to them, during the transition from water into methanol–water mixtures, the solvate sphere of the Li^+ ion becomes more strongly stabilized than that of the Na^+ ion, this being the consequence of its stronger electrostatic effect due to its smaller radius. On the other hand, the stabilization of the Rb^+ ion is lower than that of the Cs^+ ion having a larger radius, which may be attributed to the non-electrostatic effect exerted on the solvent. Feakins, Willmott and Willmott [38c] also studied the solvation conditions of bivalent ions in methanol–water mixtures. They have established that the coordination number of bivalent ions is higher than that of monovalent ions with identical radii. The Ba^{2+} ion is better stabilized in mixtures of methanol and water than the K^+ ion, even if calculated for unit charge. In the interaction of Sr^{2+} and Ba^{2+} ions with the coordinated water molecules, the electrostatic forces are predominating. The solvation phenomena of LiBr in water–acetone mixtures were studied by Nilsson and Beronius [38d], and the hydration of $HClO_4$ in water–dioxan mixtures was investigated by Gusev [38e].

5.2.4.4 EFFECT OF HYDRATION ON THE COMPRESSIBILITY AND ULTRASONIC ABSORPTION OF SOLUTIONS

The compressibility of water decreases on dissolving an electrolyte in it. This phenomenon can be explained by the strongly compressed state of the water molecules in the primary hydrate spheres of ions due to the strong local electric field (this is the electrostriction known also with respect to some other phenomena). Thus, external pressure compresses only 'free' water not participating in hydrate spheres. This phenomenon can be utilised in determining the hydration numbers, as shown in the investigations of Passynski [39], Barnartt [40], Litvinenko [41] and others. Ultrasonic absorption measurements make a higher accuracy possible in the determinations, since ultrasound brings about a high-frequency alternating increase in pressure, during which the absorption of energy also depends on the compressibility of the medium.

In order to obtain the relationship between compressibility and the hydration number of the solution, let us start with assuming that the volume V of the solution contains an overall number of n_w moles of water, and n_h moles of this are involved in the hydrate spheres of ions with no contribution to compressibility. If the molar volume of pure water is V_w^0, one has:

$$\left(\frac{\partial V}{\partial P}\right)_T = \frac{\partial}{\partial P}\left\{V_w^0(n_w - n_h)\right\}. \tag{5.2.14}$$

The measured compressibility becomes:

$$\beta = -\frac{n_w - n_h}{V}\frac{\partial V_w^0}{\partial P} =$$
$$= (n_w - n_h)\frac{V_w^0}{V}\beta_w^0, \tag{5.2.15}$$

where β_w^0 is the compressibility of pure water. Further, if n_e is the number of moles of the electrolyte dissolved in the solution, the hydration number is:

$$h = \frac{n_h}{n_e} = \frac{n_w}{n_e}\left\{1 - \frac{\beta V}{\beta_w^0 n_w V_w^0}\right\}; \tag{5.2.16}$$

while in dilute solution:

$$h = \frac{n_w}{n_e}\left\{1 - \frac{\beta}{\beta_w^0}\right\}. \tag{5.2.17}$$

Table 5.7 shows the hydration numbers of some electrolytes deduced from the compressibility data, as summarized by Robinson and Stokes,

Table 5.7

Hydration numbers of some electrolytes calculated from compressibility, activity and diffusion

Electrolyte	Hydration number calculated from		
	compressibility	activity	diffusion
LiBr	5—6	7·6	5·6
LiCl	6	7·1	6·3
NaI	6—7	5·5	3·0
NaBr	6—7	4·2	2·8
NaCl	7	3·5	3·5
KI	6—7	2·5	0·3
KBr	6—7	2·1	0·3
KCl	7	1·9	0·6
KF, NaF	8—9	—	—
$Sr(ClO_4)_2$	13	15	—
$MgCl_2$	16—17	13·7	—
$BaCl_2$	16—17	7·7	—
$LaCl_3$	—	18	—
$AlCl_3$	31—32		

Literature on page 491

together with the data calculated from activity and diffusion coefficient measurements for comparison. It is seen that these methods yield results in rough agreement with each other for only few electrolytes. The high hydration number obtained for KCl is particularly striking.

The weak point in calculating the hydration numbers from ultrasonic measurements is the assumption that the ion is incompressible together with the hydrate sphere around it. According to Dubinina and Kudryav-tsev [42] this assumption is not generally correct, but the compressibility of the hydrate sphere can be estimated by investigating the acoustic proper-ties of saturated solutions at equilibrium with the solid salt. The appli-cability of ultrasound measurements in calculating hydration numbers has been analysed in detail by Allam and Lee [43]. One should be cautious when interpreting the results, since, in addition to the simple interactions between ions and water molecules, some other interactions function, too. Therefore, the conclusions of Desnoyers, Verrall and Conway [44], drawn from ultra-sonic interferometric measurements, stating that hydration number de-creases with increasing temperature and concentration, should be applied cautiously though these results are supported by the NMR spectroscopic studies of Hindman [45a].

Up-to-date techniques of measuring ultrasonic velocity have extended the possibility of calculations of the compressibility of solutions to 5 MHz. On the basis of our recent knowledge on compressibility and the vibration potential of ions, Bockris and Saluja [45b] have analyzed in detail the solvation con-ditions of ions and the modern interpretation of the solvation and hydra-tion numbers. Water molecules are considered solvated, if they have had enough time to orientate out of the bulk water and join the ion in the stationary diffusional motion with maximum interaction. The attachment of such a water molecule to the ion is accompanied by a potential energy change of 20 kcal/mole. According to this model, too, the effect of the ion extends beyond the primary solvate sphere. There is a farther (secondary) sphere, where the 'non-solvated' water molecules are also coordinated to the ion to a certain extent. However, the molecules located here have not enough time to trans-orientate from the bulk to a position near the ion in such a degree as to produce an interaction sufficient for combining with the ion to form an entity. Thus, these water molecules will be left behind when the ion moves further; they do not belong to the solvate sphere and are not involved in the solvation or hydration number, either. It is pointed out that the solvation number of the ion defined in this way (SN) is not identical with, but lower than, the coordination number (CN). In the first solvent layer around the ion there are two kinds of solvent molecules: the coordination number is the number of all the molecules being in contact with the ion, while the solvation (hydration) number represents the molecules which remain associated with the ion also during its motion. In the case of small ions, SN and CN do not differ greatly, but for large ions SN tends to zero.

Bockris and Saluja's model also includes a time factor in the definition of the solvation number. This number will depend partly on the average period of time for which the ion stays at a given point during diffusion $(\tau_{ion, wait})$, and partly on the time required by the dipole molecules of the solvent to trans-orientate from the water structure to the ion ($\tau_{\text{water, orient}}$). When $\dfrac{\tau_{\text{water, orient}}}{\tau_{\text{ion, wait}}} \to \infty$, then $\dfrac{\text{SN}}{\text{CN}} \to 0$, when however, $\dfrac{\tau_{\text{ion, wait}}}{\tau_{\text{water, orient}}} \to \infty$, then $\dfrac{\text{SN}}{\text{CN}} \to 1$. The hydration numbers of some ions calculated according to Bockris and Saluja are listed in Table 5.2; $\varDelta$, and the ratio of the hydration and coordination numbers are shown in Table 5.8.

In his comments on the theory proposed by Bockris and Saluja, Desmoyers [45c] pointed out that in determining the solvation number, the procedure based on the difference between the compressibilities of the solution and the solvent is more reliable than that based on compressibility. He emphasized that in respect of the compressibility and heat capacity, the destruction or strengthening of the liquid structure by ions, that is, 'structural solvation' is the predominating factor. According to him, the interaction of non-solvated coordinative water molecules with the ion is not equivalent to the interaction altering the liquid structure, since the former takes place mainly after the alteration of the structure. On the other hand, structural solvation corresponds to the transition of the water molecules from the commonly bound state into the structural state altered under the effect of ions.

Table 5.8

Ratio of the hydration and coordination numbers of some ions, and the waiting and reorientation periods according to Bockris and Saluja

Ion	$\dfrac{\text{SN}}{\text{CN}}$	$\tau_{\text{ion, wait}}$	$\tau_{\text{water, orient}}$
Li^+	0.67	$4.38 \cdot 10^{-11}$ s	$1.48 \cdot 10^{-11}$ s
Na^+	0.67	$3.37 \cdot 10^{-11}$ s	$2.40 \cdot 10^{-11}$ s
K^+	0.57	$2.87 \cdot 10^{-11}$ s	$3.78 \cdot 10^{-11}$ s
Rb^+	0.47	$2.70 \cdot 10^{-11}$ s	$4.43 \cdot 10^{-11}$ s
Cs^+	0.30	$2.70 \cdot 10^{-11}$ s	$5.48 \cdot 10^{-11}$ s

Measuring the ultrasonic absorption over a broad temperature range and using a model representing the dynamical structure of associated liquids, Breitschwerdt and Wolz [45d] have drawn conclusions concerning the solvation numbers, as well as the thermodynamical parameters and relaxation times of electrolyte solutions.

5.2.4.5 OPTICAL CONSEQUENCES OF IONIC HYDRATION

The interactions between ions and solvent molecules resulting in hydration alter the energy levels of electrons which appears in the absorption spectra as well. A rather large number of studies have dealt with the absorption spectra of electrolyte solutions. The majority of these investigations, however, have been carried out to reveal the chemical equilibria in which ions participate, and not to determine the hydration number. Although the fact of hydration can be concluded from the shift of absorption maxima and from the broadening of the absorption bands in the ultra-violet and

Literature on page 491

visible region, these methods are unsuitable for determining the hydration number unequivocally. The absorption studies are usually connected with the hydration of transition metals, since the absorption bands of solutions of simple ions are in the far ultra-violet region where the absorption of the solvent hinders the measurements.

The interaction of ions with water has also been investigated by Raman spectroscopy. According to Walrafen [46], the intensity of certain Raman lines increases when dissolving KBr and NaBr, similar to the effect of the decrease in temperature, which indicates an ordering effect of the field of ions on water molecules. On the other hand, the intensity of other Raman lines–the intensity of which also decreases with increasing temperature–is reduced under the effect of electrolytes, which is probably related to the breaking or loosening of the hydrogen bonds by the field of ions. The relationship between the intensity of the Raman lines of various metal nitrates and the hydration of the cation has been studied by Vollmar [47a], who has established that the smallest change in the spectrum is brought about by NH_4^+ ions, in accordance with some other conclusions, indicating that this ion fits into the structure of water without any appreciable deformation effect.

On the basis of laser–Raman spectroscopic investigations, Janz, Balasubrahmanyam and Oliver [47b] have concluded that four to six 'strongly bound' water molecules are bound to the ions in the most dilute aqueous solutions of $Ca(NO_3)_2$, $Cd(NO_3)_2$, and $Zn(NO_3)_2$. This number decreases to two at a concentration of about 0.5 mole l^{-1} and to one at about 5 mole l^{-1}. This phenomenon has been explained by the fact that part of the strongly bound water molecules are removed from the hydrate spheres because of ion pair formation in more concentrated solutions, i.e. in this case, ions forming the pairs are in immediate contact with each other and no solvent molecules separate them.

Studying infra-red spectra, Ivanova and Zolotarev [48a] have drawn some conclusions about the state of water molecules in the first and second hydration spheres of cations.

By analysing the deviations of hydration numbers calculated from infrared absorption and colligative properties, Bonner and Woolsey [49] have concluded that spectroscopic data reflect only the interactions between ions and solvent molecules, while the hydration numbers deduced from colligative properties also involve the mutual interaction of ions, in addition to this. Their study again confirms that Li^+ ions strongly polarize the water molecules and attract their oxygen atoms, while the polarization effect of Cs^+ ions is smaller –evidently due to their more shielded nuclear charge. Several arguments support that Cs^+ ions have a structure-making effect on water, similar to that of dissolved noble gas atoms and tetra-alkyl ammonium ions. It can be ascribed to the difference in the strength of hydration that ion pair formation takes place to a lower extent in solutions of Li^+ ions (the water molecules around the ion hinder this) than in those of Cs^+ ions in which no polarized water molecules protect the cations from ion pair formation. The orientation effect of anions on water molecules is much smaller than that of Li^+ ions, and relatively the strongest polarization is caused by F^- and OH^- ions. Cl^- ions form ion pairs to a smaller degree than I^- ions, due to their stronger hydration. This is in agreement with the fact that the solvation determined by means of light absorption decreases with increasing concentration in LiCl solutions, because the concentration of free water molecules is reduced. On the other hand, the hydration of CsCl increases with increasing concentration, since ion pairs are formed due to association; these are larger than the separate ions and they exert a pronounced structure-making effect around them.

Hydration can also be concluded from the refractive index of electrolyte solutions. The method of Fajans and Joos [50] elaborated for this purpose utilizes the fact that the molar refraction of ideal solutions consists additively of the molar refraction of the solvent and the solute. The deviations from additivity give information on the interaction of the solvent with the solute. By an adequate treatment, the ionic refraction of gas ions can be calculated, and the molar refraction of solutions containing these ions can be measured. According to the empirical results, this latter case differs from the values calculated on the basis of the additivity of the molar refraction of water and the ionic refraction in dilute solution. The difference arises partly from the interaction between cations and water molecules which decreases refraction, i.e. it makes the electron shell more rigid by means of deformation, and partly from the interaction of the anions with water molecules which increases refraction, i.e. loosens the electron shell, as pointed out by Fajans and Joos. In more concentrated solutions, a contribution to this is given by the interaction of cations with anions, which also alters refraction due to the deformation of electron shells. This latter effect increases with increasing concentration.

From changes in the infra-red spectrum of water under the effect of dissolved electrolytes, it can be concluded that cations and anions bind water molecules in different orientations. The bending vibration:

$$O \Big\langle \begin{matrix} H \\ \updownarrow \\ H \end{matrix}$$

of the water molecule is altered by cations, but the stretching vibration of the $O-H$ bond remains unaltered. On the other hand, anions influence the vibration frequency of the $O-H$ bond, but they have no effect on the bending vibration. This also confirms the conclusions drawn from other phenomena that cations form their inner hydrate spheres with the oxygen atom of the water molecule adjacent to them, whereas the water molecules are bound to anions by their hydrogen atoms, forming hydrogen bonds:

$$Me^+ \cdots O \Big\langle \begin{matrix} H \\ H \end{matrix} \qquad X^- \cdots H-O \Big\rangle_H .$$

Literature on page 491

From the change of the vibration frequency, some information can be gained on the strength of the bond, which is also influenced by the shape of the ion in the case of complex ions [51a].

From investigations of the infrared spectra of aqueous solutions of strong acids, Grankina, Chuchalin, Peshevitskii, Kuzin and Khranenko [51b] have concluded that the H_3O^+ ion is hydrated by three molecules of water, and the water molecules in the complex $(H_3O . 3H_2O)^+$ can be partly or completely replaced by di-n-butyl phosphate molecules or perchlorate ions.

On the basis of studies of the near-infrared spectra of Na, Mg, Zn and Al perchlorate and Na tetrafluoroborate solutions, Subramanian and Fischer [51c] have established that all these salts exert a particular destroying effect on the structure of water; in this effect the anions play the predominant role, and their action is similar to that of increasing temperature. In the opinion of these authors, the perchlorate and fluoroborate ions are not in hydrated state. After investigating the infrated spectra of numerous salts, Andreev and Smirnova [51d] have assumed that bi- and trivalent metal ions are attached to the OH groups of water molecules. The strength of this bond depends mainly on the charge of the cation and the chemical nature of the anions in the outer hydration sphere.

Conclusions regarding the hydration number of alcohols have been drawn by Spink and Wyckoff [51e] on the basis of the infrared spectra of dilute aqueous solutions. Accordingly, when extrapolated to infinite dilution, the apparent hydration number of methanol, ethanol, 2-propanol and t-butanol are 1.0, 1.3, 1.5 and 4.0, respectively. Thus, the larger the non-polar group, the stronger the hydration. With increasing concentration of the alcohol the apparent hydration number decreases, except that of methanol, which remains unchanged. From the results the authors concluded that the OH group binds 1 molecule of water in alcohols, and its association does not change with the concentration. The higher apparent hydration number of the other alcohols can be attributed to the strengthening of the water structure around the alkyl group, and this appears as hydration ('structural hydration'). When the concentration is increased, the liquid structure strengthened around the alkyl groups becomes loosened.

Some NMR spectroscopic studies on electrolyte solutions have also contributed to the data from which the hydration of ions can be concluded. For example, Knapp, Waite and Malinovski [52a] have observed that the resonance shift of water protons is altered by dissolved electrolytes, which can be attributed to the fact that the protons of water molecules in the hydrate spheres have an environment different from that in pure water. According to their calculations, the total hydration numbers of $NaClO_4$, HCl and $HClO_4$ are 3·4, 3·0 and 2·6, respectively.

On the basis of proton shifts, Vogrin, Knapp, Flint, Anton, Highberger and Malinovski [52c] found the hydration number of alkali metal ions to be about three in solutions more dilute than 5 mole l^{-1}. In more concentrated solutions the formation of ion pairs becomes more significant. According to Akitt [52d], the hydration number of the Al^{3+} ion is about six, as indicated by the chemical shift of the protons of water.

5.2.4.6 THERMODYNAMIC ASPECTS OF HYDRATION

Important factors in the hydration of electrolyte solutions are the changes in the energy accompanying the formation of the hydrate sphere, which can be obtained in some thermodynamic procedures [52b]. The steps of the processes of dissolution of an electrolyte forming ionic crystals in the solid state can be followed mentally when decomposing the crystal, first, into an ionic gas in which ions are at an average distance from each other as in a solution of concentration c (the change in the enthalpy accompanying this process is $\Delta H_{r,c}$), then—with no change in the mutual average separation the gas ions will be transported into water, which is accompanied by a change in enthalpy, $\Delta H_{h,c}^{el}$, per mole of the electrolyte corresponding to the hydration of the cations and anions, and, perhaps, to some changes in their state occurring in solution (e.g. association). The sum of these two changes in the enthalpy is the integral molar heat of dissolution ΔH_c corresponding to concentration c:

$$\Delta H_c = \Delta H_{r,c} + \Delta H_{h,c}^{el}. \tag{5.2.18}$$

In this equation, $\Delta H_{h,c}^{el}$ is, in fact, a sum of the changes in the enthalpy of complicated processes, since, in general, when gas ions become surrounded by a liquid, in addition to solvation, their mutual interaction also changes —the dielectric constant of the liquid being higher than that of the gas— non-dissociated molecules or ion pairs can be formed, these can interact again with ions and solvent molecules, etc.

The conditions become somewhat simpler when the change in the enthalpy of the formation of infinitely dilute aqueous solutions is concerned. Under such conditions, ΔH_r, the energy required to decompose the crystal, is the lattice energy, while the energy of hydration, ΔH_h^{el}, corresponds only to the interaction of cations and anions with the solvent. Of course, this interaction is still a complicated process as it implies, in addition to the binding of water molecules to ions, the energy effects of the changes caused by ions in the structure of water, too; however, these can be included in the concept of hydration. Thus, the integral molar heat of dissolution for the formation of an infinitely dilute solution is:

$$\Delta H_0 = \Delta H_r + \Delta H_h^{el}. \tag{5.2.19}$$

The decomposition of the crystal lattice into gas ions is a highly endo-thermal process (e.g. for NaCl it is $\Delta H = 164$ kcal mole^{-1}), and the lattice energy corresponding to this can be calculated by means of thermodynamic cycles from quantities measurable experimentally or from general theoretical conclusions [53]. Owing to the uncertainties in the assumptions on which the calculations are based, however, the values calculated by different methods deviate from each other by a few per cent. The integral heat of dissolution of electrolytes that can be determined experimentally in a

Literature on page 491

Table 5.9

The enthalpy of hydration of some electrolytes at 25 °C

Electrolyte	Enthalpy of hydration/kcal mole⁻¹
LiF	−244
LiCl	−211
LiBr	−202
LiI	−163
NaF	−216
NaCl	−184
NaBr	−177
NaI	−167
KF	−198
KCl	−165
KBr	−158
KI	−149
RbF	−191
RbCl	−158
RbBr	−151
RbI	−142
CsF	−184
CsCl	−151
CsBr	−145
CsI	−136

direct and accurate way (with an error of about $0 \cdot 1$ per cent) is $\Delta H = -40$ to $+15$ kcal mole⁻¹ according to experiments. Consequently, the hydration of electrolytes is a highly exothermal process, and the heat of hydration is in the same order of magnitude as lattice energy but with opposite sign.

According to equation (5.2.19), the overall enthalpy of hydration of an electrolyte can be calculated reliably within the limits of error given above (see Table 5.9). However, the enthalpy of hydration of cations and anions cannot be separately calculated unequivocally since, according to the universal law known as the electroneutrality condition in all the real macroscopic processes, electrically equivalent positive and negative ions take part, thus the experimental data always refer to the whole electrolyte and not separately to cations or anions. The enthalpy of hydration of the electrolyte can be divided into the enthalpies of hydration of the cation and the anion only by more or less questionable assumptions.

Extended studies have dealt with the theoretical calculation of the energy changes accompanying the dissolution of gas ions (i.e. hydration). The investigations of Lange and Mishchenko [54] revealed that the potential difference at the free surface of liquids ('surface potential' χ) gives rise to theoretical difficulties in the deduction regarding separated cations and anions; the magnitude of this quantity is not known reliably. Nevertheless, in the course of the transition of identically charged ions from the gas phase into the liquid phase, some work also needs be done against the surface potential, which contributes to the effect of ions on water molecules in the bulk of the liquid phase. The change in the energy corresponding to this is usually called the *enthalpy of chemical hydration*, the sum of this and the energy of the transition through the surface potential is the *real hydration enthalpy* of gas ions.

With respect to the surface potential of liquids, at present one should be satisfied with some estimates. It can be assumed that dipole molecules are orientated at the surface of water in such a way that protons point to the bulk of the liquid. According to the estimates of Mishchenko and Kvyat [55a], $\chi = -0 \cdot 3$ to $+0 \cdot 1$ V and the energy barrier at the surface of water is about 6–7 kcal mole⁻¹.

Owing to the uncertainties in our knowledge of the surface potential, the calculations for identically charged ions are also uncertain. In the case of electrolytes containing electrically equivalent ions, this uncertainty is eliminated, since the transition of positive and negative ions across the same interface gives the same work with opposite sign.

The results of NMR measurements on the preferential solvation of ions dissolved in mixtures of water and ethanol were analyzed by Covington, Newman and Lilley [55b] from the thermodynamical point of view; as a result, they point out the connection between the variation of the chemical potential accompanying the transfer of ions from water into the solvent mixture, and the spectroscopic information regarding this process. The free-enthalpy changes of the same processes have been investigated by Wells [55c], and he inferred that the process causing these changes can be divided into two parts: (1) the transfer of the tetrahedral aquo complex from water into the methanol–water mixture; (2) the subsequent substitution of a water molecule in the tetrahedron by a methanol molecule. He discussed the deviations in the solvation of ions by H_2O and CH_3OH, taking into account the structural changes occurring in the bulk solvent.

The distribution of the enthalpy of hydration between ions. Since the enthalpy of hydration of a binary electrolyte consists of the enthalpies of hydration of the cation and the anion, a distribution for the two kinds of ions can be achieved, even when –as a consequence of the facts discussed above–it can be realized at present only conditionally. It is an empirical fact, that, extrapolating the difference of the enthalpy of hydration of electrolytes containing the same anion, but different cations (e.g. KCl—NaCl, KBr—NaBr), it is independent of the anion (moreover, it is approximately independent of the solvent, too), and the same is valid for the difference of the enthalpy of hydration of electrolytes containing the same cations and different anions (e.g. KCl—KBr, NaCl—NaBr). It follows from this that the enthalpy of hydration of an electrolyte is really an additive quantity obtained from those of the two kinds of ions in sufficiently high dilution. Thus, if at least the hydration enthalpy of one single ion could be determined separately either experimentally or theoretically, that of the other ions could be successively calculated from the enthalpy of hydration of corresponding electrolytes. Since this procedure cannot be followed yet even in approximation, the relatively most reliable method of obtaining the distribution is to start with a binary electrolyte about which it can be supposed that the hydration enthalpy of its cation is identical to that of its anion.

According to the studies of Lange and Mishchenko [54] and Kaustinskii, Drakin and Yakushevskii [56], it is justifiable to assume that the enthalpy of hydration of the Cs^+ ion is identical in good approximation with that of the I^- ion, i.e. the enthalpy of hydration of CsI (134 kcal mole^{-1}) can be divided into two equal parts for the two types of ions. Although the Pauling crystallographic radii of these ions are not identical ($r_{Cs^+} = 1 \cdot 69$ Å, $r_{I^-} = 2 \cdot 16$ Å), it can still be supposed that this difference is cancelled since the water molecules are arranged around the cation and anion in orientations of opposite direction. The enthalpies of hydration calculated on the basis

Literature on page 491

31

Table 5.10

The enthalpy of hydration of some ions calculated on the basis of the
equation: $\Delta H_{h,Cs}^{+} = \Delta H_{h1}^{-}$

Ion	$-\Delta H_{h,i}$ kcal · g-ion^{-1}	Ion	$-\Delta H_{h,i}$ kcal · g-ion^{-1}
H_3O^+	110	F^-	116
Li^+	127	Cl^-	84
Na^+	101	Br^-	76
K^+	81	I^-	67
Rb^+	75	OH^-	122
Cs^+	67	ClO_3^-	69
Ag^+	117	ClO_4^-	54
NH_4^+	78	NO_3^-	74
Mg^{2+}	467	MnO_4^-	59
Ca^{2+}	386	CN^-	83
Ba^{2+}	320	$HCOO^-$	99
Fe^{2+}	467	CH_3COO^-	101
Fe^{3+}	1056	HCO_3^-	91
Al^{3+}	1125	CO_3^{2-}	332
		SO_4^{2-}	243

of this assumption for other ions (see Table 5.10) are not in contradiction
with any known facts. This distribution is in better accordance with the
experiments than that suggested by Bernal and Fowler [57], according
to whom the enthalpy of hydration of K^+ and F^- ions can be taken as
equal, since the crystallographic radii of these ions are approximately
the same (1·33 Å). Namely, this distribution disregards the fact that water
molecules can approach the anions to a shorter distance than cations,
owing to the non-spherically symmetrical shape of the water molecules.

Several studies have dealt with the theoretical calculation of the enthalpy
of hydration (or, in general, that of solvation) on the basis of some particular models since the time of Born's work [58], however, there has been
no success in solving the problem in a satisfactory way from all points of view.
Recently, a critical survey has been published about these efforts by Mishchenko and Poltoratskii [59a]. The energy of hydration has been studied by
Burton and Daly [59b] using calculations based on the MO theory. Recently
Rosseinsky [59c] called attention to doubts in connection with the interpretation of the enthalpy of hydration of monatomic ions with respect to
hydration.

The limit of total hydration. The integral heat of dissolution of electrolytes
comprises the energy of hydration–in a broader sense of the word–and the
lattice energy. The decomposition of the crystal lattice is an endothermal
process, while hydration as a whole is an exothermal one. The dissolution
of the crystal is either endothermal or exothermal, depending on the relative
magnitude of these two quantities.

Hydration is a complicated process involving several part-processes
that are both endothermal and exothermal. So far, no detailed and satisfactorily quantitative theory on its energetics has been outlined, which is not

surprising on the basis of the above discussions. However, the semi-quantitative theory given by Mishchenko *et al.* [60] based on detailed analyis of the part-processes yields an appreciable contribution to the clarification of the energetical conditions of hydration, and it has led to new information mainly with respect to concentrated solutions. According to this, exothermal part-processes of hydration ('exoeffects') are the interactions between ions and the permanent dipoles of adjacent water molecules, the polarization of water molecules by the field of ions, the dispersion interaction between ions and water molecules and the effect of ions surrounded by the primary hydrate sphere on the farther water molecules. On the other hand, endothermal part-processes are the mutual repulsion between molecules in the hydrate sphere and the mutual repulsion between ions and the molecules in the primary hydrate layer, owing to the partial overlapping of their electron shells. These energy terms vary with the concentration of the solution and the energy change appearing as an electrostatic consequence of the change in the mutual average distance of ions contributes to it. This is revealed by the concentration dependence of the integral heat of dissolution.

Starting from the most dilute solutions of strong electrolytes, according to the very accurate thermochemical measurements of Lange [61], the initial change of the heat of dissolution with increasing concentration corresponds to the Debye–Hückel theory (its 'exothermicity' decreases), but the individual effects that become successively predominating with increasing concentration already start to be observable at low concentrations. In medium concentrations, the concentration dependence of the heat of dissolution of various electrolytes is very different. In concentrated solutions, however, the decrease in the absolute value of the heat of dissolution predominates (i.e. $\Delta H_{\text{dissolution}} \to 0$).

Studying the changes in the heat of dissolution of electrolytes as a function of concentration, Mishchenko [60] concluded that, in the concentration range from the most dilute solutions to the saturation of electrolytes of good solubility, there is a state which exhibits a turning point with respect to the properties and produces a certain change in their nature. This is the state of that solution in which the total amount of water present as solvent is bound in the primary solvate spheres of ions, but primary hydration is still not limited by the lack of water. This is the *limit of total hydration* which depends on the coordination number of ionic hydration. From investigations on the enthalpy of hydration, Mishchenko came to the conclusion that the sum of the coordination numbers of the cations and anions is 12–26 in dilute solutions. Provided that the coordination number does not change with concentration–as long as there is a sufficient amount of water present–the limit of total hydration of various electrolytes is in solutions of a molarity of 2·1–4·6.

At the limit of total hydration, where the whole amount of water present in the solution is bound in the primary hydrate spheres, the structure of the solution is changed. In more dilute solutions, there is 'free' water between hydrated ions; at the limit of total hydration and in more con-

Literature on page 491

centrated solutions, however, no such water can be found. From the limit of total hydration onwards, the hydrate spheres of ions are in immediate contact with one another in solution. The solutions of various electrolytes can be regarded as being in identical states with respect to their structure at the limit of total hydration, and in a certain sense, the theory of concentrated solutions can start with them.*

The concentration dependence of the heat of dissolution indicates that the enthalpy of hydration decreases with increasing concentration. Analysing the empirical facts in detail, Mishchenko has concluded that, in solutions more dilute than the limit of total hydration, the coordination number of hydration is probably unaltered, only the bonds between ions and water molecules are weakened, which can be called *energetical dehydration*; of course, the change in the water structure due to the presence of ions also contributes to it. In solutions more concentrated than the limit of total hydration, the coordination number of hydration also decreases with increasing concentration–due to a certain lack of water *(coordination dehydration)*.

The observations of Baron and Mishchenko [62], stating that in solutions of NaBr dissolved in a mixture of water and methanol, methanol molecules are built into the solvate sphere of the ions only when the concentration of the electrolyte is higher than the limit of total hydration, which is also confirmed by the spectroscopic investigations of Mishchenko and Pominov [63], are in accordance with the interpretation of the limit of total hydration given above.

Although current knowledge on the structure of solutions is rather deficient, it is still probable that the structure of solutions more concentrated than the limit of total hydration is more similar to the structure of the corresponding ionic crystal than to that of the dilute solution. The structure of concentrated solutions can be approximated starting with the crystal lattice of electrolytes which is extended and deformed by the water molecules inserted between the ions and completely bound by them. Figure 5.13 shows the schematic structure of a KBr solution in a concentration corresponding to the limit of total hydration, as published by Mishchenko, provided that the

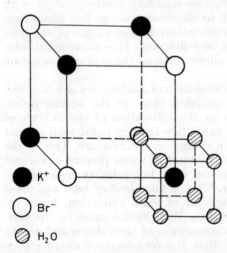

Fig. 5.13 Schematic structure of KBr solution at the limit of total hydration provided that the coordination number of both ions is eight. (For the sake of better repesentation the water molecules are shown only around one ion)

K⁺
Br⁻
H₂O

* There are no common structural properties in saturated solutions, because these are characterized only by the identity of the free enthalpy of the crystalline and dissolved state, which is not accompanied by similarity in structure.

hydration numbers of cations and anions both equal eight. In reality, of course, the lattice-like structure is smeared appreciably by the thermal motion. For the same reason, the transition between the solution-like and crystal-like structures takes place, not sharply, but gradually at the limit of total hydration.

With further increase in concentration from the limit of total hydration, the extent of coordination dehydration is also increasing because of the lack of free water, in addition energetical dehydration, and water molecules are rearranged according to the specific properties (hydrophility) of ions. If the electrolyte forms crystalline hydrates, the structure of the solution becomes more and more similar to the lattice structure of that loosened by the thermal motion, with increasing concentration.

The correlation between the entropy of dissolution and hydration. As entropy is the measure of thermodynamic probability and of the order of the state of a given system, the entropy change accompanying the dissolution of electrolytes also gives information on the direction of the change in the water structure brought about by the dissolved electrolyte.

On the basis of the general thermodynamic correlation between enthalpy, free enthalpy and entropy, the partial molar change in the entropy, $\Delta \bar{S}_w$, of water in the course of dissolution can be calculated from the partial molar enthalpy of water ($\bar{H}_w$) and from the change of the partial molar free enthalpy accompanying dissolution ($\Delta \bar{G}_w$):

$$\Delta \bar{S}_w = \frac{\bar{H}_w - \Delta \bar{G}_w}{T}. \tag{5.2.20}$$

The change in free enthalpy can be calculated from the vapour pressure of water and that of the solution (p_0 and p, respectively):

$$\Delta \bar{G}_w = \mathrm{R}T \ln \frac{p}{p_0}. \tag{5.2.21}$$

Dissolution–like any mixing of two components–is bound to give an entropy change even when the mixture formed is ideal, i.e. the structure is not altered by the addition of the second component to the first one, since dissolution is accompanied by increases in disorder.

In the course of formation of ideal solutions, the partial molar change in the entropy of water $\Delta \bar{S}_{w,id}$ is, according to the general laws of thermodynamics:

$$\Delta \bar{S}_{w,id} = - \mathrm{R}T \ln x_w \tag{5.2.22}$$

(if x_w is the molar fraction of water in the solution). In real solutions, the change in entropy corresponding to the structural changes caused by interactions of the components ('excess entropy', $\Delta \bar{S}'_w$) also contributes to it. Thus, the total change of the entropy of water in the course of dissolution is:

$$\Delta \bar{S}_w = \Delta \bar{S}_{w,id} + \Delta \bar{S}'_w. \tag{5.2.23}$$

Literature on page 492

The excess entropy corresponding to the structural changes can be calculated from this.

Some significant aspects of the concentration dependence of the excess entropy caused by the dissolution of electrolytes in water have been pointed out by Frank and Robinson [64], and later Mishchenko *et al.* [65] studied this phenomenon in detail. The dependence of the excess entropy of water on concentration and temperature varies with the nature of the dissolved electrolyte. In general, the value of $\Delta \bar{S}'_w$ is shifted towards negative values with increasing temperature which indicates that the rise in temperature helps the ordering of water molecules by ions. $\Delta \bar{S}'_w$ is negative in the whole concentration range in NH_4Cl solutions above 50 °C and in $CoCl_2$ solutions above 20 °C, i.e. the structure of water is more ordered in these solutions than in pure water. Thus, at higher temperatures, the ordering effect overcompensates the structure-breaking effect even in solutions of NH_4^+ ions with a low heat of hydration ($\Delta H_h = -78$ kcal mole^{-1}). On the other hand, dissolution of ions at temperatures about the freezing point has such a strong structure-breaking effect that even the ordering effect of Co^{2+} with a large heat of hydration ($\Delta H_h = -499$ kcal mole^{-1}) cannot compensate it either, and $\Delta \bar{S}'_w$ remains positive in all concentrations.

Isotherms of excess entropy are shown in Fig. 5.14 on some typical electrolytes. The resultant of two opposite effects is reflected in $\Delta \bar{S}'_w$: partly the structure-breaking effect of dissolved ions on pure water (the decrease in the order due to this), and partly, the ordering of water molecules under the hydration effect of ions. The effect on the structure of pure water depends strongly on temperature, while the ordering effect of the field of ions is almost independent of this. For example, the isotherm of $CoCl_2$

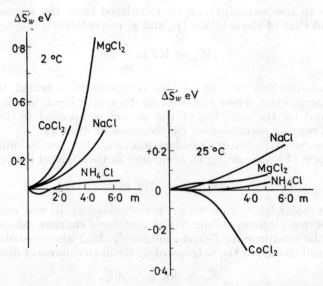

Fig. 5.14 The partial molar excess entropy of water at 2 °C and 25 °C as a function of the molality of the solution

at 25 °C is almost a mirror image of the isotherm corresponding to 2 °C (see Fig. 5.14): close to the freezing point, the disorganizing effect of the strongly hydrated Co^{2+} ions predominates, but, at 25 °C (where water structure is rather disorganized by the thermal motion), the ordering effect of the field of ions prevails already. The order of the various ions depends on temperature: at 2 °C, the order of the structure-breaking effect is $Na^+ <$ $< Mg^{2+} < Co^{2+}$, while at 25 °C, the order is opposite. Thus, in a certain sense, water is different as a solvent near the melting point and at higher temperatures.

The isotherms of NH_4Cl are of particular interest, indicating that this salt has a small thermodynamic effect on water, which hardly depends on temperature. This can be attributed to the property of the NH_4^+ ion –noted already with respect to several matters–that its partial molar volume is almost identical with that of water, and that of Cl^- ions does not differ appreciably from this, either. Fajans and Johnson [66] have already pointed out this fact in connection with the discovery that NH_4^+ ions show the least influence on entropy, viscosity, infra-red absorption, and electron diffraction of water among the salts investigated at 35 °C. It is probable that the NH_4^+ ion fits well into the tetrahedral structure of water with its four protons. Probably this is the main reason that the effect of the NH_4^+ ion differs from that of the Rb^+ ion that is similar to it in size. Both NH_4^+ and Cl^- ions fit well into the structure of water with almost no deformation. The tetrahedral configuration of the NH_4^+ ion also helps fitting into the water structure, hence the structure-breaking effect is small. Unlike this, $NaNO_3$ and KNO_3 have a large structure-breaking effect. This is probably due to the coplanar triangular shape of the NO_3^- ion that results in a poor fitting into the tetrahedral structure of water.

With respect to the solutions of fluorides of various cations of noble gas configuration, Magnusson [67] has concluded, investigating the thermodynamics of the interaction between ions and between ions and water molecules, that the ionic bond is delocalized and extended over the water molecules in the coordination sphere. In solutions of these electrolytes, covalent bonds also contribute to hydration.

Vdovenko, Gurikov and Legin [68a] have studied the thermodynamic state of water in the electrostatic field of ions in infinitely dilute solutions. Provided that there is an equilibrium between the two types of water structure–tetrahedral and tight packing–they have carried out some calculations regarding the change in this equilibrium due to the field of ions. The conclusion drawn was that, at normal temperatures, the equilibrium is shifted in the direction of the ice-like structure i.e. water is more or less stabilized.

In a detailed thermodynamical treatment of the solvation of monoatomic ions, Abrosimov [68b] elucidated the change in the liquid structure of water due to ions. He derived a relationship for the free enthalpy and enthalpy changes which correspond to the transition of positive hydration into negative one with the increase of the size of the ions.

Literature on page 492

Investigating the temperature dependence of some thermodynamic functions of the hydration of electrolytes, Lebed and Alexandrov [69a] published some detailed tables on the entropy and energy of hydration.

Studying the change of entropy accompanying the transition of an electrolyte from H_2O into D_2O, Krishnan and Friedman [69b] have confirmed the suggestion that there are two rather concentric layers around the dissolved ions. In the inner one, the orientation effect of the electric field of ions on the solvent molecules is predominating (primary hydration), while in the outer one, the orientation effect is negligible, and the structure-breaking effect (that decreases the stabilization produced by means of hydrogen bonds) prevails (secondary hydration). Primary hydration decreases in the column of alkali metals from the Li^+ ion to the Cs^+ ion, and in the column of halide ions, from the F^- to I^- ion. They have assumed that the structure-breaking effect is higher by about 5 per cent in D_2O than in H_2O, under comparable conditions.

In the thermodynamic treatment of solvation taking place in solvent mixtures, Padova [70] has concluded that the governing factor is the partial molar free enthalpy of solvation of the components, and this determines the preferred component of the solvent mixture in the solvation of ions. In most of the solvent mixtures, hydration with water is preferred to solvation of the other component. Some solvent components which probably form compounds with the salts are exceptions.

Krestov, Klopov and Patsatsiya [71] have made a detailed analysis of the thermodynamic aspects of the structural changes caused by the solvation of KCl dissolved in ethanol–water mixtures. On the basis of investigations on the change in the entropy of solvation and the entropy of dissolution, they have concluded that the structure-making effect of ethanol on water has a maximum in a solution of mole fraction 0·15 with respect to ethanol. In solutions of higher concentrations, the water structure is progressively destroyed and a mixed structure of water and alcohol is formed instead. The transition from water structure to a mixed structure is revealed clearly in the properties, however, the transition from the mixed structure into the structure of pure alcohol takes place smoothly with no marked change in the properties. This is related to the fact that the water structure is more ordered and different from the mixed structure than the latter from that of ethanol.

The effect of solvation of K^+ and Cl^- ions on the solvent depends on the concentration of ethanol. From pure water to a mole fraction of 0·15, the negative hydration of these ions is strengthened with increasing ethanol concentration. Beyond this concentration limit, however, negative hydration is weakened, which can be attributed to the fact that ethanol destroys the liquid structure in this range. The negative solvation effect of K^+ and Cl^- ions is transformed into a positive solvation effect at a mole fraction of 0·67 and 0·35 in mixtures of ethanol and water, respectively. With respect to this, the authors also extend the concept of *structural temperature* to solutions of non-electrolytes: this is the temperature at which the structure of pure water is identical with the structure of water in the solution. The structural temperature is either higher or lower than the real temperature of the solution, depending on the conditions. Starting from pure water, ethanol decreases the structural temperature of the mixture up to the maximum stabilization, whereas it increases that of solutions of higher concentration.

Ionic hydration influences solubility equilibria as well. Recently, Buslaeva, Dubnikova and Samoilov [72] studied the correlation of the nearby hydration of anions and the salting-out effect.

In discussing the process of solvate formation, Lange [73] has established that the solvation of a particle, that is, its transition from the gaseous phase into the liquid phase can be regarded as composed of two steps: the formation of a suitable cavity in the liquid, then the transition of the particle from the gaseous phase into the cavity and its 'adhering' to the surrounding liquid molecules. The energy liberated in the latter step (the 'energy of adhering') is, according to Lange, more suitable for the estimation of the interaction between the dissolved species and the solvent, than the solvation energy considered in the usual sense.

REFERENCES

to Section 5.2

[1a] As a review see, e.g., K. P. MISHCHENKO and T. M. POLTORATSKII, Voprosy termodinamiki i stroeniya vodnykh i nevodnykh rastvorov elektrolitov, p. 50 (Leningrad, 1968); N. A. IZMAILOV, Elektrokhimiya rastvorov, p. 166 (Moskva, 1966); O. Ya. SAMOILOV, Struktura vodnykh rastvorov elektrolitov i gidratatsiya ionov (Moskva, 1957). J. F. COETZEE and C. D. RITCHIE, Solute–Solvent Interactions, (London, 1969); B. E. CONWAY, Modern Aspects of Electrochemistry (1964); D. R. ROSSEINSKY, Chem. Rev. 1966, 65, 467; G. ZUNDEL, Hydratation and Intermolecular Interaction (New York, 1969); H. STREHLOW, W. KNOCHE and H. SCHNEIDER, Ber. Bunsenges., 1973, 77, 760; K. G. BREIT-SCHWERDT and H. WOLZ, Ber. Bunsenges., 1973, 77, 1000; E. LANGE, Ber. Bunsenges., 1973, 77, 1010.

[1b] I. F. HINTON and E. S. AMIS, Chem. Review, 1971, 71, 627.

[1c] H. STREHLOW and A. SCHNEIDER, Pure and Applied Chem., 1971, 25, 327; H. STREHLOW, W. KNOCHE and H. SCHNEIDER, Ber. Bunsenges., 1973, 77, 760.

[1d] A. L. VAN GEET, J. Amer. Chem. Soc., 1972, 94, 5583.

[2] O. YA. SAMOILOV, Struktura vodnykh rastvorov elektrolitov i gidratatsiya ionov, p. 76 (Moskva 1957); Disc. Faraday Soc., 1957, 24, 141.

[3] With respect to this see: E. GLUCKAUF, Chemical Physics of Ionic Solutions (ed. by B. E. CONWAY and R. G. BARRADAS) p. 67 (New York, 1966).

[4] D. M. RITSON and J. B. HASTED, J. Chem. Phys., 1948, 16, 11.

[5a] A. M. AZZAM, Z. Elektrochem. (1954) 58, 889.

[5b] G. V. KOKOVINA, O. YA. SAMOILOV and P. S. YASTREMSKII, Zhurn. Strukt. Khim., 1970, 11, 532.

[6] J. P. HUNT and H. TAUBE, J. Chem. Phys., 1950, 18, 757; 1951, 19, 602; A. C. RUTENBERG and H. TAUBE, J. Chem. Phys., 1952, 20, 825.

[7] Z. LUZ and R. G. SCHULMAN, J. Chem. Phys., 1965, 43, 3750.

[8a] C. C. McDONALD and W. D. PHILLIPS, J. Amer. Chem. Soc., 1963, 85, 3736.

[8b] O. YA. SAMOILOV, P. S. YASTREMSKII and A. P. TARASOV, Zh. Strukt. Khim., 1973, 14, 600.

[8c] M. O. TERESHKOVICH, M. I. VALENCHUK, A. V. KUPRIK, S. A. VOLKOVA and N. A. SKIDAN, Zh. Strukt. Khim., 1972, 13, 785.

[8d]   I. SMID, *Usp. Khim.*, 1973, **42**, 799.
[9a]   J. H. HILDEBRAND, *Disc. Faraday Soc.*, 1953, **15**, 9.
[9b]   YU. V. YERGIN and L. I. KOSTROVA, *Zh. Strukt. Khim.*, 1969, **10**, 971.
[9c]   A. K. LYASHCHENKO, *Zh. Strukt. Khim.*, 1970, **11**, 138.
[9d]   M. I. EMELYANOV, E. A. NIKOFOROV and G. V. POPOV, *Elektrokhim.*, 1972, **8**, 587.
[9e]   M. I. EMELYANOV and V. P. YAGODAROV, *Zh. Strukt. Khim.*, 1973, **14**, 919.
[9f]   A. S. SOLOVKIN, *Zh. Strukt. Khim.*, 1972, **13**, 514.
[9g]   K. GIESE, *Ber. Bunsenges.*, 1973, **76**, 493.
[9h]   U. KAATZE, *Ber. Bunsenges.*, 1973, **77**, 447.
[10a] KH. UEDAIRA and KH. UEDAIRA, *Zh. Fiz. Khim.*, 1968, **42**, 3024.
[10b] O. YA. SAMOILOV, I. B. RABINOVICH, Z. V. VOLOSOVA and A. F. BORINA, *Zh. Strukt. Khim.*, 1970, **11**, 207.
[10c] B. S. KRUMGALZ, *Zh. Strukt. Khim.*, 1972, **13**, 774; *Elektrokhim.*, 1972, **8**, 552.
[11]   O. YA. SAMOILOV and D. S. SOKOLOV, *Izv. AN SSSR, otd. Khim.*, 1957, **3**, 257.
[12]   V. M. VDOVENKO, YU. V. GURIKOV and E. K. LEGIN, *Zh. Strukt. Khim.*, 1966, **7**, 819.
[13]   P. S. YASTREMSKII and O. YA. SAMOILOV, *Zh. Strukt. Khim.*, 1963, **4**, 844.
[14]   R. A. HORNE, R. A. COURANT and D. S. JOHNSON, *Electrochim. Acta*, 1966, **11**, 987.
[15]   N. A. NEVOLINA, O. YA. SAMOILOV and A. L. SEIFER, *Zh. Strukt. Khim.*, 1969, **10**, 203.
[16a] I. V. RADCHENKO and A. I. RYSS, *Zh. Strukt. Khim.*, 1965, **6**, 771.
[16b] R. E. BURTON and J. DALY, *Trans. Faraday Soc.*, 1970, **66**, 1281, 2408; 1971, **67**, 1219.
[17]   E. R. NIGHTINGALE, Chemical Physics of Ionic Solutions (ed. by B. E. CONWAY and R. G. BARRADAS), p. 87 (New York, 1966).
[18]   G. ENGEL and H. G. HERTZ, *Ber. Bunsenges.*, 1968, **72**, 808.
[19a] E. A. NIKIFOROV, *Zh. Strukt. Khim.*, 1969, **10**, 137.
[19b] YU. V. ERGIN and L. I. KOSTROVA, *Zh. Strukt. Khim.*, 1972, **13**, 999.
[19c] E. BERECZ and M. ACHS-BALLA, *Acta Chim. Acad. Sci. Hung.*, 1973, **77**, 267.
[20]   R. A. ROBINSON and R. H. STOKES, Electrolyte Solutions, p. 123 (London, 1959).
[21a] C. A. KRAUS, *Ann. N. Y. Acad. Sci.*, 1949, **51**, 789; H. M. DAGGETT, E. J. BAIR and C. A. KRAUS, *J. Amer. Chem.Soc.*, 1951, **73**, 799.
[21b] M. D. MONICA and L. SENATORE, *J. Phys. Chem.*, 1970, **74**, 205.
[21c] K. SCHWABE, *Z. Phys. Chem.*, 1971, **247**, 113.
[22]   R. H. STOKES, *J. Amer. Chem. Soc.*, 1964, **86**, 979, 982, 2337.
[23]   H. L. FRIEDMAN, Ionic Solution Theory (New York, 1962).
[24]   A. J. RUTGERS and Y. HENDRIKX, *Trans. Faraday Soc.*, 1962, **58**, 2184.
[25]   R. A. HORNE and J. D. BIRKETT, *Electrochim. Acta*, 1967, **12**, 1153.
[26a] F. ACCASCINA, A. D'APRANO and R. TRIOLO, *J. Phys. Chem.*, 1967, **71**, 3469.
[26b] N. I. GUSEV, *Zh. Fiz. Khim.*, 1973, **47**, 96, 1218.
[27]   R. M. FUOSS, *Proc. Nat. Acad. Sci.*, 1959, **45**, 807.
[28]   R. H. BOYD, *J. Chem. Phys.*, 1961, **35**, 1281.
[29]   R. ZWANZIG, *J. Chem. Phys.*, 1963, **38**, 1603.
[30]   H. S. FRANK, Chemical Physics of Ionic Solutions (ed. by B. E. CONWAY and R. G. BARRADAS) p. 60 (New York, 1966).
[31]   H. EYRING, *J. Chem. Phys.*, 1936, **4**, 283.
[32a] J. N. AGAR and J. C. R. TURNER, *Proc. Roy. Soc.*, 1960, **A 255**, 307.
[32b] O. YA. SAMOILOV and P. S. YASTREMSKII, *Zh. Strukt. Khim.*, 1971, **12**, 379.
[32c] YU. V. ERGIN and L. I. KOSTROVA, *Zh. Strukt. Khim.*, 1970, **11**, 806.
[33]   G. BUCHBÖCK, *Z. Phys. Chem.*, 1906, **55**, 563.
[34]   E. W. WASHBURN, *Z. Phys. Chem.*, 1909, **66**, 513.
[35]   T. ERDEY-GRÚZ ,A. HUNYÁR, É. POGÁNY and A. VÁLI, *Hungarica Acta Chimica*, 1948, **1**, 7; T. ERDEY-GRÚZ, and A. HUNYÁR: *Hungarica Acta Chimica*, 1948, **1**, 27; A HUNYÁR, Thesis (Budapest, 1937); É. POGÁNY: *Magy. Kém. Folyóirat*, 1942, **48**, 85; A. VÁLI, Thesis, (Budapest, 1945); A. HUNYÁR, Thesis (Budapest, 1937); *J. Amer. Chem. Soc.*, 1949, **71**, 3552.
[36]   L. G. LONGSWORTH, *J. Amer. Chem. Soc.*, 1947, **69**, 1288.

[37a] C. H. HALE and T. DE VRIES, J. Amer. Chem. Soc., 1948, **70**, 2473.
[37b] K. H. KHOO, Faraday Transact. I., **1973**, 1313.
[37c] N. I. GUSEV, Zh. Fiz. Khim., 1971, **45**, 1164, 2238, 2243.
[38a] J. N. SUGDEN, J. Chem. Soc., **1926**, 174.
[38b] D. FEAKINS and P. J. VOICE, Faraday Transact. I., **1972**, 1390.
[38c] D. FEAKINS, A. S. WILLMOTT and A. R. WILLMOTT, Faraday Transact. I., **1973**, 122.
[38d] A. M. NILSSON and P. BERONIUS, Z. phys. Chem. N. F., 1972, **79**,83.
[38e] N. I. GUSEV, Zh. Fiz. Khim., 1971, **45**, 2238.
[39]   A. PASSYNSKI, Acta Physico-chimica USSR, 1938, **8**, 385.
[40]   S. BARNARTT, Quart. Rev., 1953, **7**, 84.
[41]   I. V. LITVINENKO, Zh. Strukt. Khim., 1963, **4**, 830.
[42]   E. F. DUBININA and B. B. KUDRYAVTSEV, Zh. Fiz. Khim., 1957, **31**, 2191.
[43]   D. S. ALLAM and W. H. LEE, J. Chem. Soc., **A 1966**, 426.
[44]   J. E. DESNOYERS, R. E. VERRALL and B. E. CONWAY, J. Chem. Phys., 1965, **43**, 243.
[45a] J. C. HINDMAN, J. Chem. Phys., 1962, **36**, 1000.
[45b] J. O'M. BOCKRIS and P. P. S. SALUJA, J. Phys. Chem., 1972, **76**, 2140.
[45c] J. E. DESMOYERS, J. Phys. Chem., 1973, **77**, 576.
[45d] K. G. BREITSCHWERDT and H. WOLZ, Ber. Bunsenges., 1973, **77**, 1000.
[46]   G. E. WALRAFEN, J. Chem. Phys., 1962, **36**, 1035.
[47a] P. M. VOLLMAR, J. Chem. Phys., 1963, **39**, 2263.
[47b] G. J. JANZ, K. BALASUBRAHMANYAM and B. G. OLIVER, J. Chem. Phys., 1969, **51**, 5723.
[48a] L. V. IVANOVA and V. M. ZOLOTAREV, Zh. Strukt. Khim., 1968, **9**, 385.
[48b] I. KAMPSCHULTZE-SCHEUNING and C. ZUNDEL, J. Phys. Chem., 1970, **74**, 2363.
[49]   O. D. BONNER and G. B. WOOLSEY, J. Phys. Chem., 1968, **72**, 899, O. D. BONNER, J. Phys. Chem., 1968, **72**, 2512.
[50]   K. FAJANS and G. JOOS, Z. Phys., 1924, **23**, 1.
[51a] Cf. E. R. NIGHTINGALE, Chemical Physics of Ionic Solution (ed. by B. E. CONWAY and R. G. BARRADAS) p. 87 (New York, 1966).
[51b] Z. A. GRANKINA, L. K. CHUCHALIN, B. I. PESHEVITSKII, I. A. KUZIN and S. P. KHRANENKO, Zh. Fiz. Khim., 1969, **43**, 2442.
[51c] S. SUBRAMAIAN and H. F. FISCHER, J. Phys. Chem., 1972, **76**, 84.
[51d] S. N. ANDREEV and M. F. SMIRNOVA, Zh. Fiz. Khim., 1972, **46**, 1793.
[51e] C. H. SPINK and J. C. WYCKOFF, J. Phys. Chem., 1972, **76**, 1660.
[52a] P. S. KNAPP, R. O. WAITE and E. R. MALINOVSKI, J. Chem. Phys., 1968, **49**, 5459.
[52b] As a review see e.g. J. E. DESNOYERS and C. JOLICOEUR, Modern Aspects of Electrochemistry (ed. by J. O'M. MOCKRIS and B. E. CONWAY), Vol. 5, p. 1., 1969.
[52c] B. F. J. VOGRIN, P. S. KNAPP, W. L. FLINT, A. ANTON, G. HIGHBERGER and E. R. MALINOVSKI, J. Chem. Phys., 1971, **54**, 178.
[52d] J. W. AKITT, J. Chem. Soc., 1971, 2865.
[53]   With respect to the calculation of the lattice see e.g. T. ERDEY-GRÚZ and G. SCHAY, Elméleti fizikai kémia (Theoretical Physico-chemistry), Vol. I, p. 362, (Budapest, 1962); K. P. MISHCHENKO and G. M. POLTORATSKII, Voprosy termodinamiki i stroenyia vodnykh i nevodnykh rastvorov elektrolitov, p. 61 (Leningrad, 1968).
[54]   E. LANGE and K. P. MISHCHENKO, Z. Phys. Chem., 1930, **A. 149**, 1.
[55a] K. P. MISHCHENKO and E. I. KVYAT, Zh. Fiz. Khim., 1954, **28**, 1451.
[55b] A. K. COVINGTON, K. E. NEWMAN and T. H. LILLEY, Faraday Transact., I. 1973, 973.
[55c] C. F. WELLS, Faraday Transact., I., 1973, 984.
[56]   A. F. KAUSTINSKII, S. I. DRAKIN and B. M. YAKUSHEVSKII, Zh. Fiz. Khim., 1953, **27**, 433.
[57]   J. D. BERNAL, and R. H. FOWLER, J. Chem. Phys., 1933, **1**, 515.
[58]   M. BORN, Z. Phys., 1920, **1**, 45.
[59a] K. P. MISHCHENKO and G. M. POLTORATSKII, Voprosy termodinamiki i stroeniya vodnykh i nevodnykh rastvorov elektrolitov, p. 83 (Leningrad, 1968).

[59b] R. T. Burton and J. Daly, *Trans. Faraday Soc.*, 1970, **66**, 1281.
[59c] D. R. Rosseinsky, *Electrochim. Acta*, 1971, **16**, 23.
[60] K. P. Mishchenko, *Zh. Fiz. Khim.*, 1952, **26**, 1736; K. P. Mishchenko and A. M. Sukhotin, *Zh. Fiz. Khim.*, 1953, **27**, 26.
[61] For review see e.g. E. Lange: in The Structure of Electrolytic Solutions (ed. by W. J. Hamer) p. 135 (New York, 1959).
[62] N. M. Baron and K. P. Mishchenko, *Zh. Obshch. Khim.*, 1948, **18**, 2067.
[63] K. P. Mishchenko and I. S. Pominov, *Zh. Fiz. Khim.*, 1957, **31**, 2026.
[64] H. S. Frank and A. L. Robinson, *J. Chem. Phys.*, 1940, **8**, 933.
[65] K. P. Mishchenko and A. M. Ponomareva, *Zh. Obshch. Khim.*, 1956, **26**, 1296; K. P. Mishchenko and I. F. Yakovlev, *Zh. Obshch. Khim.*, 1959, **29**, 1761; K. P. Mishchenko and E. A. Podgornaya, *Zh. Obshch. Khim.*, 1961, **31**, 1743.
[66] K. Fajans and O. Johnson, *J. Amer. Chem. Soc.*, 1942, **64**, 668.
[67] L. B. Magnusson, *J. Chem. Phys.*, 1963, **39**, 1953.
[68a] V. M. Vdovenko, Yu. V. Gurikov and E. K. Legin, *Zh. Fiz. Khim.*, 1968, **42**, 390.
[68b] V. K. Abrosimov, *Zh. Strukt. Khim.*, 1973, **14**, 211.
[69a] V. I. Lebed and V. V. Aleksandrov, *Elektrokhim.* , 1965, **1**, 1359.
[69b] C. V. Krishnan and H. L. Friedman, *J. Phys. Chem.*, 1970, **74**, 2356.
[70] J. Padova, *J. Phys. Chem.*, 1968, **72**, 796.
[71] G. A. Krestov, V. I. Klopov and K. M. Patsatsiya, *Zh. Strukt. Khim.*, 1969, **10**, 417.
[72] M. N. Buslaeva, V. T. Dubnikova, O. Ya. Samoilov, *Zh. Strukt. Khim.*, 1969, **10**, 8.
[73] E. Lange, *Ber. Bunsenges.*, 1973, **77**, 1010.

AUTHOR INDEX

Abdel-Hamid, A. A. 325, 372, 380
Abdel-Salam, M. M. Sh. 325
Abraham, M. 68
Abrosimov, V. K. 487
Accascina, F. 312, 314, 362, 363, 370, 372, 464
Achs-Balla, M. 458, 459
Ackermann, Th. 279, 282, 377, 380
Adami, L. H. 52
Adamson, A. W. 180, 240
Adler, B. J. 81
Agar, J. N. 468
Ageno, M. 51, 99, 108, 110, 377
Aggan, A. M. el 372
Agishev, A. Sh. 234
Ahn, M. K. 169
Akhmetzyanov, K. G. 105
Akitt, J. W. 293, 478
Albright, J. G. 180, 192, 229
Alder, B. J. 71
Alekseieva, L. S. 239
Alexander, K. F. 145
Alexandrov, V. V. 488
Allam, D. S. 443, 474
Ambrus, J. H. 299
Amis, E. S. 359, 362, 363, 366, 377
Andersen, J. E. 24
Anderson, D. K. 173
Andrade, E. N. 84, 94, 98, 102, 130
Andreiev, G. A. 234
Andreiev, S. N. 223, 478
Andrussow, L. 112
Anfinsen, C. B. 228
Angell, C. A. 134, 325
Anton, A. 478
Anway, A. R. 321
Arrhenius, S. 94
Arvia, A. J. 210
Asmus, E. 121, 129
Atkinson, G. 312, 321, 437
Auty, R. P. 31
Azzam, A. M. 446

Baborosky, G. 443
Bader, I. 136, 288
Badiali, J. P. 313
Bakulin, E. A. 267
Balasubrahmanyam, K. 476
Balthazár-Vass, K. 368, 381
Baranowski, B. 171

Bard, J. R. 366
Barnartt, S. 472
Barnes, W. H. 27
Baron, N. M. 484
Barone, G. 62
Barrett, R. E. 61
Bartoli, F. J. 363
Bates, R. G. 305
Batten, R. C. 102
Baur, W. H. 74
Bearman, R. J. 151
Beauregard, D. V. 61
Beck, J. 324
Beers, Y. 27
Beke, Gy. 35, 317
Benck, R. F. 132
Ben-Naim, A. 68
Benson, S. W. 64, 345
Berecz, E. 135, 136, 324, 458, 459
Bergquist, M. S. 64, 72
Bernal, J. D. 19, 20, 21, 27, 32, 33, 34, 45, 53, 66, 281, 443, 446, 451, 482
Beronius, P. 472
Bezman, I. I. 372
Bezrukov, O. Ph. 228
Bhatia, R. N. 202
Biancheria, A. 228
Bierly, T. 432
Birch, J. N. 363
Birkett, J. D. 268, 464
Bjerrum, N. 27, 28, 187, 292, 311, 420, 426, 428, 431, 443
Blanckenhagen, P. 169
Blumenfeld, L. A. 322
Blumentritt, M. 319
Bockris, J. O'M. 71, 282, 372, 381, 385, 442, 443, 474, 475
Bogoliubov, N. N. 415
Bonner, O. D. 38, 51, 61, 74, 476
Borina, A. F. 453
Born, M. 86
Boyd, G. E. 69
Boyd, R. H. 268, 271, 465
Bradley, D. C. 372
Brady, G. W. 34, 71
Bragg, W. H. 27
Breitschwerdt, K. G. 68, 475
Breslau, B. R. 127
Bressel, R. D. 134
Broadwater, T. L. 283, 367

Broersma, S. 155
Bruinink, J. 289
Brush, S. G. 85, 86
Buchanan, T. J. 37, 70
Buchböck, G. 263, 468
Buckingham, A. D. 69, 70, 71, 419
Buijs, K. 38, 49
Burnelle, L. 73
Burnham, S. 35, 36
Burns, D. T. 122
Burton, R. T. 454, 463, 482
Buslaeva, M. N. 489
Byers, C. H. 151

Cachet, H. 313
Campbell, A. N. 313
Carman, P. C. 156, 225, 227, 325
Castel, J. F. 359
Chamberlein, J. 36
Chambers, J. F. 324
Chang, Pin. 168, 218
Chapman, S. 84
Chemla, M. 196, 239
Cheng, P. Y. 123
Choppin, G. R. 38, 49
Christensen, J. J. 350
Chuchalin, L. K. 478
Churaev, N. V. 99
Clough, S. A. 27
Cobble, J. W. 240
Cohan, N. V. 74
Cole, R. H. 31
Collie, C. H. 48, 72, 419
Collins, F. C. 153, 217, 342
Connick, R. E. 65
Conway, B. E. 69, 74, 281, 282, 372, 381,
    385, 474
Copeland, C. S. 64, 345
Carmo, M. 365
Courant, R. A. 102, 286, 333, 454
Coulson, C. A. 73
Covington, A. K. 293, 481
Cox, W. M. 113, 114
Crank, J. 145, 156, 176, 177, 226
Criss, C. M. 120
Cross, P. C. 35, 36
Csillag, K. 168
Cuddeback, R. B. 41, 221
Cukrowski, A. S. 171
Cunningham, G. P. 354
Currie, D. J. 325

Dahl, P. 372
Dahlberg, D. B. 63
Dahms, H. 245, 246, 390
Daly, J. 454, 463, 482
Danford, M. D. 34, 39, 40, 42, 44
D'Aprano, A. 307, 308, 372, 464
Darmois, E. 278, 443
Datevskii, V. T. 55

Davidson, W. L. 29
Davies, D. B. 102, 103
Davies, C. W. 129, 308, 326, 363
Davies, J. T. 64
Davis, C. M. 50
Davis, H. T. 298
Day, M. C. 310
Debye, P. 294, 322, 399, 419
Demeter, K. 168
Dennison, D. M. 27
Derjaugin, B. V. 99
Desnoyers, J. E. 69, 474, 475
Deubner, A. 129
Devell, L. 223, 232
Dexter, A. R. 103
Diamond, R. M. 69, 70, 71, 433, 438
Dierksen, G. H. F. 38
Dismukes, E. B. 334
Dnieprov, G. F. 372
Dole, M. 111, 317
Dolgaya, O. M. 333
Donato, J. D. 307, 308
Doran, M. 131
D'Orazio, L. A. 72
Douglass, D. C. 224, 226, 234
Drakin, S. I. 481
Drickamer, H. G. 41, 221
Drost-Hansen, W. 102
Drücker, K. H. 324
Dubinina, E. F. 474
Dubnikova, V. T. 489
Dudziak, K. H. 98
Duer, W. C. 305
Dullien, F. A. L. 180
Dumargue, P. 203
Duncan, A. B. 73
Duncan, J. F. 307
Dunlop, P. J. 171, 180, 206, 208
Dye, J. L. 316
Dymond, J. H. 81, 181

Eagland, D. 123
Ebeling, W. 358, 424
Edelman, I. S. 42, 45
Edwards, O. W. 198
Eicher, L. D. 81
Eigen, M. 65, 279, 282, 288, 321, 377,
    380, 406, 437
Einstein, A. 122, 123, 149
Eirich, F. 97
Eisenberg, H. 349
Eley, D. D. 59
Ellerton, H. D. 171
van der Elsken, J. 169
Emelyanov, M. I. 452
Emmerich, A. 136
Engel, G. 72, 456, 457
Enskog, D. 84
Erdey-Grúz, T. 108, 110, 212, 214, 229,
    230, 244, 264, 284, 285, 313, 359, 360,

363, 364, 366, 368, 370, 373, 374, 375, 376, 381, 382, 384, 385, 386, 388, 469
Ergin, Yu. V. 457, 468
Eucken, A. 45, 46, 66, 116, 282
Evans, D. F. 69, 116, 126, 271, 278, 283, 336, 354
Evans, M. W. 60, 66, 67, 401
Everett, D. H. 65
Ebert, G. 132
Ewell, R. H. 35, 96, 97
Eyring, E. M. 321
Eyring, H. 21, 35, 41, 42, 87, 90, 96, 97, 100, 101, 103, 106, 107, 130, 158, 161, 163, 164, 167, 218, 274, 276, 281, 467

Faber, M. P. 316
Fabry, T. L. 311, 370
Fajans, K. 67, 477, 487
Falk, M. 37
Falkenhagen, H. 128, 187, 298, 304, 305, 306, 319, 322, 358, 415
Fang, J. R. 372
Feakins, D. 472
Fenerli, G. N. 135
Fernández-Prini, R. 268, 304, 412
Ferse, A. 325
Fialkov, Yu. Ya. 107
Fick, A. E. 170
Finkelstein, B. N. 116
Fischer, H. F. 478
Fischer, L. 267
Fisher, I. Z. 105, 112
Flint, W. L. 478
Fodor-Csányi, P. 230, 244
Ford, T. A. 37
Forslind, E. 38, 43, 64, 67, 72, 75
Fort, R. J. 108
Foster, N. G. 363
Fourier, J. B. J. 170
Fowler, R. H. 27, 32, 33, 34, 45, 53, 66, 281, 414, 443, 446, 451, 482
Fox, J. J. 35
Franck, E. U. 98, 338
Frank, H. S. 44, 45, 47, 49, 60, 66, 67, 72, 116, 119, 194, 401, 415, 416, 420, 422, 423, 465, 466, 467, 486
Freed, S. 71
Freeman, J. G. 293
Freise, V. 180
Frenkel, D. 169
Freundlich, H. 215
Fricke, H. 356
Friedman, H. L. 60, 297, 463, 488
Friedrich, V. J. 168, 284
Fromherz, H. 293
Frontall, C. 99, 108, 377
Frumkin, A. N. 245, 390
Frysinger, G. R. 333, 345, 438
Fuoss, R. M. 129, 155, 166, 188, 190, 211,

295, 298, 302, 305, 306, 308, 311, 316, 324, 325, 370, 415, 422, 430, 434, 465

Galinker, V. S. 135
Garland, C. W. 198
Gaysin, N. K. 235
van Geet, A. L. 444
Giber, J. 115, 231, 317
Gierer, A. 45, 282
Giese, K. 452
Gillen, K. T. 224
Gillespie, R. J. 67
Gilkerson, W. R. 310
Gingold, M. P. 99
Girgis, Y. M. 371
Gjaldbaek, J. C. 63
Glietenberg, D. 210
Glueckauf, E. 73, 401, 419, 423, 431
Godbole, E. W. 238
Godzik, K. 372
Goffredi, M. 314, 362, 363, 372, 380
Golden, S. 63
Goldenberg, N. 377
Goldfeld, M. G. 322
Goldschmidt, H. 372
Goldschmidt, V. M. 451
Goncharov, V. V. 239
Gorbanev, A. I. 70
Gorbunov, B. Z. 62
Gordon, A. R. 155, 325, 336
Gossman, A. 244
Gosting, L. J. 208, 209, 211
Gosting, L. C. 192, 236, 238
Govindaswamy, S. 133
Grankina, Z. A. 478
Greco, P. W. 363
Green, H. S. 86
Greif, L. J. 415
Griffin, R. G. 366
Gronwall, T. H. 406, 415
de Groot, S. R. 170, 204
Grotthus, Th. 280, 282, 286
Grunwald, E. 369
Grüneisen, E. 111
Gubbins, K. E. 202, 207
Guggenheim, E. A. 23, 414
Gurikov, Yu. V. 39, 40, 66, 193, 194, 221, 415, 454, 487
Gurney, R. W. 66, 73, 112, 114, 120, 121
Gusev, N. I. 285, 465, 470, 471, 472
Guttman, C. 63
Güntelberg, E. 414

Haas, C. 222
Haase, R. 265, 324
Hafez, A. M. 313
Haggis, G. A. 37, 70, 370
Hagler, A. T. 49
Halban, H. 282
Hale, C. H. 215, 264, 469

Halpern, J. 246
Hamann, S. D. 286
Hammes, G. G. 371
Hammonds, C. N. 310
Hankins, D. 55
Harned, H. S. 70, 135, 180, 192, 197, 212, 236, 238, 334, 438
Harpst, J. A. 192
Harris, F. E. 70
Hartley, G. S. 145, 155, 156, 176, 177, 186, 226
Hashitani, T. 237
Hasselle-Schuermans, N. 299
Hasted, J. B. 36, 37, 48, 70, 72, 370, 419, 431, 445
Hechler, J. J. 68
Hendrikx, Y. 443, 463
Herskovits, T. T. 123
Hertz, H. G. 63, 72, 219, 415, 456, 457
Herzog, R. O. 152
Hessler, K. 267
Hidvégi, J. 108, 363
Highberger, G. 478
Hildebrand, J. H. 63, 81, 181, 449
Hills, G. J. 341, 342, 344
Hindman, J. C. 55, 70, 73, 474
Hinton, I. F. 444
Hirschfelder, J. 163
Hirschfelder, J. O. 87
Hoch, N. J. R. 224
Hofacker, G. L. 372
Holderith, J. 115
Holt, E. 192
Hooyman, G. J. 204
Horányi, Gy. 324
Horne, R. A. 98, 102, 117, 268, 278, 286, 293, 333, 345, 346, 362, 438, 454, 464
Hornig, D. F. 37, 49, 74
Howard, B. 344
Hranenko, S. P. 478
Hsia, K. L. 302
Hudson, R. M. 197
Huffman, E. O. 198
Huggins, M. L. 281
Humeau, P. 203
Hunt, J. P. 65, 447
Hunyár, A. 212, 214, 469
Hutton, P. 370
Hückel, E. 126, 281, 294, 399, 419
Hümbelin, R. 76

Illig, R. 152
Inzelt, G. 35, 230
Irani, R. R. 180
Irish, E. M. 161, 164
Isono, T. 127
Ivanova, L. V. 476
Izatt, R. M. 350

James, J. C. 308
Jancsó, L. 35
Jander, G. 443
Janz, G. J. 68, 180, 298, 476
Jaskichev, V. S. 239
Jensen, S. J. Knak, 169
Jobling, A. 342
Johnson, D. S. 98, 102, 333, 362, 454
Johnson, O. 67, 487
Jolicoeur, C. 60
Jones, G. 111, 317
Jones, M. M. 326
Jones, R. J. 233
Joos, G. 319
Justice, J. C. 302

Kaatze, U. 452
Kaimakov, E. A. 324
Kaminsky, M. 72, 112, 113, 114, 117, 119
Kampschulte-Scheuing, I. 434
Kaposi, O. 317
Kapustinsky, A. F. 373
Kariakin, A. V. 63
Karl, D. J. 316
Kartzmark, E. M. 313
Katchalsky, A. 349
Kaustinskii, A. F. 481
Kay, R. L. 69, 126, 271, 336, 354, 367
Kegeles, G. 228
Kelbg, G. 298, 305, 306, 414, 415
Kelly, T. M. 123
Kendall, J. 106
Kennedy, J. W. 227, 240
Kepert, D. L. 307
Kessler, Yu. M. 70, 304
Kett, T. K. 173
Khalifa, M. 325
Khalil, F. Y. 313
Khanh, T. C. 389
Khoo, K. H. 469
Kim, H. 192
Kim, S. K. 319
King, C. J. 151
Kiprianov, V. A. 198, 424
Kirby, P. 372
Kirkwood, J. G. 84, 86, 371, 415
Kivelson, D. 169
Klein, G. P. 27
Klopov, V. J. 488
Klotzin, M. P. 180
Knapp, P. S. 478
Knoche, W. 371, 444
Kochnev, I. N. 62
Koefoed, J. 49
Koeller, R. C. 41, 221
Kohlrausch, F. 335
Kokovina, G. V. 449
Kondratiev, V. P. 337
Kor, S. K. 437

Korson, L. 102
Kortüm, G. 292, 350, 372
Kostrova, L. J. 451, 457, 468
Kraeft, W. D. 308, 358
Kraus, C. A. 271, 311, 354, 434, 461
Kraus, K. A. 241
Kremp, D. 304, 306, 414
Kreschek, G. C. 68
Kreshtov, G. A. 73, 488
Krienke, H. 424
Krishnan, C. V. 488
Krüger, D. 215
Kucheryavenko, K. S. 234
Kudar, H. 152
Kudish, A. I. 102
Kudriavtsev, B. B. 474
Kudryavtseva, I. V. 129
Kuecker, J. F. 126
Kugler, E. 108, 285, 360, 363, 368, 374, 375, 376, 381, 385, 388
Kulkarni, M. V. 212
Kuprik, A. V. 448
Kurant, R. A. 51
Kurmgalz, B. S. 114, 129, 271, 453
Kutschker, A. 210
Kuzin, I. A. 478
Kvyat, E. I. 480

Laforgue, A. 389
Laforgue-Kantzer, D. 389
Laity, R. W. 193, 212
Lakshmanan, G. 137
Lakshminarayanan, G. R. 68, 180
LaMer, V. K. 406, 415
Lamm, O. 180, 205
Lange, E. 480, 481, 483, 489
Lantelme, F. 239
Latimer, W. M. 60, 66, 69, 71
Launsbury, M. 324
Lawrence, A. S. C. 342
Lawyer, C. 257
Lebed, V. I. 488
Lee, W. H. 443, 474
Legard, R. J. 246
Legin, E. K. 66, 193, 194, 454, 487
Leighton, P. A. 35, 36
Leist, M. 298
Lengyel, S. 76, 115, 169, 231, 241, 284, 317
Lennard-Jones, J. 26, 73, 74, 84
Lentz, B. R. 47
Lestrade, J. C. 313
Lévay, B. 229, 230, 244
Levich, V. G. 245, 390, 424
Levien, B. J. 316
Levine, S. 433
Levitskaia, N. F. 371
Levy, H. A. 29, 34, 39, 40, 42, 44
Li, J. C. M. 168, 218
Liashchenko, A. K. 67

Lielmez, J. 202
Lifson, S. 349
Lilley, T. H. 293, 481
Lin, J. 283
Lindenbaum, S. 69
Linton, H. 282, 372, 381, 385
de Lisi, R. 362, 372, 380
Lisnianskii, L. I. 24
Litovitz, T. A. 50
Litvinenko, I. V. 472
Liu, C. T. 321
Loflin, T. 151
Lohse, E. 324
Longsworth, L. G. 154, 264, 469
Lopushanskaia, A. I. 173, 206
Lorenz, P. B. 187
Lown, D. A. 287
Lozhkina, L. G. 304
Lozhnikov, V. A. 371
Luck, W. A. P. 38, 50, 63, 74
Lumry, R. 246
Lundgren, J. O. 279
Luz, Z. 447
Lyashchenko, A. K. 452
Lyons, P. A. 192, 197, 212

Maass, O. 372
Macedo, P. B. 299
MacInnes, D. A. 349
de Maeyer, L. 288
Magnusson, L. B. 75, 487
Majthényi, L. 284, 285, 364, 383
Malenkov, G. G. 43, 55
Malinowski, E. R. 478
Malpass, V. E. 129
Malvinova, V. A. 317
Mamajek, R. C. 363
Mangold, K. 338
Marchi, R. P. 41, 42
Marcinkowsky, A. E. 241
Marchiano, S. L. 210
Marinin, V. A. 207
Markova, L. S. 239
Marsh, R. E. 43
Marshall, W. L. 345
Martin, A. E. 35
Martinov, G. A. 294
Marum, E. 372
Masterton, W. L. 63, 432
Mastroianni, M. J. 120
Matheson, A. J. 102, 103
Mathieu, I. P. 324
Maximova, I. N. 337
Maxwell, J. C. 84
Mayer, G. E. 68, 180, 415
McCall, D. W. 226, 234
McDuffie, G. E. 363
McGlashan, M. L. 23
McKoy, V. 43
McLaughlin, E. 95, 151, 219

Megaw, H. D. 29
Meixner, J. 170
Mikhailov, I. G. 73
Mikhailov, V. A. 111, 365
Mikhail, S. Z. 372
Miller, A. A. 49
Miller, D. G. 171, 187, 192, 206, 211
Miller, I. F. 127
Miller, M. L. 131, 324, 325
Miller, S. 240
Millero, F. J. 102
Mills, R. 117, 124, 223, 227, 238, 240, 241, 244
Minc, S. 112
Minton, A. P. 27, 36
Miruchulava, I. A. 298
Mishchenko, K. P. 76, 129, 441, 480, 481, 482, 483, 484, 486
Mitchell, A. G. 111, 365
Moelwyn-Hughes, R. H. 70
Monica, M. D. 463
Monk, C. B. 233, 293, 308, 433
Monroe, K. P. 106
Moore, W. R. 108
Morgan, J. 34, 35
Mori, Y. 312
Moskovitz, J. V. 55
Moulik, S. P. 124, 125, 358
Moynikan, C. T. 299
Mozumder, A. 433
Mulliken, R. S. 26
Muradova, G. A. 63
Musbally, G. M. 202
Müller, G. T. A. 197
Müller, H. 406
Myers, B. R. 333, 345, 438

Naberuzhin, Yu. I. 63, 286, 62
Nagaraja-Rao, K. 137
Nagy-Czakó, I. 230, 285, 366, 368, 370, 381, 383, 384
Nakanishi, K. 181
Namjot, A. Yu. 60
Nancollas, G. H. 307, 308, 425, 437
Narten, A. H. 39, 44, 54
Nernst, W. 170
Neronov, Yu. I. 221
Neumann, H. P. 143
Nevolina, N. A. 32, 98, 454
Newitt, D. M. 342
Newman, D. S. 257
Newman, K. E. 481
Némethy, G. 49, 60, 68
Nielsen, J. M. 240
Nightingale, E. R. 114, 119, 121, 126, 132, 455, 456
Nikiforov, E. A. 234, 452, 457
Nilsson, A. M. 472
Nir, S. 158
Northey, H. L. 288

Nothnagel, K. H. 244
Noyes, R. M. 69, 70, 71

Oakes, J. 365
O'Donnel, I. J. 208
Ohm, G. S. 170
O'Konski, C. T. 70
Oliver, B. G. 68, 476, 313
O'Neil, J. R. 52
Onsager, L. 129, 155, 166, 170, 171, 187, 188, 190, 204, 211, 295, 297, 298, 300, 305, 306, 308, 316, 319, 320, 321, 324, 325, 350, 422
O'Reilly, D. E. 219
Orentlicher, M. 52
Oscarson, J. L. 350
Ostroff, A. G. 116
Ottar, B. 229, 218
Outhwaite, C. W. 406
Owston, P. G. 28
Owen, B. B. 135, 334, 335, 336
Ozasa, T. 181
Padova, J. 488
Palei, P. N. 285
Pamfilov, A. V. 173, 206, 333
Parker, A. J. 312
Parker, H. W. 192
Passeron, E. J. 271
Passynski, A. 472
Patsatsiya, K. M. 488
Patterson, A. 321
Pauling, L. 27, 36, 42, 43, 44
Pearson, D. 345
Pedersen, L. G. 362
Penot, F. 203
Perram, J. W. 53
Perrault, G. 283
Perrin, F. 152
Peshchevitskii, B. I. 478
Peterson, S. 29
Petrea, I. C. 105
Phillips, H. O. 241
Phillips, W. D. 447
Pikal, M. J. 198, 244
Pilling, G. 123
Pin Chang 168
Pinchukov, Yu. E. 29
Pinkerton, J. M. M. 105
Pinsker, G. Z. 21
Pisarzhevskij, M. 335
Pitts, E. 298, 305, 310, 316
Pitzer, K. S. 69, 71
Powell, R. E. 60, 96, 100, 106, 167
Platteeuw, J. C. 43
Podestá, J. J. 210
Podolsky, R. J. 231
Pogány, É. 469
Poirier, J. C. 415, 416
Polestra, F. M. 228
Polissar, M. J. 242

Polson, A. 154
Poltoratskii, G. M. 482
Pominov, I. S. 484
Pople, J. A. 26, 35, 36, 46, 73, 74
Popov, G. V. 452
Posey, F. A. 437
Povarov, Yu. M. 70
Prigogine, I. 170, 204
Provencher, S. W. 350
Prue, J. E. 304, 412
Purppacher, H. R. 93, 224

Quist, A. S. 345
Quist, A. W. 44, 45

Rabinovich, I. B. 453
Radchenko, I. V. 454
Raffel, H. 153, 217
Rahman, A. 219
Ramana-Murti, M. V. 368
Rayleigh, J. W. S. 84
Reff, I. 271
Reich, A. 108
Reinfelds, G. 192
Remy, H. 443
Renkert, H. 338
Résibois, P. 298
Rice, S. A. 54
Richards, R. E. 288
Rideal, E. K. 64
Riesenfeld, E. H. 443
Righellato, E. C. 308
Rilj, N. V. 29, 46
Ritson, D. M. 48, 72, 419, 431, 445
Rizhkov, E. M. 312
Rizk, H. A. 371
Rizzo, E. 62
Roberts, N. K. 288
Robinson, A. L. 486
Robinson, C. V. 42, 45
Robinson, R. A. 42, 70, 199, 305, 358,
    406, 413, 417, 419, 431, 432, 438, 461,
    473
Rogov, V. A. 63
Rosseinsky, D. R. 482, 431
Roseveare, W. E. 96, 100, 106, 167
Rosskopf, F. 136
Rothman, L. S. 27
Rotowska, A. 112
Rowlands, L. G. 233
Rozenthal, D. K. 433
Rozental, O. M. 76
Rozenstok, Yu. L. 149
Röntgen, W. C. 31
Ruff, I. 168, 169, 247, 284, 391
Rutgers, A. J. 443, 463
Ryss, A. I. 454

Sadek, H. 313
Salomon, M. 281

Saluja, P. P. S. 443, 474, 475
Samoilov, O. Ya. 31, 33, 38, 40, 41, 42,
    45, 46, 53, 64, 65, 71, 73, 98, 119, 220,
    233, 239, 242, 243, 324, 442, 444, 448,
    449, 450, 452, 453, 454, 455, 468, 489
Sändig, R. 308
Sandved, K. 406, 415
Sarkisov, G. N. 55
Satoh, T. 131
Sauer, F. 180
Sauermann, P. F. 324
Schachmann, H. K. 123
Schaaf, H. 126
Scheraga, H. A. 47, 49, 60, 68
Schiavo, S. 312, 370
Schneider, H. 68, 444
Schober, C. 257
Schönert, H. 173, 176, 206, 210
Schreiber, P. 317
Schulman, R. G. 447
Schwabe, K. 324, 325, 463
Schwarz, M. 425
Schwarzenbach, G. 425
Sedlacek, J. 244
Seifer, A. L. 32, 98, 454
Senatore, L. 463
Senior, W. A. 50
Shakhaparonov, M. J. 303
Shamim, M. 210
Shedlovsky, T. 305, 349
Shkodin, A. M. 371
Shoolery, J. N. 71
Shraiber, L. S. 68
Shropshire, J. A. 192, 212
Shuiskii, S. J. 286
Shull, C. G. 29
Sidebottom, D. P. 316
de Sieno, R. P. 363
Simha, R. 97, 122, 123, 124
Symons, M. C. R. 54
Sinanoglu, O. 43
Sirnikov, Yu. P. 70, 73
Siver, P. Ya. 206
Skidau, N. A. 448
Slansky, C. M. 69, 71
Smid, I. 426, 448
Smirnova, M. F. 478
Smith, J. A. 288
Smith, J. E. 334
Smithson, J. M. 437
Snowden, B. S. 116
Solovkin, A. S. 452
Sokolov, D. S. 454
Spink, C. H. 478
Spinnler, J. F. 321
Spiro, M. 265, 316, 365
Stackelberg, M. V. 43, 210
Stanley, E. M. 102
Stearn, E. A. 161, 164, 281
Steckel, F. 102

Steel, B. J. 334
Stein, L. H. 225, 227
Stein, L. Z. 156
Stein, W. D. 158
Stillinger, F. H. 55, 415, 433
Stockmayer, W. H. 198
Stokes, J. M. 324, 356
Stokes, R. H. 42, 70, 117, 124, 150, 158,
    180, 188, 190, 197, 199, 238, 241, 298,
    304, 314, 316, 323, 324, 356, 358, 406,
    413, 417, 419, 431, 432, 438, 461, 463,
    473
Strehlow, H. 444
Subramanian, S. 478
Sugden, J. N. 471
Sukhotin, A. M. 312
Suryanarayana, C. V. 131, 132, 134, 324
Sutherland, W. 149, 151, 153
Svirmickas, A. 55
Swain, C. G. 278
Swift, T. J. 65
Szilágyi-Győri, E. 230, 244

Tait, M. J. 298
Tamamushi, R. 127
Tamás, J. 115, 231, 232, 241, 317
Tamm, K. 437
Tanaka, K. 237
Tarasov, A. P. 448
Tate, J. F. 326
Taube, H. 65, 437, 447
Taylor, H. S. 160
Terehova, S. D. 68
Tereskovich, M. O. 448
Tham, M. K. 207
Than, A. 363
Thirsk, H. R. 287
Thomas, L. 372
Thomas, G. O. 363
Thompson, P. T. 420, 422, 423
Tikhii, S. P. 68
Toan, N. H. 363
Tong, S. 198
Tourky, A. R. 371, 372, 380
Triolo, R. 314, 464
Troshin, V. P. 267, 317, 324
Tsvetkova, L. B. 206
Turner, J. C. R. 468
Turq, P. 239
Tyagai, V. A. 135

Uedaira, H. 119, 453
Ujszászy, K. 232
Ulbricht, H. 306, 414
Ulich, H. 443
Ushakova, E. M. 228

Vand, V. 50, 123
Valenchuk, M. I. 448
Valleau, J. P. 306

Vaslow, F. 70
Váli, A. 469
Vdovenko, V. M. 66, 193, 194, 454, 487
Verhoek, F. H. 372
Venkatesan, V. K. 131, 133, 324
Verrall, R. E. 69, 474
Verwey, E. J. W. 27
Vértes, A. 241, 317
Vértes, G. 135
Vitagliano, B. 62
Vitagliano, V. 197
Vituccio, T. 126
Vogelhut, P. O. 52
Vogrin, B. F. J. 478
Voice, P. J. 472
Volkova, S. A. 448
Vollmar, P. M. 67, 476
Volmer, M. 279
Volosova, Z. V. 453
Voroncov-Veljaminov, E. M. 228
de Vries, T. 215, 264, 469

van der Waals, J. H. 43, 169
Waite, R. O. 478
Walden, P. 154, 335, 353, 372
Waldron, R. D. 70
Walker, R. D. 202
Wail, T. T. 37
Walrafen, G. E. 37, 476
Wang, J. H. 42, 45, 46, 153, 223, 224, 227,
    228, 231, 240
Wardlaw, W. 372
Warren, B. E. 34, 35
Washburn, E. W. 263, 468
Wassermann, A. 342
Wear, J. O. 366
Weidemann, E. G. 287
Weiss, A. 244
Wells, C. F. 481
Wen, W. Y. 47, 60, 66, 72
Wenck, H. 350
Wendorff, J. 132
Wendt, R. P. 174, 198, 210
Werblan, L. 112
Weres, O. 54
While, J. W. 69
White, R. J. 433
Whorton, R. 363
Wicke, E. 279, 282, 377, 380, 406
Wien, M. 317
Williams, J. W. 297, 371
Williams, R. J. P. 437
Willmott, A. S. 472
Wilski, H. 372
Wilson, W. 319
Wirtz, K. 45, 220, 282
Wishaw, B. F. 180, 197, 323
Woessner, D. E. 116
Wolf, D. 102

Wolf, R. 324
Wolfenden, J. H. 113, 114
Wollan, E. O. 29
Wolz, H. 475
Wood, M. 55, 72
Woolf, L. A. 206, 211, 238, 240, 241, 372
Woolsey, G. B. 38, 51, 61, 74, 476
Wycckoff, J. C. 478
Wynne-Jones, W. F. K. 111, 365

Yadav, C. C. 368
Yagodarov, V. P. 452
Yakushevskii, B. M. 481
Yashkichev, V. I. 46, 47, 53
Yastremskii, P. S. 53, 448, 449, 454, 468
Yemelianov, M. I. 234, 235

Yergin, Yu. V. 451
Yermakov, V. I. 322
Yevseiev, A. M. 322
Yokoi, M. 321
Young, R. P. 117, 346, 362
Yuzhkevich, V. F. 337

Zafar, M. S. 36
Zaitseva, A. M. 105, 112
Zatsepina, G. N. 54, 280, 288, 289, 374
Zawoyski, C. 126
Zdanovskii, A. B. 136
Zolotarev, V. M. 476
Zundel, G. 287
Zwanzig, R. 268, 271, 465
Zwolinski, B. J. 81

SUBJECT INDEX

ab initio calculations 47
absolute mobility 183, 253
— reaction rate theory 281
— — — — of the transport of electricity 274
— value of the diffusion coefficient 218
— — — — — rate 160
absorption of ultrasound 104
— spectra 475
accelerating of self-diffusion 233
— — ions 266
acid-base reaction 26
activated jumps 231, 242
— state 91
activation energy 181
— — as the measure of hydration 243
— — of conduction 332
— — — — in non-electrolyte solutions 388
— — — diffusion 161, 193, 194
— — — exchange of water molecules 65
— — — ionic migration 341
— — — jumps 220
— — — migration 276
— — — protonic conduction 286
— — in solutions
— — — displacement 91
— — — — hole formation 100
— — — — viscous flow 81, 95, 100
— — — — — in non-electrolyte solutions 360
— — — rotation 119
— — — self-diffusion 46, 224, 239, 243
— — — structural 101
— — of transfer in hole 100
— — transport phenomena 102
activation enthalpy of water 102
— entropy of viscous flow 103
— volume 342
activity 146
— coefficient 146, 416, 417
— — of concentrated solutions 420
— — — ions 407
— — from the interaction between ions 409
addition-elimination mechanism 278
additivity of coefficient 113
— — enthalpy of hydration 481
— — fluidity 106

— — the equivalent conductivity 325
adhering of ions 489
adsorption of non-electrolyte 471
— — water on ions 444, 455
alcohol-water solutions 283, 285, 312, 313, 361
amphiphilic ions 71
Andrade equation 103, 130, 131
anisotropy of diffusion 323
anomalous conduction 278, 288
— liquids 31, 101
— properties of water 31
— water 99
aperipheric hydration 455
apparent diffusion coefficient 168
Arrhenius equation 80, 102, 130, 181
— theory 396
associated liquids 101, 228
— molecules 45
associating electrolytes 398
association 24, 31, 41, 69, 70, 116, 182, 196, 257, 265, 291, 294, 302, 307, 309, 331, 337, 347, 370, 397, 398, 424
— and complex ion formation 436
— — dissociation 424
— constant 309, 311, 362, 429
— — three particles 434
— theories 45
asymmetric electrolytes 433
autoprotolysis 288
average chemical potential 185
— diffusion coefficient 148

Bachinskii rule 82, 99
back-diffusion 214
B coefficient, *see* coefficient *B*
bending of the hydrogen bonds 35
— vibration 477
binary electrolyte 184
Bingham theory 82
bipolar conduction 260
Bjerrum theory 311, 420, 426, 431
Bockris-Saluja theory 474
Boltzmann distribution 400
bond angle 26
broken hydrogen bonds 36
bulk viscosity 270

cages 44, 63, 87
Cailletet-Mathias rule 21

cavities 20, 21, 30, 32, 33, 38, 40, 44, 59, 63, 64, 67, 87, 98, 111, 220, 222, 365, 383, 451, 453, 454, 463
cell 220, 243
central ion 399
characteristic distance 237, 296, 404, 406
charge numbers 184
— transport by means of electrons 390
chemical hydration 442
— potential 68, 145, 170, 173, 185, 186, 417
— — of dissolved electrolytes 407
classification of the structural theories 53
closed associates 110
cluster 62
— model 47, 60
— theory 76
coefficient B 114
— — and the entropy 121
— — — — mobility 120
— — — — size of ions 117, 118
— — — — temperature 117
collisions of ions 306
common diffusion 141
— — coefficient 145
complexes between water and dioxan 371
complexes, formation of 110
— ion formation 437
components 203
compressibility 342, 472, 474
concentration dependence of conductivity 258, 300, 302
—, effect of hydration 416
— exponent 112
— gradient 142
— potential 133
conduction coefficient 170, 171
conductivity 114, 189, 193, 256, 276, 289, 291, 300, 317, 325, 331, 339
— and hydration 470
— at high-temperatures 337
— in magnetic field 322
— — solvent mixtures 359
— mass of ions 267
— of concentrated solutions 302, 323
— — electrolyte mixtures 325
— — hydrogen ions in non-electrolyte solutions 374
— — hydroxide ions in non-electrolyte solutions 384
— — ions 355, 431
— — more than two ions 260
conductivity of weak electrolytes 259
— — — —, solutions 347
— under high field strength 317
contact ion pairs 307
continuity equation 255
convection 143
conventional activity coefficient 417
convective diffusion 143

cooperative hydrogen bonds 27, 47
coordination dehydration 484
— — number 17, 19, 21, 28, 33, 34, 36, 39, 474, 483
— — of hydration 442, 448
— — — H_3O^+ 68
— sphere 17, 36, 73
coulombic forces 395, 399
— interactions of ions 294
counter-diffusion 214
covalent molecules 293, 424
cross-effects 13, 149, 171, 182, 183, 204, 207
crystal-like structure of liquids 18
crystallization 76
crystallographic radius 269, 270, 412
cube root law 323, 420
— — — of conductivity 259
cubic expansion coefficient 332
current 260
— density 255, 256, 260
cyclic polymerization 47
cylindrical diffusion 148

Debye–Falkenhagen effect 321, 322
— –Hückel–Onsager theory 273, 294
— –Hückel limiting law 420
— — — — of the activity coefficient 410
— — theory 128, 187, 399, 405, 416, 419, 426, 470
— — —, concentration limit 421
defects 222
deformation of ionic atmosphere 295
— polarization 269
degree of association 257, 289, 291, 428
degree of dissociation 256, 257, 289, 291, 292, 347
dehydration 484
density of liquids 32
dependence of conductivity on concentration 258
depolimerization 116
desalination 325
deuterium bonds 32
Devell-Wang theory 232
development of the steady state in diffusion 146
deviations from the limiting law 302
De-Waele theory of viscosity 83
diameter of ions 296, 304, 410, 412, 470
— — molecules 88, 220
dielectric constant 352
— — and conductivity 310
— — in vicinity of ions 431
— — of non-associated polar liquids 52
— — — solutions of non-electrolytes 362
— — — water 52
— — — — adjacent to ions 445

— frictional effect 465
— relaxation 222, 268, 452
differential diffusion coefficient 148
diffusion 13, 141
— against concentration gradient 145
— and isotope exchange 243
— — phenomenological coefficient 173
— — shape of molecules 152
— coefficient 145, 150, 151, 161, 164, 174, 177, 180, 189, 199, 204, 276, 277
diffusion coefficient and concentration 148
— — — ionic size 193
— — — the structural change in solutions 193
— — of a binary electrolyte 186
— — — hydrogen ions 210
— — — ion pairs 197
— — — the individual ions 193
— constant 142
— in concentrated solutions 175
— — in multicomponent systems 204
— — non-ideal mixtures 154, 163
— of electrolytes 181, 184
— — — in concentrated solutions 199
— — — — solutions of non-electrolytes 212
— — — mixtures 207
— — hydrated ion 195
— — hydrogen ion 215
— — ion in solvent mixtures 181
— — large molecules 158
— — mixtures 203
— — non-electrolytes 141
— potential 172
— rate 160
diffusion relation with polar coordinates 148
— through interface 146
— without activation energy 181
dilatating liquids 82
dilution 258
— law 310, 347
dioxan-water solutions 285, 310, 313, 366, 382
dipole-dipole, interactions 22, 61
— moment 22
— — in liquid water 52
— — of the H_2O 27
disorder in liquids 18
dispersion forces 22
— of conductivity 321
dissociation 116, 196, 257, 265, 290, 320, 331, 337, 347, 396, 397
dissociation and association 424
— constant 291, 320, 347, 429
— under high field strength 319
distribution function 17, 88
— — theories of liquids 16
— of ions 399

— — solvent molecules around ions 446
donor-acceptor bonds 65, 279
double layers 255
driving force of diffusion 145
Dufour effect 172
dwelling time 453
— — of water around ions 449
— — of a water molecule 65, 450
dynamic viscosity 104

effect of a magnetic field on the diffusion 202
— — dissolved ions on the structure of water 64
— — field strength 317
— — hydrogen ions on self-diffusion 234
— — ionic charges on viscosity 125
— — isotopes on self-diffusion 223
— — non-electrolytes on the conductivity 351
— — organic ions on the structure of water 68
— — solute molecules on the structure of water 59
— — solutes on self-diffusion 227
effective diffusion coefficient 168
— radius 32, 40
— rigid volume 127
Einstein effect 118, 122
— equation 122, 124, 125
—-Nernst equation 225
—-Stokes law 155
electric conduction 13
— current 172
— — density 255
— field strength 253, 297
— force 253
— interactions in tracer diffusion 239
— potential gradient 255
electrochemical double layers 255
— potentials 184
electrode processes 255
electron diffusion in solutions 245
electrolytic conduction 253
electroneutrality condition 182, 255, 400
electronic conduction in solutions 390
electrophoretic effect 295, 299, 301, 318, 322, 336
— — on diffusion 188, 191, 200, 236
electrostatic interaction 128, 188, 350, 399
— — of ions 294
— — under high field strength 318
— theory of ion-water interaction 75
— — — strong electrolytes 395
electrostatics 399, 402
electrostriction 64, 73, 340, 472
elementary processes in ionic migration 273

— step of self-diffusion 222
— — — conduction 274
elimination-addition mechanism 278
empirical law of conductivity 259
energetical dehydration barrier 484
energy barrier 90, 181, 216, 281
— of adhering 489
— — hole formation 87
— — hydration 488
— — interaction 22
Engel–Hertz theory 457
enthalpy of activation 341
— — — of viscosity 130
— — chemical hydration 480
— — dissolution 479
— — evaporation 87, 96
— — hydration 69, 479, 480, 482
— — — of electrolytes 480
— — — — ions 480, 482
— — solutions 63
— — transfer 63
enthalpy, real of hydration 480
entropy of hydration 69, 121, 488
— — activation of viscosity 96, 130
— — dissolution 59
— — — and hydration 485
— — melting 100
— — mixing 217
equivalent conductivity 114, 189, 193, 256, 276, 289, 290, 291, 300, 325, 331, 339
— — of ions 257, 262
ethyloxonium ion 279
evaporation 87
excess entropy 485
— viscosity 485
exchange mechanism of the hydrate sphere 278
— of water molecules around ions 449
— — — — in the hydrate sphere 447
— rate of molecules 65
excluding ions from conduction 262
exoeffects 483
expansion 332
— of e⁻ˣ 402
experiments indicating hydration 458
extra-conductivity 283
extrapolation of limiting conductivity 259
extrapolation of transference number 262
Eyring equation for diffusion 158
— theory 21, 97, 100, 101, 218, 231, 273, 274, 278
— — of diffusion 156, 167
— — — viscosity 83, 90

Falkenhagen theory 305
Faraday number 257
Fick law 142, 147, 174, 180, 209
field strength dissociation 320

— — effect 317
flexibility of the hydrogen bonds 35
flickering cluster model 47, 60
flow 141, 142, 170
fluctuations 216, 303
fluidity 80, 106
fluidized vacancies 21
flux 80, 141, 142, 170, 176, 183, 185, 204, 253
— due to electron transfer 246
— of ions 255
force acting on the ions 253, 266
forced diffusion 141
forces beetwen ions 419
— — — and solvent molecules 419
Forslind theory 43
Frank–Wen theory 278
free energy of transfer 64
— enthalpy 170
— — barrier 156
— — of activation of viscosity 96
— — — ionic migration 274
— — — solvation 488
— — — the transfer 472
— ions 397
— volume 82, 87
frequencies of jumps 220
Fricke equation 357
frictional coefficient 150, 152, 180, 211
— force 266, 290
— resistance 150, 152, 154, 182, 188, 266, 268, 270
Friedman theory 297
Fuoss–Boyd–Zwanzig theory 465
— –Kraus theory 311
— –Onsager equations 190, 245, 308, 309, 312, 430
— — theory of mixtures 325
— –Shedlovsky theory 311

gas hydrate 44
Gibbs–Duhem equation 173
globules 169
Gordon relation 155
Gosting theory 208
gradient of concentration 142
— — enthalpy 146
Grotthus mechanism 280, 282, 286

half-life of water molecules in the hydrate sphere 447
Hall effect 258
Hartley–Crank equation 179
— — theory 226
— –Nerst equation 186
heat conduction 13, 146
— expansion coefficient 343
— flow 171
— of evaporation 87, 95, 102
— — dissolution 59, 479, 483

— — transfer 467
heavy water 32, 53
— — solutions 116, 463, 468, 488
high density regions in liquids 20
hindering effect 227
Hittorf method 262
— -type transference numbers 263
hole 33, 81, 87, 90, 96, 99
— formation 87, 100, 160, 221
— movement 87
hydrate sphere 194, 213, 269, 304, 444, 448
— water 444
hydrated ion 448
hydration 53, 64, 69, 73, 105, 113, 116, 124, 125, 133, 193, 199, 228, 231, 234, 243, 270, 294, 307, 312, 324, 352, 369, 395, 412, 432, 433, 441, 452, 458
— and activity coefficient 470, 473
— — conductivity 470
— — compressibility 472
— — diffusion 473
— — entropy of dissolution 485
— — ionic mobility 460, 466
— — optical properties 475
— — pressure 453
— — solubility 471
— — temperature 453
— — the translational movement of ions 448
— — ultrasonic absorption 474
hydration by chemical bonds 447
— effect on self-diffusion 228, 232
— enthalpy 479
— —, additivity 481
— entropy 121
— in D_2O 463, 468, 488
— — mixtures 488
— — water-dioxan mixtures 464
—, influence on concentration 416
— limit 482
— number 61, 124, 199, 202, 213, 233, 263, 270, 419, 443, 458, 460, 462, 464, 468, 470, 472, 473, 478
— of anions 70
— — cations 69
— — H_3O^+ 235
— — water 444
— sphere 488
hydrodynamic migration in non-electrolyte solutions 353
— — of oxonium ions 283
— mobility of hydroxide ion 280
— — — the oxonium ion 279
— theory 14
— — of anomalous conduction 288
hydrodynamic theory of diffusion 149
— — — electrolytic conduction 266
hydrogen bonds 22, 26, 29, 32, 35, 37,
44, 46, 48, 51, 60, 62, 67, 70, 72, 73, 74, 99, 116, 215, 282, 352, 354
— — in ion pair formation 433
— ion 193, 202, 269, 278, 372
hydrophilic water molecules 40, 45
hydrophobic cations 354
— hydration 60, 271
— ions 433
— water molecules 39, 40, 45
hydrophobicity of hydrogen ion 288
hydroxide ions 269, 278, 280, 372

ice 27
iceberg 60, 119, 401, 456, 457
— effect 442
— theory 59
ice-like structure of water 33, 42, 66
— molecules 45
ideal electrolytes 293
— solutions 408
imperfection theory 43
impulse 79
— flow 80
independent diffusion of ions 193
— migration of ions 257, 290
individual ionic mobility 266
— tracer diffusion coefficient 236
induced dipole 22
infra-red spectrum 74, 477
inner-sphere association 437
— — ion pair formation 437
interactions 61
— between the dissolved species and the solvent 489
— — — ions and the solvent 115, 441
— of ions 187, 196, 258, 395, 396, 399, 421
— — the ionic charges 188
— with the solvent 416
interdiffusion coefficient 145, 177, 179, 226
— of isotopes 217
intermolecular forces in liquids 21
internal friction 79
— pressure 133, 343
— structure 17
interstitial model 53
interstitial solutions 44
intrinsic concentration exponent 112
— diffusion coefficient 177, 178
— — flux 177
— self-diffusion coefficient 217
— transference numbers 263, 264
ion cloud 128
— –dipole force 445
— hydrate 447
— interactions 421
— — with the solvent 416
— pairs 196, 265, 257, 291, 292, 307, 314, 347, 397, 398, 424, 428, 476

— pair formation 133, 337, 426, 437
— radius 271
— triplet 397, 434
— -water interaction 75
ionic atmosphere 128, 187, 294, 399, 401, 422
— collisions 306
— conductivity 257
— diameter 304, 412, 420
— Hall effect 258
— migration 253, 341
— mobility 185, 253, 318, 331
— pairs 74
ionic radius 406, 451, 460
— sizes 304
— strength 300, 406
— — fraction 135
ionization constant 348
— of water 321
irrotational bond 70
isobar energy of activation 342
— parameters 342
isochore energy of activation 342
— parameters 342
isothermal diffusion 141
isotope exchange 243
isotopic composition and hydration 453

Jones–Dole equation 112, 113, 118, 125, 126, 127, 129, 132
— — — of transference number 317
jumps 90, 157, 164, 166, 216, 220, 231, 242, 276

Kendall–Monroe equation 106
kinematic fluidity 136
kinetic factors in hydration 449
— -statistical theory of electrolyte solutions 303
— theory of diffusion 156, 187
— — — electrolytic conduction 273, 289
— — — viscosity 85
— term 306
Kohlrausch law of independent migration of ions 257, 290
— square root law 258

Lamm theory 180
Laplace operator 147
laser light 192
— -Raman spectroscopy 476
lattice energy 479
— imperfections 18
— -like structure of liquids 18
— — — — water 31, 53, 281
— theory of liquids 16, 18
law of electroneutrality 182, 204
— — mass action 429
Lennard-Jones equation 95

— — potential 23, 219
lifetime 194
— of clusters 48
— — the hydrate sphere 234
limit in measuring the accurate conductivity 258
— of total hydration 76, 438
limiting conductivity of ions 257, 262, 355
— equivalent conductivity 257, 259
— law 325
— — for tracer diffusion 238
— — of conductivity 300
— mobility 257, 355
— value of the equivalent conductivity of weak electrolytes 259
linear diffusion 141
— law 170
liquid as damaged solid 449
— structure 16, 216
— water properties 31
— — structure 31
localized hydrolysis 438
London forces 22
lone-pair electrons 26
long-range hydration 71
loosening of the structure of water 66

macroscopic viscosity 270
magnetic field and conductivity 322
main diffusion coefficient 208
mass action law 429
— -dependence of self-diffusion 223
— flow 142, 171
— flux 142, 175, 204
— transfer 254
— transport due to electric potential 255
maximal hydration number 458
mean activity coefficient 186, 410, 414, 415
— chemical potential 185
mechanical model of the structure of liquids 19
mechanism of ionic migration 278
methyloxonium ion 279
microviscosity 125, 202, 270, 353, 467
migration 141, 253, 260, 341
— of hydrogen ion 288
— — hydroxide ion 289
— — large monovalent ions 271
migration velocity of ions 267, 275
mixing of electrolytes 193
mobility 150, 182, 185, 189, 193, 253, 262, 264, 267, 289, 318, 325, 331, 341, 348, 357, 395
— and hydration 460
— — mass of ions 267
— in non-electrolyte solutions 365
— of hydroxide ion 384

— — molecules 80
— — tracer ions 236
— — water molecules 235, 395, 448
— — — — adjacent to ions 450
molar conductivity 256, 260
— friction 180
molecular orbital 14, 26
— theories of diffusion 169
molecules 257, 424
monomeric water molecules 37
M.O. theory 26
moving boundary method 262
multicomponent systems 204
mutual diffusion coefficient 177, 179, 199

nearby hydration 195, 442, 444, 468
negative hydration 72, 117, 196, 395, 444
    448, 451, 457, 487
— solvation 488
— Wien effect 321
Nernst–Einstein equation 150, 225
— equation 186, 238
— –Hartley equation 186
— theory of diffusion 170
net transference number 265
Newtonian liquids 80
Newton's law 266
— viscosity law 79
n-m potential 24
NMR measurements 481
— spectroscopy 474, 478
non-associating electrolytes 398
— -dissociated molecules 257, 290, 292
non-electrolytes, effect on conductivity
    351
— — solutions 372
— -Newtonian liquids 82
normal liquids 16
number of activated jumps 220
— — components 203
— — independent phebomenological
    coefficient 204
number of water molecules in the hydrate
    sphere 447

obstruction effect 227, 232
— theory 356, 357
Ohm's Law 256, 260, 267, 317
Onsager dissociation theory 320
— –Fuoss equation 190, 245, 309, 312
— — theory of mixtures 325
— –Gosting–Harned equation 236, 238
— limiting law 238, 300, 302, 304, 310,
    325
— linear law 170
— Law 325
— –Provencher theory 350
— reciprocity relationship 171
— theory 236
open associates 110

optical consequences of hydration 475
optimum size of ions 461
orper in liquids 18
orientation effect on viscosity 118
— of water molecules around ions 446
— polarization 116, 269
— time 475
osmotic pressure 170
— term 306
Ostwald dilution law 291, 310, 347
— -theory of viscosity 83
outer sphere association 437
— — ion pair formation 437
oxidation-reduction process 390
oxonium ion 269, 279, 372

packing factor 152
pairformation 74, 337
pairwise diffusion 182
Panchenkov theory 168
partial molar free enthalpy 407
— — — — of the solute 417
Pauling's radii 270
peripheric hydration 455
phenomenological and diffusion coefficient
    173
— coefficient 170, 171, 183, 187, 204,
    211
— diffusion coefficient 174
— laws 170
— theories 13
physical hydration 442
Pisarzhevskii–Walden rule 272, 277, 335
Pitts equation 316
Podolsky theory 231
Poisson equation 400, 421
polarity of O—H bonds 67
polarizability 22
polarization 116, 269
— of water molecules 71
polarized diffusion 193
polarizing power 65
polyhedral cavities 20
— vacancy theory 19
polywater 99
positive hydration 71, 117, 196, 395, 444,
    448, 451, 455, 458, 487
— solvation 488
potential barrier 91, 156
— electrolytes 293, 396
— energy of interaction 23
— — — water molecules 66
preface 10
pressional diffusion 141
pressure dependence of conductivity 339
— — — hydration 453
— — — viscosity 98
primary hydration 71, 444, 445
— nearby hydration 442
proton 193, 279

— acceptor 26
— affinity 380
— — of water 285, 290
— donor 26
— exchange mechanism 280
— transfer 46, 280
protonic jumps 280
prototropic conduction 215, 235, 240, 280, 293, 352, 359, 385, 453
— — in non-electrolyte solutions 372
— mechanism in diffusion 240
prototropic migration 46
pseudo-binary systems 212
pseudonuclei 20
pseudoplastic liquids 82

quantum mechanics 23
quartz-tridymite theory 33
quasi-crystalline structure 449

radial distribution function 17
radii of hydrated ions 373
radius of gaseous ions 463
— — ions 271, 406, 460, 461
— — migrating ions 268
— — the ionic atmosphere 296, 404, 406, 411
— — — molecules 32
Raman spectra 74, 476
random mixing statistics 53
rate constant 157, 164, 275
— — for ionic migration 277
— -determining step in prototropic conduction 281
— of diffusion 142, 157
— — electron transfer 245
— — exchange of molecules 65
— — flow 170
— — freezing 76
— — ion formation 255
— — the exchange of water molecules 447
— — transport 157
ratio of broken hydrogen bonds 36
— — free ions 397
rational activity coefficient 417
real solutions 408
reciprocity relation 171, 211
recombination of ions 350
reduced mass 162
refractive index 477
relative mobility of ions 257
relay-like proton transfer 386
— — — transport 372
relaxation effect 295, 297, 299, 301, 305, 318, 322, 336, 350
— — on diffusion 188, 236
— time 128, 222, 224, 266, 295, 465
remote hydration 442
reorientation period 452

repulsive forces 22
residence time of water molecules 65, 220, 228
resistence 317
retarding of self-diffusion 233
rheology 82
rheopectic liquids 84
rigid volume 127
rigidity of molecules 23
ring formation in water 34
Robinson–Stokes method 461
rotation of molecules 102
— — water molecules 42, 221, 283, 345

salting out 471
Samoilov theory 43, 194, 220
saturated solutions 484
secondary hydration 63, 71, 442, 455, 458
— sphere 474
— viscosity 104
self-diffusion 45, 80, 93, 210, 216, 224, 233, 243
— — coefficient 80, 195, 217, 220, 226, 231, 241, 243
— — — of supercooled water 224
— — — — water 223
— — in mixtures 225
— — of ions 242
— — — liquids 235
— — — water 219
— — — — in electrolyte solutions 231
shape of the solute molecules 152
shear force 79, 90
Shedlovsky extrapolation equation 305
short-range hydration 53
— — interactions 413
Simha equation 122
sliding friction 150
softening the ice-like structure 64
solubility of non-electrolytes 471
solvation 21, 116, 175, 227, 228, 235, 265, 307, 312, 366, 369, 395, 441, 458, 472, 489
solvation effect on self-diffusion 228
— in mixtures 488
— number 152, 443, 474
— of bivalent ions 472
solvent mixtures 359
— structure of solutions 116
Soret effect 172
specific conductivity 256, 260, 332, 339, 357
— rate 91, 144, 157, 275
spectral properties 475
spherical diffusion 148
sphere model of liquids 19
square root law 293, 410, 420
— — — of conductivity 258
statistical theory of electrolytes 294

steady state in diffusion 146
Stokes–Einstein equation 151
— — law 155
— frictional law 150
Stokes' Law 158, 268, 270, 273, 318, 335, 354, 375, 460, 464
Stokes radius 268, 270, 462
— –Robinson theory 419
strength of acids 345
strengthening of the ice-like structure 60
— — — structure of water 66
stretching vibration 477
strong electrolytes 258, 291, 294, 397, 398
structural activation energy 101, 131
— cavities 38
— effect on viscosity 118
— hydration 478
— migration 287
— relaxation 68, 105
— solvation 475
— state 17
— temperature 62, 63, 66, 488
— theories 53
structure-breaking effect 66, 71, 72, 116, 454, 456, 478, 486, 488
— — — on self-diffusion 228
— — ions 467
— -making effect 71, 442, 455, 456, 476, 478
— — ions 467
— of concentrated solutions 324, 484
— — electrolyte solutions 64, 196
structure of heavy water 32, 53
— — ice 27, 54
— — liquids 17
— — non-electrolyte solutions 59
— — normal liquids 16
— — the water molecule 25
— — solution 483
— — water 31, 53, 352
supporting electrolyte 236
surface tension 471

temperature coefficient of B 117
— — of the transference number 364
— dependence of excess entropy 486
— — of conduction 331
— — — conductivity in dioxan solutions 368
— — — hydration 453
— — — ionic mobility 272, 334
— — — self-diffusion 223
— — — structure of liquids 21
— — — the transference number 333
— — — — viscosity 81, 97, 98, 130
— gradient 146
ternary electrolyte solutions 135
tetrahedral arrangement in ice 28
theories, validity of them 56
— of diffusion 149

— — electrolytic conduction 266, 273, 289
— — electrostatic interaction 289
— — ion pair formation 426
— — prototropic conduction 278
— — Samoilov 38
— — self-diffusion 217
— — strong electrolytes 128
— — structural cavities 38
— — the bent bond 35
— — — unified structure 52
— — viscosity 84
thermal diffusion 141
— expansion 332
— fluctuation 87
— motion 81
— — of the D_2O molecule 53
— wandering 80
thermodynamic aspects of hydration 479
— force 170
thermodynamic theory 184
— — of diffusion 170
thermodynamics of electrolyte mixtures 326
— — irreversible processes 170
— — the diffusion 180
thermodynamical data on hydration 76
thermostatics 170
thixotropic liquids 84
thickness of the ionic atmosphere 404, 406, 411, 422
Thomas equation 127
three component system 206
— dimensional diffusion 147
torsional vibration of water molecules 42
total flux 253
— hydration 76, 483
tracer diffusion 209, 217
— — and viscosity 240
— — as the function of concentration 244
— — coefficient 236
— — of hydrogen ions 240
— — — ions 235
— — — proton 240
transfer diffusion 168
— components 253
transference number 260, 262, 344
— — and concentration 314
— — — hydration 468
transference number in non-electrolyte solutions 364, 382, 386, 468
— — of hydrogen ion 284, 382
— — — hydroxide ion 285, 386
— — — some ions 263
— — — the cation 315
transient states in diffusion 146
transition state of migration 278
translational mobility of water molecules 270

— movement of ions 448
— velocity 253
transmission coefficient 91
transorientation time 475
transport by free volume 325
— coefficient 14
— of electric charge 245
— — impulse 80, 84
— — mass 253
— — volume 144
— phenomena 13, 253
— processes 13, 253
true electrolytes 293, 396, 398
tunnel effect 281, 287
twelve-six potential 23

unified structure 52
unipolar conduction 260
ultrasonic measurements 474
unit cell of the ice lattice 29

vacancy 18, 33, 87, 278, 449
— theory 19
valence angle 26
— of ions 184
validity of theories 56
van der Waals forces 22, 399
Vand equation 123
velocity gradient 80
— of diffusion 142
— — displacement 267
— — ions 253, 275, 297
velocity of sound in liquids 89
vibration frequency 478
— of water molecules 42
viscosity 72, 79, 80, 90, 153, 154, 335,
    352, 353, 356, 357, 467
— and hydration 124
— — self-diffusion coefficient 229
— — shape of particles 124
— — tracer diffusion 240
— at constant pressure 99
— — — volume 99
— coefficient 80, 90
—, component of 117
— law 79

— of aqueous solutions 113
— — concentrated solutions 130
— — dioxan-water mixtures 371
— — electrolyte solutions 111
— — heavy water solutions 116
— — liquid mixtures 105
— — non-electrolytic solutions 108,
    122, 360
— of solutions 113
— of ternary electrolyte solutions 135
— of water 102
—, secondary 104
viscoelastic liquids 84
viscous flow 13, 79, 84, 158, 216
viscous flux 80
voltage-dissociation 320
volume flow 176
— flux 177
— fraction 122, 136
— transfer 176
— viscosity 104
— — of solutions 112
voluminousity of liquids structure 20

waiting time 475
Walden product 151, 273, 277, 287, 335,
    336, 353, 360, 375, 385
— rule 272, 277, 335, 353, 358
Wang–Devell theory 232
water–alcohol solutions 283
— — solvent mixtures 361
— binding of anions and cations 477
— -dioxan mixtures 382
— — solvent mixtures 366
— hydrate theory 43
— molecule 25
water polymer 282
— structure 336, 340, 442
weak electrolytes 290, 347, 397, 398
Wien effect 317
— — in viscosity 129
— —, negative 321
Wilson theory 319

Zdanovskii equation 136
zero-point frequency 22